Chemical Skills

Fourth Edition

Edward I. Peters
Professor Emeritus of Chemistry
West Valley College

William T. Scroggins
Professor of Chemistry
Chabot College

McGraw-Hill, Inc.
New York St. Louis San Francisco Auckland Bogotá Caracas
Lisbon London Madrid Mexico Milan Montreal New Delhi
Paris San Juan Singapore Sydney Tokyo Toronto

This book is printed on recycled paper containing a minimum of 50% total recycled fiber with 10% postconsumer de-inked fiber.

Chemical Skills

2 3 4 5 6 7 8 9 0 MAL MAL 9 0 9 8 7 6 5 4 3 2

ISBN 0-07-049562-9

The editors were Karen M. Hughes and Kirk Emry;
the production supervisor was Denise L. Puryear.
Malloy Lithographing, Inc., was printer and binder.

Library of Congress Cataloging-in-Publication Data

Peters, Edward I.
Chemical skills / Edward I. Peters, William T. Scroggins.—4th ed.
p. cm.
Includes index.
ISBN 0-07-049562-9
1. Chemistry—Problems, exercises, etc. I. Scroggins, William.
II. Title
QD42.P46 1992
540′.76—dc20 91-26632

TABLE OF CONTENTS

TABLE OF CONTENTS

TABLE OF CONTENTS

TABLE OF CONTENTS

PREFACE

The purpose of Chemical Skills is to be to students a "tutor" that is always available, at any time and at any place, during the early stages of learning chemistry. The course's reputation as a "difficult" course is well known. We who teach it know that the "difficulty" often lies not in chemistry, but in weak mathematical skills or not knowing how to translate a problem into a workable calculation setup. A thousand-page textbook does not have space to develop basic skills for its readers. Nor does it have room to guide students gradually through the problem-solving methods that arise in the course. This book is designed to overcome both of these obstacles to starting out successfully in chemistry.

Chemical Skills addresses first poor mathematical skills. Part I, the first five chapters, ranges from a review of basic arithmetic and algebra to graphing and logarithms. These chapters will be used selectively, depending on the level of the math background, whether or not graphing skills are needed in the laboratory, and whether or not the course reaches a need for logarithms.

Chapters 2 and 3 cover mathematics and its application to measurement. Chapter 4 introduces an approach to solving chemistry problems that is used throughout the book. It includes the usual advice found in most textbooks, but we formalize that approach by "analyzing" each problem. It also "explains" quickly and in a minimum of space how example problems are solved. More about that in a moment.

Part II, chapters 6 through 13, begins with the basics of atoms, elements, chemical formulas, equation writing, and calculations based on chemical formulas and equations. Part II then concludes with a more advanced treatment of atoms, and the fundamentals of gas laws and thermochemistry. Lack of understanding of these basic chemical skills presents a barrier to successful problem solving which threatens students' survival in the chemistry course itself. Our purpose is to remove this barrier.

Part II, the remaining seven chapters, guides the student to skill in solving the problems that commonly appear a bit later in the course. In all of the above areas, Chemical Skills offers a more gradual, learning-oriented development than is possible in a textbook. Nearly all students benefit from such an approach, and for many it means survival itself.

In Chemical Skills, principles that underlie a particular skill are presented in the usual textbook format, but example problems are programmed. Rarely, when a student wants help in learning how to "do" something—solve a problem, write an equation, or perform any other act that employs what we have called a "chemical skill"—do we simply "show" him or her how to do it. More often we ask questions that will guide the student into doing it himself or herself, trusting that the act of doing will produce learning. That is what we have done in writing the programmed examples in this book. The success of this method is attested to by the large number comments received from users of earlier editions of this book and of other books that use programmed examples.

We referred above to a "formalized" problem solving approach. Most textbooks offer brief suggestions on how to solve problems. Writing down what is given and what is wanted is usually the first step. Many books do not complete that first step in their worked out examples. (We do it constantly.) Procedures vary after the given and wanted quantities are identified. In this book we recommend next a determination of which of two problem solving strategies are required by a specific example:

1. **Dimensional analysis** is used whenever the given and wanted quantities are related by one or more proportionalities. The necessary conversion factors are identified and a unit path is plotted.
2. **Algebra** is used if the relationship is fixed by a mathematical equation. In most cases, the equation is solved for the wanted quantity first. With more complicated equations, the given quantities—*including units*—are substituted into the original equation and the result calculated.

We are not so naive as to believe a student will write down the given and wanted quantities, followed by an equation or a list of conversion factors and a unit path for each problem. Nor do we want them all to be written, *unless the student is unable to get started on a particular problem.* (We sometimes ask that they be written when a new topic is introduced.) But we do want the student *always to think of these steps*, not necessarily as sequential "steps," but rather as a means of analyzing the problem. To encourage this, we show the analysis of nearly every problem example. Usually, when the steps are written, the pattern for solving the problem is obvious and no further explanation is needed.

Perhaps you would like to see how this approach is developed. It first appears in Section 2.3, on page 36. Note particularly the last paragraph in the section. In Section 2.4 we introduce two other abbreviations that are used throughout the book. For problems solved by dimensional analysis, FACTOR identifies conversion factors and PATH refers to a unit path. A reminder of the dimensional analysis procedure appears in the middle of page 58, and again near the top of pages 149. The symbol EQN is first used on an equation-type problem on page 44. A reminder appears at the bottom of page 65, just before the next application for temperature conversions.

We have placed our description of the structure of Chemical Skills in a special "To the student" section entitled HOW TO LEARN CHEMISTRY FROM THIS BOOK. It follows this preface. Note particularly the instructions on how to solve programmed examples on page iv.

You might find it interesting to see the programmed format as a student sees it by a few minutes of role playing. If so, tear out one of the periodic tables to be used as a shield and reference source for atomic masses. A glance at the text just above Example 8.5 (page 152) will show you the form in which molar mass calculations have been developed. Then try the example, looking up the atomic mass on the periodic table and writing the needed information in the book. Example 8.8 (page 155) gives you a chance to apply molar mass to a mol → g conversion just as a student will do it for the first time. Continuing through several examples, here or elsewhere in the book, shows how the student is guided into learning by doing.

Among the nicer things about writing a chemistry book are the comments and suggestions that come from instructors who review the manuscript. It is reassuring when a review praises what has been written; it is, perhaps, even more valuable when it does not. We are challenged to compare our time-honored ways with something someone else thinks is better. Sometimes we adopt the new method, and sometimes we stay with the old. Either way, we benefit from the challenge and the re-evaluation. We want to express our sincere appreciation to the following who have contributed to this edition in this way:

Ildy Boer	County College of Morris, NJ
Robert Ouellette	Ohio State University
Jean Shankweiler	El Camino College
Eric Snyder	Arcadia, California
Danny White	American River College

In addition, we greatly appreciate the assistance and encouragement of Karen Hughes, our faithful editor at McGraw-Hill.

Edward I. Peters   William T. Scroggins

To the student:

HOW TO LEARN CHEMISTRY FROM THIS BOOK

[1]**tutor** *n*: . . . a private teacher . . .

[2]**tutor** *vt*: . . . to teach or guide usu. individually in a special subject for a specific purpose . . .

Webster

Meet your tutor: this book that you hold in your hands. This is not as impersonal as it seems. This book expresses in print what we, the authors, have given to students like you when they have been stuck on some problem or skill as they are beginning to learn chemistry. Even though we cannot meet you personally, we hope you will let us help you as we have helped others in one-to-one tutoring sessions.

A tutor can help you to learn, but it is you who must do the learning. All the studying that you do and all the assignments you complete are wasted if you do not *learn* the material. Demand of yourself that you *learn* how to do what is required.

This book has been designed to help you learn. It does this by guiding you into putting into practice what you are learning while you are learning. If you use the book as it is intended—if you work the examples as suggested on the next page—you will learn more chemistry in less time. Your reward will be better test scores and more time for fun and games. Enjoy both; you will have earned them. Here are some of the features of this book that will help you to learn:

Prerequisites Most chapters open with a list of things you are expected to know or be able to do before you study the present chapter. These are things which we expect that you learned in an earlier chapter. Section references are given so you can check back if you need a refresher in a particular skill. And do check back when necessary. You will not learn something new if you have not already mastered what you need for the new learning.

Chemical Skills This is a list of things you will learn how to do when you study this chapter. Use this list as a preview of the chapter so you will know where to focus your attention as you reach different topics in your study. When you finish that topic, return to the particular chemical skill and ask yourself, "Can I do that?" If the answer is truly yes, go on. If not, go back to the section and study it some more. Solve more end-of-chapter problems. Do whatever else is necessary for you to be able to look again at that chemical skill and say with confidence, "I can do that." Then, and only then, have you *learned* that topic.

Text and Examples Here's where the actual learning takes place. We'll look at this item more closely in a moment.

End-of-Chapter Questions and Problems The only way you can be sure you have learned how to do something is to do it. The end-of-chapter questions and problems give you that opportunity. About two-thirds of the questions are answered in the back of the book. Answers to problems include complete calculation setups. Your instructor has answers to the remaining questions. The answered questions appear first, and they are separated from the unanswered questions by a bar:

┼┼┼┼┼┼┼┼┼┼┼┼┼┼┼┼┼┼┼┼┼┼

Additional Problems You will occasionally want to challenge your chemical skills with more advanced problems. The questions in this section require you to apply your knowledge to new situations which may not have been covered with an example in the chapter.

Chapter Test This is a list of questions such as those which might appear on a test based on the chapter. We recommend that you answer these questions under test conditions. That is, use only a pencil, paper, calculator, and periodic table. Do not look back to the text or examples for help. Answers are in the back of the book.

Now let's look more closely at the **Text and Example** section of each chapter, the place where the real learning occurs. To save you time, or to learn more in less time, we strongly urge you to use this section in the following way. To do this you will need four things: a pencil, a calculator, an opaque shield, and a periodic table for references. If you must buy a calculator, you will be interested in the discussion of calculators in Chapter 1. We have combined the shield and the periodic table as a tear-out card elsewhere in this book. On the side opposite the periodic table is a summary of the instructions you are about to read.

After being given the theory or technique behind a particular skill, you will practice that skill immediately in blank spaces in the book itself. These practice spaces are identified as **Examples**, set apart like this:

When you reach such a point, you should glance down the page until you find a pair of T-bars on each side:

This is where the opaque shield is placed—just below the T-bars so it covers the printing beneath that point. Read the question or example. Usually, it is followed by some comment or suggestion about how to proceed. Follow that suggestion, writing the answer or solving the problem in the space provided.

When this is done, move the shield down to the next set of T-bars, or the end of the example if there are no more T-bars. The first thing you expose on moving your shield down is the correct response to the question you have just answered. Compare your answer with the one in the book. If they are the same, as they will be most of the time, proceed as the book directs. If your answer and the book answer are not the same, find out why. Usually the comments that accompany the answer will be the only explanation you will need. If not, restudy the text material preceding the problem, or any other earlier material you may not have understood that is responsible for your incorrect answer. When you have corrected whatever needs to be corrected, proceed to the rest of the example. If you have reached the end of the example, it will close like this:

The importance of your active participation in solving example problems this way cannot be stated strongly enough. It is the key to *learning* chemistry from this workbook. Remember to demand that you master each step in the book. Satisfying that demand is the difference between merely *doing* your homework and *learning* chemistry—and the only thing that counts is what you learn.

Edward I. Peters　　　　William T. Scroggins

PART I. MATH AND PROBLEM SOLVING SKILLS

1. CALCULATORS AND MATH REVIEW

1.1 INTRODUCTION

Chemistry is a quantitative science. You, as a student, will use solutions prepared from chemicals measured in the storeroom, and you will measure out quantities of different chemicals in performing experiments. Part of studying chemistry is learning how to calculate the quantities that are needed and the quantities that will be produced. To do this, you must know how to perform certain math operations. In most cases this will mean knowing how to use your calculator.

If you are about to buy a calculator, you will find many from which to choose. A "scientific" calculator that is acceptable for lower division chemistry courses will be able to do all of the following:

	MATH OPERATION	KEY SYMBOLS
(a)	addition, subtraction, multiplication, and division;	+, −, ×, ÷
(b)	work in exponential notation;	EE or EXP
(c)	do base 10 logarithms and antilogs;	log x, INV log x or 10^x
(d)	raise any base to any power;	y^x
(e)	find reciprocals (inverses);	1/x
(f)	find squares and square roots; and	x^2, $\sqrt{x}$
(g)	work with natural (base e) logarithms;	ℓn x, INV ℓn x or e^x

A calculator with the above capabilities will also have trigonometric functions that are used in physics courses. One or more memory storages are also desirable.

Two types of operating systems are in common use: the Algebraic Operating System (AOS), used on most brands of calculators, and the Reverse Polish Notation (RPN), used mainly on Hewlett-Packard calculators. Note that the order of operations (addition, multiplication, etc.) is the same in both operating systems, but there is a difference in the order in which numbers and operations are keyed. Calculators also differ in the form of the display, particularly in the number of digits displayed. In this chapter, displays will show up to eight digits.

This chapter is limited to the mathematical operations and corresponding calculator techniques required to solve the problems in this book. After each math operation is presented, specific directions will be given for the calculator techniques to solve problems using that math operation. We suggest that you perform each operation as you read it. If any operation cannot be performed as described, consult the instruction manual that came with your calculator.

1.2 ENTERING A NUMBER

To introduce a number into your calculator, simply press the number keys in their proper order. This includes the decimal. If the number is negative, press the +/- key (sometimes identified by CHS or some other symbol) after the last digit. The number will appear in the display window of the calculator as it is entered.

Some calculators permit you to "fix" the number of digits that are displayed after the decimal point. Or you can allow the decimal to "float," that is, to show only the minimum digits needed to display the answer.

Most calculators can display eight or more digits. Very small numbers, with more than eight digits to the right of the decimal, such as 0.000000834, may have some of the digits lost in the display. This may cause the number to appear as 0.0000008, for example.

Numbers that are quite large or quite small should be entered in **exponential notation**, also known as **scientific notation**. This shows the number as the product of a coefficient, N, and an exponential, 10 raised to some integral (whole number) power, x, as $N \times 10^x$. In the standard form of exponential notation, N is equal to or more than 1, but less than 10. If you enter N outside of this range, the calculator adjusts the display to this form. Exponential math is discussed in detail in Section 1.4.

The procedure for entering a number in exponential notation is:

1. Type the coefficient, N. If the number is negative, type +/-.
2. Type EE (or whatever key is used for entering exponents).
3. Type the exponent. If the exponent is negative, type +/-.

Note: Most calculators will display an entered number in exponential notation only if (a) the calculator has been changed from "FIX" mode to "SCI" or "EXP" mode, or (b) if the entered number is too large or too small to fit the eight-digit display.

EXAMPLE 1.1: Enter the following numbers into your calculator:

10^4; 5.6709×10^{-8}; -9.87×10^6; -5.43 10^{-2}.

10^4

Enter	Press	Display
1	EE	1 00
4		1 04

5.6709×10^{-8}

Enter	Press	Display
5.6709	EE	5.6709 00
8	+/-	5.6709-08

-9.87×10^6

Enter	Press	Display
9.87	+/-, EE	-9.87 00
6		-9.87 06

-5.43×10^{-2}

Enter	Press	Display
5.43	+/-, EE	-5.43 00
2	+/-	-5.43-02

Did you get 100,000 or 10^5 instead of 10,000 or 10^4? Enter exponential numbers that don't have coefficients as "1×10^x" not as "10×10^x." Were your displays in decimal rather than exponential notation? Were digits missing? When some calculators run out of display space for a very small decimal number, they show only the zeros after the decimal, or perhaps the first one or two nonzero digits. The calculator carries the other digits, but hides them from view. Correctly calculated answers are displayed incorrectly. Be sure to change to exponential notation so that what you see is acceptable.

1.3 INVERSE

A common math operation is to take the **inverse** or **reciprocal** of a number. For any number x the inverse of that number is 1/x, sometimes written x^{-1}. (In the expression x^{-1} the "−1" is the exponential of x. Exponentials are explained in the next section.)

To find an inverse on a calculator, enter the number whose inverse you desire, and push the 1/x key. For example, the inverse of 40 is 1/40 = 0.025.

1.4 POWERS AND ROOTS: EXPONENTIALS

An **exponential** is a number in which a base, y, is raised to a power, x, as in y^x. Our number system is a base 10 exponential system. For example, $1000 = 10^3$ and $0.01 = 1/100 = 10^{-2}$. In the expressions 10^3 and 10^{-2}, 10 is called the **base**, and the superscripts 3 and −2 are the **exponents**, or **powers**, to which the base is to be raised. Two types of operations involve exponentials: raising a number to a power and taking the root of a number.

Raising a Number to a Power To square a number, multiply that number by itself. The square of 3 is $3 \times 3 = 3^2 = 9$. If an exponent is a **positive integer**, it represents the number of times the base is to be multiplied by itself. When a base is multiplied by itself three times, the result is the cube of the number. For example, "2 cubed" is the same as "2 to the third power" or: $2^3 = 2 \times 2 \times 2 = 8$. If an exponent is a **negative integer**, it is the inverse of the power of that integer: $5^{-2} = 1/5^2 = 1/25 = 0.04$.

Finding the Root of a Number If one mathematical operation is exactly the opposite of another, they are called **inverse operations**. The inverse of squaring a number is finding the square root of a number. The square root of X is the number that, when multiplied by itself, will give X as the product. A positive number has two square roots, one positive and one negative. The square root of 9 is +3 or −3, because each number, when multiplied by itself, gives 9 as the product: (+3)(+3) = 9, and (−3)(−3) = 9. There are no occasions in this book to use the negative root of any number. Hereafter we will disregard such roots without comment.

There are two ways to write the extraction of a root of a number. One is to use the square root symbol, $\sqrt{\ }$. If the cube root is desired, or any other n-th root, that number is written as a superscript in front of the root symbol: $\sqrt[3]{\ }$ for the third root, and $\sqrt[n]{\ }$ for the n-th root.

A more general way to represent the extraction of a root is to write it as an exponential. The base is the number whose root is to be found, and the exponent is the fraction 1/n, where n is the desired root. The exponent may be written either as a typical fraction, or as a decimal fraction. Accordingly, the square root of 25 and the fifth root of 32 are

$\sqrt{25} = 25^{1/2} = 25^{0.5} = 5$

$\sqrt[5]{32} = 32^{1/5} = 32^{0.2} = 2$

If your calculator has a "y^x" key, that key can be used to find the value of any base raised to any power. The general procedure is

1. Enter the base, y, the number that is to be raised to a power.
2. Press y^x.
3. Enter the exponent, x, the power to which the base is to be raised.
4. Press =.

Note that, in the answers to the examples that follow, the display on your calculator may not be the same as that shown in the book. When working with exponentials, some calculators display all answers in exponential notation, while others use exponential notation only if the number of digits is more than the calculator can display. Also, the number of digits the calculator can show varies among different brands.

EXAMPLE 1.2: Calculate: 123^2 = ________; $\sqrt[3]{123}$ = ________;

$5.67^{0.25}$ = ________; $(8.91 \times 10^{-2})^{3.4}$ = ________.

123^2

Enter	Press	Display
123	x^2	15129
	(or y^x, 2, =)	

$\sqrt[3]{123} = 123^{1/3}$

Enter	Press	Display
123	y^x	123
0.33333333	=	4.9731899

$5.67^{0.25}$

Enter	Press	Display
5.67	y^x	5.67
0.25	=	1.543106

$(8.91 \times 10^{-2})^{3.4}$

Enter	Press	Display
8.91	EE	8.91 00
2	+/−, y^x	8.91-02
3.4	=	2.689-04

1.5 ADDITION, SUBTRACTION, MULTIPLICATION, AND DIVISION

Addition and Subtraction When one quantity is to be added to another, they are written with a plus sign between them, as 12.84 + 6.93. The result of an addition is the **sum**, or **total**.

When one quantity is to be subtracted from another, they are written with a minus sign between them, as in 12.84 − 6.93. The result of a subtraction is the **difference**. In effect, subtraction is the same as the addition of a negative number: 12.84 + (−6.93) = 12.84 − 6.93. Unless there is a reason for doing otherwise, the addition of negative numbers will always be written as a subtraction.

The subtraction of a negative number is the same as the addition of a positive number. The operation: 14.28 - (-2.71) is the same as the operation: 14.28 + 2.71. Unless there is a reason for doing otherwise, the subtraction of a negative number will always be written as an addition.

Multiplication Quantities that are to be multiplied by each other are called **factors**. Multiplication may be indicated in several ways. One way is to write a multiplication sign, ×, between them, as 12.84 × 6.93. Another way is to enclose the factors in parentheses, as (12.84)(6.93). In algebra, where letters are used to represent numbers, factors may be written side-by-side: ab means a × b or (a)(b). Sometimes a raised dot is used to indicate multiplication: ab = a•b.

When a numerical factor is multiplied by a letter factor, the numerical factor is called the coefficient of the letter factor. In 12.3m, 12.3 is the **coefficient** of m. The result of a multiplication is called the **product**. The sign, positive or negative, of a product is governed by the following rule:

> A product is positive if both factors have the same sign—both positive or both negative; a product is negative if one factor is positive and the other factor is negative.

Division In almost all mathematical problems you will encounter, a division such as 12.84 ÷ 6.93 will be written as a fraction such as

$\frac{12.84}{6.93}$ or 12.84/6.93 The number above the line is the **numerator**. The number below the line is the **denominator**.

The result of a division is a **quotient**. The sign of a quotient is set by the following rule:

> A quotient is positive if the numerator and denominator have the same sign, either positive or negative; a quotient is negative if the numerator and denominator have different signs.

The general procedure on an AOS calculator for the one-step addition, subtraction, multiplication, or division of two numbers, A and B, is

1. Enter A.
2. Press required function key, +, −, × or ÷.
3. Enter B.
4. Press =.

EXAMPLE 1.3: Calculate: 12 + 345 = ________; 12 - 345 = ________;

12 × 345 = ________; 12 ÷ 345 = ________.

12 + 345

Enter	Press	Display
12	+	12
345	=	357

12 − 345

Enter	Press	Display
12	−	12
345	=	−333

12 × 345

Enter	Press	Display
12	×	12
345	=	4140

12 ÷ 345

Enter	Press	Display
12	÷	12
345	=	0.0347826

1.6 LOGARITHMS AND ANTILOGARITHMS

Math operations with **logarithms and antilogarithms**, commonly called logs and antilogs, are sufficiently complicated to merit their own chapter, Chapter 5. Logs and antilogs are found on a calculator by entering a number and just pushing one key. The procedure for performing these operations is outlined below.

1. Enter the number x.
2. Press log x for the logarithm of x. Press 10^x (or INV log x) for the antilogarithm of x.

Natural logarithms, which use the base "e" instead of base 10, are found with the calculator keys ℓn x and e^x (or INV ℓn x). See Chapter 5 for a discussion of natural logarithms.

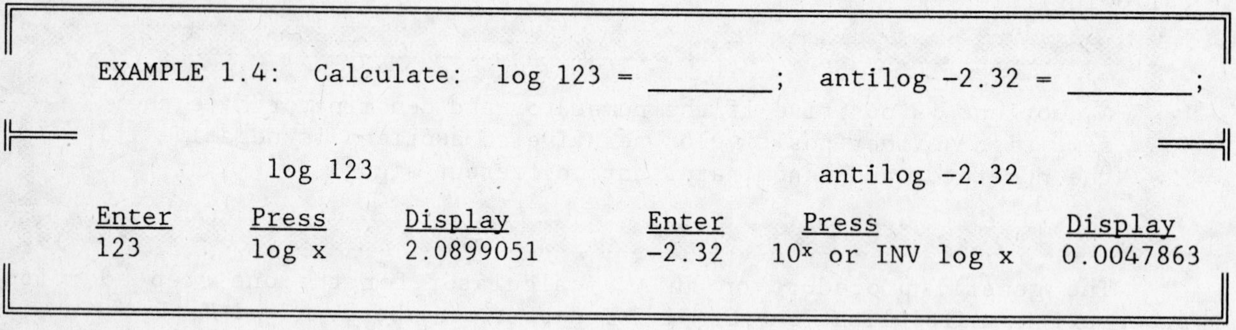

EXAMPLE 1.4: Calculate: log 123 = ________; antilog −2.32 = ________;

log 123

Enter	Press	Display
123	log x	2.0899051

antilog −2.32

Enter	Press	Display
−2.32	10^x or INV log x	0.0047863

1.7 CHAIN CALCULATIONS

A **chain calculation** is one in which two or more operations are performed, one after the other. The order in which the operations are done may be critical. Most AOS calculators perform simple calculations in the proper order when numbers and operations are entered exactly as they appear in the calculation setup. (This is not true with RPN calculators.) The calculator automatically multiplies and/or divides before adding and/or subtracting. For example, when your calculator computes 0.5054 × 78.92 + 0.4946 × 80.92, it first multiplies 0.5054 by 78.92. Then 0.4946 and 80.92 are multiplied. Finally the two products are added together.

The order in which the calculations are done may be changed by inserting parenthe tiply 0.5054 by the sum of 78.92 and 0.4946, and then multiply by 80.92. The sequence of operations for both calculations is shown below.

$0.5054 \times 78.92 + 0.4946 \times 80.92$

Enter	Press	Display
0.5054	×	0.5054
78.92	+	39.88617
0.4946	×	0.4946
80.92	=	79.9092

$0.5054 \times (78.92 + 0.4946) \times 80.92$

Enter	Press	Display
0.5054	×	0.5054
	(	0.5054
78.92	+	78.92
0.4946	)	79.4146
	×	40.136139
80.92	=	3247.8164

The numerical results are, of course, very different!

The use of parentheses allows you to alter the sequence in which the calculator usually does operations. An example is $4(0.10 + 3.00)^2$.

	Enter	Press	Display
Start with 4 times	4	×	4
Enter left parenthesis		(	0
Then enter .1 and add	.1	+	0.1
3 to it, ending with)	3	)	3.1
Square (0.10 + 3.00)		x^2	9.61
Press = to get result		=	38.44

EXAMPLE 1.5: Evaluate the expressions below.

$$\frac{84.0 - 80.0}{80.0} \times 100 = \_\_\_\_\_\_ \qquad 4.74 + \log \frac{0.10}{2.0} = \_\_\_\_\_\_$$

Enter	Press	Display
84	−	84
80	=	4
	÷	4
80	×	0.05
100	=	5

Enter	Press	Display
.1	÷	0.1
2	=	0.05
	log, +	−1.301030
4.74	=	3.438970

The most common chain calculation is the evaluation of an expression that has two or more factors in the numerator and two or more factors in the denominator. The factors may be entered in any order in such a problem. For example, the evaluation of ab/cd may be done in any of the following ways:

a × b ÷ c ÷ d	a ÷ c × b ÷ d	a ÷ d × b ÷ c
a × b ÷ d ÷ c	a ÷ c ÷ d × b	a ÷ d ÷ c × b
b × a ÷ c ÷ d	b ÷ c × a ÷ d	b ÷ d × a ÷ c
b × a ÷ d ÷ c	b ÷ c ÷ d × a	b ÷ d ÷ c × a

EXAMPLE 1.6: Use three different sequences to evaluate $\frac{72 \times 96}{18 \times 16}$

96 ÷ 16 ÷ 18 × 72

Enter	Press	Display
96	÷	96
16	÷	6
18	×	0.3333333
72	=	24

96 ÷ 18 × 72 ÷ 16

Enter	Press	Display
96	÷	96
18	×	5.3333333
72	÷	384
16	=	24

72 × 96 ÷ 16 ÷ 18

Enter	Press	Display
72	×	72
96	÷	6912
16	÷	432
18	=	24

1.8 FRACTIONS

The division of N by D (D ≠ 0), written as N/D, is a **fraction** in which N is the numerator and D is the denominator. 1/2, 2/3, 5/9, and 17/13 are fractions. In a fraction, if the...

...numerator is less than the denominator, the fraction is less than 1;

...numerator equals the denominator, the fraction equals 1;

...numerator is more than the denominator, the fraction is more than 1.

Arithmetic of Fractions Calculators cannot perform most required manipulations of fractions. Most often you must do fraction arithmetic yourself by hand. We will show how to add, subtract, multiply, divide, and cancel fractions. Each example will begin with acutal numbers and then express a general rule using letters (a, b, c, etc.). Rules for these operations are summarized at the end.

Fractions with the same denominator are added or subtracted by adding or subtracting the numerators.

$$\frac{2}{6} + \frac{3}{6} = \frac{2+3}{6} = \frac{5}{6} \text{ and } \frac{6}{7} - \frac{2}{7} = \frac{6-2}{7} = \frac{4}{7} \qquad \text{so} \qquad \frac{a}{c} \pm \frac{b}{c} = \frac{a \pm b}{c}$$

The product of two fractions is obtained by multiplying the numerator and denominator of both fractions.

$$\frac{1}{2} \times \frac{3}{4} = \frac{1 \times 3}{2 \times 4} = \frac{3}{8} \qquad \text{so} \qquad \frac{a}{c} \times \frac{b}{d} = \frac{ab}{cd}$$

If a number is divided by a fraction, the result is equivalent to multiplying that number by the inverse (reciprocal) of the fraction.

$$\frac{2}{3/4} = 2 \times \frac{4}{3} = \frac{8}{3} \qquad \text{so} \qquad \frac{a}{b/c} = a \times \frac{c}{b} = \frac{ac}{b}$$

$$\frac{1/3}{2/7} = \frac{1}{3} \times \frac{7}{2} = \frac{7}{6} \qquad \text{so} \qquad \frac{a/b}{c/d} = \frac{a}{b} \times \frac{d}{c} = \frac{ad}{bc}$$

If the same number is in both numerator and denominator, it can be **canceled**.

$$\frac{4 \times \not{3}}{\not{3} \times 2} = \frac{4}{2} \qquad \text{so} \qquad \frac{a \times \not{b}}{\not{b} \times c} = \frac{a}{c}$$

1. Addition and subtraction of fractions requires that the fractions have a common denominator. Adding or subtracting fractions with a common denominator is performed by adding or subtracting the numerators.

$$\frac{a}{c} \pm \frac{b}{c} = \frac{a \pm b}{c}$$

2. Multiplication of fractions is done by multiplying both the numerators and denominators.

$$\frac{a}{c} \times \frac{b}{d} = \frac{ab}{cd}$$

3. Dividing by a fraction is equivalent to multiplying by the inverse (reciprocal) of the fraction.

$$\frac{a}{b/c} = a \times \frac{c}{b} = \frac{ac}{b} \quad \text{or} \quad \frac{a/b}{c/d} = \frac{a}{b} \times \frac{d}{c} = \frac{ad}{bc}$$

4. When the same factor appears in both the numerator and the denominator, it can be **canceled**.

$$\frac{a \times \not{b}}{\not{b} \times c} = \frac{a}{c}$$

EXAMPLE 1.7: Perform the following calculations without a calculator.

$\frac{6}{5} - \frac{2}{5} =$ ______ $\frac{1}{2} \times \frac{2}{3} =$ ______ $\frac{3/16}{1/2} =$ ______ $\frac{6}{2/3} =$ ______

$\frac{6}{5} - \frac{2}{5} = \frac{4}{5}$ $\frac{1}{2} \times \frac{2}{3} = \frac{1}{3}$ $\frac{3/16}{1/2} = \frac{3}{16} \times \frac{2}{1} = \frac{3}{8}$ $\frac{6}{2/3} = 6 \times \frac{3}{2} = 9$

Decimal Equivalents of Fractions It is often convenient to change a fraction to its decimal equivalent. This is done by dividing the numerator by the denominator. Thus 2/5 = 2 ÷ 5 = 0.4. Fractions that do not have a common denominator may be changed to their decimal equivalents, and then added or subtracted as decimal fractions. This is an operation your calculator can do!

EXAMPLE 1.8: Perform the following subtraction with a calculator. Write the decimal equivalent of each fraction in the space provided. Then perform the indicated subtraction on those decimal equivalents.

$\frac{7}{8} - \frac{3}{5} =$ ________ - ________ = ________

$$\frac{7}{8} - \frac{3}{5} = 0.875 - 0.6 = 0.275$$

Listed below are the decimal equivalents of a few common fractions. Some fractions have a "run-on" decimal equivalent, such as 2/3 = 0.6666666.....

1/8 = 0.125	3/8 = 0.375	5/8 = 0.625
1/5 = 0.2	2/5 = 0.4	2/3 = 0.6666666...
1/4 = 0.25	1/2 = 0.5	3/4 = 0.75
1/3 = 0.3333333...	3/5 = 0.6	4/5 = 0.8
		7/8 = 0.875

Fractional quantities greater than 1 may be written as **mixed numbers** or as **improper fractions**. A mixed number consists of an integer and a fraction, such as 3¾. An improper fraction has a numerator that is larger than the denominator, such as 15/4. These two fractions are both equal to 3.75:

$$3\tfrac{3}{4} = \frac{12}{4} + \frac{3}{4} = \frac{15}{4} = 3.75$$

1.9 ALGEBRA

Algebra can be thought of as arithmetic in which letters are used to represent **variables**, quantities that may have different values. Variables, and perhaps numbers as well, appear in an **equation** which represents a mathematical relationship among the variables. In this section we will use numbers and letters to illustrate the algebraic operations you will most commonly encounter in chemistry. The operations will first be shown with numbers and then generalized with letters.

Order of Operation When two numbers are added, they may be added in any order. For example, adding 3 to 4 is the same as adding 4 to 3:

$$4 + 3 = 7 \quad \text{and also} \quad 3 + 4 = 7 \quad \text{so} \quad a + b = b + a \tag{1.1}$$

This is also true of multiplication: when numbers are multiplied, the factors may be taken in either order. The product is 12 for both 4 × 3 and 3 × 4.

$$4 \times 3 = 12 \quad \text{and also} \quad 3 \times 4 = 12 \quad \text{so} \quad a \times b = b \times a \tag{1.2}$$

The order of numbers may NOT be reversed in subtraction and division. The result of 8 − 4 is not the same as 4 − 8, nor is 8 ÷ 4 equal to 4 ÷ 8.

Grouping Terms may be grouped in addition. In 4 + 3 + 2, the sum is 9 no matter which two numbers you add first.

$$(4 + 3) + 2 = 7 + 2 = 9 \quad \text{and also} \quad 4 + (3 + 2) = 4 + 5 = 9 \quad \text{so}$$

$$a + b + c = (a + b) + c = (a + c) + b = (b + c) + a \tag{1.3}$$

This property includes subtraction, which is the addition of a negative number. 4 − 3 + 2 = 3 no matter how your combine the numbers.

$4 - 3 + 2 = 1 + 2 = 3$ and also $4 - 3 + 2 = 4 - 1 = 3$ so

$$a - b + c = (a - b) + c = (a + c) - b = (c - b) + a \quad (1.4)$$

Be careful when using your calculator to avoid using parentheses which can change the meaning of a subtraction.

$(4 - 3) + 2 = 1 + 2 = 3$ BUT NOT $4 - (3 + 2) = 4 - 5 = -1$ (NOT 3!)

Factors may also be grouped in multiplication. The product of $4 \times 3 \times 2$ is 24 no matter which two numbers you multiply first. This property includes division, which is simply multiplication by an inverse.

$(4 \times 3) \times 2 = 12 \times 2 = 24$ and also $4 \times (3 \times 2) = 4 \times 6 = 24$

$(4 \times 3) \div 2 = 12 \div 2 = 6$ and also $4 \times (3 \div 2) = 4 \times 1.5 = 6$ so

$$(a \times b) \times c = a \times (b \times c) \text{ and also } (a \times b) \div c = a \times (b \div c) \quad (1.5)$$

There is a special "grouping" possible when the sum of two numbers—a **binomial**—is to be multiplied by a single factor. To evaluate $4(3 + 2)$, the "inside to outside" procedure is to complete first the operation within parentheses, and then multiply by 4. This gives $4(3 + 2) = 4(5) = 20$. The other approach is to multiply each term in parentheses by the common factor, and then add. This process is called **binomial expansion**.

$4(3 + 2) = (4 \times 3) + (4 \times 2) = 12 + 8 = 20$ so

$$a(b + c) = ab + ac \quad (1.6)$$

Exponents If two or more exponentials to the same base are to be multiplied, the product is that same base raised to the power equal to the sum of the exponents.

$$2^2 \times 2^3 = 2^{2+3} = 2^5 = 32 \quad \text{so} \quad a^m \times a^n = a^{(m+n)} \quad (1.7)$$

The exponent of a division of exponentials to the same base is the numerator exponent minus the denominator exponent.

$$4^5 \div 4^2 = \frac{4^5}{4^2} = 4^{5-2} = 4^3 = 64 \quad \text{so} \quad a^m \div a^n = \frac{a^m}{a^n} = a^{(m-n)} \quad (1.8)$$

EXAMPLE 1.9: Simplify each expression below. For those with number bases calculate the numerical value of the expression.

$(4^{-4})(4^9) =$ ______ = ______ $\qquad \frac{3^{-2}}{3^7} =$ ______ = ______

$(x^2)(x^5) =$ ______ = ______ $\qquad (x^{-7})(x^4) =$ ______ = ______

$\frac{x^8}{x^5} =$ ______ = ______ $\qquad \frac{x^{-2}}{x^{-6}} =$ ______ = ______

$(4^{-4})(4^9) = 4^{-4+9} = 4^5 = 1024$ $\qquad$ $\dfrac{3^{-2}}{3^7} = 3^{-2-7} = 3^{-9} = 0.0000508$

$(x^2)(x^5) = x^{2+5} = x^7$ $\qquad$ $(x^{-7})(x^4) = x^{-7+4} = x^{-3}$

$\dfrac{x^8}{x^5} = x^{8-5} = x^3$ $\qquad$ $\dfrac{x^{-2}}{x^{-6}} = x^{-2-(-6)} = x^4$

Did you use the y^x key to evaluate 4^5 and 3^{-9}? If not, review Section 1.4. The calculator key sequences are: 4, y^x, 5, = and 3, y^x, 9, +/−, =.

Exponentials may be moved from a numerator to a denominator, or vice versa, by changing the sign of the exponent.

$$3^{-2} = \frac{1}{3^2} = \frac{1}{9} \qquad \text{so} \qquad a^{-m} = \frac{1}{a^m} \qquad (1.9)$$

Equation 1.9 offers a second approach to dividing exponentials. The exponential in the denomenator may be moved to the numerator by changing its sign. The problem is completed by adding the two exponents. Applying this to Equation 1.8 gives:

$$\frac{4^5}{4^2} = 4^5 \times 4^{-2} = 4^3 = 64$$

If an exponential is to be raised to a power, the exponent of the result is the product of the starting exponent times the power to which the exponential is raised.

$$(3^2)^{2.5} = 3^{2\text{x}2.5} = 3^5 = 243 \qquad \text{so} \qquad (a^m)^n = a^{mn} \qquad (1.10)$$

If any base is raised to the zeroth power, the exponential is equal to 1:

$$6^0 = 1 \text{ and } 25^0 = 1 \qquad \text{so} \qquad a^0 = 1 \qquad (1.11)$$

EXAMPLE 1.10: For each problem, first simplify the expression then calculate the answer for those with number bases.

$(2^2)^3 =$ ______ $=$ ______ $\qquad$ $(x^8)(x^5)(x^{-12}) =$ ______ $=$ ______

$\dfrac{1}{3^4} =$ ______ $=$ ______ $\qquad$ $\dfrac{x^2}{(x^5)(x^{-3})} =$ ______ $=$ ______

$(2^2)^3 = 2^{2\text{x}3} = 2^6 = 64$ $\qquad$ $(x^8)(x^5)(x^{-12}) = x^{8+5-12} = x^1 = x$

$\dfrac{1}{3^4} = 3^{-4} = 0.0123$ $\qquad$ $\dfrac{x^2}{(x^5)(x^{-3})} = (x^2)(x^{-5})(x^3) = x^0 = 1$

Calculator sequences: for 2^6: 2, y^x, 6, =; for 3^{-4}: 3, y^x, 4, +/−, =.

Occasionally, it is necessary to raise the product of two factors to a power. The answer is found by raising each factor to that power and then multiplying the results:

$$(2n)^3 = (2)^3(n)^3 = 8n^3 \quad \text{so} \quad (ab)^n = (a)^n(b)^n \tag{1.12}$$

There is a similar relationship for the division of two factors:

$$\left(\frac{a}{4}\right)^3 = \frac{a^3}{4^3} = \frac{a^3}{64} \qquad \left(\frac{a}{b}\right)^n = \frac{a^n}{b^n} \tag{1.13}$$

Order of Operation:	$a + b = b + c$	(1.1)
	$a \times b = b \times a$	(1.2)
Grouping:	$(a + b) + c = a + (c + b)$	(1.3)
	$(a - b) + c = (a + c) - b$	(1.4)
	$(a \times b) \times c = a \times (b \times c)$	(1.5)
	$(a \times b) \div c = a \times (b \div c)$	(1.5)
Binomial Expansion:	$a(b + c) = (a \times b) + (a \times c)$	(1.6)
Exponents:	$a^2 \times a^3 = a^{2+3} = a^5$	(1.7)
	$a^5 \div a^2 = a^{5-2} = a^3$	(1.8)
	$a^{-2} = 1/a^2$	(1.9)
	$(a^2)^{2.5} = a^{2x2.5} = a^5$	(1.10)
	$a^0 = 1$	(1.11)
	$(2n)^3 = (2)^3(n)^3 = 8n^3$	(1.12)
	$\left(\frac{a}{4}\right)^3 = \frac{a^3}{4^3} = \frac{a^3}{64}$	(1.13)

− a Substitute: $x = 7 - 2 = 5$

Substitute: $x + 2 = 7$ Solve for x: $x = 7 - 2 = 5$

In most cases, you will find it easier to solve simple equations first, substitute the numbers, and then calculate x. With more complicated equations, it will often be easier to substitute the numbers first, then solve for x. In either case, to arrive at an equation with x alone on one side, the letters or constants in the equation must be moved to the other side by a series of algebraic operations. The key idea in this procedure is

Whatever is done to one side of an equation must also be done to the other side.

You can add or subtract the same quantity, multiply by the same quantity, divide by the same quantity (except 0), invert both sides (if each side has only one term), raise both sides to the same power, or extract the same root. As long as the same thing is done to both sides of the equation, the result will still be a valid equation.

Procedure 1: Addition and Subtraction

$x + 2 = 5$	$x - 3 = 4$	Add/subtract the quantity
$x + 2 - 2 = 5 - 2$	$x - 3 + 3 = 4 + 3$	on the same side as x.
$x = 3$	$x = 7$	Combine terms.
$a + b = c$	$a - b = c$	
$a + b - b = c - b$	$a - b + b = c + b$	
$a = c - b$	$a = c + b$	

In practice, this procedure is sometimes called **transposing**, in which a term is moved from one side of the equation to the other with a change in sign. Above, + 2 was moved from the left side to the right, where it became − 2; and − 3 was transposed from left to right, where it became + 3.

EXAMPLE 1.11: Solve each of the following equations for x.

$x + 35 = 42$ $\qquad$ $x - 16 = 18$

$x = 42 - 35 = 7$ $\qquad$ $x = 18 + 16 = 34$

The numerical quantity was transposed in both equations. The net effect was subtracting 35 from both sides of the first equation, and adding 16 to both sides of the second.

The next example is your first occasion to apply a mathematical principle to a real physical situation. There will be many other occasions; in chemistry, that is the only kind of problem there is. Much of this book is dedicated to helping you develop logical thought patterns by which to analyze a problem and interpret it mathematically. The interpretation will lead to a calculation setup most of the time. Other times, as in this example and all others in this section, there will be an equation to solve. You will be given the values of all variables except one. Your task will be to solve for the numerical value of that unknown.

EXAMPLE 1.12: A mixture of helium and neon exerts a total pressure, P, of 4.84 atmospheres. Find the partial pressure of helium, p_{He}, if the partial pressure of neon, p_{Ne} is 1.22 atmospheres. The relationship is $p_{Ne} + p_{He} = P$. Begin by solving the equation for p_{He}.

$p_{He} = P - p_{Ne}$

The equation is solved for p_{He} by subtracting p_{Ne} from both sides.

You are now ready to substitute the given numbers and calculate the answer. Do this in the space below.

$p_{He} = P - p_{Ne} =$

$p_{He} = P - p_{Ne} = 4.84 \text{ atm} - 1.22 \text{ atm} = 3.62 \text{ atm}$

You probably did not show the units in your calculation, and we will not emphasize them in this chapter. You should recognize, however, that measurement units are an essential part of both calculation setups and answers in chemistry problems.

Ocasionally, the unknown x appears as the subtracted quantity.

Procedure 2: Unknowns With Negative Signs

$7 - x = 3$		$a - x = b$	
$7 - x + x = 3 + x$	Add x	$a - x + x = b + x$	Add X
$7 = 3 + x$	Combine	$a = b + x$	Combine
$7 - 3 = 3 - 3 + x$	Subtract 3	$a - b = b - b + x$	Subtract b
$4 = x$	Combine	$a - b = x$	Combine

EXAMPLE 1.13: Solve the following equation for x: $25 - x = 55$

$25 - 55 = x$ Add x, subtract 55

$-30 = x$ Combine terms

Procedure 3: Multiplication and Division

$5x = 75$	$\frac{x}{7} = 3$	
$\frac{5x}{5} = \frac{75}{5}$	$\frac{x}{7} \times 7 = 3 \times 7$	Divide by quantity which multiplies x. Multiply by quantity which divides x.
$x = 15$	$x = 21$	Cancel.
$ax = b$	$\frac{x}{a} = b$	
$\frac{ax}{a} = \frac{b}{a}$	$\frac{x}{a} \times a = b \times a$	Divide by quantity which multiplies x. Multiply by quantity which divides x.
$x = b/a$	$x = ab$	Cancel.

Solving $5x = 75$ is done by dividing both sides of the equation by 5. Similarly, $x/7 = 3$ is solved by multiplying both sides by 7. When term containing x is both multiplied and divided by a number, do both steps at once.

$$\frac{3x}{4} = 15 \quad \text{so} \quad \frac{3x}{4} \times \frac{4}{3} = 15 \times \frac{4}{3} \quad \text{and} \quad x = 20$$

The term $3x/4$ is the same as ¾x, a fraction multiplied by x. Note that the above process is done by **multiplying by the inverse of the fraction**.

EXAMPLE 1.14: In each of the following, solve for x.

$$3x = -27 \qquad \frac{x}{6} = 5 \qquad \frac{4x}{3} = 16$$

$$x = \frac{-27}{3} = -9 \qquad x = 6 \times 5 = 30 \qquad x = \frac{3}{4} \times 16 = 12$$

In the first problem divide both sides by 3. In the second, multiply both sides by 6. In the third, the fractional coefficients calls for the "invert and multiply" rule; multiply by 3/4, the inverse of 4/3. In all three problems you use the inverse function. When x is multiplied by a number, divide both sides by that number. When x is divided by a number, multiply both sides by that number.

Density is a measure of the mass of a unit volume of a substance. The relationship between the three variables is

$$d = \frac{m}{V} \tag{1.14}$$

where d is the density, m is the mass, and V is the volume. This equation can be used to solve the next example.

EXAMPLE 1.15: Calculate the mass of 85.0 cubic centimeters (cm^3) of metal if its density is 9.00 grams per cubic centimeter (g/cm^3).

This time your unknown is m in Equation 1.14. Solve the equation for m, substitute the given values, and calculate the answer all in one step.

$$m = V \times d = 85.0\ cm^3 \times 9.00\ g/cm^3 = 765\ g$$

EXAMPLE 1.16: The relationship between distance traveled (d), speed (s), and time (t), is expressed in the equation d = st. Use this equation to calculate the average speed of a car that travels 54 miles in 1.2 hours.

$$s = \frac{d}{t} = \frac{54\ \text{miles}}{1.2\ \text{hours}} = 45\ \text{miles/hour}$$

If the unknown is in the denominator, it must be moved to the numerator by one or more operations. Then the problem may be solved by the operations already described.

Procedure 4: Unknowns in the Denominator

$\frac{24}{x} = 6$		$\frac{16}{x} = \frac{4}{5}$	
$24 = 6x$	Multiply by x.	$16 = \frac{4}{5}x$	Multiply by x.
$x = \frac{24}{6} = 4$	Divide by 6.	$x = 16 \times \frac{5}{4} = 20$	Invert 4/5 and multiply.
$\frac{a}{x} = b$		$\frac{a}{x} = \frac{b}{c}$	
$a = bx$	Multiply by x.	$a = \frac{b}{c}x$	Multiply by x.
$x = \frac{a}{b}$	Divide by b.	$x = a \times \frac{c}{b} = ac/b$	Invert b/c and multiply.

An alternative way to solve the right-hand problem above is to "cross-multiply" and then divide by the coefficient of x:

$\frac{16}{x} = \frac{4}{5}$ Cross multiply: $4 \cdot x = 5 \cdot 16$ or $4x = 80$ Divide by 4: $x = 20$

EXAMPLE 1.17: Solve the following equations for x.

$$\frac{72}{x} = 8 \qquad \frac{64}{3x} = \frac{8}{27}$$

$8x = 72$ so $x = 9$ $\quad 3x \cdot 8 = 64 \cdot 27$ so $x = 72$

Calculator: 72, ÷, 8, = and 64, ×, 27, ÷, 3, ÷, 8, =.

Sometimes x will occur in a binomial (two term) expression.

Procedure 5: Binomial Containing the Unknown

$\frac{8}{3 + x} = 2$		$\frac{a}{b + x} = c$	
$8 = 2(3 + x)$	Multiply by (3 + x)	$a = c(b + x)$	Multiply by (b + x)
$8 = 6 + 2x$	Binomial expansion	$a = cb + cx$	Binomial expansion
$2 = 2x$	Subtract 6	$a - cb = cx$	Subtract cb
$x = 1$	Divide by 2	$x = (a - cb)/c$	Divide by c

EXAMPLE 1.18: Solve each of the following equations for x.

$$\frac{12}{x - 40} = \frac{2}{3} \qquad 25(4 - x) = 125$$

$3 \cdot 12 = 2(x - 40)$	Cross multiply.	$4 - x = 5$	Divide by 25.
$18 = x - 40$	Divide by 2.	$4 - 5 = x$	Add x; Subtract 5.
$x = 58$	Add 40.	$x = -1$	Combine.

EXAMPLE 1.19: The *mole* is the chemical unit for amount of substance. The *mole fraction*, X_M, of methanol in a mixture with ethanol is given by the equation

$$X_M = \frac{n_M}{n_M + n_E}$$

Given that $n_M = 1.5$ moles of methanol and that $X_M = 0.25$, what is n_E, the moles of ethanol?

This problem is easier to solve if you first substitute the numbers and then solve for n_E.

$0.25 = \frac{1.5}{1.5 + n_E}$	Substitute.
$0.25(1.5 + n_E) = 1.5$	Multiply by $n_M + 1.5$.
$0.375 + 0.25n_E = 1.5$	Expand the binomial.
$0.25n_E = 1.5 - 0.375 = 1.125$	Subtract 0.375.
$n_E = 1.125/0.25 = 4.5$ moles of ethanol	Divide by 0.25.

Sometimes an unknown appears in two terms. The terms are first combined, and the problem is then solved as before.

EXAMPLE 1.20: Solve this equation for x: $5x - 3 = 2x - 9$

Before you can combine terms, you must get all terms that contain the unknown on one side of the equation and all the numerical terms on the other side. Rearrange the equation that is in the space at the right.

$5x - 2x = -9 + 3$

You must subtract 2x from and add +3 to both sides of the equation.

$5x - 3 = 2x - 9$

$5x - 2x = -9 + 3$

Now combine terms and solve.

$3x = -6$ Combine terms.

$x = -2$ Divide by 3.

Procedure 6: Powers and Roots

$x^2 = 225$

$x = (225)^{1/2} = 15$

$\sqrt[5]{x} = 3$ (or $x^{1/5} = 3$)

$x = 3^5 = 243$

EXAMPLE 1.21: Solve for x in each case.

$\sqrt[3]{x} = 6$ $4x^2 = 64$ $(4x)^2 = 64$

The first problem below must be rewritten in exponential form before Procedure 6 can be applied. To solve the other problems, you will have to use other procedures as well as Procedure 6.

$\sqrt[3]{x} = x^{1/3} = 6$

$(x^{1/3})^3 = 6^3$

$x = 216$

$4x^2 = 64$

$x^2 = 64/4 = 16$

$x = \sqrt{16} = 4$

$(4x)^2 = 64$

$4x = \sqrt{64} = 8$

$x = 2$

In the first problem, the cube root of x—the third root—is the same as x raised to the 1/3 power. The problem is then solved by raising both sides to the third power . . . . In the second problem, you can divide both sides of the equation by the coefficient of x^2 before taking the square roots of both sides. Alternately, you can take the square root of both sides of the original equation, giving $2x = 8$, and then solve for x . . . . In the third problem, you first take the square root of both sides of the equation.

EXAMPLE 1.22: Gases diffuse at a rate that depends on the masses of their molecules. In an experiment, the distances that oxygen and an unknown gas diffused in the same time were measured. The values obtained were $t_{O_2} = 24$ seconds, $t_X = 72$ seconds, and the mass of oxygen molecules is known as 32. What is the mass of the unknown gas X? The relationship is

$$\frac{t_X}{t_{O_2}} = \sqrt{\frac{M_X}{M_{O_2}}}$$

This complicated equation is best solved by substituting the given values and then isolating the unknown.

$$\frac{72}{24} = \sqrt{\frac{M_X}{32}} \qquad \text{Square both sides.}$$

$$3^2 = \frac{M_X}{32} \quad \text{so} \quad M_X = 32 \times 9 = 288$$

1.11 THE QUADRATIC EQUATION

A quadratic equation is one in which the unknown is raised to the second power, as in the second and third problems of Example 1.21. Sometimes you find a problem in which the unknown is squared in one term, appears to the first power (exponent 1, or no exponent) in another term, and is absent in a third term that is a real number. There are methods by which some quadratic equations can be solved, but only one method that solves them all. It is necessary to rearrange the equation so it has the form $ax^2 + bx + c = 0$. The solution is then given by the **quadratic formula**:

$$x = \frac{-b \pm \sqrt{b^2 - 4ac}}{2a} \qquad (1.15)$$

The ± in the formula indicates that there are two values of x that satisfy the equation. These are the two roots described earlier. The conditions of the problem usually make it possible to eliminate one of the roots.

EXAMPLE 1.23: Solve the following equations for x:

$$x^2 - 7x + 12 = 0 \qquad\qquad 2x^2 + x - 10 = 0$$

$x^2 - 7x + 12 = 0$	$2x^2 + x - 10 = 0$
$a = +1, \quad b = -7, \quad c = +12$	$a = +2, \quad b = +1, \quad c = -10$
$x = \dfrac{-(-7) \pm \sqrt{(-7)^2 - (4)(12)}}{2}$	$x = \dfrac{-1 \pm \sqrt{(1)^2 - (4)(-10)}}{2(2)}$
$x = \dfrac{7 \pm 1}{2} = \dfrac{8}{2} = 4 \text{ or } \dfrac{6}{2} = 3$	$y = \dfrac{-1 \pm 81}{4} = \dfrac{-1 \pm 9}{4} = \dfrac{8}{4} = 2 \text{ or } \dfrac{-10}{4} = -\dfrac{5}{2}$

1.12 COMMON MISUSES OF CALCULATORS

This section is added to call attention to, and hopefully prevent, six ways that calculators are commonly misused.

Unnecessary Zeros In scientific writing, decimal numbers smaller than 1 are usually written with a zero in front of the decimal point. The decimal equivalent of 1/4 is .25, which is written 0.25. Even though your calculator will show such values as 0.25, it will accept .25 as a valid entry.

The rules of significant figures sometimes require recording numbers with tail-end zeros to the right of a decimal point, such as 1.30 rather than 1.3. If you must perform a calculation with a 1.30, you will be equally correct by entering 1.3.

Neither of these unnecessary uses of zeros causes an incorrect answer.

Calculating Numerator Products and Denominator Products Separately If a fraction has two or more factors in both the numerator and denominator, you may think first of calculating the numerator and denominator separately, then—in a separate operation—dividing them to get the answer. Example 1.6 is: (72 × 96)/(18 × 16). You may do this in three steps: 72 × 96 = 6912; 18 × 16 = 288; then 6912 ÷ 288 = 24. Or the answer may be obtained in one step by the calculator sequence: 72 × 96 ÷ 18 ÷ 16 = 24. The three-step method takes as many as 21 keystrokes. The one-step method takes only 12. Plus, the one-step method does not require that you write down and re-enter numbers—a process which may produce keystroke errors.

Though the longer and unnecessary procedure nearly doubles the opportunity for keying errors, it will yield a correct answer if performed properly.

Unnecessary Use of Equal Sign Unless you are interested in an intermediate result from a calculation setup, the equal sign need not be used except at the end. For example, = could be pressed after each multiplication or division step in Exercise 1.6, but exactly the same numbers would appear on the display and in the same order. The keying sequence in our first setup for solving the problem was 96 ÷ 16 ÷ 18 × 72. Students who use unnecessary equal signs would key 96 ÷ 16 = ÷ 18 = × 72 =. The two extra keystrokes accomplish nothing.

This calculator misuse does not produce an incorrect result.

Incorrect Interpretation of a Denominator Product A fraction that has two or more factors in the denominator, such as

$$\frac{24}{3 \times 4}$$

is frequently interpreted and entered into the calculator as 24 ÷ 3 × 4 = 32 instead of 24 ÷ 3 ÷ 4 = 2. If there are two or more factors in a denominator, each factor must be used as an independent divisor, not a multiplier. It is correct to interpret the setup as 24/(3 × 4) = 24/12 = 2. Here the product of the denominator factors is used as a divisor.

This calculator misuse produces an incorrect result.

Roundoffs in Display If an answer begins with four, five, or six zeros between the decimal point and the first nonzero digit, a calculator with a limited display may show only the first one or two nonzero digits. For example, a popular eight-digit display calculator gives the following:

$$\frac{2.84}{5{,}700{,}000} = 0.0000004$$

If 0.0000004 had been written as an answer, it would have been incorrect by nearly 20%. The same calculator in exponential notation mode gives the answer as 4.9825×10^{-7}. Some calculators flip automatically to exponential notation when decimal numbers become too long to fit into the display. If yours does not, be sure to change to exponential notation.

If a final answer is shortened by the number of digits the calculator can display, it is usually not acceptable.

Recording the Full Calculator Readout as the Answer Calculators give long answers that suggest very precise results. They are just that for pure numbers. But when the numbers include measured quantities, it is seldom that results have meaning beyond the first three or four digits. This is discussed in Sections 3.7 through 3.12. Until that time, we may show full calculator readouts. Beginning in Chapter 3, all answers are rounded off to a reasonable number of digits.

1.13 THE MOST COMMON CALCULATOR ERROR: NOT THINKING

Don't trust the answer your calculator gives you. Examine every problem you solve. Do some mental arithmetic. Estimate what the answer is going to be. Be sure the answer you get is about what it should be. If it isn't, find out why. You will probably find that your calculator did its job properly, but you did not. None of us are perfect typists on a calculator.

This example may seem foolish, but every chemistry teacher has seen its equivalent many times. If the problem is 27 ÷ 3, the answer is given as 81. When questioned, the student defends the answer: "That's what the calculator said." Of course it did—because what was typed into it was 27 × 3. Challenge every answer: *Is it reasonable?*

Closely related to unreasonable answers is overuse of a calculator. Probably most students would not use a calculator to find 27 ÷ 3. But how about 86 ÷ 2? Unfortunately, many students today would use a calculator to find an answer that is no more complicated than 8/2 = 4 and 6/2 = 3, or 43. And some would even report the result as 172!

Don't reach for a calculator to solve a problem you can do in your head. You will make fewer errors if you use your intelligence rather than a machine. And always be sure your answer is reasonable.

CHAPTER 1 CALCULATOR EXERCISES

Here you will find exercises with which to check yourself on the calculator techniques described in this chapter. The number of problems is small; you will have ample opportunity for practice with the problems section which begins on the next page—and in later chapters!

The exercises below include all of the kinds of calculations that are performed by chemistry students. If you learn how to perform them now, you will be able to concentrate on the chemistry when the same methods are met later. All answers are given, so your task is simply to repeat the calculations on your calculator and confirm the results.

1) $1/456 = 0.0021930$; $456^2 = 207936$; $\sqrt{456} = 21.354157$; $\log 456 = 2.6589648$

2) $2.42^{6.46} = 301.61106$; $0.0569^{0.648} = 0.1560676$;

$47.9^{0.196} = 2.1347438$; $0.368^{2.11} = 0.1213218$

3) antilog $13.49 = 3.0903 \times 10^{13}$; antilog $-8.21 = 6.166 \times 10^{-9}$

4) $$\frac{(6.626 \times 10^{-34})(1.5 \times 10^{15}) + 0.5(9.110 \times 10^{-31})(6.0 \times 10^{7})^2}{6.626 \times 10^{-34}} = 2.476296 \times 10^{18}$$

5) $$\left(\frac{3[(8.314) \times (25 + 273)]}{2.016 \times 10^{-3}}\right)^{0.5} = 1920.1206$$

6) $-\log (5.09 \times 10^{-12}) = 11.293$

7) $$15 + \left[\frac{0.023 \times (5.0)^2}{2.5}\right] \times \left[2.5 - (5.0)(0.16)\right] \div \left[(5.0)(0.08206)\right] = 63.102608$$

8) $$\frac{47.9}{47.9 + 97.3} \times 100 = 32.988981$$

9) $65 \times 1.24 \times (78.2 - 24.9) = 4295.98$

10) $$4.86 \times \frac{298}{305} \times \frac{1.14}{1.39} = 3.8944196$$

11) $$\text{antilog } \frac{63 \times 10^3}{2.303 \times 8.314 \times (273 + 55)} = 1.075048 \times 10^{10}$$

CHAPTER 1 PROBLEMS

1) a) $5.07 - 27.23 + 14.28 - 9.91 + 36.73 =$

b) $66.72 + (-87.54) + 14.05 - 2.04 =$

c) $29.02 - 14.57 + 95.79 - (-7.19) =$ d) $(2.57)(-6.09)(-8.43) =$

e) $\frac{(15.3)(-3.92)(-7.64)}{(2.43)(26.8)(-4.21)} =$

2) a) $9^3 =$ b) $\sqrt{64} =$ c) $91.2^{0.47} =$ d) $\sqrt[4]{560} =$

e) $3^5 =$ f) $12.9^{2.46} =$ g) $27^{1/5} =$ h) $\sqrt[3]{0.306} =$

3) Do not use a calculator on Problem 3. Express the answer as a fraction.

a) $\frac{2}{3} + \frac{5}{8} =$ b) $\frac{5}{2} + \frac{6}{7} + \frac{13}{28} =$ c) $\frac{12}{5} - \frac{4}{3} =$ d) $\frac{7}{8} - \frac{5}{4} - \frac{3}{10} + \frac{17}{20} =$

4) Do not user a calculator on Problem 4. Express the answer as a fraction.

a) $\frac{9}{16} \times \frac{8}{21} \times 24 \times \frac{7}{9} =$ b) $\frac{7}{8} \div \frac{2}{3} \times \frac{1}{5} =$

c) $\frac{12}{4} \times \frac{9}{1/3} \times \frac{6}{18} =$ d) $\frac{2/3}{4/5} \div \frac{28/27}{7/18} =$

5) Simplify each of the following:

a) $(x^4)(x^{-9}) =$ b) $(x^4)^3 =$ c) $\frac{5^{-6}}{5^4} =$

6) Solve each equation for x:

a) $-3x + 2 = -2x - 1$ b) $5x - 2 + 4x = 2x + 12$ c) $3(x - 2) = 12$

d) $\frac{5}{x} = 8$ e) $\frac{16}{x - 4} = \frac{1}{4}$ f) $\frac{4x}{5} - 7 = 5$ g) $-5(2x - 1) - 18 = 3x$

h) $5 + 2(3x - 4) = 7(2x - 3)$ i) $x^2 = 15 + 2x$ j) $5x^2 + 2 = 7x$

7) Solve each of the following equations for the variable indicated:

a) $I = prt$; solve for r b) $A = \frac{1}{2}(b + B)h$; solve for h

c) $y = mx + b$; solve for x d) $s = \frac{1}{2}(a + b + c)$; solve for b

e) $c^2 = a^2 + b^2$; solve for b f) $PV = (g/M)RT$; solve for M

8 - 11: Solve each equation for the unknown variable, as in Problem 7 above. Substitute the given values and calculate the answer.

8) For all triangles, $A = \frac{1}{2}bh$. If the area of a triangle, A, is 46.7 square centimeters and the height, h, is 5.86 cm, find the length of the base, b.

9) The area, A, of a circle is found by the equation $A = \pi r^2$, where r is the radius. If a circle has a radius of 3.79 cm, find the area. $\pi = 3.1416$

10) IQ, or intelligence quotient, is defined as the ratio of a person's mental age, M, to the age of that person in years, Y, expressed as a percentage: IQ = (M/Y) × 100. How old is a student if her mental age is 14 years and she has an IQ of 120?

11) Use the formula speed = distance/time to calculate how long it will take an airplane to fly 1275 miles at 510 miles per hour.

++++++++++++++++++++++

12) a) $-14.3 + 22 + (-3.62) - 7.71 =$ b) $459 - 770 + 126 - 34 - 506 =$

c) $(32.9)(-1.04)(-32.9)(-65.5) =$ d) $\dfrac{(75.6)(56.4)(-406)}{(126)(23)(-8.58)(-0.468)} =$

13) a) $4^7 =$ b) $\sqrt[3]{3.95} =$ c) $0.816^{0.354} =$

d) $67.4^{1/4} =$ e) $\sqrt[8]{646,000} =$

14) Do not use a calculator on Problem 14. Give each answer as a fraction.

a) $\dfrac{6}{5} - \dfrac{8}{9} =$ b) $\dfrac{2}{3} + \dfrac{5}{9} + \dfrac{7}{4} - \dfrac{7}{18} =$

c) $\dfrac{11}{16} - \dfrac{83}{24} + \dfrac{5}{12} =$ d) $\dfrac{7}{13} + \dfrac{15}{39} =$

15) Do not use a calculator on Problem 15. Give each answer as a fraction.

a) $8 \times \dfrac{78}{13} \div \dfrac{12}{35} =$ b) $\dfrac{22}{17} \div \dfrac{44}{51} \div \dfrac{11}{28} \div \dfrac{7}{9} =$

c) $\dfrac{2/3}{102} \times \dfrac{68}{5/18} =$ d) $\dfrac{0.12}{0.07} \div \dfrac{48}{2/3} =$

16) Simplify each of the following:

a) $(x^{-3})(x^{-5}) =$ b) $(3^2)^4 =$ c) $\dfrac{7^5}{7^4} =$

17) Solve each equation for x:

a) $4x + 17 = 2x - 7$ b) $5(x + 6) = x\left(12.5 - \dfrac{19}{2}\right)$ c) $7x = -100 + 16$

d) $\dfrac{x}{3} = \dfrac{5 + 7}{9}$ e) $\dfrac{x + 10}{2} + 15 = 5x - 7$ f) $\dfrac{17}{3x} = 47 - 81$

g) $2x^2 - 15 = 3x$ h) $14 + 5x = 3x^2$ i) $4x^3 = 127$

18) Solve each equation for the variable requested:

a) $K = C + 273$; solve for C

b) $1.8C = F - 32$; solve for C

c) $G = H - TS$; solve for S

d) $M = g/(MW \times V)$; solve for V.

19) Newton's second law of motion states that force, F, is equal to mass, m, times acceleration, a: $F = ma$. What is the mass of an object if a force of 592 kg•m/s^2 produces an acceleration of 9.76 m/s^2?

20) The behavior of a constant amount of gas is described by the equation at the right. Calculate V_2 if P_1 is 740 torr, V_1 is 3.28 liters, T_1 is 295 K, P_2 is 759 torr, and T_2 is 260 K.

$$\frac{P_1V_1}{T_1} = \frac{P_2V_2}{T_2}$$

21) The total pressure of a gaseous mixture is the sum of the partial pressures of the individual gases in the mixture. In an oxygen-nitrogen-helium system, the partial pressure of nitrogen is 390 torr, the partial pressure of helium is 120 torr, and the total pressure of the system is 650 torr. Write the equation for the partial pressure of oxygen, substitute the given values, and calculate the missing quantity.

ADDITIONAL PROBLEMS

22) Solve for T:

$$\left[125 + \frac{(1.36)(132)^2}{(25.0)^2}\right]\left[25.0 - (0.0318)(132)\right] = (132)(0.0821)T$$

23) Solve for n:

$$6.165 \times 10^{14} = 3.288 \times 10^{15}\left[\frac{1}{n^2} - \frac{1}{4^2}\right]$$

24) Solve for T:

$$\ell n\left[\frac{270}{760}\right] = \frac{44{,}000}{8.314}\left[\frac{1}{318} - \frac{1}{T}\right]$$

25) Solve for P:

$$0.318 = 0.205 - \frac{0.05916}{2}\log(1.5/P^2)$$

CHAPTER TEST

You may use a calculator unless instructed otherwise.

1) Perform each of the calculations indicated:

a) $28.63 + 4.095 + (-6.453) - 72.6 - (-1.586) =$

b) $(12.79)(9.44)(0.365) =$

c) $\frac{(-4.54)(404)(92.4)}{(47.8)(0.889)(-9.86)(-87.7)} =$

d) $\sqrt[3]{209} =$

e) $32.4^{0.715} =$

Complete parts f and g without a calculator. Give each answer as a fraction.

f) $\frac{2}{3} \times \frac{15}{32} \times 4 \div \frac{25}{7} =$

g) $\frac{4}{5} + \frac{1}{6} - \frac{1}{2} =$

2) Simplify each of the following expressions:

a) $(z^7)(z^{-9}) =$

b) $\frac{y^3}{y^{-5}} =$

c) $(4^8)(4^{-5}) =$

3) Solve each of the following equations for x:

a) $7 - x + 5 = 3 + 2x$

b) $5 + \frac{6x}{5} + 2 = -17$

c) $\frac{4}{x + 6} = 5 - 4$

d) $\frac{6abx}{2c} = \frac{4d}{e}$

e) $(2x)^3 = 64$

f) $5x^2 = 320$

g) $\sqrt[4]{5x} = 2$

h) $2x^2 - 5x - 7 = 0$

4) Solve the following equation for K: $F = \frac{Kq_1q_2}{r^2}$

5) The mass of several different metals that are deposited by electroplating may be calculated from the equation 3220 m = It(AM), in which m is the mass in grams, I is the current in amperes, t is time in minutes, and AM is the atomic mass of the metal. The atomic mass of copper is 63.5. How many minutes will be required to deposit 8.39 grams of copper at a current of 7.46 amperes?

2. CHEMICAL CALCULATIONS

PREREQUISITES

1. If you have not read HOW TO LEARN CHEMISTRY FROM THIS BOOK at the beginning of this book, please do so now. In this chapter—and through the entire book—you will be solving example problems by the procedure described there.

2. Section 1.4 covered the properties of exponents which will be specifically applied in Section 2.1 on exponential notation.

CHEMICAL SKILLS

1. Exponential-decimal conversion. Convert a number written in conventional decimal form to exponential notation; convert a number written in exponential notation to conventional decimal form.

2. Exponential mathematics. Add, subtract, multiply, and divide numbers expressed in exponential notation.

3. Dimensional analysis. Identify the starting and wanted quantities in the statement of a problem. Beginning with the starting quantity, set up and solve the problem by the dimensional analysis method.

4. Calculating percentage. Given the amount of any part of a mixture or group and the total amount of all parts in the mixture or group, calculate the percentage of the given part.

5. Using percentage. Given the percentage of any part in a mixture or group, and either the quantity of that part or the total quantity in the mixture or group, calculate the other.

2.1 EXPONENTIAL NOTATION

Scientists often work with very large or very small numbers. Calculations with these numbers are much easier if they are expressed in **exponential notation**, which is also known as **scientific notation**. Under this system, any number may be written in the form $N \times 10^m$, where N is the **coefficient** and m is a whole number. N is usually from 1 to less than 10 ($1 \le N < 10$). If, as the result of calculations, N falls outside of that range, it may be restored to "standard" form by adjusting the decimal point and the exponent.

To change a conventional number to exponential notation, first relocate the decimal so it appears immediately behind the first nonzero digit. Count the number of places the decimal was moved. Write the exponential factor as 10 raised to the power equal to that number. If the decimal was moved to the left, the exponent is positive, and no sign is necessary. If the decimal was moved to the right, the exponent is negative, so a minus sign must be inserted.

Consider the number 784,000. In exponential notation this is 7.84×10^e, where e is the exponent to be determined. To set the coefficient between 1 and 10 it was necessary to move the decimal five places left:

7 8 4 0 0 0 .
↑ 5 4 3 2 1 ↓

The five places means the absolute value of e is 5. Moving the decimal left makes e a positive number. Hence $784{,}000 = 7.84 \times 10^5$.

To change 0.000406 to exponential notation, the decimal must move four places to the right:

0 . 0 0 0 4 0 6
↓ 1 2 3 4 ↑

The absolute value of e is 4, the same as the number of places the decimal moved. Because the decimal was moved to the right, the exponent is negative. Hence $0.000406 = 4.06 \times 10^{-4}$.

The left positive/right negative rule for deciding whether the exponent is positive or negative is a simple rule, but one that is easily forgotten if it is not used regularly. The basic fact to remember about exponents is that *a negative exponent means a small number (less than one) and a positive exponent means a large number (greater than one).*

If two numbers, both in exponential notation, are compared, how can we determine which is larger? In terms of positive and negative numbers, *larger means more positive*, and *smaller means more negative*. It is obvious that +5 is larger than +3, and thus that 10^{+5} is larger than 10^{+3}. But it is not so obvious that -3 is larger than -5. It is clear that -3 is more positive than -5, however, and thus 10^{-3} is larger than 10^{-5}. Turning the last comparison around, -5 is more negative than -3, and therefore -5 is smaller than -3. The application of these ideas will appear shortly.

EXAMPLE 2.1: Write 0.0000000674 in exponential notation.

Begin by writing the coefficient with the decimal in its proper place. Then write the × 10, but do not insert an exponent at this time.

6.74 × 10

The decimal is in its proper place behind the first nonzero digit, 6.

Now count the number of places the decimal was moved. Insert that number as the exponent of 10. Do this without regard to the sign of the exponent.

$6.74 \times 10^{\ \ 8}$

The decimal moved eight places to change 0.0000000674 to 6.74.

Now decide on whether the exponent is positive or negative. Is the number 0.0000000674 large (___) or small (___)? Will the exponent be positive (___) or negative (___)? Check the right answers, and then insert a plus sign or a minus sign in front of the exponent 8 above. (Normally a + sign is not necessary, but if that is the right answer, put it in this time only.)

6.74×10^{-8}

The number 0.0000000674 is quite small. When expressed in exponential notation, its exponent must be negative.

Often in working with problems in exponential notation, a calculated answer has the form of exponential notation except that the coefficient is not between 1 and 10. The decimal point must therefore be moved and the exponent changed accordingly.

Suppose a calculated result is 264×10^5, and you wish to write this value with a coefficient of 2.64: $264 \times 10^5 = 2.64 \times 10^e$. What is e? If you were asked to write 264 in exponential notation, you would undoubtedly have no trouble writing 2.64×10^2. So we substitute and then combine exponentials:

$$264 \times 10^5 = (2.64 \times 10^2) \times 10^5 = 2.64 \times (10^2 \times 10^5) = 2.64 \times 10^7$$

EXAMPLE 2.2: A calculated result is 0.000378×10^{-9}. Rewrite the result with a coefficient of 3.78.

Begin by substituting the exponential notation form of 0.000378. Then multiply the exponentials to get the final expression.

$$0.000378 \times 10^{-9} = 3.78 \times 10^{-4} \times 10^{-9} = 3.78 \times 10^{-13}$$

To change a number written in exponential notation to conventional decimal form, you literally perform the indicated multiplication. If the exponent is positive, you are multiplying the coefficient by a large number. The product will be larger than the original coefficient, so the decimal is moved to the right the number of places indicated by the exponent. If the exponent is negative, you are multiplying the coefficient by a number less than 1—actually dividing the coefficient by a large number. The answer will be smaller than the starting coefficient; the decimal is moved to the left the number of places equal to the absolute value of the exponent.

EXAMPLE 2.3: Express 7.24×10^4 and 1.29×10^{-6} in conventional form.

7.24×10^4 = ____________ 1.29×10^{-6} = ______________

$7.24 \times 10^4 = 72{,}400$; $1.29 \times 10^{-6} = 0.00000129$

In each case the number of places the decimal is moved corresponds to the absolute value of the exponent.

Calculations in Exponential Notation The rules for addition, subtraction, multiplication, and division of numbers in exponential notation are summarized below.

1. When multiplying numbers in exponential notation, the coefficients (the numbers in front of the powers of 10) are multiplied in the usual way, and the exponents are *added*.
2. When dividing numbers in exponential notation, the coefficients are divided normally, and the denominator exponent is *subtracted* from the numerator exponent.
3. Adding and subtracting numbers in exponential notation, when both numbers are to the same power of 10, is performed by adding/subtracting the coefficients and keeping the same power of 10.

Of course, your calculator knows the above rules, too. Whenever you do problems with exponentials, your calculator performs these operations without thinking. But that's just the problem. If you make a mistake keying in a number (which is more complicated for exponentials), you may obtain a wildly incorrect answer. Always be sure your answer is reasonable. Develop the skill with exponential math that is necessary for you to estimate the final result.

Multiplying exponentials means adding their exponents. Dividing exponentials means subtracting their exponents. To multiply 4.5×10^8 by 1.4×10^{-3}, we multiply 4.5×1.4 in the conventional way, and then multiply $10^8 \times 10^{-3}$ by adding their exponents. *Don't reach for that calculator yet!* Think first! Figure out the approximate answer in your head. How? Well, 4.5×1.4 is perhaps 5 or 6 something. Multiplying $10^8 \times 10^{-3}$ means adding 8 and -3 to get 5. So the answer is about 6×10^5. *Now* use the calculator to get

$$(4.5 \times 10^8)(1.4 \times 10^{-3}) = 6.3 \times 10^5$$

If you had difficulties getting this result on your calculator, review the material in Sections 1.2 and 1.5.

EXAMPLE 2.4: Estimate the product of the multiplication below. Then find the product on your calculator.

$(9.6 \times 10^{-4})(2.5 \times 10^{-5})$ = ______________(est) = ______________(calc)

$9.6 \times 2.5 \approx 10 \times 2 \approx 20$. $10^{-4} \times 10^{-5} = 10^{-9}$.
Estimated answer: $20 \times 10^{-9} = 2 \times 10^{-8}$. By calculator, 2.4×10^{-8}.

In division of exponentials, the power of 10 of the denominator is subtracted from that of the numerator. Let's estimate the result when we divide 8.5×10^{-4} by 3.4×10^{-7}. 8.5 divided by 3.4 is between 2 and 3. To divide 10^{-4} by 10^{-7} means subtracting -7 from -4: (-4) - (-7) = -4 + 7 = +3. The exponent in the answer is thus 10^3. Our estimate is about 2×10^3. By calculator:

$$\frac{8.5 \times 10^{-4}}{3.4 \times 10^{-7}} = 2.5 \times 10^{3}$$

EXAMPLE 2.5: First estimate, and then calculate, the answer to the following division problem:

$\frac{2.25 \times 10^{3}}{6.00 \times 10^{-6}}$ = ______________(est) = ______________(calc)

$2.25 \div 6.00 \approx 2/6 \approx 0.3$ something. $10^3 \div 10^{-6} = 10^9$.
Estimated answer: 0.3×10^9 or 3×10^8. By calculator, 3.75×10^8.

Multiplication and division of several factors are handled in exactly the same way. Estimate the coefficient by multiplying/dividing all the coefficients given. Determine the exponents of the answer by adding the exponents of all the terms multiplied and subtracting the exponents of those divided. If a number in conventional form is mixed in with numbers in exponential notation, treat it simply as a coefficient—which it is, $\times 10^0$.

EXAMPLE 2.6: Estimate, then calculate:

$\frac{(6.3 \times 10^{-12})(3.6 \times 10^{4})(0.75)}{(4.2 \times 10^{-5})(1.5 \times 10^{8})}$ = ______________(est) = ______________(calc)

Numerator: $6 \times 4 \times 0.75 \approx 18$; Denominator: $4 \times 1.5 \approx 6$; $18/6 \approx 3$.
Summation of exponents: -12 + 4 - (-5) - 8 = -11.
Estimated answer: 3×10^{-11}. By calculator, 2.7×10^{-11}.

In addition and subtraction, remember that all exponentials must be the same before the coefficients are added or subtracted. A calculator takes care of all this. In estimating a sum or difference, the terms with the larger exponents will account for nearly all of the result. For example, the sum of $2.4 \times 10^4 + 3.6 \times 10^4 + 4.5 \times 10^5$ might be estimated at about 5×10^5, not much larger than the third term alone. After equalizing the exponentials, the actual sum is

$$0.24 \times 10^5 + 0.36 \times 10^5 + 4.5 \times 10^5 = 5.1 \times 10^5$$

EXAMPLE 2.7: Estimate the sum and difference below, and then use your calculator to find the answers. Remember that, with negative exponents, 10^{-6} is larger than 10^{-7}.

(a) $3.12 \times 10^{-4} + 6.09 \times 10^{-3} + 5.48 \times 10^{-5}$ = ____________(est)

= ____________(calc)

(b) $6.21 \times 10^{-9} - 8.19 \times 10^{-10}$ = ____________(est) = ____________(calc)

(a) 10^{-3} is the larger exponent. When exponents are adjusted, the 3 in the first term falls into the first decimal column, and the 5 in the third term is in the second decimal column. Estimated answer, about 0.3 + 6, or about 6.3×10^{-3}. By calculator, 6.4568×10^{-3}.

(b) $(6.2 - 0.8) \times 10^{-9}$ is about 5.4×10^{-9}. By calculator, the answer is 5.391×10^{-9}.

2.2 PROPORTIONALITY

Direct Proportionalities If two variable quantities are related in such a way that when one is doubled, the other is doubled; when one is tripled, the other is tripled; or when one is reduced by 35%, the other is reduced by 35%; these quantities are said to be **directly proportional** to each other. The relationship is called a **direct proportionality**. In a direct proportionality, when one quantity changes, the other quantity changes in the same direction.

When you travel at constant speed in a car, the distance you travel is directly proportional to the time you drive. The proportionality may be expressed mathematically as

$$d \propto t \tag{2.1}$$

where d is distance, t is time, and $\propto$ is the symbol for proportionality. For every proportionality, there is a multiplier, known as the **proportionality constant**, by which the proportionality may be changed into an equation. Proportionality constants often have physical meaning. In this case, the proportionality constant is speed, s:

$$d = st \tag{2.2}$$

By solving the equation for the proportionality constant, the units in which it is measured, if any, appear:

$$s = \frac{d}{t} = \frac{\text{miles}}{\text{hour}} = \text{miles per hour} \tag{2.3}$$

Not all proportionality constants have units. One of the first of these constants you met in arithmetic was pi, π. It is a proportionality constant in the equation for the circumference (C) of a circle,

$$C = \pi d \tag{2.4}$$

in which d is the diameter. The circumference is directly proportional to the diameter. Solving the equation for π,

$$\pi = \frac{C}{d} = \frac{\text{length unit}}{\text{length unit}} = \text{dimensionless number} \tag{2.5}$$

Equations 2.3 and 2.5 show that *the ratio of variables is constant in a direct proportionality*. (A ratio is a way of expressing the quotient of two numbers. It is usually written as a fraction, as C/d above. It may also be shown by two dots, as C:d.)

Inverse Proportionalities If two quantities are related in such a way that when one is doubled, the other is halved; or when one is reduced to one-third of its starting value, the other is tripled; the quantities are **inversely proportional** to each other. The relationship is called an **inverse proportionality**. In an inverse proportionality, when one quantity increases, the other quantity decreases; they change in opposite directions.

If 48 minutes are required for a pump to remove the oil from a large tank, two identical pumps should be able to empty the tank in 24 minutes, three pumps in 16 minutes, and four pumps in only 12 minutes. The time (t) is inversely proportional to the number of pumps (p). Expressed mathematically,

$$t \propto \frac{1}{p} \tag{2.6}$$

A proportionality constant (k) changes the relationship to an equation:

$$t = k \times \frac{1}{p} = \frac{k}{p} \tag{2.7}$$

Solving for the proportionality constant gives

$$k = pt = (\text{pumps})(\text{minutes}) = \text{pump-minutes} \tag{2.8}$$

In this case the units of the proportionality constant are pump-minutes. Equation 2.8 shows that *the product of the variables is constant in an inverse proportionality*.

Multiple Proportionalities When one quantity is proportional to each of two or more other variables, it is proportional to the product of those variables. In a manufacturing operation, for example, the number of units produced may be proportional (a) to the number of workers and (b) the number of hours worked. If one worker can produce X units per hour, three workers can produce 3X units per hour. In 7 hours one worker can produce 7X units, and three workers can make 3 × 7 = 21X units. Thus

$$\text{units} \propto \text{workers} \qquad \text{units} \propto \text{hours} \qquad \text{units} = X\,(\text{workers})(\text{hours})$$

where X is the proportionality constant in units per worker-hour.

There are no examples or chemical skills associated with this section. It is included at this point because the ideas it presents underlie many topics throughout the book, beginning in the next section. The main ideas are as follows:

1. In a direct proportionality, when one variable increases, the other variable increases also.
2. The ratio of variables is constant in a direct proportion.
3. In an inverse proportion, when one variable increases, the other variable decreases.
4. The product of variables is constant in an inverse proportionality.
5. If a quantity is proportional to each of two or more variables, it is proportional to the product of those variables.
6. Any proportionality may be changed to an equation by introducing a proportionality constant.
7. A proportionality constant may or may not have units.

2.3 CHEMICAL PROBLEM SOLVING

Before we begin using the mathematical tools you have been studying, we will describe a general approach to problem solving that is applied throughout this book. All problems in chemistry must first be interpreted and translated into some sort of mathematical setup. Then the rules of algebra and arithmetic may be applied to calculate the answer. The last part is easy. The interpretation and translation sometimes are not. The main purpose of this section is to help you develop a way to interpret the problem and put it into a solvable mathematical form. We recommend five steps for solving every problem in this book. They are:

1. Identify (perhaps list) what is given.

At the beginning we suggest that you actually write down what is given in a problem. Sometimes you may not use everything that is given, but write it down anyway. Sometimes you are not given everything you need to know. A problem will not remind you, for example, that there are 60 minutes in an hour. In this chapter we will identify **GIVEN** quantities by name.

2. Identify what is wanted.

Problems are usually quite clear in telling you what to calculate, but it still helps to write it down so you know exactly where you are going. We will talk about the quantity you are asked to **FIND** in this chapter.

3. Identify and write the mathematical connection between the given quantity (or quantities) and the wanted quantity.

This step is the heart of chemical problem solving. We analyze the given and wanted quantities with respect to the theories, principles, and relationships that have been developed in the text and translate them into an equation or calculation setup. Most chemical problems fit into one of two classes—and sometimes a combination of them.

Algebraic Equation. When the given and wanted quantities make up an algebraic equation we strongly recommend that you solve the equation algebraically for the wanted quantity. Then substitute the known values, *with units*, and calculate the answer. In the next chapter you will find that the units become a major aid in checking the correctness of your algebra. This advantage is lost if you substitute numbers directly into the original equation.

Dimensional Analysis. If there is a proportional relationship between the given and wanted quantities, the problem is most easily solved by **dimensional analysis.** The setup begins with a *starting quantity*, which is often the *only* given quantity in the problem. The starting quantity is multiplied by one or more conversion factors that yield the desired result. Again the setup and result include both numbers and units.

4. *Calculate and write down the answer.*
The real work was done in Step 3. At this point you and your calculator decide what the numerical result is, and you alone complete the step by writing down the units. *The units are an essential part of every answer.*

5. *Confirm that the answer is reasonable in both size and units.*
No problem should be considered finished until Step 5 is completed. Is the answer reasonable? Does algebraic treatment of units in the calculation setup yield the correct units in the answer? (You will see what this means in the next section.) If it does not, the setup is wrong. Has your use of the calculator yielded the correct numerical answer? This question is answered by estimating the answer with rounded numbers—mentally, if possible—and comparing the estimated answer with that reached with your calculator.

Frankly, we don't expect that you will approach each problem by counting out Step 1, Step 2, . . . Step 5. But if you *think* about those steps consciously in this chapter and the next, you will develop a routine that will improve your problem solving skills. In the solved examples that follow, we will keep reminding you of the method. Listing GIVEN, FIND, and other symbols we will develop later is the quickest way for us to "explain" how we solved a problem.

2.4 THE DIMENSIONAL ANALYSIS METHOD FOR SOLVING PROBLEMS

Dimensional analysis is one of several names for the problem-solving method you will use in most of this book. It is also known as the *unity factor method*, the *factor-label method*, and by other names. The method will be introduced with a problem:

Problem: How many inches are in 4 feet?

Before you began to read this sentence, you probably calculated the answer mentally: 48 inches. Our interest is not in the answer, however, but in how you reached it. We will complicate this simple example by using it to establish procedures that later will make complicated problems simple.

The Starting Quantity (GIVEN). The given information in the problem is 4 feet. A starting quantity is always an amount of something that can be counted or measured. The starting quantity always includes the units in which it is counted or measured, such as 3 dozen eggs, 18 marbles, 7 feet, or 8 pounds. The starting quantity belongs to a particular problem; it is not useful in any other problem except by coincidence. The starting quantity is the quantity on which any particular problem is based. Much of your success in solving problems depends on your ability to identify the starting quantity.

The Wanted Quantity (FIND). Not only must you know where to start, you must also know where you are going! What quantity have you been asked to find? Unlike the starting quantity, the wanted quantity does not appear in the problem as a given number and unit. The quantity to find above is "how many inches?" Only the unit has been specified: "? inches" will be your answer, where you must calculate the number that will replace the question mark.

Conversion Relationships. The **conversion relationship** in the above problem is 12 inches = 1 foot. It gives the relationship between the units of the given quantity, feet, and the units of the quantity to find, inches. A conversion relationship is an equality or a proportionality between two measured or defined quantities. Conversion relationships may be expressed in equations, as above, or in phrases, such as 12 inches per foot, 3000 cars a month, 8-hour workday, or 45 miles per hour. A conversion relationship does not belong to a particular problem, but may be used in any problem in which its conditions are satisfied.

Some conversion relationships are permanent because they are fixed by definition, such as the 12 inches = 1 foot. The relationship is true for any problem. Its units measure the same thing, length. The relationship is a true equality. Other relationships are temporary, useful for just this problem because two quantities are directly proportional within the conditions of the problem. For example, at a constant speed of 45 miles per hour, or 45 miles/hour, distance is proportional to time.

The Conversion Factor (FACTOR). To use a conversion relationship, it is expressed in a fractional form called a **conversion factor**. Two conversion factors may be written for every relationship. They are reciprocals of each other. The conversion factors for 12 inches = 1 foot are

$$\frac{12 \text{ inches}}{1 \text{ foot}} \text{ and } \frac{1 \text{ foot}}{12 \text{ inches}} \quad \text{or, using abbreviations,} \quad \frac{12 \text{ in.}}{1 \text{ ft}} \text{ and } \frac{1 \text{ ft}}{12 \text{ in.}}$$

(Note: The abbreviation for inch, in., is the only symbol for a unit that is followed by a period.)

The Unit Path (PATH). The **unit path** is the step or series of steps in which units of the given quantity are changed to the units of the quantity to find. Each step involves multiplying by a conversion factor. The unit path is "feet → inches" when changing 4 feet to inches. Multiplying 4 feet by the conversion factor in the form 12 inches/1 foot yields 48 inches, which has the units of the quantity to find, "? inches":

$$4 \text{ ft} \times \frac{12 \text{ in.}}{1 \text{ ft}} = 48 \text{ in.}$$

A conversion factor derived from a true equality, such as 12 inches = 1 foot, does not change the physical meaning of the quantity given; only the units are changed. The distance "4 feet" is *exactly* the same as the distance "48 inches." Conversion factors derived from temporary relationships do change the meaning of the given quantity. Two hours at 45 miles per hour change hours to miles in

$$2 \text{ hours} \times \frac{45 \text{ miles}}{1 \text{ hour}} = 90 \text{ miles}$$

Problem: How many feet are in 36 inches?

Again you probably have the answer already. The given quantity is 36 inches, the quantity to find is ? feet, and the unit path is inches → feet. Almost by instinct, you know you must divide the given quantity, 36 inches, by the conversion factor, 12 inches/1 foot. Remember that when you divide by a fraction, you multiply by its inverse. In this case, this is the same as multiplying by the conversion factor in the form 1 foot/12 inches.

$$36 \text{ in.} \times \frac{1 \text{ ft}}{12 \text{ in.}} = 3 \text{ ft}$$

Get used to thinking of dividing by a fraction as multiplying by its inverse. You will have many occasions to perform this operation.

Problem: Simplify the algebraic expression $4a \times \dfrac{12b}{a}$.

Most students will begin solving this problem by canceling the "a" factor that appears in both numerator and denominator. Then they multiply the numbers to get 48b. This is shown side-by-side with the setup and solution of the question "How many inches are in 4 feet?"

$$4a \times \frac{12b}{a} = 4\cancel{a} \times \frac{12b}{\cancel{a}} = 48b \qquad 4 \text{ ft} \times \frac{12 \text{ in.}}{1 \text{ ft}} = 4 \cancel{\text{ft}} \times \frac{12 \text{ in.}}{1 \cancel{\text{ft}}} = 48 \text{ in.}$$

This shows that units in a problem may be canceled in exactly the same way algebraic variables are canceled. This is the major feature of the dimensional analysis method of solving problems, which may now be stated as a three-step procedure:

1. Identify and write down the quantity given. Then identify and write down the quantity to find.
2. Develop the unit path, a logical series of steps from the given quantity to the quantity to find. You must know the conversion factor between the units in each step of the path.
3. Following the unit path, multiply the given quantity by a conversion factor for each step of the unit path so the unwanted units cancel, leaving only the units of the quantity to find.

$$\left(\text{given quantity}\right) \times \left(\begin{matrix}\text{one or more} \\ \text{conversion factors}\end{matrix}\right) = \text{quantity to find}$$

Your thought process in Step 2 will vary. When the units are very familiar, as in the inch-foot conversions above, you will probably think of the unit path first and the conversion factors second. With unfamiliar units and more complicated problems you are more apt to try to identify conversion factors you know and then figure out how to connect them to solve the problem. Generally, that is what we will do. In your thinking, however, be flexible, ready to go in whatever direction seems logical for the problem at hand.

We caution you not to perform the second and third steps mechanically, without thinking why you are doing what you do, blindly juggling conversion factors until the units "come out right." You can get right answers this way, but little or no understanding of the problem. It is better to think about each step and figure out whether you should multiply by a given conversion factor or its inverse. If you do this thoughtfully, each step will be logical and will make sense. Calculation to any point in the setup will have physical meaning and the units will identify that meaning.

What happens if your setup is wrong? As you check to see if your answer is reasonable, you will find that the units of your answer are "nonsense units." For example, if, in finding the number of feet in 36 inches, you had multiplied by 12 inches per foot instead of 1 foot per 12 inches, your result would have been

$$36 \text{ in.} \times \frac{12 \text{ in.}}{1 \text{ ft}} = 432 \text{ in.}^2/\text{ft}$$

"Inches squared per foot" are nonsense units. Whenever you get an answer with nonsense units, go back to the setup, find the error and correct it.

EXAMPLE 2.8: How many minutes are in 2 weeks?

Identify the given quantity, the quantity to find, and the conversion factors you need to solve the problem. Write them below.

GIVEN: 2 weeks FIND: ? minutes
FACTORS: 7 days/week; 24 hours/day; 60 minutes/hour

Beginning with the given quantity, we wish to set up conversion factors to change weeks to minutes. We do not have a direct conversion from weeks to minutes, so several steps will be required. Can you develop a unit path that begins with weeks, passes through a series of time units included in the above conversion factors, and ends in minutes? Summarize your procedure by writing the unit path below.

PATH: weeks → days → hours → minutes

Now set up the first of the conversion factors. Begin with the given quantity. Then select the conversion factor that will change weeks to days. Write it as a multiplier of the given quantity.

$$2 \cancel{\text{weeks}} \times \frac{7 \text{ days}}{1 \cancel{\text{week}}} \times \frac{\qquad}{\qquad}$$

Notice that the unit "weeks" cancels in the setup so far. If you had used the wrong conversion factor, 1 week/7 days, there would have been no cancellation. You would have been left with nonsense units, weeks²/day, at this point.

The numerical answer so far is 14 days. We really have no interest in this intermediate answer, so we simply extend the setup for the next step along the unit path. This means multiplying by the conversion factor that converts days to hours. Insert the numbers and units in the space provided.

$$2\ \text{weeks} \times \frac{7\ \text{days}}{1\ \text{week}} \times \frac{24\ \text{hours}}{1\ \text{day}} \times \underline{\qquad\qquad} =$$

Calculated this far you would have the number of hours in 2 weeks, another intermediate answer we aren't interested in.

Following the unit path, set up the remaining conversion factor. You will then have the units of the wanted quantity. Calculate the answer.

$$2\ \text{weeks} \times \frac{7\ \text{days}}{1\ \text{week}} \times \frac{24\ \text{hours}}{1\ \text{day}} \times \frac{60\ \text{minutes}}{1\ \text{hour}} = 20{,}160\ \text{minutes}$$

EXAMPLE 2.9: A student has a part-time job in which she works 4-hour shifts. Her pay scale is $5 per hour. How many shifts must she work in order to earn $220?

The statement of the problem includes a given quantity and some conversion factors. Identify them, as well as the quantity to find.

GIVEN: 220 dollars FIND: ? shifts
FACTORS: 5 dollars/hour; 4 hours/shift

Now think through the problem. There is no direct conversion between dollars and shifts. But there is a way that requires more than one step. Write the unit path.

PATH: dollars → hours → shifts

You know how many dollars each hour is worth, and you also know how many hours are in each shift. "Hours" is the connecting link between dollars and shifts.

This time see if you can complete the entire setup of the problem. Start with the given quantity. Think whether you will multiply by each conversion factor or divide by it (multiply by its inverse). Calculate the answer. Be sure it is reasonable in both numbers and units.

$$220 \text{ ~~dollars~~} \times \frac{1 \text{ ~~hour~~}}{5 \text{ ~~dollars~~}} \times \frac{1 \text{ shift}}{4 \text{ ~~hours~~}} = 11 \text{ shifts}$$

You have seen that cancellation of units is a major feature of the dimensional analysis method. It is a part of most problem setups in this book. Nevertheless, we will no longer draw the cancellation lines in the printed setups. The main reason for this is that the *act* of drawing them, rather than seeing them already drawn, helps you to learn dimensional analysis. When you examine a printed setup, therefore, draw the cancellation lines yourself. Doing so will direct your attention to the very thing you should see in each step of the problem setup.

The dimensional analysis method is a powerful problem-solving tool. It lets you see the whole problem at once as a series of logical steps that make up the unit path. If you understand these steps, you understand the problem. With each problem you solve, demand that understanding of yourself. It may take a little more time at the beginning, but your success in solving chemistry problems will be assured.

2.5 PER: A RATIO AND A DIVISION PROBLEM

EXAMPLE 2.10: If you used 16.0 gallons of gasoline when driving 376 miles, what was your gas mileage over that distance?

This section begins with a familiar calculation to point out that you are already acquainted with what is to be presented. The problem asks the number of miles a car traveled for each gallon of gasoline burned on a particular trip. Calculate the answer, and be sure to include its units.

$$\frac{376 \text{ miles}}{16.0 \text{ gallons}} = 23.5 \text{ miles/gallon, or miles per gallon}$$

In the above example you divided a measured number of miles by a measured number of gallons. If you extended the "miles/gallon" units to words, you read the ratio as "miles *per* gallon." The arithmetical operation, division, and the use of the word *per* are the things to notice here. *Per* implies division. Speed in miles per hour is obtained by dividing the distance traveled by the time required. A farmer might calculate crop yield in bushels per acre by dividing bushels harvested by acres planted. A manufacturer calculates production in units per day by dividing total units by working days. Many examples could be given.

Per also suggests a comparison between two kinds of units and a proportional relationship between them. The mileage found in Example 2.10 can be used as a conversion factor to solve other problems in which this same mileage is obtained.

EXAMPLE 2.11: At 23.5 miles per gallon, how many miles can you travel on a full tank of 22.0 gallons?

List the given quantity, quantity to find, conversion factor, and unit path. Set up and solve the problem.

GIVEN: 22.0 gallons FIND: ? miles
FACTOR: 23.5 miles/gallon
PATH: gallons → miles

$$22.0 \text{ gallons} \times \frac{23.5 \text{ miles}}{1 \text{ gallon}} = 517 \text{ miles}$$

EXAMPLE 2.12: At the 23.5 miles per gallon calculated in Example 2.10, how many dollars will you spend on gasoline during a 658-mile trip if the price of gasoline is \$1.35 per gallon?

GIVEN: 658 miles FIND: ? dollars
FACTORS: 23.5 miles/gallon; \$1.35/gallon
PATH: miles → gallons → dollars

$$658 \text{ miles} \times \frac{1 \text{ gallon}}{23.5 \text{ miles}} \times \frac{\$1.35}{\text{gallon}} = \$37.80$$

2.6 PERCENT: PARTS PER HUNDRED

Cent refers to *hundred*. There are 100 cents in a dollar, and 100 years in a century. *Per cent* therefore means *per 100*. When we say X percent of a mixture is component A, it means there are X parts of A per 100 parts of all things in the mixture. "Parts" may be measured in grams, liters, any other unit, or simply counting units.

Percentage problems fall into two categories. In one you are given information about the composition of a mixture and asked to calculate the percentage of one item in the mixture. This is done by substituting into an equation that is developed in the next paragraph. In the other category, the percentage is given and you are asked to find either the quantity of the mixture or of a part of the mixture. This can *usually* be done by dimensional analysis.

Finding Percentage The percentage of one part of a mixture may be figured out as a "ratio and proportion" problem. Imagine that you and a friend shared the driving in the 376-mile trip of Example 2.10, and that you drove 141 of those miles. The *part quantity* you drove was 141 miles; the *total quantity* was 376 miles. The ratio of your part quantity to the total quantity is 141 miles/376 miles. That ratio is to be expressed as parts per 100 total. Therefore

$$\frac{141\text{ miles}}{376\text{ miles}} = \frac{X}{100}$$

where X is the parts per hundred you drove—the percentage of the driving done by you. Solving for X,

$$X = \frac{141\text{ miles}}{376\text{ miles}} \times 100 = 37.5\%$$

These ideas may be summarized in a defining equation for percent, which has exactly the same form:

$$\text{percent} = \frac{\text{part quantity}}{\text{total quantity}} \times 100 \qquad (2.9)$$

The next example is solved by the above algebraic equation, rather than by dimensional analysis. There are two given quantities, rather than just one as was the case before. The conversion factor and unit path are replaced by the equation (EQN).

EXAMPLE 2.13: Calculate the percentage of salt in a solution prepared by dissolving 10.0 grams of salt in 40.0 grams of water.

Write the GIVEN and FIND quantities. Follow with the equation. Substitute the given values and solve the problem. Do it carefully. What is the part quantity? What is the total quantity?

GIVEN: 10.0 g salt, 40.0 g water FIND: % salt

$$\text{EQN: } \%\text{ salt} = \frac{\text{g salt}}{\text{total grams}} \times 100 = \frac{10.0\text{ g salt}}{50.0\text{ g solution}} \times 100 = 20.0\%$$

The total quantity is the grams of salt *plus* the grams of water, not just the grams of water.

Using a Given Percentage If the percentage of one part of a mixture is *known* and you are given either the quantity of that part or the total quantity of the mixture, you can find the other by a dimensional analysis calculation. This is based on the "parts per hundred" idea of percentage. If X percent of a mixture is A, the following equation may be written:

$$X\% = \frac{\text{X parts A}}{\text{100 parts mixture}} \tag{2.10}$$

In the sense of X parts of A per 100 parts of mixture, Equation 2.10 is a conversion factor between parts of A and parts of mixture. For example, if there are 176 people in a large lecture section and 62.5% of them are women, how many women are in the class? The given quantity is 176 people. The quantity to find is ? women. The unit path is people → women. The percentage gives the needed conversion factor of 62.5 women/100 people. Therefore

$$176 \text{ people} \times \frac{62.5 \text{ women}}{100 \text{ people}} = 110 \text{ women}$$

EXAMPLE 2.14: How many grams of salt and how many grams of water would you use to prepare 85.0 grams of a solution that is 12.0% salt?

Start by finding the grams of salt. Remember the steps: given quantity, quantity to find, conversion factor, unit path, and then solve the problem.

GIVEN: 85.0 g solution FIND: ? g salt
FACTOR: 85.0 g salt/100 g solution
PATH: g solution → g salt

$$85.0 \text{ g solution} \times \frac{12.0 \text{ g salt}}{100 \text{ g solution}} = 10.2 \text{ g salt}$$

The total mass of the solution will be 85.0 grams. The mass of salt in the solution will be 10.2 grams. How many grams of water must be used? Set up and solve.

85.0 g solution - 10.2 g salt = 74.8 g water

EXAMPLE 2.15: A mineral of copper is called *bluestone*; its copper content is 25.5%. In a certain experiment, 561 grams of copper were recovered from a pure sample of the mineral. What was the mass of the sample?

The dimensional analysis method should now be a habit: GIVEN, FIND, PATH, FACTOR, SOLUTION.

GIVEN: 561 g copper FIND: ? g sample
FACTOR: 25.5 g copper/100 g sample
PATH: g copper → g sample

$$561 \text{ g copper} \times \frac{100 \text{ g sample}}{25.5 \text{ g copper}} = 2200 \text{ g sample}$$

The dimensional analysis approach to solving percentage problems is no doubt different from what you have learned previously. Students generally find it easier than conventional methods, and they make fewer errors. It will be used whenever the occasion arises in this book.

CHAPTER 2 PROBLEMS

1) Write each of the following numbers in exponential notation:

a) 91,100 b) 0.000000075 c) 6400 d) 0.000165 e) 0.0816 f) 935

2) Convert each of the following numbers in exponential notation to conventional decimal form.

a) 2.24×10^{-5} b) 9.3×10^{2} c) 4.20×10^{4}
d) 2.95×10^{-3} e) 7.35×10^{-2} f) 8.18×10^{-12}

3) Rewrite each number with a coefficient between 1 and 10:

a) 278×10^{3} b) 92×10^{-4} c) 0.611×10^{5}
d) 43.9×10^{-1} e) 0.0165×10^{-2} f) 0.0641×10^{6}.

4) Complete the following calculations and, if appropriate, express the result in exponential notation:

a) $(6.66 \times 10^{6})(4.94 \times 10^{4}) =$ b) $(9.30 \times 10^{4})(4.39 \times 10^{-12}) =$

c) $(6.25 \times 10^{-8})(5.81 \times 10^{-3}) =$ d) $\dfrac{8.56 \times 10^{3}}{9.72 \times 10^{2}} =$

e) $\dfrac{5.37 \times 10^{9}}{4.41 \times 10^{-3}} =$ f) $\dfrac{6.68 \times 10^{-8}}{3.67 \times 10^{-5}} =$

g) $\dfrac{(9.80 \times 10^{2})(1.2 \times 10^{1})(7.34 \times 10^{-4})}{(4.3 \times 10^{-5})(3.42 \times 10^{3})} =$

h) $\dfrac{(5.69 \times 10^{5})(6.51 \times 10^{-4})(2.91 \times 10^{2})}{(8.63 \times 10^{9})(6.1)(1.49 \times 10^{-3})} =$

5) Perform the following additions and/or subtractions. Express the result with a coefficient between 1 and 10.

a) $1.76 \times 10^{9} + 4.65 \times 10^{10} =$ b) $7.15 \times 10^{-2} - 2.89 \times 10^{-3} =$

c) $9.16 \times 10^{8} + 1.64 \times 10^{6} + 5.28 \times 10^{7} =$

d) $2.72 \times 10^{-4} + 8.16 \times 10^{-6} - 6.09 \times 10^{-5} =$

6) A bicyclist can ride a bicycle at an average speed of 1320 feet per minute. Use dimensional analysis to calculate the hours needed to travel 25 miles. (There are 5280 feet in 1 mile.)

7) A barbecue is being planned for 175 people. The ice cream sandwiches for desert cost \$1.68 per dozen. It is assumed that the average number of sandwiches eaten per person will be 1.25. What will it cost to buy ice cream sandwiches for the barbecue? Solve the problem by dimensional analysis.

8) A public school district furnishes pencils to its elementary school children. The pencils the secretary orders from the district warehouse each year are packaged in boxes that contain one gross (12 dozen = 1 gross) pencils. The average use of pencils is 9.3 pencils per student over the school year. If the school's enrollment is 812 students, what is the minimum number of boxes of pencils that should be ordered? Use dimensional analysis.

9) What will be the cost of gasoline for a round trip between San Francisco and Los Angeles in an automobile that averages 24.0 miles per gallon? The price of gasoline is \$1.44 per gallon, and the one-way distance between the cities is 403 miles. Solve the problem by dimensional analysis.

10) If an investor pays \$2925.00 for 225 shares of stock in a chemical manufacturing company, what is the price per share of stock?

11) An equipment manufacturer finds that the mass of one gross (144) cartons of a certain replacement part weighs 2088 pounds. Calculate the weight per carton.

12) A gold-bearing ore was found to contain 6.48 grams of gold. What is the percentage of gold in the ore if the mass of the sample was 75.0 grams?

13) Calculate the percentage of ammonia in a solution prepared by dissolving 13.2 grams of ammonia in 225 grams of water.

14) A hydrochloric acid solution is 37.0% hydrogen chloride. How many grams of hydrogen chloride are present in 542 grams of solution?

15) When a salt solution weighing 175 grams is evaporated to dryness, 44.0 grams of salt remain. What percentage of the original solution was salt?

16) A silver ore has been analyzed and found to be 23.4% silver. How many grams of ore are needed to obtain 50.0 grams of silver?

17) A source of iron is the iron oxide found in magnetite ore. If a particular ore is 27.0% iron oxide, how much of the oxide can be obtained from 520 kilograms of that ore?

18) How many grams of an 8.50% sucrose solution must be poured out to get 40.4 grams of sucrose?

19) Water is 11.2% hydrogen and 88.8% oxygen by weight. If 75.5 grams of hydrogen are produced by decomposing water, how many grams of water were decomposed? Also, how many grams of oxygen were produced?

++++++++++++++++++++++

20) Write each of the following numbers in exponential notation:

a) 707,000 b) 0.00563 c) 6,460,000,000 d) 0.0000000000134

21) Write each of the following numbers in conventional decimal form:

a) 1.48×10^{-5} b) 8.50×10^{12} c) 7.98×10^{-22} d) 9.37×10^{7}

22) Rewrite the following numbers so the coefficient is between 1 and 10:

a) 56.1×10^{9} b) 0.00115×10^{-8} c) $68\ 640 \times 10^{-2}$ d) 0.000389×10^{14}

23) Complete each of the following calculations:

a) $(5.63 \times 10^{25})(7.76 \times 10^{-8}) =$ b) $(4.80 \times 10^{13})(6.20 \times 10^{31}) =$

c) $\dfrac{1.97 \times 10^{26}}{9.07 \times 10^{13}} =$ d) $\dfrac{2.69 \times 10^{6}}{7.87 \times 10^{-11}} =$

e) $\dfrac{(6.39 \times 10^{-4})(6.42 \times 10^{7})}{(7.24 \times 10^{5})(8.08 \times 10^{8})(7.41 \times 10^{-30})} =$

24) Perform the following calculations:

a) $3.12 \times 10^{-5} + 1.62 \times 10^{-5} + 5.24 \times 10^{-6} =$

b) $1.36 \times 10^{6} - 9.42 \times 10^{5} =$

c) $2.95 \times 10^{-3} - 9.32 \times 10^{-3} + 4.16 \times 10^{-2} =$

25) A fruit stand advertises peaches for sale at $5.00 per twenty pound box. If you count 45 peaches in the box, what is the cost of each peach?

26) The tread on a certain automobile tire wears by one thousandth of an inch (0.001 inch) for each 2600 miles driven. After how many months must a tire with 0.007 inch of tread be replaced if it is driven an average of 375 miles per week? (The average month has 4.33 weeks.)

27) A petroleum storage tank 35.5 feet in diameter has a capacity of 617 gallons per inch of depth. How many hours will it take a pump operating at 239 gallons per minute to lower the depth of oil in the tank from 11.4 feet to 2.1 feet?

28) There were 1629 babies born in a municipal hospital in 1982. Calculate the average rate of births per month.

29) During a half-hour period, 247 automobiles pass a checkpoint on a freeway. What was the average flow of traffic in cars per minute?

30) Out of 11,300 units coming off a production line, 724 are rejected by quality control. What is the reject percentage?

31) As a result of a promotional campaign, a new magazine received 132,338 "subscriptions" which the customer may cancel after receiving the first issue. Past experience indicates that 63.9% of those receiving the first issue will continue their subscriptions. How many permanent readers is this?

32) The expected response to a direct mail survey is 17%. How many questionnaires must be mailed to produce at least 500 replies?

33) A Christmas bird count conducted by the National Audubon Society reported 231 sightings of a certain bird in 1981. In 1982 only 197 of these birds were seen. What is the percentage decline in sightings?

34) It has been observed that 14% of the automobiles entering an airport parking lot remain for more than 24 hours. If 84 cars arriving on a certain day are still there 24 hours later, estimate the total number of cars using the lot that day.

ADDITIONAL PROBLEMS

35) A student measures the weight of an object several times and obtains: 1.18 g, 1.28 g, 1.26 g, 1.22 g and 1.16 g. a) What is the average value? b) A number often used to measure the precision of such data is the average deviation, d. Compute d as the average of the absolute values of the differences between each number and the average. c) Also used to report precision is the relative average deviation, RAD. Compute RAD as the percentage which the deviation is of the average.

36) Express the number of seconds in a year in exponential notation. 1 minute = 60 seconds, 1 hour = 60 minutes, 1 day = 24 hours, and 1 year = 365.25 days.

37) Identify whether each of these pairs of variables are directly or inversely proportional. Assume that other factors do not influence the relationship. a) gallons of gasoline consumed and miles traveled in a vehicle, b) body weight and time spent exercising, c) weight of a ball and distance traveled when thrown, and d) volume of water in a pot and the amount of heat needed to bring it to a boil.

38) Percentage error is defined as $[(e - t)/t] \times 100$ where e is the experimental value and t is the true value of the quantity measured. What is the percentage error if a student measures the mass of an object to be 2.5 grams when its actual weight is 2.0 grams?

CHAPTER TEST

1) Express in exponential notation the numbers given in conventional decimal form; express in conventional decimal form the numbers given in exponential notation.

a) 7,835,000 b) 0.0000218 c) 3.19×10^{-4} d) 5.03×10^{5}

2) Calculate the following: $\dfrac{(2.89 \times 10^{5})(5)(3.08 \times 10^{-6})(1.09 \times 10^{-4})}{(7.21 \times 10^{3})(5.82 \times 10^{-6})(3)}$

3) Calculate the following: $6.21 \times 10^{4} + 3.06 \times 10^{3} + 5.54 \times 10^{5}$

4) Calculate the following: $7.19 \times 10^{-3} - 8.24 \times 10^{-4} =$

5) How many seconds does it take a race horse to run six furlongs at 40.7 miles per hour if: 1 furlong = 40 rods; 5.5 yards = 1 rod; and 5280 feet = 1 mile?

6) A sample of a chemical compound is analyzed and found to contain 2.31 grams of sodium, 1.40 grams of nitrogen, and 4.80 grams of oxygen. Calculate the percentage of sodium in the compound.

7) The newspaper account of a local election states that the 18,421 votes cast represented 67.8% of the registered voters. How many registered voters are in that district?

8) The registrar of a college reported that this year's enrollment is only 91.2% of last year's total of 7419. What is this year's enrollment?

3. MEASUREMENT

PREREQUISITES

1. Section 2.4 describes dimensional analysis, which is used in making conversions within the metric system of units. Density problems are also solved by the dimensional analysis method.

2. Sections 1.10 and 1.11 review algebraic manipulations that appear in temperature conversions.

3. Section 2.1 on exponential notation is used extensively in working with the metric system.

CHEMICAL SKILLS

1. Significant figures. Count the number of significant figures in a given quantity.

2. Rounding off. Round off a given quantity to a specified number of significant figures.

3. Math operations with significant figures. Add, subtract, multiply, and divide given quantities and express the result in the proper number of significant figures.

4. Metric unit conversions. Given a mass, length, or volume measurement in any metric unit, convert that quantity to any other metric unit.

5. English-metric unit conversion. Given a mass, length, or volume measurement in any metric or English unit, and an English-metric conversion table, convert any other English or metric unit.

6. Unit conversion with squared or cubed units. Given any squared or cubed measurement, convert to any other squared or cubed unit.

7. Temperature conversion. Given a temperature in degrees Celsius, degrees Fahrenheit, or kelvins, convert to the other two scales.

8. Density. Given two of the following, calculate the third: (a) mass of a sample of a substance; (b) volume occupied by that sample; and (c) density of the substance.

3.1 METRIC UNITS

In the sciences, metric units of measurement are used almost exclusively. These units are being further refined and formalized as **SI units, the System of International Units**. In this book we will use the units most commonly found in recent textbooks, while calling attention to alternatives when appropriate. You should be aware of these alternatives and be flexible enough to adopt whatever is required by your instructor.

The **metric system** is a decimal system in which large and small units for the same measurement differ by multiples of 10. A particular unit is assigned to each kind of measurement. The powers of 10 that modify the size of the unit are identified by a series of prefixes that are shown in Table 3.1. The common metric conversion factors at the bottom should be memorized.

Measurement of Mass Technically, the **mass** of an object is a measure of that object's resistance to a change in its state of motion, a property called **inertia**. The **weight** of an object is a measure of the attractive force it experiences in a gravitational field. In a constant gravitational field, such as on the surface of the earth, weight is proportional to mass. Both depend on the amount of matter in the sample. In most chemical considerations the distinction between the terms is not important, and they are used interchangeably. Accordingly, to "weigh" something means to measure its mass; if an object "weighs" 4 kilograms, its mass is 4 kilograms.

The standard unit of mass is the **kilogram**, the mass of a platinum-iridium cylinder stored at the International Bureau of Weights and Measures in France. In the laboratory, mass is measured on a balance in grams; 1000 grams = 1 kilogram. A kilogram is equivalent to about 2.2 pounds.

Measurement of Length The standard unit of length in the metric system is the **meter**, which is slightly more than 3 feet. The more common smaller units are the centimeter (1/100 meter) and the millimeter (1/1000 meter). Longer distances are measured in kilometers (1000 meters). In the laboratory, a metric ruler is commonly used to measure length.

TABLE 3.1: METRIC PREFIXES

Symbols and Prefixes

NUMBER OF UNITS	PREFIX	SYMBOL	NUMBER OF UNITS	PREFIX	SYMBOL
1,000,000,000 = 10^9	giga-	G	0.1 = 10^{-1}	deci-	d
1,000,000 = 10^6	mega-	M	0.01 = 10^{-2}	centi-	c
1,000 = 10^3	kilo-	k	0.001 = 10^{-3}	milli-	m
100 = 10^2	hecto-	h	0.000001 = 10^{-6}	micro-	μ
10 = 10^1	deka-	da	0.000000001 = 10^{-9}	nano-	n

Common Metric Conversion Relationships

1 kilounit = 10^3 units
1 centiunit = 10^{-2} units
1 milliunit = 10^{-3} units
1 microunit = 10^{-6} units
1 nanounit = 10^{-9} units

Measurement of Volume Many different units are commonly used for volume: the gallon, quart, and fluid ounce are the popular measures. The metric system uses the **liter**, abbreviated L, modified by the usual metric prefixes for larger and smaller quantities. A liter is just a bit larger than a quart.

Within the metric system, 1 liter is exactly equal to 1 cubic decimeter, 1 dm^3. This volume may be represented as a cube measuring 1 decimeter, or 10 centimeters, on each edge. In cubic centimeters, the volume of the cube is 10 cm × 10 cm × 10 cm or 1000 cm^3. This leads to an important relationship:

$$1 \text{ L} = 1000 \text{ cm}^3 = 10^3 \text{ cm}^3 \tag{3.1}$$

Since 1 L = 1000 cm^3, and by the metric system of prefixes, 1 L = 1000 mL, then

$$1 \text{ mL} = 1 \text{ cm}^3 \tag{3.2}$$

In other words, a cubic centimeter and a milliliter are the same volume.

In the laboratory, volume is usually measured with a graduated cylinder. Other volume-measuring devices are the volumetric flask, pipet, and buret.

3.2 UNCERTAINTY IN MEASUREMENT

A common laboratory procedure is to determine the mass of an object by weighing it on a balance. Let's suppose two types of balances are available, a centigram balance and an electronic analytical balance. The smallest measurement unit on the centigram balance is the centigram, 0.01 gram. That is why it is called a centigram balance. The analytical balance, on the other hand, measures mass to the ten-thousandth of a gram, 0.0001 gram.

You begin by weighing the object on a centigram balance. When the indicator shows that the instrument is in balance, you find that the ten-gram beam reads 30 g, the one-gram beam reads 6 g, and the tenth-gram beam reads 0.4 and the centigram indicator shows 0.05. You record the mass as 36.45 g.

The final digit, the 5 in the hundredths place, is uncertain. Measuring to the centigram on a centigram balance is pushing the balance to the very limit of its capability. As a mechanical device, it is less than perfect; no mechanical device is perfect. This first uncertain digit is called the **doubtful digit**.

If you were to remove the object from the balance, recheck the zero point of the balance, and weigh the object again, you would probably obtain the same weight—except for the doubtful digit. Successive measurements might yield 36.44 g, 36.45 g, 36.46 g, and 36.45 g.

The reported value is 36.45 g. This number has four digits. The 3 in the tens place, the 6 in the ones place, and the 4 in the tenths place, are certain. The 5 in the hundredths place is doubtful. Taken together, all the certain digits plus the first uncertain digit—the doubtful digit—make up the **significant figures** of the reported value. The quantity 36.45 g has four significant figures.

The decimal place of the last significant figure of a measured quantity indicates the **precision** of the measuring instrument. Precision is a measure of the agreement among a series of measurements. The centigram balance has a precision of about ±0.01 g.

If you now weigh the same object on the analytical balance, you get different answers! The values obtained are 36.4522 g, 36.4529 g, and 36.4527g. The reported average is 36.4526 g. This number has six significant figures. The doubtful digit is four places to the right of the decimal. The precision of the analytical balance is about ±0.0004 g. The analytical balance is a much more precise instrument than is the centigram balance.

Almost all of the values in this book are measured quantities. Therefore, they have some uncertainty. One of the goals of this chapter is to help you to recognize the uncertainty in measured quantities and to show you how to account for that uncertainty in calculated results based on those measurements.

One additional point is to be made. Let's say that you weighed the above object on a third balance and obtained weights of 39.72 g, 39.70 g, and 39.71 g. These results agree well with each other, and so are precise—at least to within ±0.01 g. However, they do not agree with the measurements made on the two other balances. The **accuracy** of the measurements is in doubt. Accuracy is a measure of how close the reported value is to the true, or accepted, value. As you can see, it is quite possible for results to be precise but not accurate!

3.3 COUNTING SIGNIFICANT FIGURES

The rules listed below will be used to count the number of significant figures in all values given in this book. It should be mentioned that not all instructors follow the same set of rules. If your instructor gives you a different set of rules, follow them!

1. The concept of significant figures applies only to measured quantities. Accordingly, counting numbers, such as 32 people, and defined quantities, such as 1000 m/km, are not measured and do not have significant figures. They are exact numbers.
2. Begin counting significant figures at the first nonzero digit.
3. Stop counting significant figures at the doubtful digit, which should be the last digit recorded.

Zeros are sometimes confusing when it comes to counting significant figures, so let's examine them more closely. Suppose you weighed an object on a centigram balance with a precision of ±0.01 gram, and the mass was 50.30 grams. Notice that the mass is recorded at 50.30 grams, not 50.3 grams. The tail-end zero must be shown because it is the doubtful digit. Beginning with the 5 and counting to the zero—the doubtful digit, the last digit shown—the measurement has four significant figures.

Now suppose you wish to express the same measurement—a measurement good to four significant figures—in kilograms. 50.30 grams is equivalent to 0.05030 kg. (We will discuss how to convert from one metric unit to another later in this chapter.) Counting significant figures again begins at the first nonzero digit, 5. Specifically, counting significant figures does not begin at the decimal point—unless the first nonzero digit is in the tenths place.

If the same measurement, still a four-significant-figure number, were to be expressed in milligrams the value would be 50,300 mg. Here a problem arises. Knowing the history of the measurement, we recognize that the zero after the 3 is the doubtful digit and is therefore counted as significant. The second zero after the 3 is necessary to locate the decimal point, but it is not significant. This violates the rule that the last digit shown is the doubtful digit. Tail-end zeros to the right of the decimal point are ambiguous when it comes to counting significant figures.

This problem is resolved simply by writing the number in exponential notation. The coefficient is written in the proper number of significant figures, with the doubtful digit being the last digit shown. Hence, the mass is 5.030×10^4 mg. If the measurement had been good to only three significant figures it would be 5.03×10^4; and if five significant figures, 5.0300×10^4. Actually, exponential notation resolves all questions about the number of significant figures in a value, large or small. The same quantity could also be expressed 5.030×10^{-2} kg rather than 0.05030 kg. There is no temptation to begin counting significant figures at the decimal point when exponential notation is used.

EXAMPLE 3.1: If the following numbers represent measured quantities or results calculated from measurements, write next to each quantity the number of significant figures:

(a) 3.82 L ______ (b) 24 m ______ (c) 0.0619 kg ______

(d) 2.1×10^4 g ______ (e) 3.4610 km ______ (f) 712,000 cm ______

(a) 3; (b) 2; (c) 3; (d) 2; (e) 5; (f) 3 to 6 (ambiguous)

In (c), 0.0619 is three significant figures, not four. Begin counting at the first nonzero digit, not at the decimal point. In (e), 3.4610 km has five significant figures. A tail-end zero to the right of the decimal is written specifically to show that it is the doubtful digit, and therefore counted as significant. In (f), 712,000 must be written in exponential notation to show the number of significant figures:

Three: 7.12×10^5 cm Four: 7.120×10^5 cm
Five: 7.1200×10^5 cm Six: 7.12000×10^5 cm

3.4 ROUNDING OFF

Calculations from measurements frequently yield results that suggest the calculated quantity is more precise than the measurements from which it originates. For example, if 407.4 grams of a metal occupy 51.6 cubic centimeters, its density is, according to a calculator,

$$\frac{407.4 \text{ g}}{51.6 \text{ cm}^3} = 7.895349 \text{ g/cm}^3$$

Assuming an uncertainty of ±1 in the doubtful digit, the mass measurement has an uncertainty of 1 part in 4074, or 0.02%; the volume uncertainty is 1 part in 516, or 0.2%; but the uncertainty in the calculated density is 1 part in

7,895,349, or 0.000001%. Calculated results cannot be more precise than the least precise of the measurements from which they begin. Therefore, answers such as the above must be rounded off to reflect the uncertainty of the data.

The rule for rounding off calculated results is:

> If the first digit to be dropped is 0, 1, 2, 3, or 4, leave the preceding digit unchanged. If the first digit to be dropped is 5, 6, 7, 8, or 9, raise the preceding digit by 1.

Example: If 38.2345 g is to be rounded off to four significant figures, the result will be expressed to the second decimal place. The first digit to be dropped—the digit now in the third decimal place—is 4. The digit before it is therefore left unchanged. The rounded off result is 38.23 g.

Example: If 0.012587 cm is to be rounded off to three significant figures, the result will be expressed to the fourth decimal place. The first digit to be dropped is the 8 now in the fifth decimal place. The preceding digit, 5, is therefore increased by 1. The rounded off result is 0.0126 cm.

Example: Round off the density calculated above, 7.895348 g/cm^3, to three significant figures. The first digit to be dropped is 5 in the third decimal place. The preceding digit must be increased by 1. This time, increasing 9 in the second decimal place by 1 makes it 10, which spills over into the first decimal place. The rounded off result is 7.90 g/cm^3. The tail-end zero must be shown, as it is the doubtful digit.

Different textbooks and instructors have different ways of handling the preceding digit when the first digit to be dropped is 5. The "best" way is a matter of opinion. No way is actually incorrect, as the last digit in the rounded off result is doubtful anyway. If your instructor prefers a rule other than the one given above, you should, of course, use it.

EXAMPLE 3.2: A calculated result yields 21.409653 on the display of a calculator. Round off this number to two, three, four, five, six, and seven significant figures.

two: ____________ three: ____________ four: ____________

five: ____________ six: ____________ seven: ____________

two: 21	three: 21.4	four: 21.41
five: 21.410	six: 21.4097	seven: 21.40965

At five significant figures, dropping 6 in the fourth decimal place raises 9 to 10 in the third decimal, which spills over into the second decimal. The tail-end zero must be shown as the doubtful digit.

3.5 SIGNIFICANT FIGURE RULE FOR MULTIPLICATION AND DIVISION

Calculation does not improve precision. As mentioned before, a result calculated from measurements can be no more precise than the least precise measurement in the calculation. This condition is satisfied by two rules, one for multiplication and division, and the other for addition and subtraction.

> The answer to a multiplication or division problem must be rounded off to the same number of significant figures as the smallest number of significant figures in any factor.

EXAMPLE 3.3: The following problem is given with the calculator readout for the answer. Assuming all factors are correctly expressed in terms of significant figures, round off the result to the proper number of significant figures.

$$\frac{239.1 \times 46.23 \times 0.00290}{16.508 \times 114.29} = 0.01699021 =$$

Examine the factors in the setup, determine the smallest number of significant figures in any factor, and then round off the result to the same number of significant figures.

0.01699021 = 0.0170 to three significant figures.

The smallest number of significant figures in any factor is three in 0.00290. The answer must therefore be rounded off to three significant figures, dropping the 9 in the fifth decimal place. This increases the fourth decimal place by 1, which spills over into the third decimal.

EXAMPLE 3.4: The density of lead is 11.3 g/cm³. Calculate the mass of 2342 cm³ of lead. Express the result in grams, and in the proper number of significant figures.

GIVEN: 2342 cm³ FIND: g FACTOR: 11.3 g/cm³ PATH: cm³ → g

$$2342 \text{ cm}^3 \times \frac{11.3 \text{ g}}{1 \text{ cm}^3} = 26{,}464.6 \text{ g} = 2.65 \times 10^4 \text{ g}$$

The volume is given in four significant figures, but the density in only three. The 1 cm^3 in the denominator is a counting number from a definition. It therefore does not contribute to the determination of significant figures in an answer. The result must be rounded off to three digits, the smallest number of significant figures in the data. Exponential notation must be used as 26,500 grams could be three, four, or five significant figures.

3.6 SIGNIFICANT FIGURE RULE FOR ADDITION AND SUBTRACTION

> The answer to an addition or subtraction problem must be rounded off to the first column that has a doubtful digit.

The procedure is to examine the numbers being added and/or subtracted, determine the largest column (tens, units, tenths, hundredths, etc.) containing a doubtful digit, and round off the answer to that column. If numbers are expressed in exponential notation, all exponentials must be to the same power.

EXAMPLE 3.5: Assuming all numbers at the right are correctly expressed in terms of significant figures, round off the sum to the proper number of significant figures.

```
     27.14
  9,224.8
      5.7567
    948.84
 ----------
 10,206.5367 = ________
```

```
     27.1|4
  9,224.8|
      5.7|567
    948.8|4
 --------------
 10,206.5|367 = 10,206.5
```

27.14 and 948.84 are doubtful in the hundredths column, 9244.8 is doubtful in the tenths column and 5.7567 is doubtful in the ten-thousandths column. The largest column with a doubtful digit is the tenths column. The answer must therefore be rounded off to tenths.

A mechanical way to determine the column to which a sum or difference must be rounded off is to draw a vertical line through the numbers, as above, so all spaces immediately to the left of the line are filled, but one or more spaces to the right are vacant. The answer must be rounded off to the line.

Example 3.5 illustrates an important point. The answer contains more significant figures (six) than any term entering into the addition. Sums often have more significant figures than are present in any term, and differences may have fewer, as in the next example. Keep this in mind when you are tempted to use for addition and subtraction the more frequently applied rule for multiplication and division. The smallest number of significant figures in any term to be added or subtracted has nothing to do with rounding off a sum or difference.

EXAMPLE 3.6: The following data are collected in an experiment:

Measured mass of beaker + dry chemical	148.77 g
Measured mass of beaker + dry chemical after some of chemical has been removed	−106.409 g
Mass of chemical removed (by difference)	42.361 g = ________g

Assuming both masses to be expressed properly with respect to significant figures, round off the answer to the correct number of significant figures. Record that rounded off result in the space provided.

```
 148.77|  g
-106.40|9 g
  42.36|1 g = 42.36 g
```

3.7 SIGNIFICANT FIGURE POLICY FOR THIS WORKBOOK

Significant figures will receive little if any comment in other chapters in this workbook. Instead we will adopt a general policy that all answers will be rounded off routinely to the number of significant figures justified by the data. Most important, answers to problems should not be reported in all of the digits shown by a calculator.

3.8 METRIC UNIT CONVERSIONS

A conversion within the metric system is simply a matter of moving the decimal point. Deciding whether to move it to the right or to the left, and how far, is not always so simple.

The method for solving such metric conversions is dimensional analysis, introduced in Chapter 2. In that chapter we developed a routine for solving problems by the dimensional analysis method. It included the identification of the GIVEN quantity, the quantity to FIND, conversion FACTORS, and a unit PATH. We suggest that you continue making these identifications, but we recognize that you will make them mentally as your skill develops. Nevertheless, we strongly recommend that you write the unit path for all problems that are solved by dimensional analysis. It really helps. Our answers will usually show the complete analysis of the problem so you can see where our numbers come from.

The problem below illustrates the method applied to metric conversions.

EXAMPLE 3.7: What is the number of grams in 5.23 kilograms?

Identify the starting and wanted quantities as GIVEN and FIND.
Obtain the needed FACTOR from the relationships in Table 3.1.
Plan the PATH by which the problem may be solved.

GIVEN: 5.23 kg FIND: ? g FACTOR: 1000 g/kg PATH: kg → g

Now write the setup in the space below, but do not calculate the answer. Save that until the next step.

$$5.23 \text{ kg} \times \frac{1000 \text{ g}}{1 \text{ kg}} =$$

Each kilogram has 1000 grams, so multiplication is required. Notice the form of the ratio: 1000 g/kg. The "1000" modifies the grams in the numerator in the same way that the "k" for "kilo" modifies the grams in the denominator.

Will you use a calculator on this multiplication? Hopefully, no. When you multiply a number by a multiple of 10, all you must do is to move the decimal that number of places that is the same as the number of times 10 is used as a factor. 1000 is 10 × 10 × 10, or 10^3. The decimal must be moved three places, which is the number of zeros in 1000 and also the exponent when the factor is expressed as an exponential. It doesn't take a calculator to move a decimal three places. Prove it by writing the answer after the = sign above.

$$5.23 \text{ kg} \times \frac{1000 \text{ g}}{1 \text{ kg}} = 5230 \text{ g} \quad \text{or} \quad 5.23 \text{ kg} \times \frac{10^3 \text{ g}}{1 \text{ kg}} = 5.23 \times 10^3 \text{ g}$$

The most common error in making metric conversions is to move the decimal point the wrong way. This can be avoided if you always write the metric prefix of your conversion factor on one side of the fraction bar and the number of units represented by that prefix on the other side. In other words, if the prefix is above the line, the number is below the line, and vice versa.

To illustrate, the prefixes and numbers from Table 3.1 that you should memorize are repeated below, applied specifically to mass units. The metric abbreviations for the prefixes and their corresponding numbers are tabulated separately at the right.

1 kilogram = 10^3 grams		k	10^3
1 centigram = 10^{-2} gram		c	10^{-2}
1 milligram = 10^{-3} gram		m	10^{-3}
1 microgram = 10^{-6} gram		μ	10^{-6}
1 nanogram = 10^{-9} gram		n	10^{-9}

It is the letter for each prefix and its matching number that must be on opposite sides of the fraction bar. In Example 3.7 the conversion factor was

$$\frac{1000 \text{ g}}{1 \text{ kg}} = \frac{10^3 \text{ g}}{1 \text{ kg}}$$

The number, 10^3, is above the line, and the prefix symbol, k, is below the line.

The following conversion factors are INCORRECT:

The number, 10^3, and k are both in numerator $\frac{10^3 \text{ kg}}{1 \text{ g}}$ or $\frac{1 \text{ g}}{10^{-2} \text{ cg}}$ The number, 10^{-2}, and c are both in denominator

Other correct conversion factors from the table are

$$\frac{10^{-3} \text{ g}}{1 \text{ mg}} \qquad \frac{1 \text{ mg}}{10^{-3} \text{ g}} \qquad \frac{10^{-2} \text{ g}}{1 \text{ cg}} \qquad \frac{1 \text{ ng}}{10^{-9} \text{ g}}$$

In each case the number of grams and the matching letter are on opposite sides of the bar.

Observe these rules in working with metric conversions and you are not apt to move the decimal point in the wrong direction. Even so, always be sure to check your result. In making a conversion, the same quantity has a smaller number of large units and a larger number of small units. In Example 3.7, a small number (5.23) of large units (kilograms) is the same as a large number (5.23×10^3) of small units (grams).

EXAMPLE 3.8: Calculate the number of grams of a pharmaceutical powder in 943 centigrams.

Begin with GIVEN-FIND-FACTOR-PATH. Then set up and solve for the answer. When it comes to moving the decimal point, you do the thinking, not your calculator.

GIVEN: 943 cg FIND: ? g FACTOR: 10^{-2} g/cg PATH: cg → g

$$943 \text{ cg} \times \frac{10^{-2} \text{ g}}{1 \text{ cg}} = 9.43 \text{ g}$$

In this case the given quantity, 943 cg, is multiplied by the conversion factor 10^{-2} g/cg. The number in the conversion factor, 10^{-2}, is in the numerator and the matching symbol, c for centi-, is in the denominator. Multiplying by 10^{-2} means making the number smaller. So the decimal must be moved to the left. Though the decimal is not written in 943, it is understood to follow the 3. Starting with 943. and moving the decimal two places left gives 9.43. The 943 cg is a large number of small units compared to 9.43 g, a small number of large units.

EXAMPLE 3.9: Convert 145 cm to kilometers.

Table 3.1 does not give a direct conversion relationship between centiunits and kilounits, but you can get from one to the other in two steps. Begin with GIVEN-FIND-FACTOR-PATH noting that two conversion factors will be needed for this problem.

GIVEN: 145 cm FIND: ? km FACTOR: 10^{-2} m/cm; 10^3 m/km PATH: cm → m → km

Begin the setup by writing the starting quantity and the first conversion factor. You may, if you wish, set up the entire problem, but we will stop at each step.

$$145 \text{ cm} \times \frac{10^{-2} \text{ m}}{1 \text{ cm}} \times \frac{\qquad}{\qquad} =$$

The setup to this point changes centimeters to meters. If we were to calculate the answer it would be the number of meters in 145 centimeters.

Our interest is not in the intermediate answer, so we will not calculate it. Instead we will extend the setup to the wanted unit, kilometers. Insert the numbers and units of the second conversion factor in the open space above. Complete the problem, expressing the answer in exponential notation.

$$145 \text{ cm} \times \frac{10^{-2} \text{ m}}{1 \text{ cm}} \times \frac{1 \text{ km}}{10^3 \text{ m}} = 1.45 \times 10^{-3} \text{ km}$$

3.9 METRIC-ENGLISH UNIT CONVERSIONS

In order to help you become familiar with metric units, we will, in this chapter only, compare them with the English units still used in the United States. Table 3.2 gives these comparisons for mass, length, and volume units. You may use Table 3.2 as a source of conversion factors between the systems. Use of English-metric conversions is a straightforward application of dimensional analysis.

It is not necessary to memorize all the conversion relationships shown in Table 3.2. We recommend that you learn just one of each quantity. These are shown in bold type in the table. Note in particular the "≡" used in the conversion 1 inch ≡ 2.54 cm. We will use this symbol for definitions. This means that 1 inch is *exactly* 2.54 cm and does *not* represent three significant figures.

TABLE 3.2: ENGLISH-METRIC CONVERSION RELATIONSHIPS

MASS	LENGTH	CAPACITY
1 pound = 453.6 g	**1 inch ≡ 2.54 cm**	1.057 quarts = 1 L
2.205 pounds = 1 kg	39.37 inches = 1 m	**1 gallon = 3.785 L**
1 ounce = 28.35 g	1 mile = 1.609 km	1 ounce = 29.57 mL

EXAMPLE 3.10: Calculate the grams of ground beef in a supermarket package labeled 1.73 pounds.

Identify the GIVEN and FIND quantities, the conversion FACTOR (from the relationships in Table 3.2), and the unit PATH needed to solve the problem.

GIVEN: 1.73 lb FIND: ? g FACTOR: 453.6 g/lb PATH: lb → g

In the space below write the given quantity at the left. To obtain the quantity you were asked to find, will you multiply the given quantity by 453.6 grams per pound, or by its inverse, 1 pound per 453.6 grams? Figure it out and complete the setup by inserting the required conversion factor. Calculate the answer.

$$1.73 \text{ lb} \times \frac{453.6 \text{ g}}{1 \text{ lb}} = 785 \text{ g}$$

If there are 1.73 pounds, and each one has 453.6 grams, you must multiply by 453.6 grams per pound to find the total number of grams.

Your calculator no doubt gave you 784.728 as the answer to the above example. We rounded it off to 785. All answers from this point in this book are rounded off to the number of significant figures allowed by the data.

EXAMPLE 3.11: The net weight of oil in a 55-gallon drum is 412 pounds. Find the mass of the oil in kilograms.

Begin with the GIVEN-FIND-FACTOR-PATH sequence. Think carefully; there is a small distraction in the problem.

GIVEN: 412 lb FIND: ? kg FACTOR: 2.205 lb/kg PATH: lb → kg

Notice that there is enough information to solve the problem without using the size of the drum. This is an example of a problem that contains more information than you need.

Now write the calculation setup and solve for the answer.

$$412 \text{ lb} \times \frac{1 \text{ kg}}{2.205 \text{ lb}} = 187 \text{ kg}$$

EXAMPLE 3.12: Convert 4.21 meters to feet.

Table 3.2 does not have a direct conversion between meters and feet, so you must do a bit of searching to find a unit path for which you know all the conversion factors. Use either or both Tables 3.1 and 3.2, plus whatever English conversions you already know, and see if you can come up with the unit path by which this problem may be solved.

GIVEN: 4.21 m FIND: ? ft FACTORS: 10^{-2} m/cm; 2.54 cm/in.; 12 in./ft
PATH: m → cm → in. → ft

Begin with the given quantity and set up the conversion factors in the order of the unit path until the quantity you wish to find is reached. Calculate the answer.

$$4.21 \text{ m} \times \frac{1 \text{ cm}}{10^{-2} \text{ m}} \times \frac{1 \text{ in.}}{2.54 \text{ cm}} \times \frac{1 \text{ ft}}{12 \text{ in.}} = 13.8 \text{ ft}$$

3.10 UNIT CONVERSIONS WITH SQUARES AND CUBES

The volume of a rectangular box is found by multiplying the length by the width by the height. If the box is 20.0 cm long, 15.0 cm wide, and 10.0 cm high its volume is

$$20.0 \text{ cm} \times 15.0 \text{ cm} \times 10.0 \text{ cm} = 3000 \text{ cm}^3$$

Notice that cubic centimeter is written cm^3, indicating centimeters of length have been used as a factor three times.

We will now find the volume of the same box in cubic inches. To do so, each length measurement must be converted to inches:

$$20.0 \text{ cm} \times \frac{1 \text{ in.}}{2.54 \text{ cm}} = 7.87 \text{ in.}$$

$$15.0 \text{ cm} \times \frac{1 \text{ in.}}{2.54 \text{ cm}} = 5.91 \text{ in.} \qquad 10.0 \text{ cm} \times \frac{1 \text{ in.}}{2.54 \text{ cm}} = 3.94 \text{ in.}$$

The volume is therefore 7.87 in. × 5.91 in. × 3.94 in. = 183 in.^3

Instead of changing each centimeter measurement to inches, we might have omitted the intermediate values and produced a single setup:

$$20.0 \text{ cm} \times \frac{1 \text{ in.}}{2.54 \text{ cm}} \times 15.0 \text{ cm} \times \frac{1 \text{ in.}}{2.54 \text{ cm}} \times 10.0 \text{ cm} \times \frac{1 \text{ in.}}{2.54 \text{ cm}} = 183 \text{ in.}^3$$

Observe that the conversion factor $\frac{1 \text{ in.}}{2.54 \text{ cm}}$ has been used as a multiplier three times. It may therefore be expressed exponentially as $\left(\frac{1 \text{ in.}}{2.54 \text{ cm}}\right)^3$, which is the same as $\frac{(1 \text{ in.})^3}{(2.54 \text{ cm})^3}$ or $\frac{1^3 \text{ in.}^3}{2.54^3 \text{ cm}^3}$. In both the numerator and the denominator you cube both the number and the unit. This is the same as cubing an algebraic expression, such as

$$(2a)^3 = 2a \times 2a \times 2a = 2 \times 2 \times 2 \times a \times a \times a = 2^3 \times a^3$$

It follows that the setup can be shortened by writing

$$20.0 \text{ cm} \times 15.0 \text{ cm} \times 10.0 \text{ cm} \times \frac{1^3 \text{ in.}^3}{2.54^3 \text{ cm}^3} = 183 \text{ in.}^3$$

The above illustration shows that:

When you convert a volume expressed in one set of cubic length units to another set of cubic length units, the conversion factor between length units is cubed—both numbers and units.

EXAMPLE 3.13: How many cubic centimeters are in 0.639 cubic feet?

Table 3.2 gives no direct conversion between centimeters and feet, so the unit path will have at least two steps. Write that path here, using steps for which you know the conversion factors.

GIVEN: 0.639 ft^3 FIND: ? cm^3 FACTORS: 12 in./ft; 2.54 cm/in.
PATH: $ft^3 \rightarrow in.^3 \rightarrow cm^3$

Now proceed with the complete setup and solution of the problem. Remember how you handle the numbers in the conversion factors.

$$0.639 \text{ ft}^3 \times \frac{(12 \text{ in.})^3}{(1 \text{ ft})^3} \times \frac{(2.54 \text{ cm})^3}{(1 \text{ in.})^3} = 0.639 \text{ ft}^3 \times \frac{12^3 \text{ in.}^3}{1^3 \text{ ft}^3} \times \frac{2.54^3 \text{ cm}^3}{1^3 \text{ in.}^3}$$
$$= 1.81 \times 10^4 \text{ cm}^3$$

EXAMPLE 13.14: Convert 0.0448 cubic meters to cubic millimeters.

You know the procedure. Set up below and solve. You will surely find exponential notation handy on this one.

GIVEN: 0.0448 m^3 FIND: ? mm^3 FACTOR: 10^{-3} m/mm PATH: $m^3 \rightarrow mm^3$

$$0.0448\ m^3 \times \frac{1\ mm^3}{(10^{-3})^3\ m^3} = 0.0448 \times 10^9\ mm^3 = 4.48 \times 10^7\ mm^3$$

Cubing the exponential 10^{-3} yields $10^{-3} \times 10^{-3} \times 10^{-3} = 10^{-9}$. Dividing by 10^{-9} is equivalent to multiplying by 10^9.

EXAMPLE 3.15: An automobile engine has an internal volume of 286 in.^{3} What is this in liters?

You're on your own on this one!

GIVEN: 286 in.^{3} FIND: ? L FACTORS: 2.54 cm/in, 1 mL = 1 cm^3, 10^{-3} L/mL
PATH: in.$^3 \rightarrow cm^3 \rightarrow mL \rightarrow L$

$$286\ in.^3 \times \frac{2.54^3\ cm^3}{1\ in.^3} \times \frac{1\ mL}{1\ cm^3} \times \frac{10^{-3}\ L}{1\ mL} = 4.69\ L$$

3.11 TEMPERATURE

The Celsius temperature scale is used for all scientific measurements. Its two fixed points are at 0°C for the normal freezing point of water and 100°C for the normal boiling point. The corresponding temperatures on the more familiar (in the United States only) Fahrenheit scale are 32°F and 212°F, respectively. The two scales are related by the equation

$$1.8 \times °C = °F - 32 \qquad (3.3)$$

If you are given one temperature and asked to find the other, we recommend that you solve Equation 3.3 for the temperature you are to find, substitute the given temperature, and calculate the answer.

Recall that, when a problem is solved by an algebraic equation, our analysis begins with one or more given quantities, the quantity to find, and the equation: GIVEN-FIND-EQN. There is no conversion factor, or unit path.

EXAMPLE 3.16: The thermostat on a hot tub is set at 102°F. Find the corresponding Celsius temperature.

Solve Equation 3.3 for °C, substitute the given value and calculate the answer.

GIVEN: 102°F FIND: ? °C EQN: 1.8 × °C = °F − 32

$$°C = \frac{°F - 32}{1.8} = \frac{102 - 32}{1.8} = \frac{70}{1.8} = 39°C$$

EXAMPLE 3.17: The temperature is −48°C on a cold arctic night. Find the Fahrenheit equivalent.

GIVEN: −48°C FIND: ? °F EQN: 1.8 × °C = °F − 32

$$°F = 1.8(°C) + 32 = 1.8(-48) + 32 = -86 + 32 = -54°F$$

The SI system of units identifies the **kelvin**, abbreviated K, as the unit of thermodynamic temperature. The kelvin is the same size as a Celsius degree. They are related by the equation K = °C + 273.15. Most temperatures are measured to the nearest Celsius degree, so we will disregard the decimal part of the above equation. It thus becomes

$$K = °C + 273 \tag{3.4}$$

Conversions are straightforward applications of the above equation.

EXAMPLE 3.18: (a) What is the Kelvin temperature at 25°C?
(b) What Celsius temperature corresponds to 224 K?

(a) GIVEN: 25°C FIND: ? K EQN: K = °C + 273

°C + 273 = 25 + 273 = 298 K

(b) GIVEN: 224 K FIND: ? °C EQN: K = °C + 273

°C = K − 273 = 224 − 273 = −49°C

3.12 DENSITY: A PHYSICAL PROPERTY WITH DERIVED UNITS

Many physical quantities are measured in units that are combinations of those already considered. Volume is an example. The unit of volume is a length unit cubed. Similarly, area is a length unit squared. Speed combines units of length and time, as in miles per hour. Units such as these are called **derived units**, because they are derived from fundamental measurement units.

If you were to weigh different volumes of a pure substance, you would discover that the mass and the volume are directly proportional to each other:

$$m \propto V \tag{3.5}$$

This is logical. Two gallons of milk weigh twice as much as one gallon. Introducing a proportionality constant, d,

$$m = d \times V \tag{3.6}$$

Solving Equation 3.6 for d gives the units of the proportionality constant and defines the property of matter called **density**:

$$\text{density} \equiv \frac{\text{mass}}{\text{volume}} \qquad \text{or} \qquad d \equiv \frac{m}{V} \tag{3.7}$$

In words, the density of a substance is its mass per unit volume. Units of density match the definition: mass units over volume units, such as grams per cubic centimeter, grams per milliliter, or grams per liter.

To find the density of a substance you measure the mass and the volume of the same sample of that substance, and then divide the mass by the volume. In other words, you insert measurements into the defining equation for density.

EXAMPLE 3.19: A block of wood measures 20.5 cm × 4.60 cm × 1.60 cm. Its mass is 71.3 grams. Calculate the density of the wood.

Unlike earlier problems, this example does not have just a single given quantity. It has four given quantities, three of which may be used to find the volume of the sample. Recall that the volume of a rectangular solid is length × width × height. Therefore the three dimensions, exactly as they appear in the problem, are a mathematical expression for volume. They may be substituted directly into Equation 3.7. The fourth starting quantity, mass, is the numerator in Equation 3.7. Make the substitutions below and calculate the density.

GIVEN: 20.5 cm; 4.60 cm; 1.60 cm; 71.3 g FIND: g/cm³ EQN: $d = \frac{m}{V}$

$$d = \frac{m}{\text{length} \times \text{width} \times \text{height}} = \frac{71.3 \text{ g}}{20.5 \text{ cm} \times 4.60 \text{ cm} \times 1.60 \text{ cm}} = 0.473 \text{ g/cm}^3$$

As was pointed out in Section 2.5 on "per" relationships, ratios such as density (mass per unit volume) can be used as conversion factors in dimensional analysis problems. When expressed as

$$d = \frac{X \text{ g}}{1 \text{ cm}^3} \tag{3.8}$$

density can serve as the conversion factor for either unit path between mass and volume, $g \rightarrow cm^3$ or $cm^3 \rightarrow g$.

EXAMPLE 3.20: How many cubic centimeters are occupied by 75.0 grams of zinc if its density is 7.14 grams per cubic centimeter?

This is a one-step conversion based on the relationship in Equation 3.8.

GIVEN: 75.0 g FIND: ? cm^3 FACTOR: 7.14 g/cm^3 PATH: g → cm^3

$$75.0 \text{ g} \times \frac{1 \text{ cm}^3}{7.14 \text{ g}} = 10.5 \text{ cm}^3$$

CHAPTER 3 PROBLEMS

1) How many significant figures are in each of the quantities listed:

a) 454 mg ____ d) 0.0680 km ____ g) 0.1536 g ____

b) 0.0353 L ____ e) 10.0 mL ____ h) 0.0060 g ____

c) 52.20 mL ____ f) 3×10^7 kg ____ i) 1.898×10^{-3} g ____

j) 2500 m ____

2) Round off the given quantity 7.758064 to the number of significant figures indicated.

one ________ three ________ five ________

two ________ four ________ six ________

3) Perform the indicated calculation and express the answer in the correct number of significant figures:

$$\frac{0.370 \times 843 \times 0.0704}{0.0042 \times 17.10} =$$

4) Complete the following operations and round off the answers to the proper number of significant figures:

a) $18.7 - 0.56 =$ ________ b) $1.59 \times 10^{-4} - 6.42 \times 10^{-5} =$ ________

5) Round off the quantity 4.106738 to the number of significant figures indicated:

one ________ three ________ five ________

two ________ four ________ six ________

6) Calculate the answer required and express that answer in the proper number of significant figures:

$$\frac{1.81 \times 10^4 \times 0.000130 \times 2.248}{0.00851 \times 87.3 \times 5.6 \times 10^3} =$$

7) Complete the following operations and express the results in the correct number of significant figures:

a) $294.696 + 10.4752 + 0.701 + 0.0086 + 41.61 + 17.009 =$

b) $(4.39 \times 10^3) + (1.26 \times 10^2) + (5.81 \times 10^4) =$

8) Complete each blank with the number of the units shown that are equal to the quantity given at the left. There is no planned order reading from left to right on any line; each quantity should be calculated independently of all others. You should not refer to Table 3.1, but you may refer to Table 3.2, unless your instructor directs otherwise.

a) 6.91 g	______ mg	______ kg	______ oz
b) 8.12 kg	______ g	______ mg	______ lb
c) 4.25 lb	______ g	______ mg	______ kg
d) 14.8 oz	______ g	______ cg	______ kg
e) 301 mm	______ cm	______ m	______ yd
f) 29.4 in	______ cm	______ mm	______ km
g) 4.19 ft	______ m	______ mm	______ cm
h) 65.2 km	______ m	______ cm	______ miles
i) 2410 ft	______ m	______ km	______ miles
j) 1.95 ft^3	______ m^3	______ cm^3	______ in^3
k) 3070 cm^3	______ m^3	______ in^3	______ mm^3
l) 7.15 L	______ mL	______ qt	______ gal
m) 119 mL	______ cm^3	______ L	______ qt
n) 0.816 gal	______ L	______ mL	______ cm^3

9) It is not uncommon for temperatures in California's Death Valley to reach 110°F during the summer. Calculate this temperature in °C and K.

10) Sodium chloride (table salt) melts at 801°C. Calculate this temperature on the Fahrenheit and Kelvin scales.

11) A student found a block of wood that did not float in water, so she decided to determine its density. The block measured 5.2 cm × 3.4 cm × 1.7 cm, and had a mass of 39.1 grams. What is the density of the wood?

12) 63.5 grams of alcohol having a density of 0.818 g/mL are required for a chemical reaction. How many milliliters of alcohol should be measured out?

13) The density of a certain metal is 7.39 g/cm³. What is the mass of 29.5 cm³ of the metal?

14) A piece of an irregularly shaped metal with a mass of 67.6 grams was found to have a volume equal to 12.8 mL, determined by water displacement. What is the density of the metal?

15) The density of a saline (salt) solution is 1.27 g/mL. What is the mass of 326 milliliters of the solution?

16) Magnesium is a light metal with a density of 1.74 g/cm³. What volume is occupied by 1.35 kg of magnesium?

++++++++++++++++++++++

17) Count the number of significant figures to which each of the following quantities is measured:

a) 37.50 g ____ b) 0.00724 L ____ c) 896.0 mm ____

d) 2.300×10^4 cm³ ____ e) 6.8×10^{-4} ____ f) 5290 L ____

18) Round off 0.034950 to the number of significant figures shown:

One __________ Two __________ Three __________

19) Perform the calculation indicated and write the answer in the correct number of significant figures.

$$\frac{(0.035)(1.260 \times 10^{-3})(5410)}{(17.91 \times 10^{-6})(0.030406)} =$$

20) Calculate the following sum: $0.0936 + (1.41 \times 10^{-3}) + 1.07 =$

21) Complete each blank with the number of the units shown that are equal to the quantity given at the left. You may refer only to Table 3.2, unless your instructor directs otherwise.

a) 23.6 cm	______ m	______ mm	______ in
b) 4.79 m	______ mm	______ km	______ ft
c) 0.542 L	______ mL	______ gal	______ cm³
d) 809 mg	______ kg	______ g	______ oz
e) 1.36 gal	______ L	______ m³	______ cm³
f) 0.0519 kg	______ g	______ cg	______ lb
g) 2640 mL	______ L	______ gal	______ cm³

22) A high pressure steam power plant delivers steam to the turbines at 1024°F. Express this temperature in Celsius degrees and kelvins.

23) Liquid nitrogen boils at −196°C. What is this temperature in kelvins and in Fahrenheit degrees?

24) Calculate the volume of 89.5 grams of platinum, which has a density of 21.5 grams per cubic centimeter.

25) The density of a motor oil is 0.814 g/mL. What is the mass of one quart (946 mL) of this oil?

26) Find the density of a certain alloy if a block 55.1 cm × 92.0 cm × 44.0 cm has a mass of 1790 kilograms.

ADDITIONAL PROBLEMS

27) A graduated cylinder which contains 47.3 mL weighs 89.62 grams. When a solid is submerged in the water, the volume level rises to 55.4 mL and the mass to 123.79 grams. a) What is the volume of the solid? b) What is the mass of the solid? c) What is its density?

28) A 5.62 gram sample of salt mixed with sand is treated with 50.00 mL of water to dissolve all the salt. A 5.00 mL portion of the water solution is evaporated and found to contain 0.48 grams of salt. How many grams of sand were in the original mixture?

29) Convert 67 lb/ft^3 to mg/mm^3.

30) Suppose that a new temperature scale is established, the "student degree," °S, in which the boiling and freezing points of water are 85°S and −35°S, respectively. Derive an equation relating °C and °S and calculate the °S equivalent to 112°C.

CHAPTER TEST

1) State the number of significant figures in each of the following:

3.80 L ____________ 0.0042 km ____________ 806.0 cm ____________

7500 mL ____________ 2.026×10^5 nm ____________

2) Round off the quantity 1.2960546 to two, three, four, five, six, and seven significant figures.

Two ____________ Three ____________ Four ____________

Five ____________ Six ____________ Seven ____________

For 3 and 4, assume that all number are correct with respect ot significant figures. Calculate the answers and round them off to the proper number of significant figures.

3) $\frac{(3.70 \times 10^{-6})(2791)}{(639.92)(0.0082)} =$

4)
$$\begin{array}{r} 106.054 \\ 35.60 \\ 0.01428 \\ 2.3311 \\ \hline \end{array}$$

5) A grain of sand weighs 0.8 milligrams. Express this in (a) grams and (b) kilograms.

6) A fisherman caught a 1.26-kilogram trout. How many pounds is this? (453.6 grams = 1 pound)

7) How many (a) centimeters and (b) millimeters are in 79.8 meters?

8) What number of meters is equal to 60.5 kilometers?

9) How many feet are in 4.92 meters? (2.54 cm = 1 inch)

10) Calculate the number of liters in 94.1 gallons. (3.785 L = 1 gal)

11) How many cubic meters are in 18.2 cubic yards? (1 meter = 39.37 in.)

12) Find the number of liters in 786 cubic centimeters.

13) What Celsius temperature is equal to 41°F?

14) Convert 82°C to (a) Fahrenheit degrees and (b) kelvins.

15) The mass of 60.0 milliliters of a nickel plating solution is 67.8 grams. What is its density?

16) What is the mass of 252 cubic centimeters of silver if its density is 10.5 grams per cubic centimeter?

4. GRAPHS OF EXPERIMENTAL DATA

PREREQUISITES

1. The direct and indirect proportionalities of Section 2.2 are repre-sented by graphs, which have distinct properties.

2. Density, introduced in Section 3.12, appears in one of the end-of-chapter problems.

CHEMICAL SKILLS

1. Plot a graph. Given a set of data showing the relationship between two variables, plot a graph of those data.

2. Read a graph. Given a graph showing how one variable changes with respect to another variable, and given the value of either variable, read the value of the other variable from the graph.

3. Slope. Given a straight-line graph, calculate the slope of the line. Identify its units and their physical significance, if any.

4. Equations from graphs. Given a straight-line graph, write the equation for the relationship between the variables.

Graphs are used to show how quantities that are studied in the laboratory are related to each other. The purpose of this chapter is to introduce you to some of the techniques that produce good graphs from real-life data, and then to interpret those graphs.

4.1 THE PRESENTATION OF DATA

A chemistry class performed an experiment to determine the relationship between the temperature of a sample of gas and the pressure it exerts if its volume is held constant. Nine constant temperature baths were set up. Each student was given a metal container with a pressure gauge attached. The students proceeded around the room, immersing the container into each bath until the container and gas reached the temperature of the bath. When the pressure became constant, it was read and recorded in a notebook. The data collected by two students—we'll call them Karen and Ric so we may refer to their data later—are recorded in Table 4.1.

Can you look at Table 4.1 and make any sense out of it? About the only thing you can say from the numbers alone is that, as the temperature goes down, the pressure goes down also—with one exception. A table of data is not an adequate presentation from which to search for relationships. A graph is much easier to analyze. We will plot the data from Table 4.1 to illustrate the preparation and analysis of a graph. There will be several detours into other examples as we go along, but we will keep returning to this experiment as our central example.

4.2 THE ANATOMY OF A GRAPH

Figure 4.1 shows the structure of a graph. The lined pattern on which the graph is drawn is known as a **grid**. The heavy intersecting lines are the **axes** of the graph. The horizontal axis is commonly called the **x axis**, and the vertical axis is the **y axis**. The axes intersect at the **origin**. A point on the graph is identified by its **coordinates**, the x value of the point followed by the y value. When written, they are enclosed in parentheses and separated by a comma: (x,y).

Each axis has a **scale**. This is a set of numbers that are distributed uniformly along the axes. They tell the values of the variables at any point. Scale numbers increase from left to right on the horizontal axis and from bottom to top on the vertical axis.

TABLE 4.1: PRESSURE EXERTED BY A GAS AT DIFFERENT TEMPERATURES
Quantity and Volume Constant
Readings of Two Students

	PRESSURE (torr)	
TEMPERATURE (°C)	Karen	Ric
100	1174	960
73	1079	860
52	1020	820
25	949	761
0	861	698
-30	758	614
-48	700	551
-79	601	571
-94	566	443

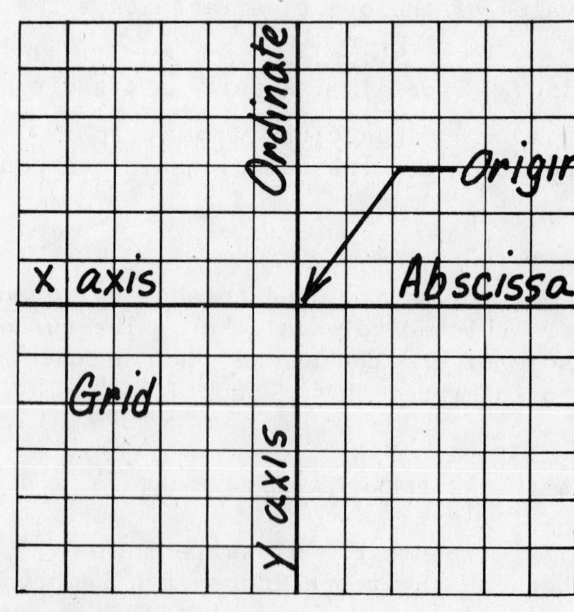

Figure 4.1

4.3 THE PROPERTIES OF A GOOD GRAPH

Before considering how to construct a graph, we will list the properties of a good graph so we can have a clear view of what we are trying to produce:

1. The graph is easy to plot and easy to read.
2. The axes are clearly labeled so the reader is able to see exactly what is plotted and its value. This includes the units in which each value is expressed.
3. To the left of a vertical axis and beneath a horizontal axis there is enough space for numbers and a label without crowding. This means that the axes are not at the edge of the grid, but inside the grid.
4. If the label on the vertical axis is printed vertically, the words are readable by turning the page one-quarter turn clockwise.
5. Scales are uniform. Scale numbers should be "round" numbers—numbers like 20, 50, 100, rather than 17, 224, etc.—set on the heaviest lines at regular intervals at least two lines apart. The axis does not have to begin at 0, but the beginning value should be shown.
6. If every fifth and/or tenth line of the grid is heavier than other lines, the scales should be such that the number of units represented by each 10-line major division is 1×10^n, 2×10^n, or 5×10^n. We will call this the **1-2-5 rule**. By this rule, the only nonzero digit in the number of units per 10-line division is 1, 2, or 5. Thus each major division might be 0.01, 1, 100, 0.2, 20, 0.005, 5, 50. (On rare occasions, 4×10^n is acceptable.)
7. Scales are selected so the graph covers as much of the page as possible. This rule may be modified if the range of values is very small, causing the smallest measured increment to spread over five or more of the smallest grid lines.
8. Plotted points are easily seen. X's, circled X's, large dots, circled dots, and circles are all commonly used, as well as some others. If two or more data sets appear on the same grid, one symbol is used for points from one data set and a different symbol for points from each of the other data sets.
9. The curve—even a straight-line curve—is drawn smoothly *among* the points rather than as a zig-zag curve that passes *through* each point. The curve has no sharp breaks, unless the relation plotted has sharp breaks. The curve does not have to touch every point.
10. The graph has a title that tells exactly what is plotted.

We will refer to this list of properties of a good graph often in the discussion ahead. In doing so, a particular property will be identified by its number in the list. For example, we might wish to emphasize the importance of the 1-2-5 rule of Item 6—and we do! Only if this rule is obeyed is a graph easily plotted and easily read (Item 1). This is because 1, 2, and 5 are the only integers that divide evenly into the 10 lines of a major division.

4.4 SELECTING THE SCALE

The most common grid used for plotting experimental data is "millimeter" paper on which the individual lines are 1 mm apart. Each fifth line is heavier than the individual lines, and each tenth line is heavier than the fifth lines. A more "open" grid with the same three levels of line thickness has 20 individual lines per inch. All illustrations in this book were plotted on 20-line-per-inch paper. We recommend that you use this grid too.*

Selecting a suitable scale is the most important step in preparing a graph. The scale determines how easily the graph is plotted, and how easily it is read (Item 1). The following steps produce a good scale:

1. Determine the range of values to be plotted. This is the highest number minus the lowest number.
2. Establish the range to be represented by each major (10-line) and/or minor (1-line) division on the graph. First get the "tentative" range, and then raise it to a number that conforms to the 1-2-5 rule (Item 6) so it will be easy to plot and easy to read (Item 1).
3. Scale the axis with appropriate round numbers on the major lines. Scaled lines should be no closer than every other line (Items 5, 6).

To illustrate how this procedure is followed, suppose the values 348, 397, 449, 504, and 551 are to be plotted along an axis drawn lengthwise on a 20-line-per-inch grid. The grid has 20 major lines. Item 3 in the properties of a good graph says that the axis begins inside the grid. Suppose we decide to have the axis begin at least two major lines from the edge of the grid, thereby leaving room for numbers and a label on the other axis. That leaves 18 major lines over which to plot the values. Following the steps in order:

1. The range to be plotted is 551 - 348 = 203.
2. The "tentative" range for each 10-line division assumes the 203 units will be distributed uniformly over all 18 major divisions. Thus each major division will be 203/18 = 11.3 units. To get the "plotting" range, this number is raised to the next number that obeys the 1-2-5 rule. In this case the number is 20; each major division is 20 units. The smallest division on the grid has a value of 2.
3. Beginning the axis with a round number that is just smaller than the first data point, 340, yields the scale that follows. The given values are plotted above the axis. The full range of plotted values does not fill the page, but it is more than half the page and therefore "covers as much of the page as possible" (Item 7). Any attempt to expand the scale to 10 units per major division would require more than the entire length of the grid, even without leaving two major lines for labels for the other axis (Item 3).

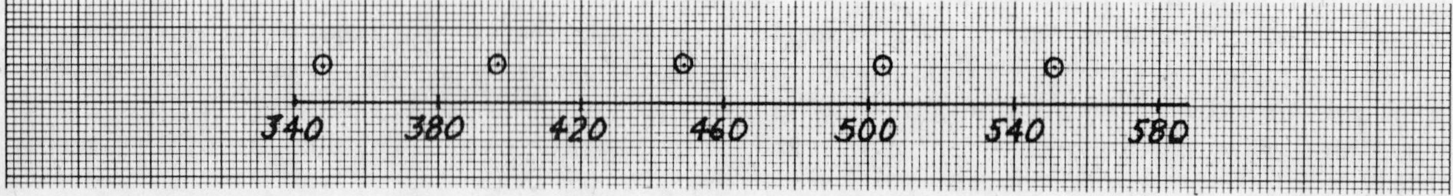

*Several sheets of 20-lines/inch graph paper are at the back of the book.

As drawn, the scaling of the axis is entirely satisfactory. However, the plotted points are located toward the left side of the axis. A better balanced graph can be produced by shifting the entire axis range of 340 to 580 to the right by two major divisions on the grid. The same goal can be reached by beginning the axis at 300 rather than 340. Both are shown below. Technically, these changes are not necessary, but they do improve the appearance of the graph.

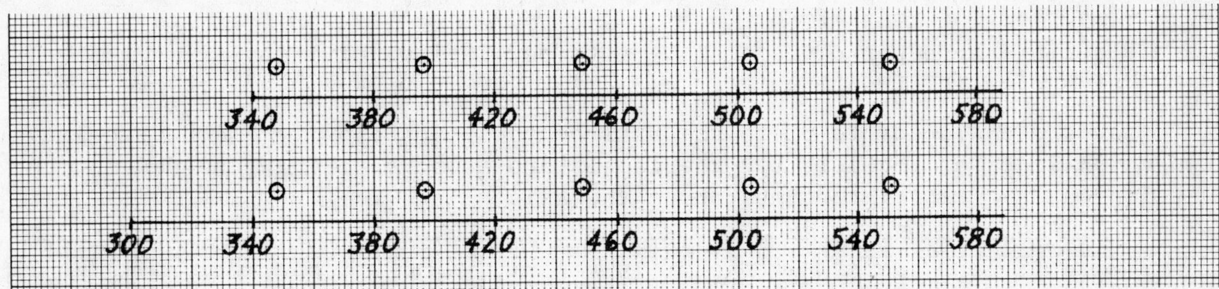

Item 5 in the properties of a good graph says the axis does not have to begin at zero. The reason is plain if we plot the same numbers on an axis that does include zero. Each major division becomes 50 units instead of 20, and the plotted points are "bunched" in one small part of the scale:

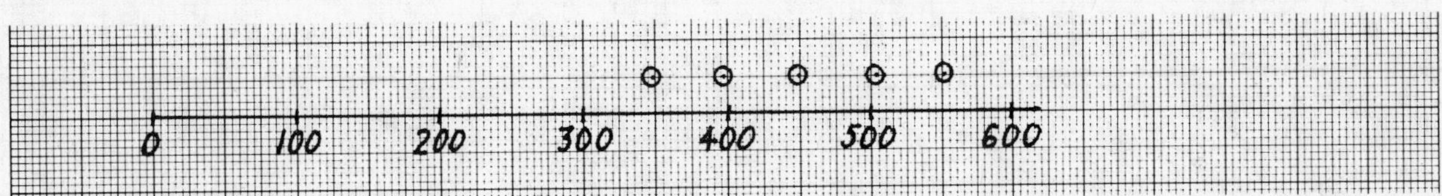

Occasionally the conditions of an experiment are such that the origin, (0,0), is a real value, actually a point on the curve, and that it is necessary to the proper interpretation of the graph. In such case the scale should begin at 0,0 even if it does crowd other plotted points.

EXAMPLE 4.1: Establish a horizontal axis that is suitable for plotting the following numbers: 20.5, 39.7, 61.0, 92.8, and 116.9. Mark the positions of the numbers above the axis.

The short side of the grid on 20-line-per-inch paper has 14 major divisions. At least two divisions should be left for the label and numbering the scale (Item 3). Thus 12 major divisions are available for the axis. Find (a) the range to be plotted and (b) the tentative range per major division.

Range to be plotted: 116.9 - 20.5 = 96.4
Tentative range per major division: 96.4/12 = 8.0

The plotting range per division is found by raising the tentative range to the first number that has 1, 2, or 5 as its only nonzero digit. What is that number? Also, with that plotting range per major division, what is the value of each line on the graph?

Plotting range per division: 10 units per 10 line division.
Value of each line: 1 unit.

You are now ready to place the axis on the grid below. Remember to leave two major lines at the left for the labels and numbers that are needed for the vertical axis. Plot the points above the axis.

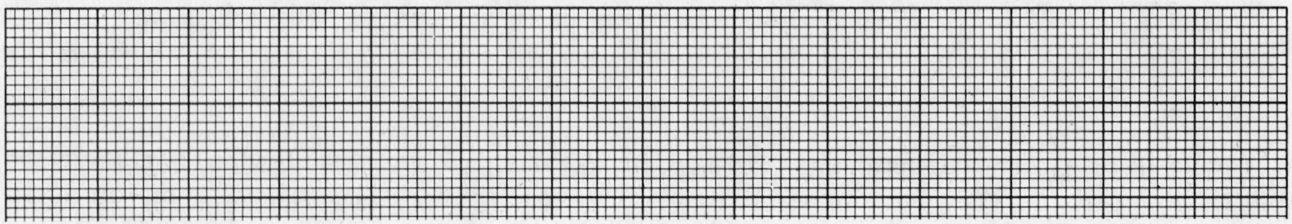

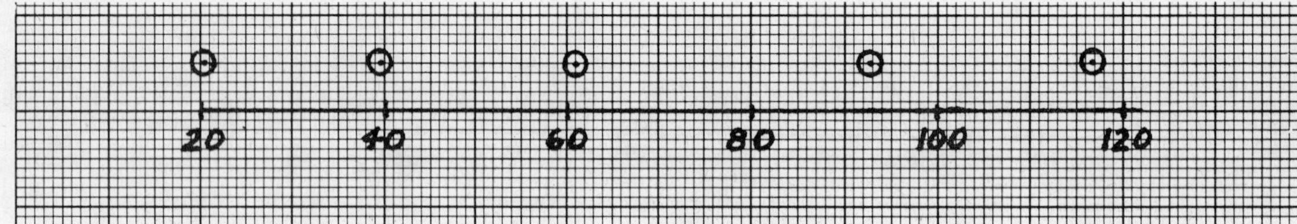

It would be quite satisfactory to begin this scale at zero, which would shift everything above two major divisions to the right.

EXAMPLE 4.2: The grid below has 20 major divisions. Leaving two lines at the left for numbers and a label, set up on that grid a scale on which the following values might be plotted:

0.000564 0.000824 0.000983 0.001075 0.001291

The numbers are a bit more awkward this time, being long decimal fractions. They can be simplified by rewriting them in exponential notation, and then plotting only the coefficients. One warning: if the coefficients are to be comparable, they must all be written to the same exponential, regardless of where the decimal happens to fall. With this precaution in mind, use the space beneath the numbers to rewrite the numbers in exponential notation.

$0.000564 = 5.64 \times 10^{-4}$
$0.000824 = 8.24 \times 10^{-4}$
$0.000983 = 9.83 \times 10^{-4}$
$0.001075 = 10.75 \times 10^{-4}$
$0.001291 = 12.91 \times 10^{-4}$

It was necessary to place the decimal after the second digit in the last two numbers in order to keep the exponential at 10^{-4}.

Now draw the axis on the grid below. Do this without guiding questions this time, following the same procedure as before. Plot the numbers above the axis to be sure they fit satisfactorily.

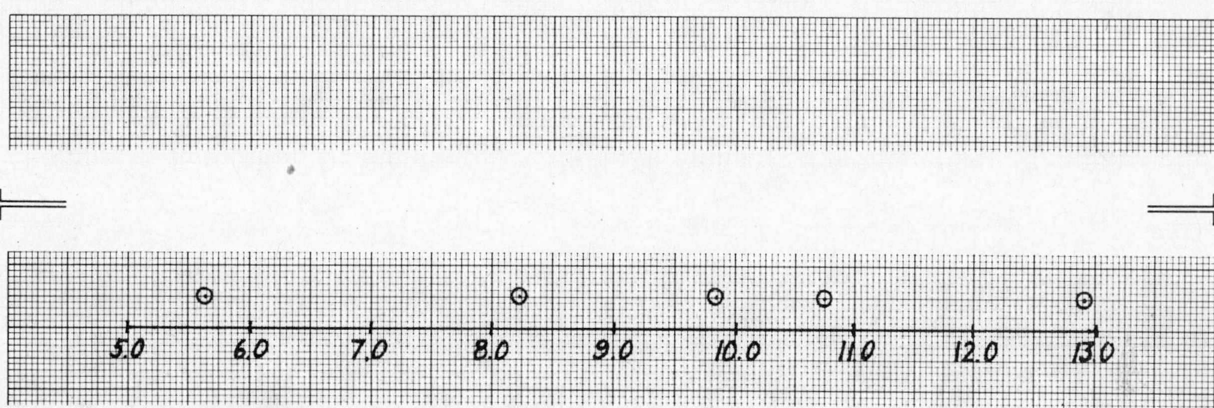

The range of numbers is 12.91 - 5.64 = 7.27. The tentative range per division is 7.27/18 = 0.40, which is increased to 0.50 units per division for plotting. This makes each line equal to 0.05 units and every other major line equal to 1 unit.

The numbers plotted are not the values they are supposed to represent; they are 10,000 times those values. This fact is stated in the label given to the axis. We will show how this is done in the next section.

So far the numbers for which you have been drawing axes have not presented any problems. This is not always the case. Sometimes any scale that satisfies the 1-2-5 rule produces a graph whose plotted points do not cover enough of the grid. Compromises must be made. You might try turning the paper 90° clockwise and plot the horizontal axis along the long dimension of the grid and the vertical axis along the short dimension. Another alternative is to test 4×10^n as the number of plotted units per 10-line division. If nothing works, it may be necessary to violate one of the properties of a good graph. That happens rarely, however. Do not give up too quickly.

The time has come to begin building the graph of the data in Table 4.1. Both curves will be plotted on the same grid. Pressure is to be plotted vertically, and temperature is to be plotted horizontally.

EXAMPLE 4.3: Develop the axes and scales to plot pressure versus temperature from the data of Table 4.1.

Let us begin with the horizontal axis, temperature. The numbers from the table are 100, 73, 52, 25, 0, -30, -48, -79, and -94. Find the range of temperatures; calculate the tentative range per major division; adjust to an acceptable plotting range; and draw the scaled axis on the grid below. Notice that, because there are some below-zero (negative) temperatures, the vertical axis will not be at the left side of the page. This means the entire width of 14 major lines is available for the temperature axis. Mark the temperatures to be plotted.

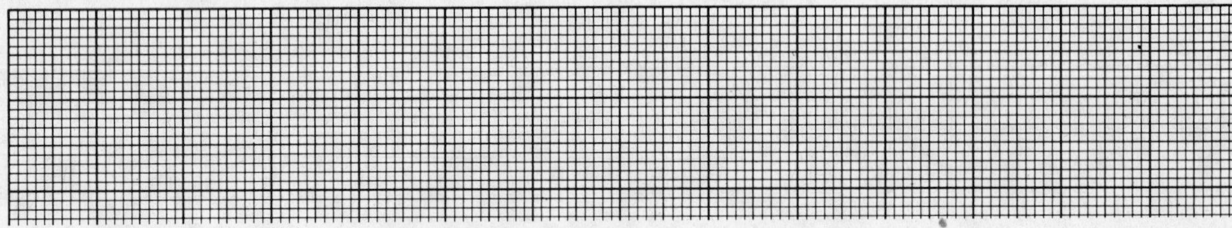

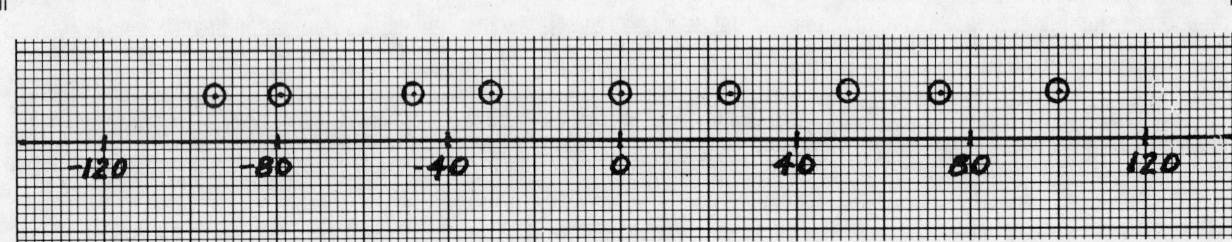

The temperature range is 100 - (-94) = 194. The tentative range over the full 14 major lines is 194/14 = 13.9, which must be adjusted upward to 20°C per major division. The temperatures run about the same distance on each side of zero, so 0°C is best placed in the center of the page.

Now draw the vertical axis on the grid below, which has been rotated horizontally for printing convenience. The total span of major divisions is 20, and at least two must be left for the label for the horizontal axis. Curves for both Karen and Ric will be plotted on the same graph, so the range is from the lowest pressure to the highest in either data set. The pressures are:

Karen:	1174,	1079,	1022,	949,	861,	758,	700,	601,	566	torr
Ric:	960,	860,	820,	761,	698,	614,	551,	571,	443	torr

From these numbers, find the tentative range per division and the plotting range, and then transfer the scale to the grid. Plot the two sets of values, using circled X'S for Karen's pressures and circled dots for Ric's.

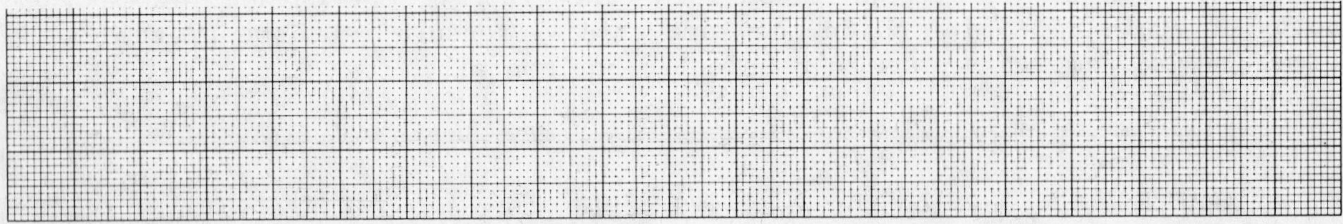

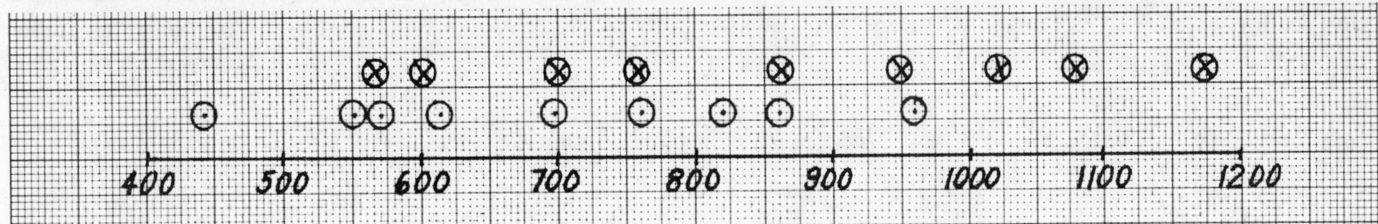

Range = 1174 - 443 = 731. Tentative division range = 731/18 = 40.6 torr per division. Adjust to 50 torr per division. Each line represents 5 torr.

A complete graph of these data will be developed as the chapter progresses. Remove a sheet of graph paper from the back of the book. Work lightly in pencil, as you will probably make corrections before your graph is completed. Copy the horizontal temperature axis from Example 4.3 along the second major line from the bottom. Then draw the vertical pressure axis upward from the 0°C mark on the temperature axis. Number the pressure axis too, but very lightly because you may wish to relocate the numbers later. When your scales are finished, turn to Table 4.1. Carefully plot all points from Karen's data set on the grid you have prepared. Use a circled X for each point, as in Example 4.3. Then do the same with a circled dot for Ric's data. Then put the graph aside temporarily. We will return to it shortly.

Two final comments before leaving this section on scaling an axis: Item 5 says scales must be uniform. Each increment of the scale must be the same as every other increment. Any departure from this principle distorts the shape of the graph. It no longer represents the relationship between the plotted variables.

On rare occasions it may be convenient to plot data in fractions. If so, recognize that 1/2, 1/3, 1/4, . . . are not uniform. In fact, they are increasingly smaller and closer to the origin. It is much better to change fractions to their decimal equivalents before plotting. If fractions are necessary, they must all have the same denominator, and it is the numerators that are plotted.

4.5 LABELS FOR THE AXES

If a graph is to be read, the quantity plotted on each axis must be clearly identified. The label must include the units in which the quantity is expressed.

There are no precise rules for writing labels, but they must accomplish the purpose above. The label for the horizontal axis is usually *printed* beneath the scale numbers. To add the label to a vertical axis, turn the paper one-quarter turn clockwise (Item 4) and print the label above the scale numbers. The units follow the quantity plotted: Mass, g. Some prefer to enclose the units in parentheses: Mass (g).

Add the label to the temperature axis on your graph of the pressure-temperature experiment. It is best not to label the pressure axis at this time because the curves pass through that axis, as already noted.

A special labeling problem appears when an axis is scaled in exponential notation. In Example 4.2 you produced a scale over which the following numbers were plotted:

$0.000564 = 5.64 \times 10^{-4}$
$0.000824 = 8.24 \times 10^{-4}$
$0.000983 = 9.83 \times 10^{-4}$
$0.001075 = 10.75 \times 10^{-4}$
$0.001291 = 12.91 \times 10^{-4}$

Suppose these numbers represent very small distances, D, expressed in meters. How would you label the axis drawn in Example 4.2? How do you interpret a scale value such as 5.64? The scale value is 10^4 times larger than D. Therefore, the scale should be labeled, "Distance, meters $\times 10^4$."

To find D, the scale value must be multiplied by 10^{-4}, the inverse of the exponential in the label. This can be very confusing to the reader. To remove this confusion, we suggest that an equation be added to the label telling the reader exactly what to do with the scale number to find the quantity it represents. The complete scale and label for Example 4.2 would then be

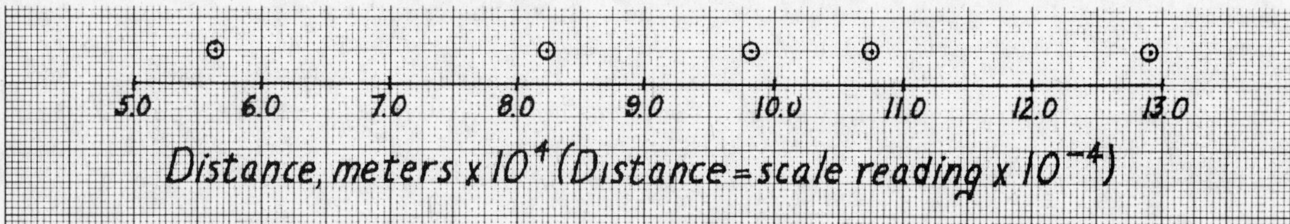

EXAMPLE 4.4: The scale below has been drawn to graph the following masses, expressed in kilograms:

4,750,000; 6,040,000; 6,980,000; 7,840,000; 9,310,000

The scale numbers represent the coefficients when the masses are written in exponential notation. Write an appropriate label for the scale.

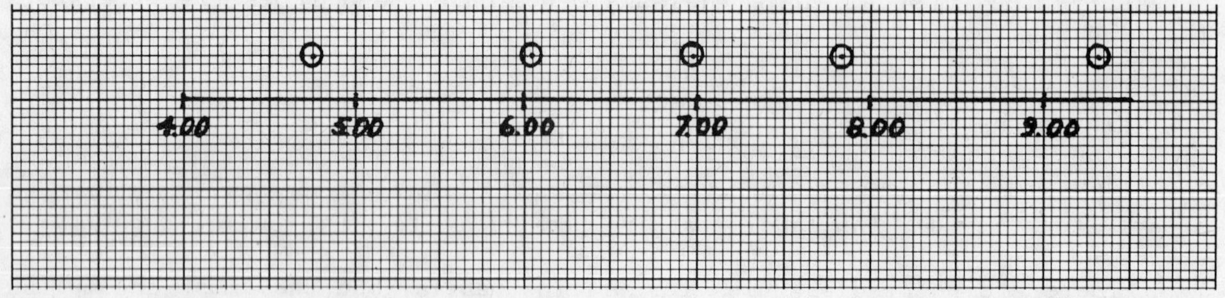

Mass, kg $\times 10^{-6}$ (Mass = scale reading $\times 10^6$)

The scale value of the first plotted point is 4.75. This is the number of kilograms $\times 10^{-6}$. The number kilograms is therefore 10^6 times as large as the scale reading: 4.75×10^6.

4.6 DRAWING A CURVE

Once the points on a graph are plotted on the grid, a curve is drawn "through" them. (In the language of graphs, the word *curve* includes a straight line.) Almost always, a curve is a "smooth" curve that passes *among* the points, touching some and perhaps missing others, rather than a wavy or zig-zag curve that passes from point to point. Most changes between variables studied in chemistry occur gradually, rather than in jumps, so the graphical representation of those changes has no sharp breaks.

It is easier to "average out" the points above and below a linear (straight line) curve if you use a transparent ruler to draw the line. Similarly, a transparent French curve helps in developing the gradual changes in a nonlinear curve. Another handy device is a flexible drawing guide that can be adjusted to fit any curve.

All measurements contain experimental errors. It follows that points plotted from measured values include these errors. One of the functions of a graph is to average out experimental errors. Thus a point on a curve gives a *more reliable* set of values than a plotted point that is not on the curve, even though the values of the plotted point have been measured.

With these ideas in mind, return to your graph of the pressure-temperature data in Table 4.1. Draw the best curve you can for each data set you have already plotted. (You may wish to make one curve solid and the other dashed. That is one way to distinguish between two or more curves plotted on the same grid.) The "best" curve probably will not touch every point, but the average distance of points above the line will be about the same as the average distance of points below the line. There is an exception to this. An outlying point—a point that does not seem to fit the curve established by the other data points—should be disregarded. Such a point usually represents a gross error in the data. It should not be allowed to distort the curve.

When your curves are drawn you will see where they pass through the vertical axis, and where space is available to write the number scale on one side or the other. Add that scale, and also label the axis in another open space.

When two or more curves are plotted on the same grid, the source of each curve must be identified. One way to do this is to place a legend in an open space on the grid. (Yet to come is a title that will be placed on the grid in another open space. Be sure to leave room.) Draw a circled X, and next to it write "Karen's data." Beneath that, place a circled dot labeled "Ric's data." If you have used a solid line for one data set, and a dashed line for the other, those lines can be used in the legend instead of the circled dot and X. Alternately, the names, Karen and Ric, might be placed alongside their respective curves.

After your graph is fully labeled, set it aside once again.

4.7 THE TITLE OF THE GRAPH

The title of a graph must state briefly, but completely, what the graph represents. It should identify the variables measured. A good title includes any conditions that have been imposed in an experiment. For example, if the rate of a chemical reaction is studied as a function of concentration, it is necessary to control the temperature, which also affects rate. The title could then be, "Rate of a Chemical Reaction as a Function of Concentration at Constant Temperature." The title might be even more specific as, "Rate of Decomposition of Hydrogen Peroxide as a Function of Concentration at 25°C."

There is no set rule governing where a title should be placed on a graph, unless such a rule is established by an instructor, employer, or other such authority. If the axis and scale can be selected so there is room above or below the graph, those are probably the best places. Otherwise the title may be worked in wherever room is available.

Now return to the pressure versus temperature graph you are plotting. The amount of gas in the container remains the same, and the container is rigid, having a fixed volume. From this information, write a title for your pressure-temperature curves and place that title on your graph.

4.8 THE TEMPERATURE VS. PRESSURE GRAPH

Your graph of pressure versus temperature is now complete. You may wish to compare it with ours, Figure 4.2. We will refer to it often in the pages ahead. You may wish to place a marker at the page so you can get to it quickly.

Your axes and ours should be the same, as they were developed in an example. There are various acceptable ways you might have identified the top curve as coming from Karen's data and the bottom from Ric's. The title could be worded several ways, but somehow it should state that pressure has been plotted against temperature. It should also note that the quantity and volume of the gas are constant.

You no doubt noticed that Ric's points do not fit on a straight line as well as Karen's do. In Ric's data, the 73°C point was distinctly off the line, and the -79°C point was way off. The point at -79°C is an example of an outlying point that was probably caused by a gross error. It should be ignored. The 73°C point is probably a routine precision variation that should be averaged with others in drawing the curve. It is noteworthy that Karen's 73°C pressure is also low, the only point that does not touch her line. If others in the class have the same point below the line, it suggests that the temperature was not really 73°C, but closer to 65-68°C.

4.9 INTERPOLATION AND EXTRAPOLATION

We now turn to several topics that, as a group, may be thought of as the use and interpretation of graphs. The simplest use to which a graph may be put is **interpolation**. To interpolate is to estimate the value of one variable from a given value of the other when both values are within the range over which measurements have been made. In short, this is "reading" a graph.

As an example, refer to Karen's curve in Figure 4.2. What pressure will the gas exert at 40°C? Moving up the 40°C line until reaching Karen's curve, and then moving across to the y axis, we read 987 torr. In reverse, at what temperature does Karen's curve predict that the gas pressure will be 635 torr? Starting from 635 on the vertical axis and moving left to Karen's curve, we cross the curve at a temperature of about -71°C.

EXAMPLE 4.5: Use Ric's curve in Figure 4.2 to estimate the temperature at which pressure is 551 torr.

According to Ric's curve in Figure 4.2, pressure is 551 torr at -52°C.

Notice that, in Table 4.1, Ric recorded a pressure of 551 torr at -48°C. Which temperature, -52°C or -48°C, do you suppose is most apt to be correct for 551 torr? Explain your choice.

-52°C. Because a curve averages out experimental errors, it is more apt to be correct than any single reading.

Interpolation can be used to predict the temperature at which Ric's gas sample will exert a pressure of 500 torr, but not Karen's. She did not use a temperature low enough for her pressure to drop to 500 torr. That pressure is outside the range of her data. But could we not estimate that value anyway? Is it not reasonable to expect that a straight line that runs from 100°C to -94°C will continue to be straight for a few more degrees? To assume that a trend continues outside of the observed range is called **extrapolation**.

EXAMPLE 4.6: Use a straight edge to extrapolate Karen's curve in Figure 4.2 until it intersects the 500-torr line. Estimate the temperature at which pressure will be 500 torr.

-113°C.

In all probability it is reasonable to predict from Karen's data that her gas pressure will be 500 torr at -113°C. Would a similar prediction be justified for 400 torr, 300 torr, 200 torr, and 100 torr? And how about 1200 torr and 1300 torr, both values higher than those she observed?

Considering the nature of gases, extrapolation to higher temperatures and lower pressures is probably quite reasonable. The gas will continue to behave as a gas. However, at some combination of low temperature and high pressure almost every gas condenses to a liquid—and its behavior becomes erratic before condensation begins. We must be cautious about extrapolation. It can lead to some interesting predictions, but they must always be confirmed in the laboratory.

Extrapolating pressure-temperature data predicts the existence of an "absolute zero" temperature. Both Karen's and Ric's curves have been redrawn and extrapolated to the horizontal axis in Figure 4.3. The extrapolated portions of the curves are in dashes. Karen's curve hits the temperature axis at -271°C, and Ric's at -268°C.

If a large number of similar experiments were to be analyzed it would be found that all gases tend to reach the temperature axis at about -273°C. Even though that point cannot actually be reached because the gases condense, there is overwhelming evidence from other experiments that this is the lowest temperature possible. You will study this absolute zero in a later chapter.

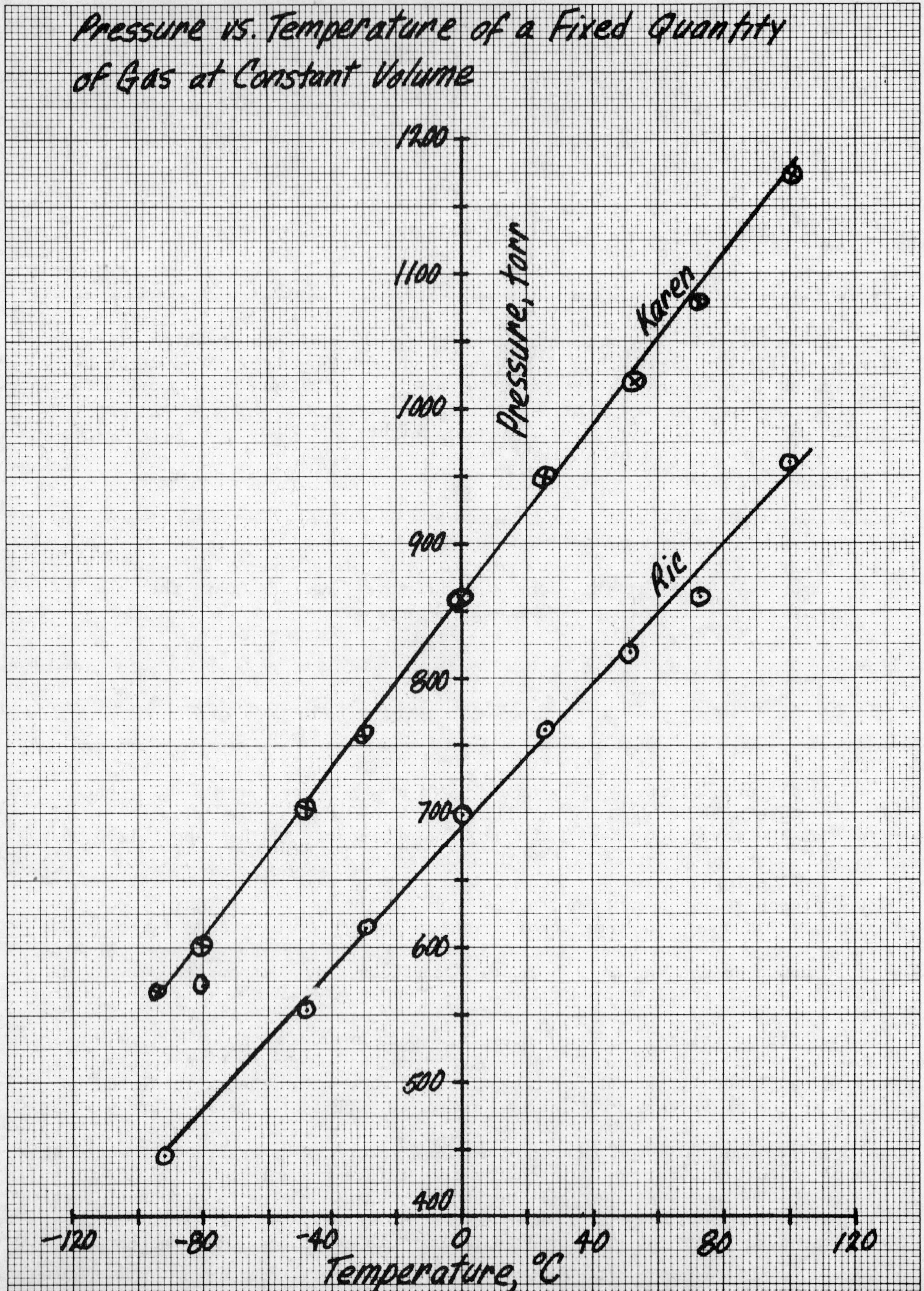
Pressure vs. Temperature of a Fixed Quantity
of Gas at Constant Volume
1200
1100
1000
900
800
700
600
500
400
Pressure, torr
Karen
Ric
-120
-80
-40
0
40
80
120
Temperature, °C

Figure 4.2

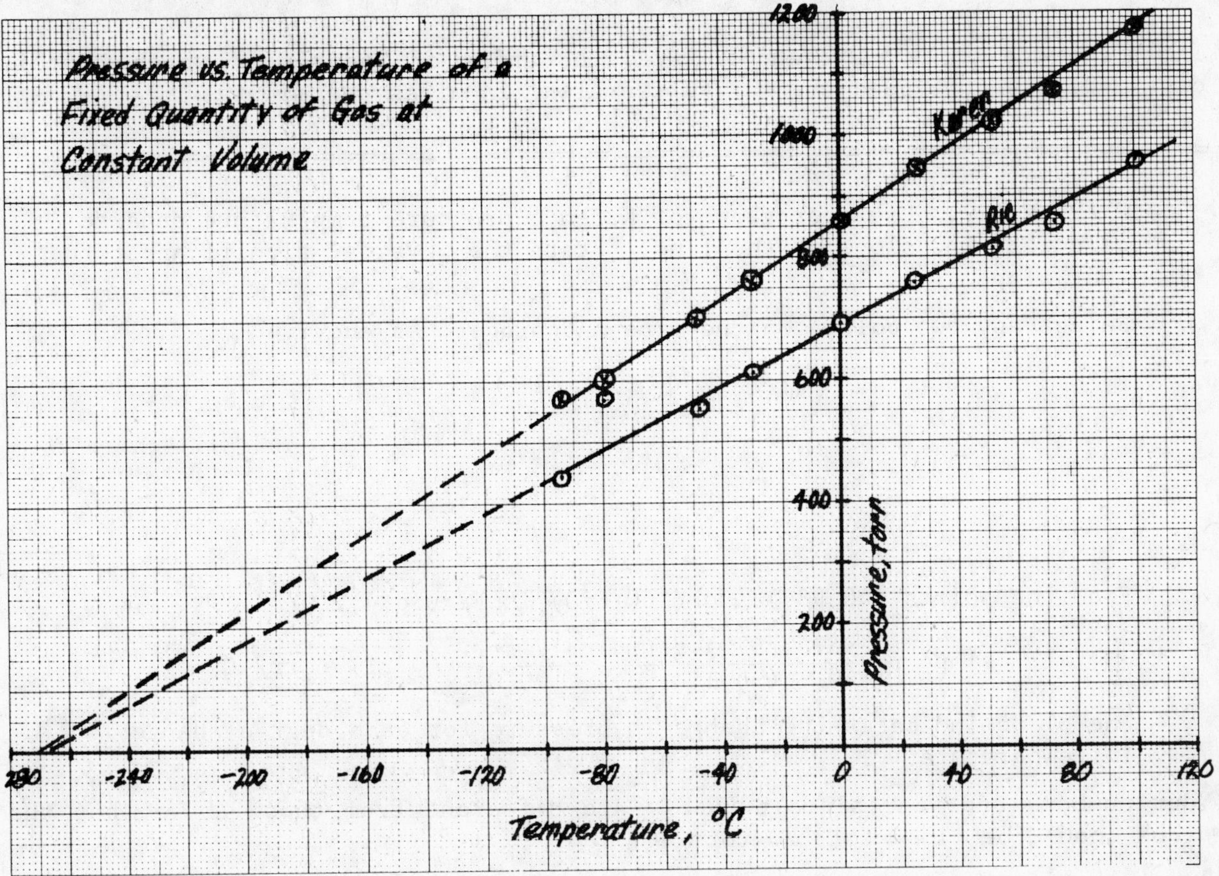

Figure 4.3

4.10 THE SLOPE OF A CURVE

In mathematical terms, the slope of a straight-line curve is the ratio of the change of y values between two points divided by the change of x values between the same points. Using the Greek delta, Δ, for change,

$$\text{slope} = \frac{\Delta y}{\Delta x} \qquad (4.1)$$

The slope of a graph of experimental data often has physical significance. For example, Figure 4.4 is a graph of distance traveled at constant speed on a highway. To find the slope of the line, select any two points *on the line*. Remember, points on the line are more reliable than plotted points. Substituting the coordinates of A and B into Equation 4.1,

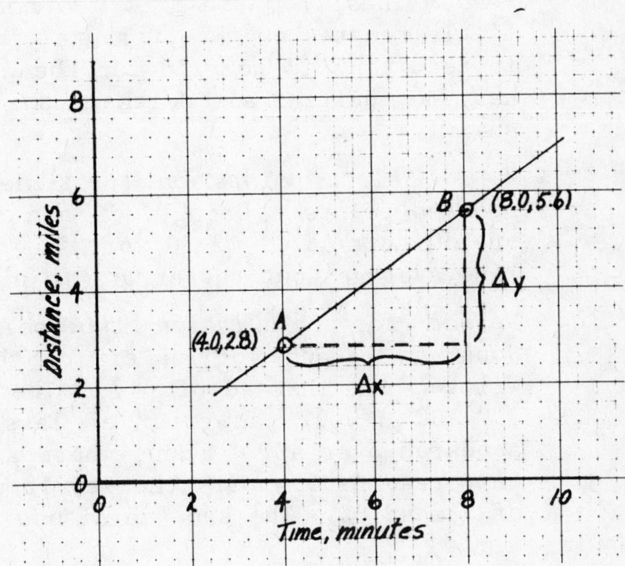

Figure 4.4

$$\text{slope} = \frac{\Delta y}{\Delta x} = \frac{(5.6 - 2.8)\ \text{miles}}{(8.0 - 4.0)\ \text{minutes}} = 0.70\ \text{miles/minute}$$

Including the units of y and x gives the units of the slope, which in this case indicate the speed the car is traveling in miles per minute.

EXAMPLE 4.7: Use your graph of Karen's pressure-temperature data to calculate the slope of the line. State the physical significance of that slope.

Select any two points on the line and proceed as above.

From the points (40,989) and (-60,670) in Figure 4.2,

$$\text{slope} = \frac{(987 - 670)\ \text{torr}}{[40 - (-60)]\ \ ^\circ\text{C}} = \frac{317\ \text{torr}}{100^\circ\text{C}} = 3.17\ \text{torr}/^\circ\text{C}$$

The slope indicates the change in pressure for each degree change in temperature.

Your numbers and your slope probably are different from those above, but the slope should be close to 3.20 torr/°C.

There are three things to keep in mind when finding the slope of a line. (If your answer to the above examples was much different from 3.20 torr/°C, you probably didn't do one of these things.) These are:

1. *Read values from the axes; do not count squares on the grid.* Counting little squares on the pressure-temperature graph yields 63.8/50.0 = 1.28 instead of 3.17 torr/°C for the answer to Example 4.7. The 1.28 answer has no units, as neither the numerator nor denominator measures anything.

2. *Select points on the line for calculation, not coordinates of plotted points.* Using 73°C and 25°C temperature points yields a slope of 130 torr/48°C = 2.70 torr/°C—quite different from 3.17 torr/°C. The line averages out the errors in plotted points.

3. *Select for calculation two points that are far from each other.* Suppose, because of an error, the numerator in Example 4.7 had been written 312 instead of 317. The calculated slope would then have been 312/100 = 3.12, an error of less than 2%. Had the same 5 torr error appeared over a 20° span, where a pressure change of 64 torr is mistakenly read as 59 torr, the result would have been 59/20 = 2.95, an error of almost 8%. The same magnitude of error in the numerator or denomina-

tor results in a smaller percentage error in a calculated slope when the points are far apart.

Not all curves are straight. The slope of a curve that is not straight changes from point to point, and therefore must be identified as the slope at a specific point. To find that slope, draw a tangent to the curve at that point and calculate the slope of the tangent by the method described above.

Some curves have negative slopes. This occurs when the variables are inversely related—when one variable increases as the other decreases. Figure 4.5 is a nonlinear graph with a negative slope. If we select the point (60,0.010) and draw a tangent at that point, it intersects the vertical axis at about 0.024 moles per liter and the horizontal axis at about 105 seconds. Using these values, the slope is

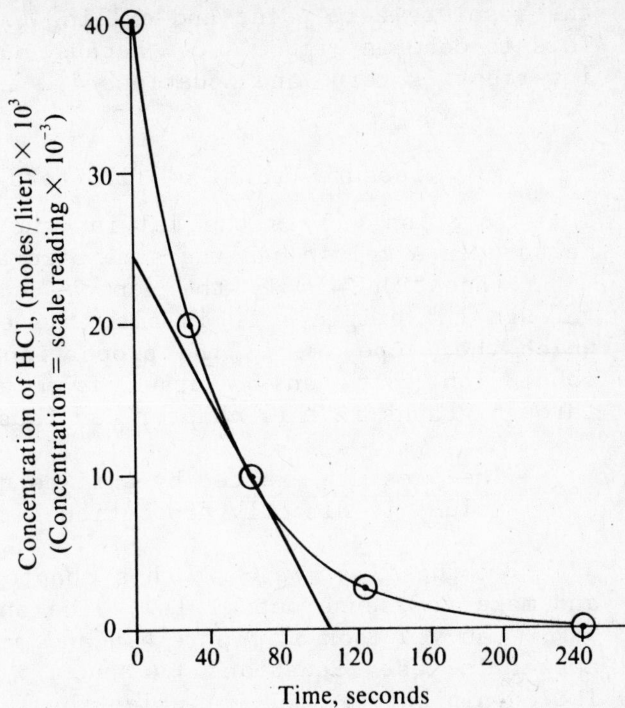

Figure 4.5 Variation of the concentration of hydrochloric acid with time for the reaction of magnesium with hydrochloric acid.

$$\text{slope} = \frac{(0 - 0.024)\ \text{mol/L}}{(105 - 0)\ \text{sec}} = -2.3 \times 10^{-4}\ \text{mol/L}\cdot\text{sec}$$

4.11 EQUATIONS DERIVED FROM STRAIGHT-LINE (LINEAR) GRAPHS

If an equation can be derived from a graph, it expresses the relationship between the variables plotted. Figures 4.2 and 4.3 can be converted into an equation by using the slope-intercept form of the equation for a straight line,

$$y = mx + b \tag{4.2}$$

In this equation x and y are the variables, m is the slope, and b is the y intercept, the value of y where the line crosses the vertical axis. The slope of Karen's line is 3.17. Examination of the graph shows that the line crosses the pressure axis at 860 torr. Substituting these values into Equation 4.2, the equation for Karen's line is

$$\text{pressure (torr)} = 3.17\ (\text{torr}/^\circ\text{C}) \times \text{temperature}\ (^\circ\text{C}) + 860\ (\text{torr}) \tag{4.3}$$

So far this gives us only the equation for Karen's particular sample of gas. Figure 4.3 shows her graph and Ric's extrapolated to the absolute zero at -273°C. The equation for Ric's line would have a different slope and a different pressure intercept. The difference between the graphs is caused by a difference in quantity of gas in the two experiments.

There is something naturally "comfortable" about a graph that starts at the origin. Suppose Figure 4.3 is redrawn by relocating the vertical axis at the absolute zero point and calling that point zero on the temperature scale. This is done in Figure 4.6. Because the line passes through the origin, the y intercept is zero, and Equations 4.2 and 4.3 simplify to

$$y = mx \tag{4.4}$$

$$\text{pressure (torr)} = 3.17\ \text{(torr/K)} \times \text{temperature (K)} \tag{4.5}$$

K in Equation 4.5 is the kelvin, the unit in which absolute temperature is measured. A kelvin has the same size as one Celsius degree.

Equation 4.4 is the general equation of a straight line that passes through the origin. It is also the equation of a direct proportionality in which the slope, m, is the proportionality constant (see Section 2.2). The conclusion that can be drawn from the experiment that has been analyzed through graphs in this chapter can thus be summarized as follows:

> The pressure exerted by a fixed quantity of a gas held at constant volume is directly proportional to absolute temperature.

On the first page of this chapter we asked, "Can you look at Table 4.1 and make any sense out of it?" The answer was no. It is now appropriate to ask, "Can you look at Figure 4.6 and make any sense out of it?" This time the answer is yes. Equations 4.4 and 4.5, plus the above summary, state the relationship between the variables that have been studied.

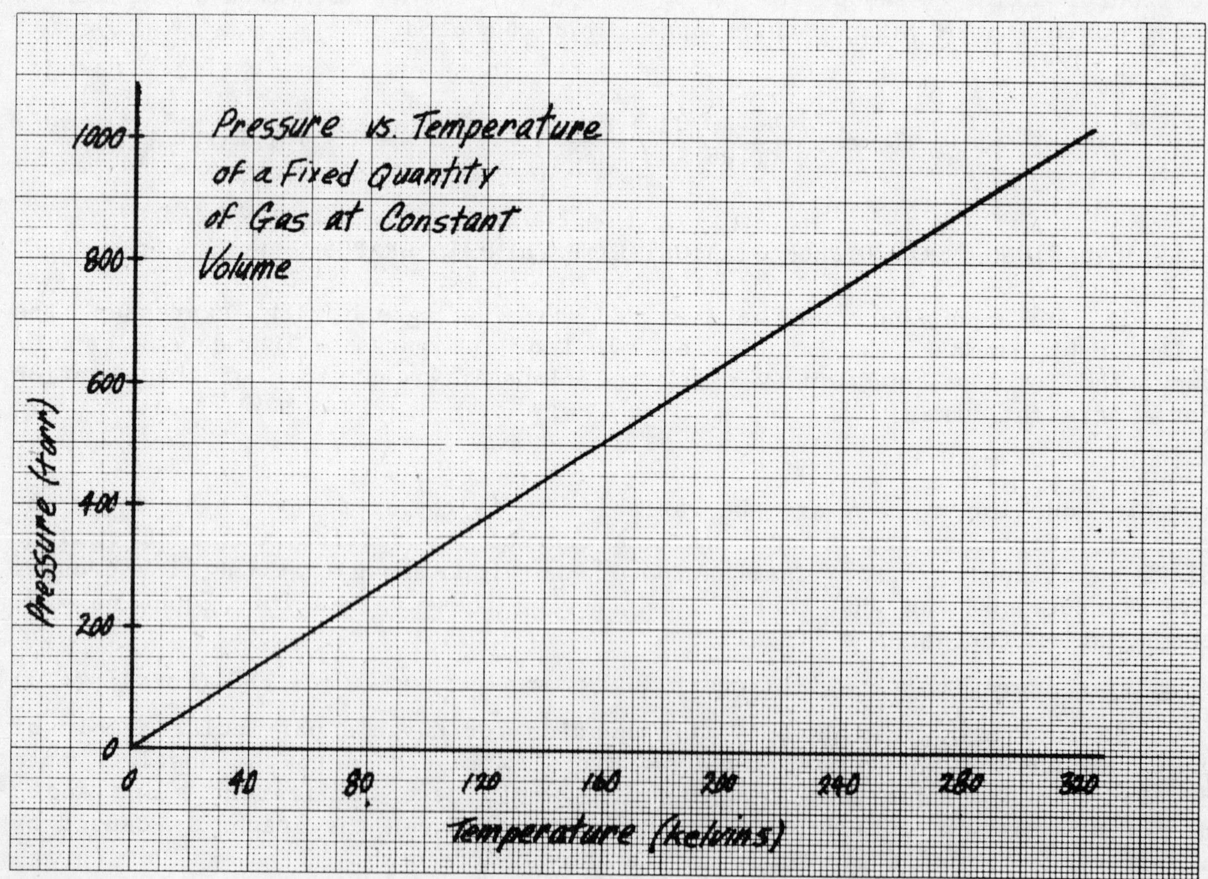

Figure 4.6

4.12 EQUATIONS DERIVED FROM NONLINEAR GRAPHS

It is unfortunate that not all graphs of laboratory data are straight lines. A straight-line graph is the easiest to analyze and to express in an equation. Recognizing this, the scientist tries to present data in a form that produces a straight-line graph. This usually involves restating one or both variables.

Figure 4.7 is an example of how a nonlinear graph can be replotted as a straight line. Figure 4.7A is a graph of an inverse proportionality for which the general equation is

$$xy = k \tag{4.6}$$

The symbol k represents a constant. If both sides of Equation 4.6 are divided by x, the equation becomes

$$y = \frac{k}{x} = k \times \frac{1}{x} \tag{4.7}$$

This has the same form as the equation of a straight line that passes through the origin (Equation 4.4 and Figure 4.6). To produce the straight line, you would plot y vertically and the reciprocal of x horizontally (Figure 4.7B). The constant k is the slope of the line.

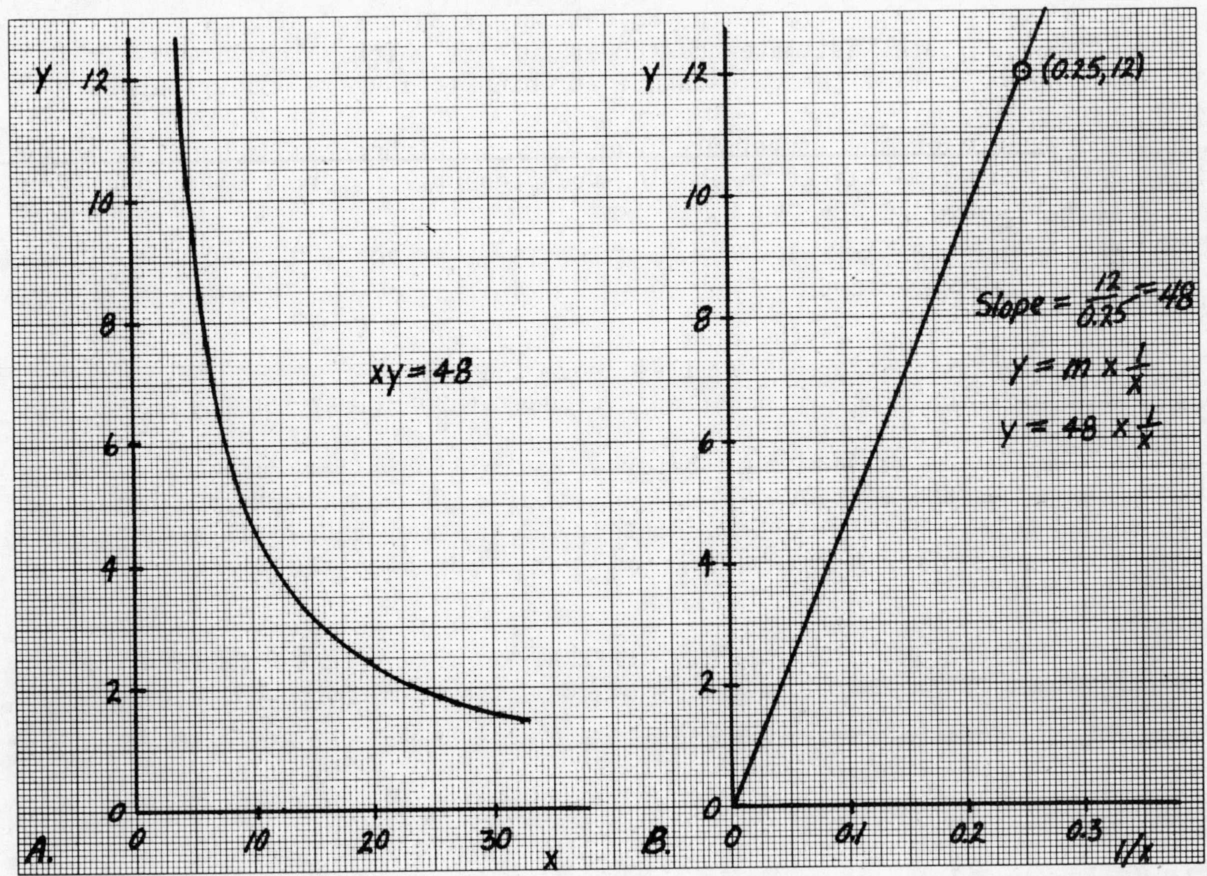

Figure 4.7

4.13 OTHER NONLINEAR GRAPHS

Some experiments produce nonlinear graphs that cannot be simplified. The graph may be linear over one section but not over another; or it may be linear over two portions, but with different slopes; or it may have no predictable behavior because of the nature of the change taking place. In cases such as these, a simple x-y graph is plotted and the best possible curve is drawn through the points. Relationships such as these cannot be summarized in a simple equation.

GRAPHING EXERCISES

1) The solubility of a compound can be expressed in several ways, one of which is moles of compound that will dissolve in 1 kilogram of water. A mole (abbreviated mol) is a quantity unit you will use often in chemical calculations later in this book. Solubility depends on temperature. Prepare a graph of the solubility of potassium nitrate, KNO_3, from the following information:

Temperature, °C	0	20	40	60	80	100
Solubility, mol/kg	1.61	2.80	5.78	11.20	16.76	24.50

a) From your graph, estimate the solubility at 86°C and at 37°C.
b) From your graph, estimate the temperature at which solubility is 7.32 mol/kg water, and the temperature at which it is 17.7 mol/kg water.

2) The pressure of the atmosphere decreases at higher altitudes. From the information below, make a graph showing the pressure in torr in the range from 10 to 25 miles above the earth.

Altitude, miles	10	15	20	25
Pressure, torr	80	24	7	2

a) At what elevation is the pressure 63 torr?
b) Estimate the pressure 12.7 miles above the earth.
c) What will the pressure be at an altitude of 30 miles? 50 miles?

3) The temperature of a mixture before, during, and after freezing is not a smooth curve, as are most of the changes studied in this chapter. Instead it appears to follow distinct paths during each of the three stages. Plot a graph of temperature in Celsius degrees over a period of 20 minutes from the following data:

TIME min	TEMPERATURE °C	TIME min	TEMPERATURE °C	TIME min	TEMPERATURE °C
0	86.9	7	79.4	14	78.9
1	84.7	8	79.4	15	78.9
2	83.1	9	79.4	16	78.8
3	81.3	10	79.3	17	78.1
4	79.2	11	79.2	18	76.4
5	79.3	12	79.1	19	75.2
6	79.7	13	79.1	20	74.2

Ideally, the freezing temperature of the solution is the temperature at the abrupt change between the first and second stages. From your graph, what do you estimate this temperature to be?

4) The equilibrium vapor pressure of a liquid changes with temperature. Plot a graph of the vapor pressure of water at temperatures well above the normal boiling point, using the data below:

Vapor Pressure, torr	11,700	25,100	48,100	84,700	139,900
Temperature, °C	200	240	280	320	360

a) Estimate the vapor pressure of water at 306°C.
b) At what temperature is the vapor pressure 17,500 torr?
c) Can you predict the temperature at which the vapor pressure is 1.47×10^5 torr?
d) What is the vapor pressure at 180°C?

5) Most coffee drinkers like their coffee hot. They know that it doesn't stay hot in certain kinds of cups, particularly outdoors on a cold day. One student, curious about how fast coffee cools when purchased from an outdoor snack wagon at a late-fall football game, did an experiment. He measured the temperature of a cup of coffee at different times. His readings were:

Time, minutes	0	1	2	4	6	8	10
Temperature, °C	93.8	59.5	51.0	36.0	29.0	24.6	20.5

Prepare a graph of these readings, and then answer the following questions:

a) If you decide you want to drink your coffee while its temperature is 40°C or higher, how many minutes do you have to finish the cup?
b) When is the rate of cooling the greatest? Estimate the rate of cooling at 1.5 minutes, and again at 7 minutes.

6) An experiment was conducted in which the volume of a fixed quantity of gas was measured at different temperatures at constant pressure. The results are tabulated below. Graph the data, and answer the questions from the graph.

Temperature, K	405	455	505	533	573	594
Volume, L	5.94	6.65	7.34	7.97	8.42	8.86

a) What will the volume be at 450 K?
b) At what temperature will the volume be 7.73 liters?
c) Calculate and state the physical significance of the slope of the curve.
d) Determine the equation of the curve.
e) Write a statement expressing the relationship between the variables. Be sure to identify any restrictions that may apply.

7) A beaker was placed on a balance and different volumes of a liquid were poured into it. The mass of the beaker plus liquid was measured each time. The following data were collected:

Mass, g	192.8	204.0	219.6	236.5	246.3
Volume, mL	23	34	45	59	68

Prepare a graph that best represents the data as collected.

a) Calculate the density of the liquid.
b) Write an equation for the curve. This may require a second graph.
c) What is the mass of the beaker?

8) Below are data from an experiment in which the distance travelled by a freely falling steel ball was measured at different times from the moment of release. Prepare a graph of these data. Then prepare a second graph of distance versus the square of time. If you wish, the second graph may be plotted on the same grid as the first. The vertical axis and distance scale would be the same, but the horizontal axis would have a second scale added to it, the square of time.

Distance, meters	0.53	1.24	2.56	4.15	6.47
Time, seconds	0.33	0.50	0.72	0.92	1.15

a) Estimate the distance travelled in 0.83 seconds.
b) Estimate the time required to fall 5.83 meters.
c) Write an equation for the relationship between time and distance fallen.
d) Use your equation to confirm your answers to a and b.

ADDITIONAL PROBLEMS

9) The change in concentration, C, depends on elapsed time, t, as shown below.

t(min):	0	100	200	300	400	500	600
C(mol/L):	0.105	0.0598	0.0340	0.0194	0.0110	0.0063	0.0036

A straight line relationship exists for one of the following: C vs t, ln C vs t, or 1/C vs t. Plot these relationships on the graph on the next page. Which one is linear? Calculate the slope of the linear plot.

10) A student collected the following data for the vapor pressure of a liquid as a function of temperature.

P(atm):	0.022	0.037	0.069	0.145	0.349
T(°C):	1.2	10.9	23.7	40.4	62.6

A linear relationship exists between log P and 1/T, where T has been converted to kelvin.

a) Convert each of the above temperatures to kelvin. Graph log P (as y) vs 1/T (as x, with T in kelvin) and compute the slope. Spread out the points for maximum precision in the determination of the slope.

b) Determine the y-intercept by redrawing the graph to show the value of log P when 1/T = 0.

c) Write the equation for the straight line. From this equation compute the value of the x-intercept.

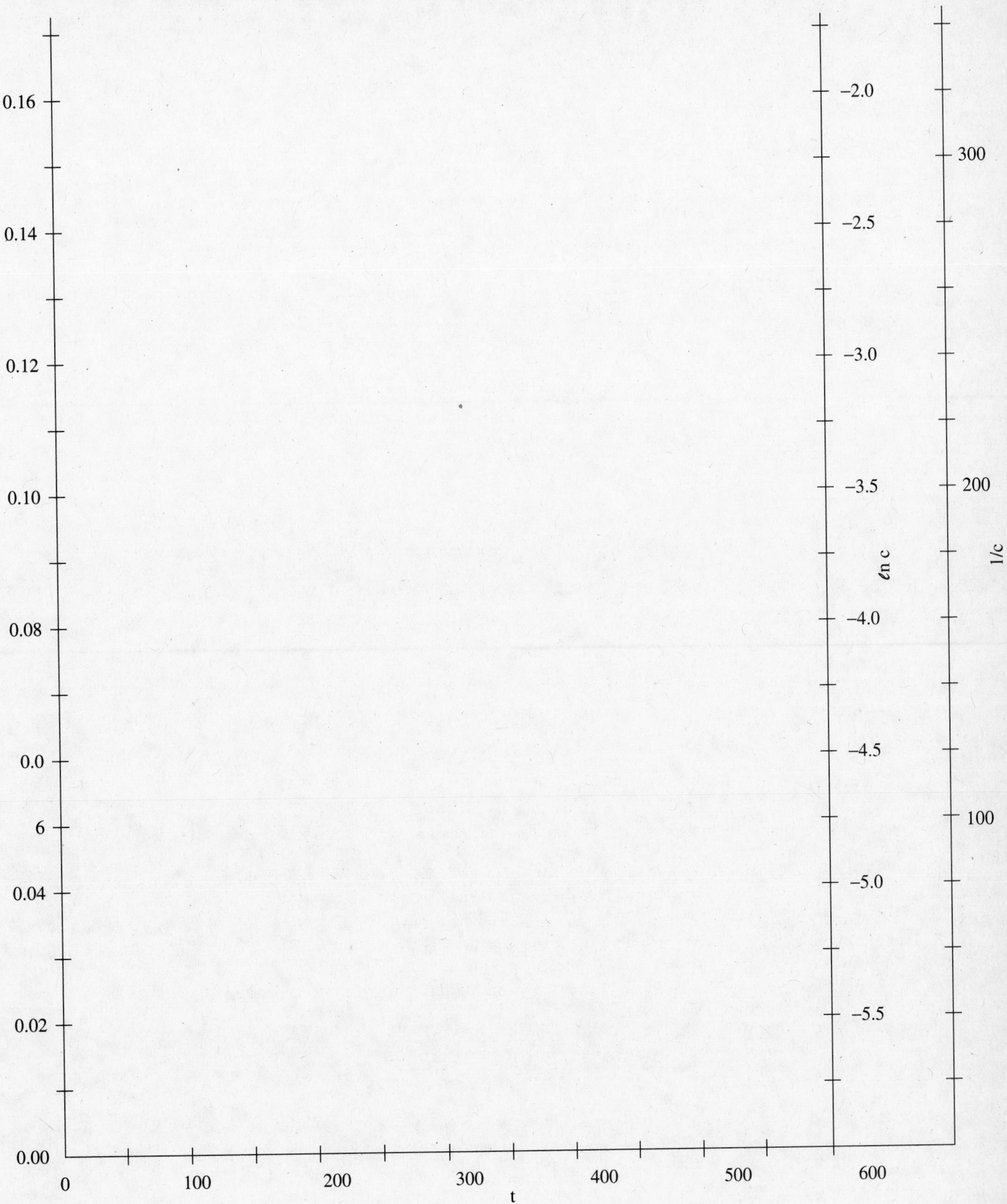
0.16
0.14
0.12
0.10
0.08
0.0
6
0.04
0.02
0.00
0
100
200
300
400
500
600
t
−2.0
−2.5
−3.0
−3.5
−4.0
−4.5
−5.0
−5.5
ℓn c
300
200
100
1/c

5. LOGARITHMS

PREREQUISITES

1. You will use calculators to perform calculations that include logrithms, as discussed in Section 1.6.

2. Sections 1.9 and 1.10 discussed algebraic operations with exponents which will be performed in this chapter.

3. Section 1.4 presented exponential notation which will be used for many numbers throughout this chapter.

4. Sections 3.3 through 3.7 gave the rules of significant figures that are applied when working with logarithms.

CHEMICAL SKILLS

1. Logarithms. Use a calculator to find the logarithm of any positive number.

2. Antilogarithms. Use a calculator to find the antilogarithm of any number.

3. Algebra of logarithms. Given an equation with a logarithmic expression and the values of all variables except one, solve the equation for the unknown variable.

Many natural laws follow exponential relationships, and the equations for such laws involve logarithms. Topics in general chemistry for which logs are used include the vapor pressure of liquids, the thermodynamic, kinetic and equilibrium behavior of chemical reactions, and the variation of electrochemi-cal cell voltage with concentration. In addition, logs provide a useful way to express very large and very small numbers. This application is most apparent in the use of the pH scale for acidity.

Most of these topics are commonly introduced in the second half of the general chemistry course, so you won't find an immediate use for the discussion of logarithms presented here. You might even wish to postpone the study of this chapter until you need to use logarithms. In this book, the only chapter that will use logs is Chapter 20 on acid-base equilibria.

One more observation: It is not the purpose to present here a complete discussion of logarithms. That is left to your mathematics classes. This chapter is aimed specifically and exclusively at logarithms as they are used in general chemistry.

5.1 LOGARITHMS OF POWERS OF 10

Table 5.1 organizes some basic number facts from which to begin the study of logarithms. The table ranges over seven orders of magnitude, from 10^3 to 10^{-3}. It may be extended indefinitely in either direction to numbers that are infinitely large or infinitesimally small.

The common logarithm of a number is the power, or exponent, to which 10 must be raised to be equal to the number.

From this definition,

$$\text{if} \quad N = 10^x \quad \text{then} \quad \log N = \log 10^x = x \tag{5.1}$$

From Table 5.1, when N = 1000 (or 10^3), log 1000 = log 10^3 = 3; when N = 0.01 (10^{-2}), log 0.01 = log 10^{-2} = -2; and so forth, throughout the table. In each case, the logarithm of the number is the exponent of 10.

TABLE 5.1: FORMS OF NUMBERS AND THEIR LOGARITHMS

DECIMAL FORM	FRACTION FORM	EXPONENTIAL FORM	LOGARITHM
1000	1000/1	10^3	3
100	100/1	10^2	2
10	10/1	10^1	1
1	1/1	10^0	0
0.1	1/10	10^{-1}	-1
0.01	1/100	10^{-2}	-2
0.001	1/1000	10^{-3}	-3

The above discussion and Table 5.1 illustrate these facts:

1. A common logarithm is an exponent of 10.
2. The logarithm of a number greater than 1 is a positive number.
3. The logarithm of 1 is zero.
4. The logarithm of a number less than 1 is a negative number.
5. The logarithm of a number that is a whole-number power of 10 is an integer—the same whole number.

If you have caught on to the discussion this far, you will be able to answer the following questions without using a calculator:

EXAMPLE 5.1: Write the logarithms of (a) 10,000,000; (b) 0.000001.

Begin by rewriting the numbers as exponentials of 10. In effect, you will express the numbers in exponential notation in the form 1×10^n.

10,000,000 = ______________ 0.000001 = ______________

10,000,000 = 10^7; log 10^7 = ______ 0.000001 = 10^{-6}; log 10^{-6} = ______

In both cases, the magnitude (size) of the exponent is simply a count of the number of places the exponent was moved.

Now you can write the logarithms in the spaces above.

---

log 10,000,000 = log 10^7 = 7; log 0.000001 = log 10^{-6} = -6

In each case, the logarithm is simply the exponent of 10.

5.2 MULTIPLICATION AND DIVISION WITH LOGARITHMS

Because logarithms are exponents, the rules for exponents in multiplication and division apply to logarithms. When exponents to the same base are multiplied, the exponents are added. Written first as a general equation, and then to the base 10,

$$a^m \times a^n = a^{m+n} \qquad 10^y \times 10^z = 10^{y+z} \tag{5.2}$$

By the definition of logarithms in Equation 5.1, y in the second equation above is the logarithm of the factor 10^y, z is the logarithm of 10^z, and y + z is the logarithm of the product. This leads to the general rule of logarithms for multiplication:

> The logarithm of the product of two or more factors is the sum of the logarithms of the factors.
>
> $$\log ab = \log a + \log b \tag{5.3}$$

When exponentials are divided, there is a subtraction of exponents:

$$\frac{a^m}{a^n} = a^{m-n} \qquad \frac{10^y}{10^z} = 10^{y-z} \tag{5.4}$$

The general rule of logarithms for division is:

The logarithm of the quotient of two numbers is found by subtracting the denominator logarithm from the numerator logarithm:

$$\log \frac{a}{b} = \log a - \log b \qquad (5.5)$$

5.3 LOGARITHMS THAT ARE NOT INTEGERS

If the logarithm of 10^0 is 0 and the logarithm of 10^1 is 1, you may well suspect that the logarithm of a number between 1 and 10 is a fraction. For example, the square root of 10 is about 3.16. The square root of 10 can be written $10^{0.5} = 3.16$. It follows that log 3.16 = 0.5.

Try this on your calculator: Enter 3.16, and then press the log key. The display will show 0.4996871, which is 0.500 if rounded off to three significant figures. This illustrates how you find the logarithm of a number with a calculator: enter the number and press the log key.

EXAMPLE 5.2: Use your calculator to find the logarithms below. It will be necessary to use exponential notation when entering the last number. Express each answer to three decimal places.

log 5.24 = ________ log 52.4 = ________ log 524 = ________

log 52,400 = ____________ log 5,240,000,000,000 = ________

$$\log 5.24 = \log (5.24 \times 10^0) = 0.719$$
$$\log 52.4 = \log (5.24 \times 10^1) = 1.719$$
$$\log 524 = \log (5.24 \times 10^2) = 2.719$$
$$\log 52{,}400 = \log (5.24 \times 10^4) = 4.719$$
$$\log 5{,}240{,}000{,}000{,}000 = \log (5.24 \times 10^{12}) = 12.719$$

In the answer to Example 5.2, all starting numbers have been written in exponential notation. Notice that the coefficient of all numbers is 5.24. Notice also that the digits after the decimal of the logarithm are all .719, which is the logarithm of 5.24. Finally, observe that the digits before the decimal in the logarithm are the same as the power of 10.

This all comes from Equation 5.2, which shows that when numbers are multiplied, their logarithms are added. The mechanics of this appears below for the logarithm of 524, followed by a general statement for all values of the exponent:

$$\log 524 = \log (5.24 \times 10^2) = \log 5.24 + \log 10^2 = 0.719 + 2 = 2.719$$
$$\log (5.24 \times 10^n) = \log 5.24 + \log 10^n = 0.719 + n = n.719$$

The logarithm of a number smaller than 1 is negative. Though the numbers change, this is actually a continuation of the series above. Specifically, for 0.524 and 5.24×10^{-15},

$$\log 0.524 = \log (5.24 \times 10^{-1}) = 0.719 + (-1) = -0.281$$
$$\log (5.24 \times 10^{-15}) = 0.719 + (-15) = -14.281$$

EXAMPLE 5.3: Find the logarithm of (a) 3.91×10^8 and (b) 4.32×10^{-9}.

$\log 3.91 \times 10^8 =$ ______ $\log 4.32 \times 10^{-9} =$ ______

$\log 3.91 \times 10^8 = 8.592$ $\log 4.32 \times 10^{-9} = -8.365$

Occasionally you will have to find the logarithm of a reciprocal, or inverse, of a number. If the number is x, its inverse is 1/x. The relationship between the log of a number and its inverse is

$$\log (1/x) = -\log x \tag{5.6}$$

A number raised to a power may be expressed as y^x. The logarithm of such an expression may be simplified as follows:

$$\log (y^x) = x \log y \tag{5.7}$$

EXAMPLE 5.4: Find the following logarithms:
(a) $\log 1/0.00313 =$ ______ (b) $\log 0.0416^{0.25} =$ ______

$\log 1/0.00313 = -\log 0.00313 = -(-2.504) = 2.504$
$\log 0.0416^{0.25} = 0.25 \log 0.0416 = 0.25(-1.381) = -0.35$

5.4 ANTILOGARITHMS

The number that corresponds to a given logarithm is called its **antilogarithm**. Reversing Equation 5.1,

$$\text{if} \quad \log N = x \quad \text{then} \quad \text{antilog } x = 10^x = N \tag{5.8}$$

If 1.772 is the logarithm of a number, then the number is $10^{1.772} = 59.2$. Similarly, the antilogarithm of $-5.487 = 10^{-5.487} = 3.26 \times 10^{-6}$.

How to find an antilogarithm on a calculator depends on the calculator you own. With some calculators you can choose between two or more methods. Examine your calculator. If it has a "10^x" key, its main purpose is to find antilogarithms. If that key is missing, there is probably an "INV" or "2ND" key that must be used with the "log" key. A "y^x" key offers another way. The use of these keys is illustrated for finding the antilogarithm of 1.772:

To Use This Key	Enter	Press	Display
10^x	1.772	10^x	59.156163
INV and log	1.772	INV and log	59.156163
y^x	10	y^x	10
	1.772	=	59.156163

EXAMPLE 5.5: Find the antilogarithms of (a) 7.39 and (b) -16.149

antilog 7.39 = ____________ antilog -16.149 = ____________

antilog 7.39 = $10^{7.39}$ = 2.5×10^{7}; antilog -16.149 = $10^{-16.149}$ = 7.10×10^{-17}

5.5 LOGARITHMS AND SIGNIFICANT FIGURES

In Example 5.2 and the discussion that followed it, we found the logarithm of 5.24×10^{n} where n had the values -15, -1, 0, 1, 2, 4, 12, and the general value n itself. The number 5.24×10^{n} has three significant figures, regardless of the value of n. The logarithms should also have three significant figures—and each logarithm satisfies this condition.

The part of each logarithm that is related to the coefficient, 5.24, is the three digits *after* the decimal point. These are 0.719 when n is 0 or positive, and 1 - 0.719 = 0.281 when n is negative. The number *in front of* the decimal point is established only by the exponent of 10, and has nothing to do with significant figures. Thus, as logarithms, -14.281, 0.719, 2.719 and 12.719 all have three significant figures. *The number of significant figures in a logarithm is the number of digits after the decimal.*

5.6 CHEMICAL CALCULATIONS WITH LOGARITHMS

There are many occasions to work with very small numbers in solving chemical equilibrium problems. These are usually written in exponential notation, but sometimes they are expressed as "p numbers." The best known of these is pH, a measure of acidity. Examples of pH will be considered in Chapter 20. In general, if N is any small number,

$$pN \equiv -\log N \tag{5.9}$$

EXAMPLE 5.6: The symbol K is commonly used for equilibrium constants. Find pK for the dissociation constant of acetic acid if K = 1.8×10^{-5}. Be sure to use the proper number of significant figures.

GIVEN: K = 1.8×10^{-5} FIND: pK EQN: pK = -log K

pK = -log K = -log (1.8×10^{-5}) = - (-4.74) = 4.74

K is in two significant figures, so pK is also in two significant figures. Because pK is a logarithm, counting significant figures begins at the decimal.

EXAMPLE 5.7: The pOH of a solution is 3.29. If pOH = -log $[OH^-]$, calculate $[OH^-]$.

GIVEN: pOH = 3.29 FIND: $[OH^-]$ EQN: pOH = -log$[OH^-]$

$[OH^-]$ = antilog -pOH = antilog -3.29 = $10^{-3.29} = 5.1 \times 10^{-4}$

Note that the equation that defines pOH must be multiplied by -1 before the antilog is found: log $[OH^-]$ = -pOH, which yields $[OH^-]$ = antilog -pOH.

Many equations that include logarithms appear in chemistry, and you must be able to solve them for any variable. The equations sometimes look frightening, but they are not really difficult. Just solve the equation for the wanted variable, substitute the given quantities, and calculate the answer. If the wanted quantity appears as a logarithm or is included in a logarithm, solve for the logarithm and then find the antilogarithm.

EXAMPLE 5.8: An equilibrium constant can be calculated from thermodynamic data by the equation $\Delta G° = -2.303RT \log K$. Find K if the other factors are: $\Delta G° = 1.39 \times 10^4$; R = 8.314; T = 523.

Rearrange the equation so log K is alone on the left side and all other factors are on the right.

GIVEN: $\Delta G° = 1.39 \times 10^4$, R = 8.314, T = 523 FIND: K
EQN: $\Delta G° = -2.303RT \log K$

$$\log K = \frac{\Delta G°}{-2.303RT} = \underline{\hspace{4cm}}$$

Now substitute the given numbers and calculate log K.

$$\log K = \frac{\Delta G°}{-2.303RT} = \frac{13{,}900}{(-2.303)(8.314)(523)} = -1.39$$

If log K is -1.39, K = antilog -1.39. Calculate K: ____________

K = antilog -1.39 = $10^{-1.39}$ = 0.041

Note that both -1.39, a logarithm, and 0.041 are written in two significant figures.

The half-life of a radioactive substance is the time required for half of a sample to "disappear," or be changed to another substance. The process is described by the equation

$$(0.301)n = \log (S/R) \tag{5.10}$$

where S is the amount of original sample and R is the amount remaining after n half-lives. The value of n can be calculated by dividing the time elapsed by the half-life.

EXAMPLE 5.9: If you have 6.43 milligrams of a radioisotope with a half-life of 1.71 hours, how many milligrams will be left 5.37 hours later?

Let's list the GIVEN-FIND-EQUATION as the first step in this problem. As a hint, there are two equations, and both are "given" in the paragraph before this example.

GIVEN: S = 6.43 mg; time elapsed = 5.37 hr; half-life = 1.71 hr FIND: R

EQN: (0.301)n = log (S/R); n = time elapsed ÷ half-life

The second equation comes from the sentence just before the example.

This time the wanted quantity is a part of a logarithm. We therefore solve first for log (S/R), and then by antilog get S/R. With S given, R can be calculated. Complete the problem.

1. $\log (S/R) = (0.301)n = (0.301)(5.37/1.71) = 0.945$
2. $S/R = \text{antilog } 0.945 = 8.81$
3. $R = S/8.81 = 6.43 \text{ mg}/8.81 = 0.729 \text{ mg}$

In Step 1, the values needed to determine n are substituted into Equation 5.10 and the value of log (S/R) is computed. For ease of calculation, it is usually best to solve for log x and then calculate x as the antilog of that number, as in Step 2. When a value for S/R has been obtained, Equation 5.10 is solved algebraically for R, the given and calculated values are substituted, and the answer calculated. This is Step 3.

5.7 NATURAL LOGARITHMS

This entire chapter has described common logarithms to the base 10. Many natural phenomena involve logarithms to the base 2.718 . . . , which is known as **base e**. To distinguish between the bases, the common logarithm of X is written $\log_{10} X$, or simply log X, and the **natural logarithm** is $\log_e X$ or ln X. The relationship between logarithms to the two bases is

$$\ln X = (\ln 10)(\log X) = 2.303 \log X \quad (5.11)$$

This is the source of the 2.303 factor that appears in so many logarithmic equations.

Most calculators have a natural log key that makes it just as easy—in fact, easier—to work with natural logarithms than base 10 logarithms. This is because the 2.303 factor is absent. For instance, the equation in Example 5.8, $\Delta G° = -2.303RT \log K$, becomes $\Delta G° = -RT \ln K$. The calculator operation is the same, except that the ln key is pressed instead of the log key.

Listed below is a summary of the logarithmic relationships covered in this chapter.

$\log 10^x = x$	(5.1)
$\log ab = \log a + \log b$	(5.3)
$\log a/b = \log a - \log b$	(5.5)
$\log 1/x = -\log x$	(5.6)
$\log y^x = x \log y$	(5.7)
$\text{antilog } x = 10^x$	(5.8)
$pN = -\log N$	(5.9)
$\ln X = 2.303 \log X$	(5.11)

CHAPTER 5 PROBLEMS

1) Write the logarithms of each of the following numbers:

a) 8.10×10^4 b) 100,000,000 c) 439,000 d) 6.11×10^{-5}
e) 0.024 f) 0.97×10^9 g) 0.000000000010 h) 36.2×10^{-4}

2) Write the antilogarithm for each of the following logarithms:

a) 0.861 b) 2.62 c) 0.1124 d) -4.057
e) 1.982 f) -0.56 g) 13.664 h) -10.367

3) Find y in each of the following:

a) $y - 6.214 = \log 32.6$ b) $3.7\ y = \log 0.0039$ c) $27y^2 = 733 \log 82$

4) Calculate y in each of the following:

a) $6.62 \log y = 8.46$ b) $\dfrac{\log y}{-304} = -5.06 \times 10^{-5}$

c) $\dfrac{-5.69 \times 10^{11}}{\log y} = 4.97 \times 10^{10}$

5) In a certain solution, pCl is 9.34. If pCl = -log $[Cl^-]$, find $[Cl^-]$.

6) The hydroxide ion concentration, $[OH^-]$, of a solution is 6.2×10^{-6} moles per liter. Calculate pOH if pOH = -log $[OH^-]$.

7) Calculate K for a redox reaction, using the equation $E° = \frac{0.0592}{n} \log K$. For this reaction, E° = 1.10 and n = 2.00.

8) If $\Delta G° = -2.303RT \log K$, find $\Delta G°$ for a reaction in which $K = 8.2 \times 10^{12}$, and T = 325. Then recalculate $\Delta G°$, using $\Delta G° = -RT \ln K$. The results should be the same.

9) Calculate the number of half-lives, n, if a 15.234 gram sample (S) decays to 14.639 grams (R) in 5.06 years. The equation is

$$0.301n = \log (S/R) = \log S - \log R.$$

10) The equation $\log \frac{k_2}{k_1} = \left(\frac{E_a}{2.303R}\right)\left(\frac{T_2 - T_1}{T_2T_1}\right)$ appears in the study of rates of chemical reaction, in which k is called the rate constant. What is k_2/k_1 when $E_a = 4.8 \times 10^4$, $T_1 = 2.9 \times 10^2$, $T_2 = 3.0 \times 10^2$, and R = 8.314?

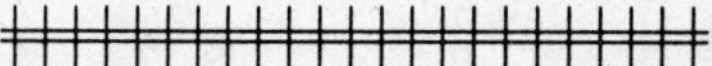

11) Find the logarithms of:

a) 617 b) 0.00354 c) 5.83×10^{-12} d) 2.72×10^6

12) Find the antilogarithms of: a) -14 b) 7.85 c) -4.116

13) Calculate y in each of the following:

a) y = log (48.7/660) b) 4.1y = 3.82 log 63.5

14) Find y in each of the following:

a) $(3.98 \times 10^{-4})(852 \log y) = 3.20$ b) $\frac{6}{\log y} = -4.26$

15) The pK of a certain equilibrium is 13.42. Find K. (pK = -log K)

16) The silver ion concentration, $[Ag^+]$, of a solution is 6.3×10^{-9}. Calculate pAg if pAg = -log $[Ag^+]$.

17) If E° = 0.069 and n = 3.00 in the equation $E = E° - \frac{0.0592}{n} \log Q$, find the value of Q that will cause E to become zero.

18) Calculate E in $\log k = \log A - \frac{E}{2.303RT}$ when k = 0.39, $A = 1.3 \times 10^{10}$, R = 8.314, and T = 293. Repeat the calculation, using $\ln k = \ln A - \frac{E}{R\,T}$. The answers should be the same.

ADDITIONAL PROBLEMS

19) Solve the equation $\Delta G° = -RT\ell nK$ for K. Compute the value of K if $\Delta G°$ = 4450, R = 8.314, and T = 298.

20) Solve the equation $E = E° = (0.05916/n)\log(P/C^2)$ for P. If E = 0.462, E° = 0.760, n = 2.00, and C = 5.00, what is P?

21) Solve $R_2/R_1 = (A_2/A_1)^a$ for a. What is a if R_2 = 0.562, R_1 = 0.397, A_2 = 0.0260, and A_1 = 0.0130?

22) Solve $\ell n(C_2/C_1) = -kt$ for C_2. If $C_1 = 4.0 \times 10^{15}$, k = 0.795 and t = 14.70, what is C_2?

CHAPTER TEST

1) Write the logarithms of: a) 2.55×10^6 b) 4.83×10^{-26}.

2) Find the antilogarithms of: a) 9.49 b) -5.93.

3) $3y = 15 \log (3.71 \times 10^{-8})$. What is the value of y?

4) Calculate y if $2.20 \log y - 3.9 = 12.5$.

5) The potential of an electrochemical cell, E, is found by the equation shown below. Calculate that potential for a silver-copper cell if E° = 0.46, n = 2.00, and Q = 4.6.

$$E = E° - \frac{0.0592}{n} \log Q$$

6. ATOMS, ELEMENTS, AND THE PERIODIC TABLE

CHEMICAL SKILLS

For this chapter you may consult a periodic table such as that printed inside the back cover of this book. DO NOT use a periodic table which has more than basic information: atomic number, atomic mass, and elemental symbol.

1. For the electron, proton, and neutron, state the symbol, charge, relative mass and location within the atom.

2. For each common element (those in Table 6-1), given the symbol, write the name, and vice versa.

3. Give the group and period numbers for each common element.

4. Identify any common element given its group and period numbers.

5. Define and use correctly the terms atomic number, mass number, isotope, atomic mass, and atomic mass unit.

6. Given the isotope symbol for an element, state the number of protons, neutrons, and electrons.

7. Write the isotope symbol for any common element given its atomic and mass numbers, or information from which they can be determined.

8. Given the relative mass and percentage abundance of the isotopes of an element, calculate its atomic mass.

9. Given the atomic mass of an element and the relative mass and percentage abundance of all its isotopes except one, calculate the relative mass and percentage abundance of the missing isotope.

10. For each common element, give its molecular formula and physical state at room conditions.

6.1 THE SCIENCE OF CHEMISTRY

In the previous five chapters we have dealt almost totally with math and problem solving. Now we begin talking about *CHEMISTRY*! What is chemistry? Simply put, chemistry is a way of looking at the world at two levels. First there is the atomic and molecular level, consisting of the tiny particles of which matter is made. Second is observing how visible quantities of matter behave and then using them to create useful products.

When chemists look around them, they see more than paper, wood, metal, air, dirt—they see complex carbohydrates (the paper and wood), metallic corrosion (the surface of the metal), molecules moving freely in space (gas), and combinations of minerals (dirt). After studying chemistry for a while, you, too, will see your environment differently. This chapter "begins at the beginning" of chemistry: atoms.

6.2 THE NUCLEAR ATOM

Atoms are made up primarily of three types of particles: protons, neutrons, and electrons. **Protons** carry a positive electrical charge and are found in the center of the atom, called the **nucleus**. The nucleus is quite small, only about 1/100,000th the diameter of the atom. **Neutrons** are not charged and are also located in the nucleus of the atom. **Electrons** are negatively charged and are spread throughout the atom. Protons and neutrons are of about the same mass, but electrons are much lighter, only about 1/1800th the mass of a proton or neutron.

The above description of the **nuclear atom** is summarized in Figure 6-1. It is important that you form a picture of the nuclear atom in your mind. This mental picture will help you organize and remember much of what we have to say about the atom.

Nuclear Atom	Particle	Symbol	Charge	Relative Mass	Location
e^-	electron	e^-	−1	1/1800	outside nucleus
	proton	p^+	+1	1	in nucleus
p^+ n (nucleus*)	neutron	n	0	1	in nucleus

*(relative size: 1/100,000th of atom)

FIGURE 6.1. THE NUCLEAR ATOM

The charges on the particles within the atom determine how the atom behaves. **Electrostatic forces** between charged particles cause them either to repel each other—push each other apart—or to attract each other—pull them closer to each other. A simple rule summarizes this interaction. *Like charges repel; unlike charges attract*. This means that electrons repel one another (e^- repels e^-). Protons also repel one another (p^+ repels p^+). Electrons and protons are oppositely charged, so they attract one another (p^+ attracts e^-). Neutrons, having no charge, experience no electrostatic attraction or repulsion.

When atoms undergo chemical changes, the most important forces involved are electrostatic. Much of chemistry can be explained by looking at the interaction among charged particles, particularly e^- and p^+.

EXAMPLE 6.1: Identify each of the following particles.

(a) charged, in nucleus ____________

(b) in nucleus, not attracted to electron ____________

(c) massive, charged ____________

(a) proton (b) neutron (c) proton

6.3 ELEMENTS AND THE PERIODIC TABLE

There are many different kinds of atoms. Over 100 different elements are known to exist. An **element** is identified by the number of protons in the nucleus. Atoms of the element hydrogen, for example, all have just one proton in their nuclei. Helium atoms all have two protons, carbon has six protons, oxygen eight, and so on. The number of protons in the nucleus of an atom is called its **atomic number**. The symbol Z is used to represent the atomic number of an element. Thus, for hydrogen, $Z = 1$, for helium, $Z = 2$, for carbon, $Z = 6$, and for oxygen, $Z = 8$. We will learn more about the importance of the atomic number in a moment.

Each element is represented by a **chemical symbol**. Elemental symbols have either one or two letters. The first letter is always capitalized, and the second letter, if present, is always in lower case—a "small" letter. We will present a method of learning chemical symbols for the elements in a moment.

The **periodic table** arranges elements according to their chemical and physical properties. We will use the periodic table again and again to help remember patterns in the chemistry of the elements and their compounds. Elements appear in columns that reflect their properties and in rows in order of their atomic numbers. A complete periodic table, showing all the known elements, appears inside the back cover of this book.

About 50 elements are used in most beginning chemistry courses. They are presented in alphabetical order, along with their atomic numbers, in Table 6.1. You should "memorize" these so that, when given any one for an element, you can write the others for that same element. This is quite a task—until we add that you may refer to a periodic table to assist you. Then it becomes surprisingly easy, and the drudgery of memorizing almost disappears.

Figure 6.2 is a partial periodic table, filled in only for those elements listed in Table 6.1. Beneath the periodic table is an alphabetical list of the same elements without symbols and without atomic numbers. This is followed by suggestions on how to learn the names and symbols of the elements and their positions in the periodic table all at the same time. Spend some time doing the things suggested. When you feel confident, try your skill on the following examples—but first put Figure 6.2 away, and refer only to a full periodic table. The table on your tear-out shield is probably your handiest reference.

TABLE 6.1: NAMES, SYMBOLS, AND ATOMIC NUMBERS OF SOME COMMON ELEMENTS

Name	Symbol	Atomic Number	Name	Symbol	Atomic Number	Name	Symbol	Atomic Number
aluminum	Al	13	gallium	Ga	31	phosphorus	P	15
antimony	Sb	51	germanium	Ge	32	potassium	K	19
argon	Ar	18	gold	Au	79	rubidium	Rb	37
arsenic	As	33	helium	He	2	scandium	Sc	21
barium	Ba	56	hydrogen	H	1	selenium	Se	34
beryllium	Be	4	indium	In	49	silicon	Si	14
bismuth	Bi	83	iodine	I	53	silver	Ag	47
boron	B	5	iron	Fe	26	sodium	Na	11
bromine	Br	35	krypton	Kr	36	strontium	Sr	38
cadmium	Cd	48	lead	Pb	82	sulfur	S	16
calcium	Ca	20	lithium	Li	3	tellurium	Te	52
carbon	C	6	magnesium	Mg	12	thallium	Tl	81
cesium	Cs	55	manganese	Mn	25	tin	Sn	50
chlorine	Cl	17	mercury	Hg	80	titanium	Ti	22
chromium	Cr	24	neon	Ne	10	vanadium	V	23
cobalt	Co	27	nickel	Ni	28	xenon	Xe	54
copper	Cu	29	nitrogen	N	7	zinc	Zn	30
fluorine	F	9	oxygen	O	8			

There are several common terms used in referring to the periodic table. You already know that the number appearing above each symbol is that element's atomic number, Z. The number below each symbol is the **atomic mass**, sometimes called atomic weight, of that element. We'll say more about atomic mass a bit later.

Each horizontal row of the table is known as a **period**. The number of elements in a period varies. The first period has only two elements, hydrogen and helium. Note that hydrogen appears in two places. Periods 2 and 3 have 8 elements, periods 4 and 5 have 18, and periods 6 and 7 have 32, although not all of the elements in period 7 are known to exist.

The elements in each vertical column of the periodic table form a **group** or **family**. The groups are numbered left to right from 1 to 18. This method of numbering is fairly recent. In the past, Roman numerals were used for groups. Many periodic tables that use this system still exist, so we will show these Roman symbols in the periodic tables in this book as well.

The first two columns, 1 and 2, are groups IA and IIA. Columns 3 through 7 are groups IIIB through VIIB. Columns 8, 9, and 10 constitute group VIIIB. The next two, columns 11 and 12, are groups IB and IIB. The last six groups, columns 13 to 18, are groups IIIA to VIIIA. Group VIIIA is also know as group 0. Got it? Remember that periodic tables appear in just about every chemistry classroom and textbook. You do not have to <u>memorize</u> the periodic table, just learn to use it.

Elements in all "A" groups (1, 2, 13-18) are often referred to as **representative elements**. In a similar way, elements in the "B" groups (3-12) are known as **transition elements** or **transition metals**.

Some families have special names. Group 1 is known as the alkali metal family. Group 2 is the alkaline earth family. Group 16 is often called the chalcogens. Group 17 elements are the halogens. Group 18 is the noble gas family.

1	2	3	4	5	6	7	8	9	10	11	12	13	14	15	16	17	18
IA	IIA	IIIB	IVB	VB	VIB	VIIB		VIIIB		IB	IIB	IIIA	IVA	VA	VIA	VIIA	VIIIA
1 H																1 H	2 He
3 Li	4 Be											5 B	6 C	7 N	8 O	9 F	10 Ne
11 Na	12 Mg											13 Al	14 Si	15 P	16 S	17 Cl	18 Ar
19 K	20 Ca	21 Sc	22 Ti	23 V	24 Cr	25 Mn	26 Fe	27 Co	28 Ni	29 Cu	30 Zn	31 Ga	32 Ge	33 As	34 Se	35 Br	36 Kr
37 Rb	38 Sr									47 Ag	48 Cd	49 In	50 Sn	51 Sb	52 Te	53 I	54 Xe
55 Cs	56 Ba									79 Au	80 Hg	81 Tl	82 Pb	83 Bi			

FIGURE 6.2. THE COMMON ELEMENTS. The elements whose symbols and positions in the periodic table you should know are listed alphabetically both in Table 6.1 and below. In addition, they appear in the partial periodic table above.

aluminum	cadmium	gallium	lead	phosphorus	sulfur
antimony	calcium	germanium	lithium	potassium	tellurium
argon	carbon	gold	magnesium	rubidium	thallium
arsenic	cesium	helium	manganese	scandium	tin
barium	chlorine	hydrogen	mercury	selenium	titanium
beryllium	chromium	indium	neon	silicon	vanadium
bismuth	cobalt	iodine	nickel	silver	xenon
boron	copper	iron	nitrogen	sodium	zinc
bromine	fluorine	krypton	oxygen	strontium	

The following procedure will help you in learning the names and symbols of the above elements.

1. *Locate in the periodic table the symbol of each element in the list. Get a mental picture of its position in the table. If you have trouble with any element, look up its atomic number in Table 6.1 and then find its position and symbol. Don't refer to Table 6.1 any more than absolutely necessary.*
2. *Repeat the process, taking the elements in irregular order, as many times as necessary until you can quickly find the position and symbol of any element in the list.*
3. *Read the symbols in the periodic table in the order of their atomic numbers and try to identify the name of each element. If you cannot, use Table 6.1 as a reference, but only as a last resort.*
4. *Repeat the process, selecting symbols at random until you can state the name of each element as soon as you see its symbol.*

The heavy stair-step line in the periodic table separates the **metals** from the **nonmetals**. The elements below and to the left of the line are the metals. Those to the upper right of the line are nonmetals. Some of the elements along the line are called **metalloids** because their properties fall in between those of metals and nonmetals. The metals in groups 3 through 12, in the "valley" of the table, are called **transition metals**. Those in the two rows below the main body of the table are the **rare earth metals**.

EXAMPLE 6.2: Fill in the name or symbol of the following elements.

(a) copper ________ (d) K ____________

(b) chlorine ________ (e) F ____________

(c) tin ________ (f) P ____________

(a) Cu (b) Cl (c) Sn (d) potassium (e) fluorine (f) phosphorus

EXAMPLE 6.3: (a) Give the group and period numbers for the following.

barium ________ iodine ________ neon ________

(b) Identify the following elements.

group VA(15), period 4 ______ group VIIIA(8), period 4 ______

(a) Ba: group IIA (2), period 6; I: group VIIA (17), period 5;
Ne: group VIIIA (18), period 2
(b) As, arsenic (atomic no. 33); Fe, iron (atomic no. 26)

6.4 ISOTOPES AND ELEMENTAL SYMBOLS

All atoms of a given element have the same number of protons. However, atoms of an element can have different numbers of neutrons. Such atoms are called **isotopes** of one another. For example, three isotopes of hydrogen are known. Common hydrogen has 1 proton and 0 neutrons in its nucleus. The second isotope of hydrogen, commonly called deuterium, has 1 proton and 1 neutron. The third isotope, called tritium, has 1 proton and 2 neutrons. All are atoms of hydrogen because they all have 1 proton in the nucleus.

All the isotopes of a given element behave about the same chemically. The greatest difference among them is their mass. Adding neutrons adds mass to the atom. The **mass number** of an isotope is the total number of protons and neutrons. For example, common hydrogen has a mass number of 1 (1 proton + 0 neutrons), deuterium has a mass number of 2 (1 p + 1 n), and the mass number of tritium is 3 (1 p + 2 n). Note that **isotopes have the same atomic number (same number of protons) but different mass numbers (different numbers of neutrons)**. The mass number is often added to the name of the element to distinguish among its isotopes. So common hydrogen is called hydrogen-1, deuterium is hydrogen-2, and tritium is hydrogen-3.

The chemical symbol of an element may be modified to show a particular isotope. The atomic number, Z, is indicated by a subscript to the left of the elemental symbol. The mass number, A, is given as a left superscript:

isotope symbol: ${}^{A}_{Z}Sy$, Z = atomic number, A = mass number, Sy = element symbol

The symbols for the three isotopes of hydrogen are thus ${}^{1}_{1}H$, ${}^{2}_{1}H$, and ${}^{3}_{1}H$. The atomic number Z is often omitted from the isotope symbol.

EXAMPLE 6.4: Complete the following chart.

	Isotope Symbol	Atomic Number	Mass Number	Number of Protons	Number of Neutrons
a)	${}^{22}Ne$				
b)		24	52		
c)				35	46

	Isotope Symbol	Atomic Number	Mass Number	Number of Protons	Number of Neutrons
a)	${}^{22}Ne$	10	22	10	12
b)	${}^{52}Cr$	24	52	24	28
c)	${}^{81}Br$	35	81	35	46

(a) The symbol ${}^{22}Ne$ tells us two things: (1) the element is neon, atomic number Z = 10 (from the periodic table) and (2) that the mass number A is 22. Z gives us directly the number of protons, 10. A is the sum of protons and neutrons, so the number of neutrons is A − Z = 22 − 10 = 12.

(b) Z = 24 so the element is chromium, Cr. A = 52 so the symbol is ${}^{52}Cr$. The number of protons = Z = 24. Neutrons = A − Z = 52 − 24 = 28.

(c) Protons = Z = 35, the element is bromine, Br. A = protons + neutrons so A = 35 + 46 = 81. The symbol is ${}^{81}Br$.

6.5 ATOMIC MASS

Atoms are quite small. Their masses are much too small to measure by ordinary methods. Nevertheless, chemists have developed a *relative atomic mass scale* to be able to compare the masses of atoms with one another. This scale measures atoms in **atomic mass units, amu** for short. The size of an amu is based on the mass of the isotope carbon-12. *The mass of a carbon-12 atom is exactly 12.00 amu.* All other atomic masses are reported relative to this mass.

The atomic mass of each element appears underneath the symbol for the element in the periodic table inside the front cover of this book. Take a look at the atomic mass of carbon as it appears in this table. You will note that the atomic mass of C is listed as 12.01 amu, not 12.00 as you may have expected. Why the difference?

The difference is due to the existence of a second isotope of carbon, carbon-13. For carbon, as well as all other elements, *the atomic mass is the average mass of all the atoms of an element as they occur in nature.* As we find carbon in our environment, 99.00% exists as C-12 (12.00 amu) and the remaining 1.00% exists as C-13 (13.00 amu—due to the extra neutron). The average mass of all carbon atoms is 12.01 amu, computed as follows:

$$\begin{aligned} 99\% \text{ of mass of C-12} &= 0.9900 \times 12.00 \text{ amu} = 11.88 \text{ amu} \\ 1\% \text{ of mass of C-13} &= 0.0100 \times 13.00 \text{ amu} = \underline{0.130 \text{ amu}} \\ & \qquad 12.01 \text{ amu} \end{aligned}$$

In the above example we used just four digits for the masses of the isotopes and for the atomic mass of carbon. The number of digits in a measured quantity reflects how well we know the value of that quantity. The atomic masses of the elements, to the maximum number of digits to which they are known, are listed in a table inside the front cover of this book. Some are known to many digits. For example, the atomic mass of F is 18.998403 amu. Others are less well known; the atomic mass of S is 32.06 amu. In this book we will *always use atomic masses to four significant figures.*

EXAMPLE 6.5: Naturally occurring boron is 20.0% ^{10}B, 10.01 amu, and 80.0% ^{11}B, 11.01 amu. What is the atomic mass of boron?

$$\begin{aligned} 0.200 \times 10.01 \text{ amu} &= 2.00\underline{2} \text{ amu} \\ 0.800 \times 11.01 \text{ amu} &= \underline{8.80\underline{8} \text{ amu}} \\ & 10.81\underline{0} \text{ amu} = 10.81 \text{ amu} \end{aligned}$$

It is possible to determine the relative mass and percentage abundance of an unknown isotope. The information needed is the atomic mass and the relative mass and percentage of the other isotope (or isotopes, if more than two exist). Suppose that we know that rubidium, atomic mass 85.47, has two isotopes. One has a mass of 84.91 and an abundance of 72.15% The mass and percentage of the other can be found as follows. Using the set up we showed before, we see three "blanks."

$$
\begin{array}{ccccc}
0.7215 & \times & 84.91 \text{ amu} & = & 61.26\underline{3} \text{ amu} \\
\boxed{} & \times & \boxed{\text{ amu}} & = & \boxed{\text{ amu}} \\
\hline
 & & & & 85.47 \text{ amu}
\end{array}
$$

The first step is to see that the sum of the two isotope percents must be 100.00%. This means that the missing isotope has an abundance of 100.00% − 72.15% = 27.85%. This appears as the decimal 0.2785 in the problem set up. Next note that the part of the atomic mass due to the unknown isotope is 85.47 − 61.26$\underline{3}$ = 24.20$\underline{7}$ amu. The set up now becomes:

$$
\begin{array}{ccccc}
0.7215 & \times & 84.91 \text{ amu} & = & 61.26\underline{3} \text{ amu} \\
\boxed{0.2785} & \times & \boxed{\text{ amu}} & = & \boxed{24.20\underline{7}\text{ amu}} \\
\hline
 & & & & 85.47 \text{ amu}
\end{array}
$$

Finally, the missing mass, when multiplied by its fraction 0.2785 must give its contribution to the average, 24.20$\underline{7}$ amu. That is, 0.2785 × ? = 24.20$\underline{7}$ amu or ? = 24.20$\underline{7}$ amu ÷ 0.2785 = 86.92 amu.

$$
\begin{array}{ccccc}
0.7215 & \times & 84.91 \text{ amu} & = & 61.26\underline{3} \text{ amu} \\
\boxed{0.2785} & \times & \boxed{86.92\text{ amu}} & = & \boxed{24.20\underline{7}\text{ amu}} \\
\hline
 & & & & 85.47 \text{ amu}
\end{array}
$$

The unknown isotope has a mass of 86.92 amu and an abundance of 27.85%!

EXAMPLE 6.6: Gallium has an atomic mass of 69.72 amu and has two naturally occurring isotopes. One has a mass of 68.93 amu and an abundance of 60.40%. What is the mass and percentage of the other?

$$
\begin{array}{ccccc}
0.6040 & \times & 68.93 \text{ amu} & = & 41.63\underline{4} \text{ amu} \\
\boxed{(1)} & \times & \boxed{(3)\text{ amu}} & = & \boxed{(2)\text{ amu}} \\
\hline
 & & & & 69.72 \text{ amu}
\end{array}
$$

(1) = 1 − 0.6040 = 0.3960
(2) = 69.72 − 41.63$\underline{4}$ = 28.08$\underline{6}$
(3) = 28.08$\underline{6}$/0.3960 = 70.92

The unknown isotope has a mass of 70.92 amu and an abundance of 39.60%.

6.6 FORMULAS AND STATES OF THE ELEMENTS

The chemical formula of a pure substance, either an element or a compound made up of two or more elements, uses elemental symbols to identify the element or elements in that substance. Subscript numbers are written after the symbols to show the number of atoms of each element present in the smallest particle of the substance. (In some cases the subscripts show the ratio of atoms of different elements in the substance.) If only one atom of an element is present, the subscript is omitted.

Usually the formula of an element is simply its elemental symbol. However, the most stable form of some elements at room temperature and pressure is as molecules rather than as single atoms. Molecules are particles consisting of two or more atoms bonded together in such a way that they behave as a single unit.

The natural, stable form of seven of the elements consists of diatomic molecules, in which each molecule has two atoms. (The prefix di- means two.) These elements are three common gases, nitrogen, oxygen, and hydrogen; and the halogens: fluorine and chlorine, both gases, liquid bromine, and solid iodine. In writing the formulas of these elements, a subscript 2 is used to indicate that the molecule has two atoms: N_2, O_2, H_2, F_2, Cl_2, Br_2, and I_2.

The diatomic-molecule formula of these substances MUST BE USED whenever you refer to these elements AS ELEMENTS, NOT CHEMICALLY COMBINED WITH ANY OTHER ELEMENT. This is how they appear in chemical equations, and it is the form in which they are used in chemical calculations. Failure to write the diatomic formulas for these elements is probably the most common mistake made by beginning chemistry students.

Sulfur, a yellow solid, and phosphorus, a white solid in its molecular form (there are other forms of phosphorus), also exist as molecules. Sulfur forms stable molecules containing eight atoms, and phosphorus molecules have four atoms. Their molecular formulas are S_8 and P_4. Because these molecular forms do not affect chemical calculations, they are not usually written this way, but simply as S and P.

It is common to include the physical state—gas, liquid, or solid—as part of the formula of an element. This is done by adding (g) for gas, (ℓ) for liquid, or (s) for solid immediately after the formula.

Most elements are solids at normal temperature and pressure (25°C and 1 atmosphere). Those that are gases include the six noble gases: He(g), Ne(g), Ar(g), Kr(g), Xe(g), Rn(g), which are monatomic (mon- or mono- means 1). In addition there are the five diatomic gases, $H_2(g)$, $F_2(g)$, $Cl_2(g)$, $O_2(g)$, and $N_2(g)$. Only two elements are liquids under ordinary conditions: monatomic Hg(ℓ) and diatomic $Br_2(\ell)$.

The formulas and states of the elements are summarized in Figure 6.3.

FIGURE 6.3. FORMULAS AND STATES OF THE ELEMENTS AT ORDINARY CONDITIONS*

12 IIB	13 IIIA	14 IVA	15 VA	16 VIA	17 VIIA	18 VIIIA
					$H_2(g)$	He(g)
			$N_2(g)$	$O_2(g)$	$F_2(g)$	Ne(g)
			$P_4(s)$	$S_8(s)$	$Cl_2(g)$	Ar(g)
					$Br_2(\ell)$	Kr(g)
					$I_2(s)$	Xe(g)
Hg(ℓ)						Rn(g)

*All elements not shown are written as monatomic solids.

EXAMPLE 6.7: For each name given below, write the formula of the element in its natural, stable form, including the state designation. For each formula given, add the state and name the element.

aluminum	______	O_2( )	______
iodine	______	Ca( )	______
hydrogen	______	Ne( )	______
sulfur	______	Na( )	______

aluminum, Al(s)	iodine, I_2(s)	O_2(g), oxygen	Ca(s), calcium
hydrogen, H_2(g)	sulfur, S(s)	Ne(g), neon	Na(s), sodium

CHAPTER 6 PROBLEMS

1) Fill in this chart.

Particle	Charge	Mass	Location
______	______	1/1800	______
______	+1	1	______
______	0	______	______

2) Give symbols for the following elements.
a) carbon b) magnesium
c) calcium d) manganese
e) chromium f) iron
g) chlorine h) iodine

3) Write the names of the following elements.
a) K b) Sb
c) P d) Sn
e) Be f) Se
g) Ba h) Te

4) a) Can you name two elements in the same family as oxygen? sodium? arsenic? silver?
b) Can you name two elements in the same period as those in part a?

5) Classify each of the elements below as metals or non-metals. For those which are metals, classify them as main group, transition metal, or rare earth.
a) Pb b) S
c) Fe d) U
e) Xe f) V

6) Define an isotope. How do isotopes of the same element differ in mass? in number of protons? in number of neutrons? in number of electrons?

7) Complete this chart.

Isotope Symbol	Atomic Number	Mass No.	No. p^+	No. n
____	30	___	___	37
^{77}As	___	___	___	___
____	___	131	53	___
____	___	___	38	52

8) Define atomic mass. Why is atomic mass reported on a relative scale? Which isotope has an exactly defined atomic mass and serves as the mass to which all others are relatively reported?

9) a) Naturally occurring silver has two isotopes. Isotope A has a relative mass of 106.9051 and an abundance of 51.82%. Isotope B has a relative mass of 108.9047. Calculate the atomic mass of silver from these data.

b) Naturally occurring iron has four isotopes.

A: 53.9396 5.82%
B: 55.9349 91.66%
C: 56.9354 2.19%
D: 57.9333 0.33%

Calculate the atomic mass of iron from these data.

10) a) Chlorine has an atomic mass of 35.453 and two naturally occurring isotopes. Isotope A has a relative mass of 34.96885 and an abundance of 75.53%. What is the mass and abundance of the second isotope?

b) Silicon has an atomic mass of 28.0855 and three isotopes.

A: 27.97693 92.21%
B: ??? ???
C: 29.97376 3.09%

What is the mass and abundance of the missing isotope?

11) Naturally occurring lithium consists of two isotopes, ^{6}Li and ^{7}Li, and has an atomic mass of 6.941. Which isotope is more abundant? What are the percent abundances of the two lithium isotopes? Their masses are 6.01512 and 7.01600, respectively.

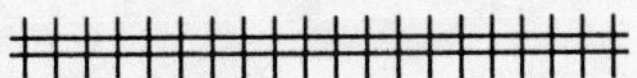

12) Compare the three subatomic particles in 1) mass, 2) charge, 3) attractive/repulsive force, and 4) position within the atom.

13) Give symbols for the following elements.

a) tin
b) argon
c) titanium
d) arsenic
e) gold
f) bromine
g) silver
h) boron

14) Write the names of the following elements.

a) F
b) Fe
c) Al
d) I
e) Cr
f) Pb
g) Cu
h) Cs

15) a) Can you name two elements in the same family as bromine? neon? phosphorus? mercury?
b) Can you name two elements in the same period as those in part a?

16) Classify each of the elements below as metals or nonmetals. For those which are metals, classify them as main group, transition metal, or rare earth.

a) Na b) Am
c) Cl d) Cu
e) C f) Au

17) Define atomic number and mass number. Where are each of these written on the isotope symbol? How do isotopes differ in atomic number? in mass number?

18) Complete this chart.

Isotope Symbol	Atomic Number	Mass No.	No. p^+	No. n
^{235}U	___	___	___	___
___	8	17	___	___
___	___	206	82	___
___	___	22	___	12

19) Define the atomic mass unit. Why is each atomic mass an average value? Do all atomic masses have the same number of significant figures?

20) Rubidium has an atomic mass of 85.4678 and two naturally occurring isotopes. Isotope A has a relative mass of 84.9117 and an abundance of 72.15%. What is the mass and abundance of the second isotope?

21) Naturally occurring neon has three isotopes.
A: 19.99244 90.92%
B: 20.99395 0.257%
C: 21.99138 8.82%
Calculate the atomic mass of neon from this data.

22) Naturally occurring chlorine consists of two isotopes, ^{35}Cl and ^{37}Cl, and has an atomic mass of 35.453. Which isotope is more abundant? What are the percent abundances of the two chlorine isotopes? Their masses are 34.969 and 36.966, respectively.

ADDITIONAL PROBLEMS

23) To the right is a portion of the periodic table with letters substituted for the elemental symbols.
a) J has x protons, so how mny protons does M have?
b) T is in the third period, so how many protons does it have?

J	L	M
Q	R	T

24) Identify these elements.
a) solid diatomic non-metal
b) metal with fewest protons
c) non-metal in group IIIA (13)
d) metal whose symbol is two vowels
e) diatomic liquid

25) Silicon, atomic mass 28.0855 amu, has three naturally occurring isotopes. ^{28}Si has a mass of 27.9769 amu at 92.21%. ^{29}Si has a mass of 28.9765 amu at 4.70%. What is the mass and abundance of the third isotope?

26) Copper, atomic mass 63.546 amu, has two naturally occurring isotopes. Their masses are 62.9298 amu for ^{63}Cu and 64.9278 amu for ^{65}Cu. What are their percentages? [Hint: If A is the fraction of the first isotope, then 1 − A is the fraction of the second isotope.]

CHAPTER TEST

1) Given symbols and names for the following elements.
   a) silicon
   b) strontium
   c) gold
   d) K
   e) P
   f) Cl

2) Complete this chart.

Isotope Symbol	Atomic Number	Mass No.	No. p^+	No. n
___	22	___	___	24
^{55}Fe	___	___	___	___
___	___	127	54	___
___	___	___	28	32

3) a) Can you name two elements in the same family as carbon? magnesium? antimony? mercury?
   b) Can you name two elements in the same period as those in part a?

4) Naturally occurring copper has two isotopes. Isotope A has a relative mass of 62.9298 and an abundance of 69.09%. Isotope B has a relative mass of 64.9278. Calculate the atomic mass of copper from these data. Calculate the atomic mass of neon from this data.

5) Magnesium has an atomic mass of 24.312 and three isotopes.
   A: 23.98504 78.70%
   B: 24.98584 10.13%
   C: ??? ???
   What is the mass and abundance of the missing isotope?

7. CHEMICAL NAMES AND FORMULAS

PREREQUISITES

In Chapter 6 you learned the names, symbols, and formulas of some commonly encountered elements and their positions in the periodic table. In this chapter you will use this information to write the formulas of chemical compounds. Recall that, when referring to a specific group in the periodic table, we will follow the traditional group designation by the recently IUPAC-approved number, such as Group IVA (14).

CHEMICAL SKILLS

You may refer to an approved periodic table, one containing only the atomic number, symbol, and atomic mass, while practicing these skills.

1. Compounds of two nonmetals. Given the name (or formula) of a compound formed by two nonmetals, write its formula (or name).

2. Monatomic ions. Given the name (or formula) of a monatomic ion formed by any representative element, or from the transition elements shown in Figure 7.1, write the formula (or name) of the ion.

3. Acids. Given the name (or formula) of any binary acid, or of an oxoacid listed in Figure 7.2, or of acetic acid, write the formula (or name) of the acid.

4. Ions derived from acids. Given the name (or formula) of any ion that could result from the ionization of an acid listed in Chemical Skill 3 above, write the formula (or name) of that ion.

5. Other common names. Given the name (or formula) of water or ammonia, or of the hydroxide or ammonium ions, or of any anion listed in Table 7.1, write the formula (or name).

6. Ionic compounds. Given the name (or formula) of any ionic compound made up of any cation and any anion identified in the chapter, write the formula (or name) of the compound.

The names and formulas of chemical substances are fixed by a system of nomenclature. If you try to memorize the names and formulas of important ions and compounds without learning the system, you will be able to reproduce those names and formulas temporarily. You will also be overwhelmed by the amount of memorizing you must do. If, on the other hand, you learn the system and how to apply it, you will accomplish four things:

1. You will reduce memorizing by an estimated 80-90%.
2. You will spend less time learning what you must learn.
3. You will retain your nomenclature skills for a much longer time.
4. You will be able to write names and formulas of substances you have never heard of before.

We cannot overemphasize this advice: LEARN THE SYSTEM.

7.1 CHEMICAL FORMULAS

There are two kinds of pure substances: elements, which we studied in Chapter 6, and **compounds**, which are made up of atoms of different elements. chemists use **chemical formulas** to represent elements and compounds when writing of chemical changes. The formula of a substance is made up of the symbols of the elements in that substance. A subscript number after a symbol shows the number of atoms of the element in the formula unit of the substance. If there is only one atom of an element in the unit, the subscript is omitted; the absence of a subscript indicates one atom in the formula unit.

A **molecular substance** consists of distinct, individual particles, or **molecules**. Some elements and some compounds are molecular. The "formula unit" of a molecular substance is the molecule itself, which is represented precisely by the chemical formula. Water, H_2O, is a molecular substance. There are two hydrogen atoms and one oxygen atom in a water molecule.

An **ionic compound** is made up of **ions** formed by the elements in the compound. Ionic compounds generally do not exist as individual molecules, but rather as crystals made up of huge numbers of ions in a definite ratio with respect to each other. The formula unit of an ionic compound consists of the symbols of the elements whose ions are in the compound, with subscripts to show the ratio between those ions. Calcium chloride is an ionic compound. Its formula is $CaCl_2$, which shows that for each ion coming from calcium there are two ions that come from chlorine.

7.2 COMPOUNDS FORMED BY TWO NONMETALS

A molecular compound that is formed by two nonmetals is a **molecular binary compound**. The system by which these compounds are named is summarized below.

The names of compounds formed by two nonmetals consist of two words:

1. The first word is the name of the element appearing first in the chemical formula, including a prefix to indicate the number of atoms of that element in the molecule.

2. The second word is the name of the element appearing second in the chemical formula, modified with the suffix *-ide*. It also

includes a prefix to indicate the number of atoms of that element in the molecule.

The first element listed in the formula is generally the element with the lowest *electronegativity*. The concept of electronegativity is beyond the scope of this book, but approximate values can be related to the periodic table. Generally, the element with the lowest electronegativity is farther to the left and lower down in the periodic table. The element listed second will thus be farther to the right and toward the top of the periodic table.

The first ten prefixes used for the number of atoms of an element in a molecule are as follows:

1	mono-	2	di-	3	tri-	4	tetra-	5	penta-
6	hexa-	7	hepta-	8	octa-	9	nona-	10	deca-

The prefix *mono-* is frequently omitted when a molecule has only one atom of an element. The "o" in *mono-* and the "a" in the prefixes from 4 to 10 are omitted if the resulting word "sounds better," usually when the next letter is a vowel. For example, a compound ending in O_5 is a *pent*oxide rather than a *penta*oxide.

EXAMPLE 7.1: Use the two rules shown above to write the formula of sulfur hexafluoride:________; and the name of P_2O_3:________________

sulfur hexafluoride, SF_6 P_2O_3, diphosphorus trioxide

For the first compound, the first rule places the symbols of the elements in the same order as they appear in the name: SF. The absence of a prefix for sulfur indicates one atom in the molecule, so there is no subscript after S in the formula. The *hexa-* prefix for fluorine indicates six fluorine atoms in the molecule. Hence, SF_6. The name of the second compound is the name of the two elements, phosphorus and oxygen, in the order in which they appear in the formula. The prefix *di-* is applied to phosphorus because there are two atoms in the molecule, and *tri-* is applied to oxygen, which is modified to end in *-ide*: diphosphorus trioxide.

EXAMPLE 7.2: Opposite each compound name below, write the formula; opposite each formula, write the name.

dinitrogen pentoxide____________ N_2O_4________________

sulfur trioxide________________ S_2F_{10}________________

carbon monoxide________________ SiS_2________________

dinitrogen pentoxide, N_2O_5	N_2O_4, dinitrogen tetroxide
sulfur trioxide, SO_3	S_2F_{10}, disulfur decafluoride
carbon monoxide, CO	SiS_2, silicon disulfide

You may wonder why the prefix, *mon-*, is used in carbon monoxide in the above example. This compound and carbon dioxide are so frequently met in chemistry that it has become customary to distinguish between them by using oxygen prefixes in both names. This is an example of "common usage" that sometimes replaces rules in a nomenclature system. Other examples are the common names for the familiar substances water, H_2O, and ammonia, NH_3.

7.3 MONATOMIC IONS

An atom is electrically neutral. It has the same number of protons, each with a +1 charge, and electrons, each with a -1 charge. If an atom gains or loses one or more electrons, it becomes a new species called an **ion**. If the atom gains electrons, the ion is negatively charged and is called an **anion**. If the atom loses electrons, its charge is positive, and it is called a **cation** (pronounced *cat'-ion*). Typical ion formations may be represented as:

Sodium atom	$Na \rightarrow Na^+ + e^-$	Sodium ion
Sulfur atom	$S + 2\ e^- \rightarrow S^{2-}$	Sulfide ion

The superscript + in Na^+ indicates that the ion has a charge of positive 1; the superscript 2- in S^{2-} represents the negative 2 charge of that ion.

Some common ions formed by a single element are shown in Figure 7.1. Most of these are monatomic, which means they are formed by a single atom. Compare the formula of each element with the formula of the ion derived from that element. The formula of the element is the symbol of the element; the formula of the ion is the symbol of the element with the ionic charge written as a superscript. *In writing the formula of an ion you must always, ALWAYS, include the charge.* Without the charge, the ionic formula is not correct.

The name of a monatomic cation is simply the name of the metal from which the ion is formed, followed by the word *ion*.

Monatomic cations are named by the above rule. Accordingly, the name of Na^+ is *sodium ion*.

Some metals form more than one ion, each with a different charge. The best known example is iron, which forms the ions Fe^{2+} and Fe^{3+}. These ions are distinguished from each other by calling them "iron two" and "iron three," respectively. Note that the number in the name is the charge on the ion. The names are written *iron(II)* and *iron(III)*.

1	2	6	7	8	9	10	11	12	13	14	15	16	17	18
IA	IIA	VIB	VIIB	←VIIIB	VIIIB	VIIIB→	IB	IIB	IIIA	IVA	VA	VIA	VIIA	VIIIA
1+	2+								3+		3-	2-	1-	0
H^+													H^-	
Li^+											N^{3-}	O^{2-}	F^-	
Na^+	Mg^{2+}								Al^{3+}		P_3^-	O^{2-}	Cl^-	
K^+	Ca^{2+}	Cr^{3+} Cr^{2+}	Mn^{3+} Mn^{2+}	Fe^{3+} Fe^{2+}	Co^{3+} Co^{2+}	Ni^{2+}	Cu^{2+} Cu^+	Zn^{2+}					Br^-	
Rb^+	Sr^{2+}						Ag^+	Cd^{2+}		Sn^{4+} Sn^{2+}	Sb^{5+} Sb^{3+}		I^-	
Cs^+	Ba^{2+}						Au^{3+} Au^+	Hg^{2+} Hg_2^{2+}		Pb^{4+} Pb^{2+}	Bi^{3+}			

FIGURE 7.1. COMMON ELEMENTAL IONS. All ions in this table are monoatomic except Hg_2^{2+}, which is diatomic. Charges of "A" group ions are listed across the top.

Notice three essential features of the written forms of these names.

1. The charge on the metal ion is written in Roman numerals.
2. The Roman numeral is enclosed in parentheses.
3. There is no space between the name of the element and the first parenthesis.

Students often overlook the parentheses, which must be included if the name is to be written correctly.

An older nomenclature system distinguishes between two ions of the same element by using the suffix -ous for the lower charge, and -ic for the higher. This system is still used on the labels of stock room chemicals. The suffixes are generally applied to the Latin names of the elements. By this method iron(II) ion is ferrous ion, and iron(III) ion is ferric ion. We will not use this method in this book.

The name of a monatomic anion is the name of the element, modified to end with an *-ide* suffix, followed by "ion."

The *-ide* suffix in the name of a monatomic anion is the same as the -ide suffix that is applied to the same elements in naming compounds formed by nonmetals. Thus the name of S^{2-} is *sulfide ion*.

The periodic table is your guide to charges on monatomic ions, particularly those formed by the representative elements. Notice that all monatomic ions from Group IA (1) have a +1 charge. Similarly, ions from Group IIA (2) have a +2 charge, and from Group IIIA (13), +3. Among the anions the charges are -3 for Group VA (15) elements, -2 for Group VIA (16) elements, and -1 for elements in Group VIIA (17). Learn the charges on monatomic ions formed by elements in these groups, rather than trying to memorize the charge on the monatomic ion from each element.

The metals in Group IVA (14) appear to form ions having charges of +4 or +2. Actually, the properties of tin(IV) and lead(IV) compounds are molecular in character, rather than ionic. This suggests that the "ions" do not exist in the same sense that tin(II) and lead(II) ions do. This notwithstanding, tin(IV) and lead(IV) compounds are commonly named as if they consisted of ions.

Most transition metals form more than one positive ion. The charge on an ion derived from a transition metal is a part of the name of the ion. Silver ion, Ag^+, zinc ion, Zn^{2+}, and nickel ion, Ni^{2+}, are exceptions, as each is the only cation normally formed by the corresponding element. Charges on these ions must be memorized.

One of the ions of mercury, Hg_2^{2+}, is diatomic. Its two positive charges come from two atoms, or +1 from each atom. It is therefore the mercury(I) ion. Note that mercury(I) ion is not Hg^+. The second mercury ion, Hg^{2+}, is called by its expected name, mercury(II) ion.

EXAMPLE 7.3: Referring only to a regular periodic table (not Figure 7.1), write the formula of each ion whose name is given, and the name where the formula is given.

potassium ion________	Br^-________
silver ion________	Zn^{2+}________
manganese(II) ion________	Fe^{3+}________
strontium ion________	Te^{2-}________

potassium ion, K^+	Br^-, bromide ion
silver ion, Ag^+	Zn^{2+}, zinc ion
manganese(II) ion, Mn^{2+}	Fe^{3+}, iron(III) ion
strontium ion, Sr^{2+}	Te^{2-}, telluride ion

Charges are not part of the names of potassium, zinc, silver, or strontium ions because each element forms only one ion. Charges must be included in the names of iron and manganese ions because each forms more than one ion.

You establish the +2 charge on strontium ion because strontium is in Group IIA (2) of the periodic table.

Te^{2-} is a monatomic anion, so its name is the name of the element, tellurium, modified with an *-ide* suffix.

7.4 ACIDS

Acids may be defined in several ways. In this chapter we identify an **acid as any substance that can donate (give up) a hydrogen ion (H^+) when in water solution**. To *donate* H^+ a substance must *contain* H in its formula. It is usually the first elemental symbol in the formula.

The water solution of an acid and the acid compound usually have the same formula. The compound hydrogen chloride, HCl, when dissolved in water is called hydrochloric acid. Its formula is also HCl. In practice, the context in which the formula appears usually shows whether you should use the acid name or the compound-of-two-nonmetals name. One way to remove all doubt is to add the state symbol to the formula, (g) if gaseous hydrogen chloride is intended, or (aq) if its water solution, hydrochloric acid, is meant. (The "aq" identifies an *aqueous* solution, from the Latin *aqua* for water.) Thus HCl by itself is usually hydrochloric acid, HCl(aq) is definitely hydrochloric acid, and HCl(g) is hydrogen chloride.

Binary Acids The name *hydrochloric acid* illustrates the rule by which binary acids are named:

> To name a binary acid, apply the prefix *hydro-* and the suffix *-ic* to a modified form of the name of the nonmetal combined with hydrogen.

One of the characteristics of chemical families is that their members form similar compounds. This is true of the binary acids of the halogens, which, using the rule stated above, are sometimes called **hydrohalic** acids. Rather than simply supplying the names of these acids, we'll give you the opportunity to learn by applying the rule, even extending it a bit beyond the hydrohalic acids.

EXAMPLE 7.4: For each of the following formulas, write the name; for each name, write the formula:

hydrofluoric acid__________________ HI__________________________________acid

hydrobromic acid___________________ H_2S________________________________acid

hydrofluoric acid, HF — HI, hydroiodic acid
hydrobromic acid, HBr — H_2S, hydrosulfuric acid

For comments about hydrosulfuric acid, read on.

H_2S is a binary acid in which hydrogen is combined with the nonmetal sulfur. The presence of two hydrogen atoms in the molecule, instead of one, does not affect the way the acid is named; you still apply the prefix *hydro-* and the suffix *-ic* to a modification of the name of the nonmetal.

You do not have to memorize the formulas of the binary acids. Their formulas can be predicted by adding enough H^+ ions to balance the charge on the common negative ion of the nonmetal. For example, S commonly forms S^{2-} (see Figure 7.1). Hydrosulfuric acid therefore takes two +1 charges from hydrogen ions to cancel the one -2 charge from the sulfide ion. With this in mind, you probably can write the names and formulas of the other two binary acids formed by the Group VIA (16) elements:

EXAMPLE 7.5: Write the names and formulas of the binary acids formed by selenium and tellurium.

selenium:______________________ tellurium:______________________

hydroselenic acid, H_2Se; hydrotelluric acid, H_2Te

Both names correspond with the rule for binary acids, and both formulas correspond with the formulas of other hydrogen compounds of Group VIA (16) elements, including water, H_2O.

Oxoacids **Oxoacids are acids that contain hydrogen, a nonmetal, and oxygen.** A common oxoacid is HNO_3, called nitric acid. This name illustrates the basis of the system used to name oxoacids.

> The name of the most common oxoacid of a nonmetal is the name of the nonmetal changed to end in *-ic*, followed by the word "acid."

Frequently, there is more than one oxoacid for a given nonmetal. This is true for nitrogen. In addition to HNO_3, nitric acid, a second acid is known: HNO_2, called nitrous acid. The second acid is distinguished from the first by a change in the ending.

> When a nonmetal forms an oxoacid with one less oxygen atom than the "-ic acid," the name of the acid with the fewest oxygens has the ending "-ous acid."

A useful device for remembering this system is to note that the ending *-ic* has the vowel "i" as in "high" for a high number of oxygens. The ending *-ous* has the vowel "o" as in "low" for a low number of oxygens.

Figure 7.2 lists the oxoacids of the nonmetals. It has the form of the five groups at the right side of the periodic table, in periodic table format, in keeping with our efforts to use the table as a memory aid. Generally, the formulas of the oxoacids of the nonmetals in a family follow a pattern. For example, if you remember that sulfur forms sulfuric and sulfurous acids, H_2SO_4 and H_2SO_3, you can reasonably predict that selenium, right below S in the periodic table, forms selenic and selenous acids, H_2SeO_4 and H_2SeO_3.

As with all predictions based on generalizations, these must be checked in the laboratory. For the two acids of selenium the predictions are confirmed; they are also confirmed when extended to tellurium, the next element in Group VIA (16). Elsewhere in Figure 7.2 you will find formulas that do not

IVA 14	VA 15	VIA 16	VIIA 17	0 18
H_2CO_3	**HNO_3** HNO_2		HOF	
	H_3PO_4	**H_2SO_4** H_2SO_3	$HClO_4$ **$HClO_3$** $HClO_2$ $HClO$	
	H_3AsO_4	H_2SeO_4 H_2SeO_3	$HBrO_4$ $HBrO_3$ $HBrO_2$ $HBrO$	
		H_2TeO_4 H_2TeO_3	HIO_4 HIO_3 HIO_2 HIO	

FIGURE 7.2. SELECTED OXOACIDS USED IN THE NOMENCLATURE SYSTEM. Names and formulas of acids shown in **boldface** should be memorized. All other names and formulas can be derived from them. These acids are:

- H_2CO_3 carbonic acid
- HNO_3 nitric acid
- H_3PO_4 phosphoric acid
- H_2SO_4 sulfuric acid
- $HClO_3$ chloric acid

fit a consistent pattern, and you will not find some you might expect. Explanation of these variances is beyond the scope of this book. The important thing to see is that the periodic table makes it possible to *predict* the existence of chemical compounds, and the nomenclature system establishes their names if they are indeed real.

Notice that, for the halogen family, four oxoacids of the same element are possible. This requires the addition of a prefix (*per-*) to distinguish between the *-ic* acid and the acid having one more oxygen atom than the *-ic* acid. Another prefix (*hypo-*) marks the difference between the *-ous* acid and the acid with one less oxygen. Fluorine forms only one oxoacid, and some of the oxoacids of the other halogens are unstable and incapable of being isolated in the free state.

In the halogen family, the prefix *per-* is added to the name of the oxoacid with the most oxygens, and the prefix *hypo-* is added to the name of the oxoacid with the fewest oxygens.

$HClO_4$ is per-chlor-ic acid
$HClO_3$ is chlor-ic acid
$HClO_2$ is chlor-ous acid
$HClO$ is hypo-chlor-ous acid

EXAMPLE 7.6: For each of the following names, write the formula; for each formula, write the name.

sulfurous acid__________ H_3PO_4__________

iodic acid__________ $HBrO$__________

bromous acid__________ HIO_4__________

sulfurous acid, H_2SO_3 H_3PO_4, phosphoric acid
iodic acid, HIO_3 $HBrO$, hypobromous acid
bromous acid, $HBrO_2$ HIO_4, periodic acid

The two common oxoacids of sulfur are H_2SO_4 and H_2SO_3. The *-ous* ending indicates the one with the lowest number of oxygens. . . . HIO_3 involves a direct substitution of iodine for chlorine in chloric acid, $HClO_3$. . . . The *-ous* suffix in bromous acid means one less oxygen than the *-ic* acid. Chloric acid is $HClO_3$, so chlorous acid is $HClO_2$; the halogen substitution then gives $HBrO_2$. . . . $HBrO$ has as its chlorine counterpart $HClO$, hypochlorous acid. It follows that the matching bromine acid is hypobromous acid. . . . As $HClO_4$ is perchloric acid, so HIO_4 must be periodic acid. (The name of the acid is pronounced *per-iodic*, not *pe-ri-od-ic*, as in the periodic table.)

7.5 NEGATIVE IONS DERIVED FROM THE TOTAL IONIZATION OF OXOACIDS

Binary acids yield monatomic ions like the chloride ion, Cl^-. Oxoacids yield **oxoanions**, negative ions that contain oxygen. Oxoanions are polyatomic; they contain more than one atom. The nitrate ion, NO_3^-, from the ionization of nitric acid is typical. It is also an example of the rule for naming polyatomic anions:

> If the name of a polyatomic oxoacid ends in *-ic*, the name of the oxoanion from its total ionization is formed by changing the *-ic* of the acid name to *-ate*. If the acid name ends in *-ous*, the anion name ends in *-ite*.

By this rule, chloric acid, $HClO_3$, yields the chlorate ion, ClO_3^-, and chlorous acid, $HClO_2$, produces the chlorite ion, ClO_3^-. As before, learn the rule so you can apply it to any acid.

EXAMPLE 7.7: Write the names of the polyatomic anions from these acids:

phosphoric acid: ____________ ion carbonic acid: ____________ ion

citric acid: ____________ ion nitrous acid: ____________ ion

phosphoric acid, phosphate ion carbonic acid, carbonate ion

citric acid, citrate ion nitrous acid, nitrite ion

In the three *-ic* acids, the *-ic* has been replaced by *-ate*. In nitrous acid, the *-ous* ending has been changed to *-ite*.

The formula of an oxoanion is directly related to the formula of the oxoacid that produces the anion.

> The formula of an oxoanion from the total ionization of an oxoacid is the formula of the parent acid minus all ionizable hydrogen—the H's written first in the acid formula. The negative charge on the anion acid is equal to the number of acidic hydrogens removed from the formula of the acid.

Examples are:

- Hypochlorous acid, $HClO$, loses one hydrogen to form hypochlorite ion, ClO^-
- Carbonic acid, H_2CO_3, loses two hydrogens to form carbonate ion, CO_3^{2-}
- Phosphoric acid, H_3PO_4, loses three hydrogens to form phosphate ion, PO_4^{3-}

If you know the name and formula of the acid, neither the name nor the formula of the anion needs to be memorized! Both may be derived from the above rules.

EXAMPLE 7.8: The formula of acetic acid is $HC_2H_3O_2$, and tartaric acid is $H_2C_4H_4O_6$. Write the names and formulas of the ions derived from the total ionization of these acids.

Don't be concerned by the complex formulas of these acids. The above rule apples to all acids. Remember that only the hydrogens written first in the acid formulas are released.

acetic acid: ____________ tartaric acid: ____________

acetic acid: acetate ion, $C_2H_3O_2^-$; tartaric acid: tartarate ion, $C_4H_4O_6^{2-}$

Acetic acid has only one H written first, so an acetate ion has a -1 charge. Tartaric acid has two acidic hydrogens, so tartarate ion has a -2 charge.

The next example gives you practice going both ways between names and formulas of polyatomic ions. The name of the parent acid can be figured out from the name of the anion. If the acid formula is known, the formula of the anion is found as in Example 7.8. If the substance is unfamiliar, don't be concerned. Just apply the rules, and you should have no problem.

EXAMPLE 7.9: For each of the following names, write the formula; for each formula, write the name.

bromite ion ____________ IO_4^- ____________

nitrite ion ____________ SO_3^{2-} ____________

tellurite ion ____________ BrO^- ____________

bromite ion, BrO_2^- — IO_4^-, periodate ion
nitrite ion, NO_2^- — SO_3^{2-}, sulfite ion
tellurite ion, TeO_3^{2-} — BrO^-, hypobromite ion

With *-ite* endings, bromite, nitrite, and tellurite ions come from bromous acid, $HBrO$; nitrous acid, HNO_2; and tellurous acid, H_2TeO_3. (The name and formula of tellurous acid correspond with sulfurous acid, sulfur and tellurium being in the same family.) Removing hydrogens from the acid formulas and assigning negative charges equal to the number of hydrogens removed gives the formulas of the anions. . . . The -1 charge on IO_4^- indicates that its parent acid has one H; this is HIO_4, periodic acid. In the ion, the *-ic* is changed to *-ate*: periodate. Similarly, SO_3^{2-} comes from sulfurous acid, H_2SO_3, which yields sulfite ion; and BrO^- comes from hypobromous acid, which produces the hypobromite ion.

Figure 7.3 shows all of the oxoanions that come from the total ionization of the oxoacids in Figure 7.2.

7.6 NEGATIVE IONS FROM THE STEPWISE DISSOCIATION OF POLYPROTIC ACIDS

Thus far we have considered only the total dissociation of polyprotic acids, acids containing two or three ionizable hydrogens. In total dissociation the acid molecule releases all of its ionizable hydrogen. The fact is that polyprotic acids lose their hydrogens one at a time, yielding intermediate ions that still contain hydrogen. For example, the dissociation of sulfuric acid by steps is

$$H_2SO_4 \xrightarrow{-H^+} HSO_4^- \xrightarrow{-H^+} SO_4^{2-}$$

The "$-H^+$" over each arrow indicates the subtraction of one hydrogen ion from the starting species. The intermediate ion, HSO_4^-, can be thought of as a

IVA 14	VA 15	VIA 16	VIIA 17	0 18
CO_3^{2-}	NO_3^- NO_2^-		OF^-	
	PO_4^{3-}	SO_4^{2-} SO_3^{2-}	ClO_4^- ClO_3^- ClO_2^- ClO^-	
	AsO_4^{3-}	SeO_4^{2-} SeO_3^{2-}	BrO_4^- BrO_3^- BrO_2^- BrO^-	
		TeO_4^{2-} TeO_3^{2-}	IO_4^- IO_3^- IO_2^- IO^-	

FIGURE 7.3 OXOANIONS DERIVED FROM THE TOTAL DISSOCIATION OF THE OXOACIDS IN FIGURE 7.2.

sulfate ion, SO_4^{2-}, with a hydrogen ion, H^+, attached to it. It is therefore given the logical name, hydrogen sulfate ion. In a similar fashion,

> The intermediate ion in the stepwise dissociation of any diprotic acid is given the name of the final ion preceded by "hydrogen."

Phosphoric acid, a triprotic acid, has two intermediates:

$$H_3PO_4 \xrightarrow{-H^+} H_2PO_4^- \xrightarrow{-H^+} HPO_4^{2-} \xrightarrow{-H^+} PO_4^{3-}$$

The HPO_4^{2-} ion is the phosphate ion with a hydrogen ion attached, so it is called the hydrogen phosphate ion. $H_2PO_4^-$ is a phosphate ion with two hydrogens attached, so it is named the dihydrogen phosphate ion. The prefix *di-* for two distinguishes $H_2PO_4^-$ from HPO_4^{2-}. This is the only place a prefix indicating number appears in the name of an ionic species in the nomenclature system used in this book. A summary of important intermediate ions is in Table 7.1.

An older nomenclature system uses the prefix *bi-* to identify intermediate ions of the diprotic acids. Most familiar is the bicarbonate ion, HCO_3^-. The $H_2PO_4^-$ ion is sometimes called the monobasic phosphate ion, and HPO_4^{2-} is the dibasic phosphate ion, where *mono-* and *di-* indicate the number of hydrogen ions lost from the neutral acid molecule. We will not use these names in this book.

TABLE 7.1: STEPWISE DISSOCIATION OF POLYPROTIC ACIDS

ACID	INTERMEDIATE ION(S)		FINAL ION
H_2CO_3 carbonic	HCO_3^- hydrogen carbonate		CO_3^{2-} carbonate
H_2SO_4 sulfuric	HSO_4^- hydrogen sulfate		SO_4^{2-} sulfate
H_2SO_3 sulfurous	HSO_3^- hydrogen sulfite		SO_3^{2-} sulfite
H_2S hydrosulfuric	HS^- hydrogen sulfide		S^{2-} sulfide
H_3PO_4 phosphoric	$H_2PO_4^-$ dihydrogen phosphate	HPO_4^{2-} hydrogen phosphate	PO_4^{3-} phosphate

EXAMPLE 7.10: For each of the following names, write the formula; for each formula, write the name. Refer only to a complete periodic table while completing this example.

hydrogen sulfide ion ____________ HSO_3^- ____________

hydrogen selenate ion ____________ HTe^- ____________

hydrogen sulfide ion, HS^- — HSO_3^-, hydrogen sulfite ion

hydrogen selenate ion, $HSeO_4^-$ — HTe^-, hydrogen telluride ion

Selenium and tellurium are in the same chemical family as sulfur, and form corresponding ions.

7.7 OTHER POLYATOMIC IONS

The Ammonium Ion, NH_4^+ The compound ammonia, NH_3, is capable of combining with a hydrogen ion to form the ammonium ion, NH_4^+: $NH_3 + H^+ \rightarrow NH_4^+$. This ion is both common and important; it must be named or written from memory. It is the only polyatomic cation normally considered in introductory chemistry.

The Hydroxide Ion, OH^- To a very tiny but very important extent, water behaves as an acid in releasing a proton in dissociation. This is seen more clearly by writing the formula of water as HOH: $HOH \rightarrow H^+ + OH^-$. The OH^- ion is the hydroxide ion, which is present in substances known as **bases** and is responsible for most of their properties. Notice that the *-ide* ending is applied to a diatomic ion in this case. The name and formula of the hydroxide ion must be memorized.

The Cyanide Ion, CN^- The cyanide ion is another diatomic ion ending in *-ide*. It may be found in the very slight dissociation of hydrocyanic acid, HCN: $HCN \rightarrow H^+ + CN^-$. If there is an occasion in this book to use the cyanide ion, you may look it up; it need not be memorized.

Tables 7.2 and 7.3 list all of the ions discussed so far, plus some of the many others that are known. We suggest you do not rely on these tables, but regard them as a "last-resort, use-only-if-absolutely-necessary" reference for ion formulas.

7.8 THE FORMULAS OF IONIC COMPOUNDS

The two nomenclature rules for ionic compounds are:

1. The name of an ionic compound is the name of the cation, followed by the name of the anion.

2. The formula of an ionic compound is the formula of the cation followed by the formula of the anion, each taken that number of times as is necessary to yield a total charge of zero. (Note: individual ionic charges are not normally included in the formula of a compound.)

Notice that in both name and formula, the cation comes first, and the anion last. We will develop the application of these rules by examples.

EXAMPLE 7.11: Write the formulas of potassium iodide, potassium bromide, and potassium nitrate.

According to Rule 1, the name potassium iodide indicates the presence of the potassium and iodide ions; potassium bromide signifies potassium and bromide ions; and potassium nitrate signals the potassium and nitrate ions. In order that you may see clearly what you are working with, write the formulas of these ions:

potassium ion, ____; iodide ion, _____; bromide ion, ____; nitrate ion, _____

Potassium ion, K^+; iodide ion, I^-; bromide ion, Br^-; nitrate ion, NO_3^-

Rule 2 requires that the formula of the compound contains the two ions in such numbers that the net charge is zero. How many potassium ions will have to be with one iodide ion to yield a net—or total—charge of zero? Similarly, how many bromide ions and how many nitrate ions are required to neutralize the +1 charge of a single potassium ion? For one potassium ion, K^+, you need . . .

_____iodide ions, I^-; _____bromide ions, Br^-; _____nitrate ions, NO_3^-.

All three answers are 1.

An iodide ion has a -1 charge. It requires a +1 charge to yield a total charge of zero. One potassium ion satisfies that requirement. Similarly the +1 charge of a potassium ion needs a -1 charge from either a bromide ion or a nitrate ion to give a total charge equal to zero.

The formulas follow from the ion formulas, K^+, I^-, Br^-, and NO_3^-. Remember, the subscript 1 is not used in a formula if the number of atoms or ions is 1. The formulas:

potassium iodide,______; potassium bromide, ______; potassium nitrate, ______

Potassium iodide, KI; potassium bromide, KBr; potassium nitrate, KNO_3

TABLE 7.2: COMMON CATIONS

IONS OF +1 CHARGE	IONS OF +2 CHARGE	IONS OF +3 CHARGE
Alkali Metals:	Alkaline Earths:	
Group IA(1)	Group IIA(2)	Group IIIA(13)
Li^+ lithium	Be^{2+} beryllium	Al^{3+} aluminum
Na^+ sodium	Mg^{2+} magnesium	Ga^{3+} gallium
K^+ potassium	Ca^{2+} calcium	
Rb^+ rubidium	Sr^{2+} strontium	Transition Elements
Cs^+ cesium	Ba^{2+} barium	Cr^{3+} chromium(III)
		Mn^{3+} manganese(III)
Transition Elements	Transition Elements	Fe^{3+} iron(III)
Cu^+ copper(I)	Cr^{2+} chromium(II)	Co^{3+} cobalt(III)
Ag^+ silver	Mn^{2+} manganese(II)	
	Fe^{2+} iron(II)	
Others	Co^{2+} cobalt(II)	
NH_4^+ ammonium	Ni^{2+} nickel(II)	
H^+ hydrogen	Cu^{2+} copper(II)	
OR	Zn^{2+} zinc	
H_3O^+ hydronium	Cd^{2+} cadmium	
	Hg_2^{2+} mercury(I)	
	Hg^{2+} mercury(II)	
	Others	
	Sn^{2+} tin(II)	
	Pb^{2+} lead(II)	

TABLE 7.3: COMMON ANIONS

IONS OF -1 CHARGE		IONS OF -2 CHARGE	IONS OF -3 CHARGE
Halogens:			
Group VIIA(17)	Oxoanions	Group VIA(16)	Group VA(15)
F^- fluoride	ClO_4^- perchlorate	O^{2-} oxide	N^{3-} nitride
Cl^- chloride	ClO_3^- chlorate	S^{2-} sulfide	P^{3-} phosphide
Br^- bromide	ClO_2^- chlorite		
I^- iodide	ClO^- hypochlorite	Oxoanions	Oxoanion
		CO_3^{2-} carbonate	PO_4^{3-} phosphate
Intermediate anions	BrO_3^- bromate	SO_4^{2-} sulfate	
HCO_3^- hydrogen carbonate	BrO_2^- bromite	SO_3^{2-} sulfite	
	BrO^- hypobromite	$S_2O_3^{2-}$ thiosulfate	
HS^- hydrogen sulfide		$C_2O_4^{2-}$ oxalate	
HSO_3^- hydrogen sulfite	IO_4^- periodate	CrO_4^{2-} chromate	
	IO_3^- iodate	$Cr_2O_7^{2-}$ dichromate	
HSO_4^- hydrogen sulfate	NO_3^- nitrate	Intermediate anion	
$H_2PO_4^-$ dihydrogen phosphate	NO_2^- nitrite	HPO_4^{2-} hydrogen phosphate	
Others	Others		
$C_2H_3O_2^-$ acetate	OH^- hydroxide	Diatomic elemental	
SCN^- thiocyanate	CN^- cyanide	O_2^{2-} peroxide	
MnO_4^- permanganate	H^- hydride		

EXAMPLE 7.12: Write the formulas of magnesium iodide, magnesium bromide, and magnesium nitrate.

The anions are the same as in the last example. The formula of the magnesium ion you can determine from the periodic table. It is ________.

Magnesium ion, Mg^{2+}

Concentrate first on magnesium iodide to see how these ions go together. Ask yourself, "How many iodide ions, I^-, each with a -1 charge, must be used to balance the +2 charge of a single magnesium ion, Mg^{2+}, yielding a total charge of zero?" Then write the formula of magnesium iodide.

Magnesium iodide, MgI_2.

The two plus charge of a single magnesium ion, Mg^{2+}, requires the -1 charges of two iodide ions, I^-, to give a total of zero. The compound is MgI_2.

The formula of magnesium bromide should follow readily, but magnesium nitrate introduces something new. You need two nitrate ions in magnesium nitrate. This is two polyatomic ions.

> When a polyatomic ion is used more than once in a formula, the entire ion symbol is enclosed in parentheses, followed by a subscript to indicate the number of times the ion appears.

Magnesium hydroxide, for example, is a combination of one magnesium ion and two hydroxide ions, OH^-. The formula is $Mg(OH)_2$. Notice that MgO_2H_2 is *not* the formula; in that form, the identity of the hydroxide ion is lost, and the compound could not be named from the formula by Rule 1. Now, in the same manner that you used for magnesium iodide, write the formulas of magnesium bromide and magnesium nitrate.

magnesium bromide ________________ magnesium nitrate ____________________

magnesium bromide, $MgBr_2$; magnesium nitrate, $Mg(NO_3)_2$

There are two things to notice in these answers. First, the formula of magnesium bromide is not $Mg(Br)_2$. The bromide ion, Br^-, is a monatomic ion, even though the symbol for bromine contains two letters. *Only polyatomic ions are enclosed in parentheses when taken more than once.* Second, don't be concerned about what seems to be a double subscript in $Mg(NO_3)_2$. The subscript 3 is a part of the formula of the nitrate ion, indicating three oxygen atoms in each nitrate ion. The subscript 2 shows that two nitrate ions, the species in parentheses, are in each formula unit of magnesium nitrate.

EXAMPLE 7.13: Write the formulas of potassium sulfate and magnesium sulfate.

Now you will need the formula of the sulfate ion. It is ________.

sulfate ion, SO_4^{2-}

Combining the sulfate ion with the potassium ion, K^+, and magnesium ion, Mg^{2+}, differs only slightly from the formula writing you have already done. The principle is the same: total charge in the formula must be zero.

potassium sulfate ________ magnesium sulfate ________

potassium sulfate, K_2SO_4; magnesium sulfate, $MgSO_4$

It takes two potassium ions, K^+, each at +1, to match the -2 charge of one sulfate ion. With magnesium sulfate, the ion charges are +2 and -2, so the formula has one of each of the ions.

The formula of aluminum oxide is a 3-2 combination: Al^{3+} for aluminum ion and O^{2-} for oxide ion. When you meet a 3-2 combination, think 6. Two aluminum ions give a total positive charge of 6: $2(+3) = +6$. Three oxide ions yield a negative charge of -6: $3(-2) = -6$. The total charge on the compound is zero with that combination of ions, yielding the formula Al_2O_3.

EXAMPLE 7.14: Write the formulas of aluminum bromide, aluminum sulfate, and aluminum phosphate.

The phosphate ion is the only new ion. Its formula is ________.

phosphate ion, PO_4^{3-}

Now, from the ion formulas Al^{3+}, Br^-, SO_4^{2-}, and PO_4^{3-}, write the formulas of the compounds.

aluminum bromide ________ aluminum sulfate ________

aluminum phosphate ________

aluminum bromide, $AlBr_3$ aluminum sulfate, $Al_2(SO_4)_3$
aluminum phosphate, $AlPO_4$

Three bromide ions at -1 each are required to match the +3 of a single aluminum ion. . . . Aluminum sulfate is a 3-2 combination, requiring two +3 charges from aluminum ion and three -2 charges from sulfate ion. . . . Aluminum phosphate comes to zero with one ion each, one at +3 and the other at -3.

To review,

> Formulas of all ionic compounds are written from the formulas of the ions, which include their charges.
>
> Ions are combined in such numbers that the total charge is zero.
>
> Polyatomic ions appearing more than once in a formula are enclosed in parentheses.

Keep these facts in mind and use them faithfully. Then you will be able to write the more demanding formulas in the next examples.

EXAMPLE 7.15: Write the formulas of the following ionic compounds:

ammonium nitrate ______ ammonium carbonate ______
iron(II) sulfate ______ iron(III) sulfate ______
sodium hydrogen carbonate ______ calcium hydrogen carbonate ______
aluminum hydrogen phosphate ______ aluminum dihydrogen phosphate ______
potassium bromate ______ sodium nitrite ______
sodium periodate ______ potassium sulfite ______

ammonium nitrate, NH_4NO_3
iron(II) sulfate, $FeSO_4$
sodium hydrogen carbonate, $NaHCO_3$
aluminum hydrogen phosphate, $Al_2(HPO_4)_3$
potassium bromate, $KBrO_3$
sodium periodate, $NaIO_4$
ammonium carbonate, $(NH_4)_2CO_3$
iron(III) sulfate, $Fe_2(SO_4)_3$
calcium hydrogen carbonate, $Ca(HCO_3)_2$
aluminum dihydrogen phosphate, $Al(H_2PO_4)_3$
sodium nitrite, $NaNO_2$
potassium sulfite, K_2SO_3

In ammonium nitrate, NH_4NO_3, the presence of nitrogen in both cation and anion does not alter the procedure for writing individual ions in the formula; the nitrogens are not combined.

In ammonium carbonate, $(NH_4)_2CO_3$, the polyatomic ion is enclosed in parentheses when taken more than one time; also in iron(III) sulfate, $Fe_2(SO_4)_3$; calcium hydrogen carbonate, $Ca(HCO_3)_2$; aluminum hydrogen phosphate, $Al_2(HPO_4)_3$; and aluminum dihydrogen phosphate, $Al(H_2PO_4)_3$.

In iron(II) sulfate, $FeSO_4$, and iron(III) sulfate, $Fe_2(SO_4)_3$, the roman numerals tell the charges on the iron ions that must be matched by the proper number of sulfate ions.

The 3-2 combination arises in iron(III) sulfate, $Fe_2(SO_4)_3$, and aluminum hydrogen phosphate, $Al_2(HPO_4)_3$.

$Ca(HCO_3)_2$, $Al_2(HPO_4)_3$, and $Al(H_2PO_4)_3$ have the *entire* intermediate polyatomic anion enclosed in parentheses; hydrogen is a part of those ions.

7.9 THE NAMES OF IONIC COMPOUNDS

In the foregoing section the names of ionic compounds have been stated with the assumption they would be self-explanatory. According to Rule 1 in Section 7.8, the name of the cation comes first, followed by the name of the anion. In order for you to state the name of a compound when given the formula, you must simply recognize the ions present. You will probably be able to do this for each compound in the following example except, perhaps, the last one.

EXAMPLE 7.16: For each formula given, write the name of the compound.

$NaOH$ ______________________ $(NH_4)_2SO_4$ ______________________

Li_2SO_3 ______________________ KNO_2 ______________________

$Ba(HCO_3)_2$ ______________________ KIO_4 ______________________

CaS ______________________ $FeCl_2$ ______________________

$NaOH$, sodium hydroxide
Li_2SO_3, lithium sulfite
$Ba(HCO_3)_2$, barium hydrogen carbonate
CaS, calcium sulfide
$(NH_4)_2SO_4$, ammonium sulfate
KNO_2, potassium nitrite
KIO_4, potassium periodate
$FeCl_2$, iron(II) chloride

Hopefully you picked up the *-ite* suffixes for lithium sulfite and potassium nitrite. Special comment on iron(II) chloride follows.

In determining the name of $FeCl_2$, you had to choose between two monatomic ions of iron. Is the compound iron(III) chloride or iron(II) chloride? The choice is based on the fact that the total ionic charge of a compound is zero. This means that the positive charge from the iron ion must be equal to the negative charges from the chloride ions. There are two chloride ions, each with a charge of -1. This gives a total chloride charge of -2, which requires +2 from the single iron ion to make the compound have zero charge. You conclude that iron(II) is present, and the compound name must be iron(II) chloride.

You must reason similarly while naming any compound that contains an element that forms more than one monatomic cation. Then be sure to include the cation charge in the name of the compound. Notice that the ionic charge is not included in the name of a compound in which the cation has only one possible charge. $CaCO_3$, for example, is calcium carbonate, not calcium(II) carbonate.

EXAMPLE 7.17: The element that is the cation in each of the following compounds may have either of two positive charges. Write the correct name of each compound.

$CoCl_3$ ______________________ Fe_2S_3 ______________________

Hg_2Br_2 ______________________ $MnSO_4$ ______________________

CuI ______________________ $Hg(NO_3)_2$ ______________________

SnF_2 ______________________ $Cr(C_2H_3O_2)_3$ ______________________

$CoCl_3$, cobalt(III) chloride
Hg_2Br_2, mercury(I) bromide
CuI, copper(I) iodide
SnF_2, tin(II) fluoride
Fe_2S_3, iron(III) sulfide
$MnSO_4$, manganese(II) sulfate
$Hg(NO_3)_2$, mercury(II) nitrate
$Cr(C_2H_3O_2)_3$, chromium(III) acetate

In $CoCl_3$, three -1 charges from the chloride ions, Cl^-, must be matched by +3 from one cobalt ion; hence cobalt(III) chloride.

In Fe_2S_3, three -2 charges from the sulfide ions, S^{2-}, require +3 from each iron ion, a 3-2 combination; hence iron(III) sulfide.

In Hg_2Br_2, two -1 charges from two bromide ions indicate a +2 from the positive ion, which is, in this case, the diatomic mercury(I) ion, Hg_2^{2+}.

In $Hg(NO_3)_2$, two nitrate ions furnish -1 each, a total of -2, to be balanced by a single mercury ion giving +2; hence mercury(II) nitrate.

Similar reasoning leads to the determination of the cation charge in the other four compounds.

CHEMICAL NOMENCLATURE EXERCISE

For each chemical name that follows, write the formula; for each formula, write the name. Use oxidation numbers when necessary to distinguish between different ions formed by the same element, but not otherwise. Using the nomenclature rules developed in this chapter, you will be able to complete most of this exercise by referring only to a periodic table. A few substances contain ions listed in Tables 7.2 or 7.3 that are not mentioned elsewhere in the chapter. Those substances are identified with an asterisk (*); you may refer to Table 7.2 or 7.3 when writing those names or formulas.

phosphorus ______________________ O_2 ______________________

zinc ion ______________________ H_2SO_4 ______________________acid

sodium bromide ____	OH^- ____
oxygen difluoride ____	$BaBr_2$ ____
potassium ion ____	Na_2O ____
silver chloride ____	$CaCl_2$ ____
iron(III) ion ____	P_2O_3 ____
zinc hydroxide ____	Na ____
copper(II) bromide ____	NH_4I ____
nitrogen ____	$Ba_3(PO_4)_2$ ____
nickel sulfide ____	Li^+ ____
calcium hydrogen carbonate ____	NO_3^- ____
manganese(II) carbonate ____	Li_3PO_4 ____
iron(III) phosphate ____	$Ni(HSO_4)_2$ ____
phosphate ion ____	C ____
nitrite ion ____	KF ____
perbromate ion ____	PbO ____
calcium hydrogen sulfate ____	HF ____acid
uranium hexafluoride ____ (Z = 92 for uranium)	Cs_2CO_3 ____
potassium sulfide ____	$CuSO_4$ ____
hydrogen phosphate ion ____	$HgBr_2$ ____
bromine ____	IO_2^- ____
mercury(I) chloride ____	Ca_3P_2 ____
bromous acid ____	Al_2S_3 ____
aluminum oxide ____	SO_4^{2-} ____
hydrogen telluride ion ____	Sr(OH)2 ____
iron(III) chloride ____	Cl_2 ____
rubidium sulfate ____	$In_2(CO_3)_3$ ____
sodium hydride ____	$H_2PO_4^-$ ____
potassium oxalate* ____	Mn^{2+} ____
hydrogen carbonate ion ____	Cl_2O ____
silicon dioxide ____	Na_2O_2* ____
calcium hydrogen telluride ____	$Mg(HSeO_4)_2$ ____
sulfide ion ____	$Cr(ClO_4)_3$ ____
ammonium phosphate ____	F^- ____
potassium thiocyanate* ____	F_2 ____

selenic acid ____________ $CaTeO_3$ ____________

strontium bromide ____________ $Ga(HCO_3)_3$ ____________

ammonium acetate ____________ Na_2CrO_4* ____________

magnesium nitride ____________ BaH_2 ____________

magnesium hydrogen selenite ____________ H_2TeO_4 ____________acid

sodium dichromate* ____________ CI_4 ____________

hydrosulfuric acid ____________ KCN* ____________

---

phosphoric acid ____________ HNO_2 ____________acid

disulfur hexafluoride ____________ N_2O_5 ____________

barium dihydrogen phosphate ____________ $Fe(NO_3)_2$ ____________

lithium ____________ Ne ____________

cobalt(III) phosphate ____________ HCl ____________acid

calcium sulfite ____________ HCO_3^- ____________

perchloric acid ____________ HBrO ____________acid

barium peroxide* ____________ $KMnO_4$* ____________

gallium sulfate ____________ SeO_4^{2-} ____________

ammonium ion ____________ HS^- ____________

ADDITIONAL PROBLEMS

Write names and formulas for all these ion combinations.

	Br^-	HPO_3^{2-}	Se^{2-}	ClO^-	$C_2O_4^{2-}$	CrO_4^{2-}
Cr^{2+}						
NH_4^+						
H^+						
Sr^{2+}						
Fe^{3+}						
Zn^{2+}						

CHAPTER TEST

You may refer to an approved periodic table, one containing only the atomic number, symbol and atomic weight of each element, in answering these questions.

For each chemical name that follows, write the formula; for each formula, write the name.

chlorine ________	Fe ________
helium ________	N_2 ________
phosphorus tribromide ________	S_2F_2 ________
disilicon hexafluoride ________	OF_2 ________
calcium ion ________	Br^- ________
sulfide ion ________	Fe^{3+} ________
iodic acid ________	HBr ________acid
sulfurous acid ________	HNO_3 ________acid
phosphate ion ________	$SO_4{}^{2-}$ ________
perchlorate ion ________	OH^- ________
hydrogen carbonate ion ________	$BrO_3{}^-$ ________
ammonium phosphate ________	Na_2S ________
iron(II) nitrate ________	Na_2CO_3 ________
potassium hypobromite ________	$Mg(H_2PO_4)_2$ ________
sodium hydrogen sulfite ________	$PbCl_2$ ________
potassium fluoride ________	$Cu(IO_4)_2$ ________

8. CHEMICAL FORMULA PROBLEMS

PREREQUISITES

1. Exponential notation, developed is Section 2.1, will be used for expressing numbers greater than 1000 or less than 0.001.

2. Dimensional analysis, introduced in Section 2.4, is used extensively in solving problems related to chemical formulas.

3. Section 2.6 on percentage calculations is used in expressing percentage composition of chemical compounds.

4. The GIVEN-FIND-FACTOR-PATH method of analyzing a problem that was introduced in Section 2.3 is used throughout this chapter.

5. Chapter 6 introduced the elements and atomic mass.

6. Chapter 7 covered chemical formulas and names, which are the starting points of many problems in this chapter. Calculations involving chemical formulas are the major topic of this chapter.

CHEMICAL SKILLS

1. Molar mass. Calculate the molar mass of a compound from its formula.

2. Avogadro's number. Given the mass or number of moles of a substance and its chemical formula, calculate the number of atoms, molecules, or ions in the sample; given the number of atoms, molecules, or ions in the sample of a substance and its formula, calculate the mass or number of moles in the sample.

3. Mass-to-mole conversions. Given the number of grams (or moles) of a substance whose formula is known, calculate the number of moles (or grams).

4. Percentage composition. Calculate the percentage composition of a compound from its formula.

5. Empirical formula. Given information from which the number of grams of each element in a sample of a compound can be determined, find the empirical formula of the compound.

6. Molecular formula. Given the molar mass of an unknown compound and information from which its empirical formula can be found, write the molecular formula of that compound.

8.1 THE MOLE

Section 6.5 introducted the atomic mass scale and gave the masses of the atoms in atomic mass units, amu. However, discussion of the tiny masses of individual atoms in a unit as small as the amu is not very practical. Industrial chemists and almost all laboratory chemists are concerned with bulk samples of chemicals that can be weighed on balances. These quantities contain inconceivably large numbers of atoms. The concept of the **mole** has been developed to bridge the gap between the tiny world of atoms and the real world of weighable amounts.

The mole is that amount of any substance that contains the same number of units as there are atoms in exactly 12 grams of carbon-12. This number of atoms is huge: experimentally it has been found that, to four significant figures, there are 6.022×10^{23} units per mole. This quantity is called **Avogadro's number** in honor of one of the earlier thinkers in the area of atomic mass.

It is easier to think of a mole in terms of this number than in terms of its formal definition. As a number, it is exactly like the word *dozen*, which you immediately recognize as 12 units of something. There are always 12 units of the "thing" per dozen, no matter what the thing is. Any time you read the word *mole*, mentally substitute the word *dozen*. Whatever the sentence means to you with the word *dozen*, it means the same thing with the word *mole*, except that the number is bigger. To illustrate, consider the following example:

EXAMPLE 8.1: How many sodium atoms are in 3 dozen sodium atoms? How many sodium atoms are in 3 moles of sodium atoms?

Even though you already know the answer to the first question, write down the arithmetic by which the problem is solved.

$3 \times 12 = 36$ sodium atoms

Now try for the number of atoms in 3 moles of sodium.

$3 \times 6.022 \times 10^{23} = 18.07 \times 10^{23} = 1.807 \times 10^{24}$ sodium atoms

The calculation method is exactly the same, but the number is bigger.

Did you recognize the unwritten dimensional analysis applications in Example 8.1? In answering the first question, the given quantity is 3 dozen sodium atoms. The number of sodium atoms is the quantity you are asked to find. By definition, there are 12 units of any "thing" per dozen. This "per" relationship establishes the conversion factor between one dozen of any "thing" called X and units of X as 12 units of X/1 dozen X. The unit path is dozens → units. This analysis is summarized as

GIVEN: 3 doz atoms FIND: number of atoms

FACTOR: 12 atoms/doz PATH: doz → units

The calculation setup is

$$3 \text{ dozen Na atoms} \times \frac{12 \text{ Na atoms}}{1 \text{ dozen Na atoms}} = 36 \text{ Na atoms}$$

The mole may be applied in a similar manner for the second question in Example 8.1. The general conversion relationship and resulting conversion factor for a mole of any "thing" called X are

$$6.022 \times 10^{23} \text{ units of X} = 1 \text{ mol X} \qquad (8.1)$$

$$\text{and} \quad \frac{6.022 \times 10^{23} \text{ units of X}}{1 \text{ mole X}}$$

For the second question in Example 8.1 the unit path is moles → atoms. The calculation setup is

$$3 \text{ mol Na atoms} \times \frac{6.022 \times 10^{23} \text{ Na atoms}}{1 \text{ mol Na atoms}} = 1.807 \times 10^{24} \text{ Na atoms}$$

The above conversion factor can be used for any conversion between moles of a substance and the number of atoms of that substance in either direction. For the unit path atoms → moles the conversion factor would be inverted in the calculation setup.

EXAMPLE 8.2: How many moles of nitrogen atoms are in a sample that contains 1.14×10^{23} atoms?

To refresh you on the GIVEN-FIND-FACTOR-PATH analysis of a problem, we suggest that you fill in the details for this example. Then set up the problem and calculate the answer.

GIVEN: FIND:

FACTOR: PATH:

GIVEN: 1.14×10^{23} N atoms FIND: moles of N atoms
FACTOR: 6.022×10^{23} N atoms/mol N atoms PATH: atoms → moles

$$1.14 \times 10^{23} \text{ N atoms} \times \frac{1 \text{ mol N atoms}}{6.022 \times 10^{23} \text{ N atoms}} = 0.189 \text{ mol N atoms}$$

As noted in Chapter 3, we realize that you will usually run through the GIVEN-FIND-FACTOR-PATH process mentally, without consciously thinking about the individual quantities. That is as it should be; you should develop that skill. But when a new kind of problem is introduced, we may ask for a full analysis to help you get started. And we will continue to use the summaries as a quick way of explaining how we solved a problem.

If you are ever doubtful about how to set up a problem, try writing down the unit path. If that doesn't clear the way, write out the whole analysis. It's one way to get started, and it may be all the help you need.

8.2 MOLAR MASS OF ATOMS

Atomic masses in amu are just the masses of single atoms. Mole quantities of atoms are large enough to be weighed on conventional balances. **The mass in grams of one mole of a substance is called its molar mass (MM).** The units of molar mass are grams per mole. This leads to the very important conversion factor between mass and moles. If MM is the number of grams in 1 mole of any substance X, this conversion factor is expressed as

$$\text{MM of X} = \frac{\text{MM g X}}{1 \text{ mol X}} \qquad (8.2)$$

The atomic mass unit (amu), molar mass, and the mole are all based on carbon-12, which has had assigned to it, in one way or another, a value of exactly 12. The result is that the atomic mass of carbon-12 is exactly 12 amu, and the molar mass of carbon-12 is exactly 12 grams per mole—the same number with different units. In fact, the atomic mass of every element in atomic mass units is numerically equal to the molar mass in grams per mole. If you know or can find one, you automatically have the other. The atomic mass of magnesium, for example, is 24.31 amu; therefore, its molar mass is 24.31 g/mol.

EXAMPLE 8.3: How many moles are in (a) 75 grams of aluminum metal and (b) 25 milligrams of neon gas.

Part (a) requires only one conversion factor. Use dimensional analysis to solve that problem first.

GIVEN: 75 g Al FIND: mol Al FACTOR: 26.98 g Al/mol Al PATH: g → mol

$$75 \text{ g Al} \times \frac{1 \text{ mol Al}}{26.98 \text{ g Al}} = 2.8 \text{ mol Al}$$

In (b) you will need two-steps to change 25 mg Ne to moles.

GIVEN: 25 mg Ne FIND: mol Ne
FACTOR: 20.18 g Ne/mol Ne; 10^{-3} g/mg PATH: mg → g → mol

$$25 \text{ mg Ne} \times \frac{10^{-3} \text{ g Ne}}{\text{mg Ne}} \times \frac{1 \text{ mol Ne}}{20.18 \text{ g Ne}} = 0.0012 \text{ mol Ne}$$

EXAMPLE 8.4: How many atoms are present in each of the quantities given in Example 8.3: (a) 75 g Al and (b) 25 mg Ne.

Now an additional conversion has been added: moles to atoms. Check Equation 8.1 if necessary. Set up and solve both problems.

The additional conversion factor is 6.022×10^{23} atoms of X/mol X.

(a) GIVEN: 75 g Al FIND: atoms Al
FACTORS: 26.98 g Al/mol Al, 6.022×10^{23} atoms Al/mol Al
PATH: g → mol → atoms

$$75 \text{ g Al} \times \frac{1 \text{ mol Al}}{26.98 \text{ g Al}} \times \frac{6.022 \times 10^{23} \text{ atoms Al}}{\text{mol Al}} = 1.7 \times 10^{24} \text{ atoms Al}$$

(b) GIVEN: 25 mg Ne FIND: atoms Ne
FACTORS: 20.18 g Ne/mol Ne, 10^{-3} g/mg, 6.022×10^{23} atoms Ne/mol Ne
PATH: mg → g → mol → atoms

$$25 \text{ mg Ne} \times \frac{10^{-3} \text{ g Ne}}{\text{mg Ne}} \times \frac{1 \text{ mol Ne}}{20.18 \text{ g Ne}} \times \frac{6.022 \times 10^{23} \text{ atoms Ne}}{\text{mol Ne}} = 7.5 \times 10^{20} \text{ atoms Ne}$$

8.3 MOLAR MASS OF MOLECULAR AND IONIC SUBSTANCES

In this section you will be combining atomic masses—or molar masses of atoms—that you will read from the periodic table. These masses are known to a varying number of decimal places, all beyond any need you might have at this time. Atomic masses in the periodic tables in this book have been rounded off to four significant figures, which is as many as we will need for beginning work in chemistry.

The molar mass of any substance is the mass in grams of one mole of units of that substance as represented by its chemical formula. The molar mass of oxygen atoms—formula O—is 16.00 g/mol. This is the bottom figure in the box for oxygen (Z = 8) in the periodic table.

Oxygen is one of the seven elements that form stable diatomic molecules. The formula of an oxygen molecule is O_2. One mole of oxygen molecules, at two atoms per molecule, has two moles of oxygen atoms. The molar mass of O_2 is therefore 2 × 16.00 g/mol = 32.00 g/mol. Similarly, the molar mass of nitrogen molecules, N_2, is 2 × 14.01 g/mol = 28.02 g/mol. The molar mass of nitrogen atoms, N, is simply 14.01 g/mol.

From these examples you can see how important it is to specify clearly the chemical species whose molar mass you would calculate. Always identify that species by its chemical formula. This is a good habit that will help you avoid many mistakes.

The molar mass of a compound is calculated from its chemical formula. One water molecule, H_2O, contains two hydrogen atoms and one oxygen atom. One mole of water molecules contains two moles of hydrogen atoms, each having a mass of 1.008 gram, and one mole of oxygen atoms, having a mass of 16.00 grams. Using the mathematical axiom "the whole is equal to the sum of its parts," the molar mass of water is

$$\begin{array}{rcl} 2(1.008\ \text{g/mol}) & = & 2.016\ \text{g/mol} \\ +\ 1(16.00\ \text{g/mol}) & = & \underline{16.00\ \ \text{g/mol}} \\ & & 18.016\ \text{g/mol} \quad = \quad 18.02\ \text{g/mol} \end{array}$$

Note the proper use of significant figures.

Hereafter we will write molar mass calculation setups on one line, just as you will no doubt approach them on your calculator. The above setup is

$$2(1.008\ \text{g/mol}) + 16.00\ \text{g/mol} = 18.016\ \text{g/mol} = 18.02\ \text{g/mol}$$

Written this way, the doubtful digit to which the result must be rounded is less apparent, but it can be identified if you examine the numbers carefully. If in doubt, you can always write the numbers to be added vertically.

EXAMPLE 8.5: Calculate the molar mass of chlorine, Cl_2; carbon dioxide, CO_2; and propanol, C_3H_7OH.

The calculations for chlorine and carbon dioxide are handled in the same way as for oxygen, nitrogen, and water. Try those two first in the space below, leaving propanol for a second step.

Cl_2: CO_2:

Cl_2: (2)(35.45 g/mol) = 70.90 g/mol
CO_2: 12.01 g/mol + (2)(16.00 g/mol) = 44.01 g/mol

The molar mass of propanol is found in exactly the same way. You must be careful to count all the atoms of each element in the molecule regardless of how the formula happens to be written. For example, in propanol, C_3H_7OH, there are three elements. How many atoms of carbon are in the molecule? Of hydrogen? Of oxygen? With that hint, calculate the molar mass of propanol.

C_3H_7OH:

C_3H_7OH: (3)(12.01 g/mol) + (8)(1.008 g/mol) + 16.00 g/mol = 60.09 g/mol

Notice that hydrogen appears twice in the formula, and the total number of hydrogen atoms in a molecule is 8. Formulas for organic compounds frequently have the same element in two or more positions.

EXAMPLE 8.6: Calculate the molar masses of sodium chloride, NaCl; magnesium bromide, $MgBr_2$; and barium nitrate, $Ba(NO_3)_2$.

You can probably handle NaCl and $MgBr_2$ without further comment. Perform those calculations here:

NaCl: $MgBr_2$:

NaCl: 22.99 g/mol + 35.45 g/mol = 58.44 g/mol
$MgBr_2$: 24.31 g/mol + (2)(79.90 g/mol) = 184.11 g/mol

Before attempting the calculation for barium nitrate, remember what the subscript 2 means in the formula $Ba(NO_3)_2$. It indicates that there are 2 moles of nitrate ions, NO_3^-, in 1 mole of $Ba(NO_3)_2$. In that case, how many moles of nitrogen atoms are in 1 mole of $Ba(NO_3)_2$? (_______) How many moles of oxygen atoms? (_______)

2 moles of nitrogen atoms and 6 moles of oxygen atoms

Each mole of nitrate ions has 1 mole of nitrogen atoms, and there are 2 moles of nitrate ions. This makes 2 × 1 = 2 moles of nitrogen atoms. Each mole of nitrate ions has 3 moles of oxygen atoms, so for 2 moles of nitrate ions there are 2 × 3 = 6 moles of oxygen atoms.

Now that you know the number of moles of each kind of atom in 1 mole of $Ba(NO_3)_2$, complete the calculation of the molar mass.

$Ba(NO_3)_2$:

$Ba(NO_3)_2$: 137.3 g/mol + (2)(14.01 g/mol) + (6)(16.00 g/mol) = 261.3 g/mol

8.4 OTHER NAMES RELATED TO ATOMIC MASSES AND THE MOLE

Chemists are not in complete agreement on the terms and definitions that have been introduced in this chapter. Probably the most common difference you are apt to encounter is the use of "weight" instead of "mass." Other common variations you are apt to meet include the following:

Molecular Mass; Formula Mass The **molecular mass** of a molecular compound is the mass of one molecule of that compound compared to the mass of one atom of carbon-12 at exactly 12 atomic mass units. Ionic compounds do not exist in individual particles such as atoms or molecules. Instead they arrange themselves in a crystal framework of positive and negative ions in the same ratio that the ions appear in the chemical formula. **The formula mass** is used to describe the mass of an imaginary particle having the same number of atoms of each element as appears in the chemical formula, again compared to the mass of an atom of carbon-12 at exactly 12 atomic mass units.

Gram Atomic Weight; Gram Molecular Weight; Gram Formula Weight These terms are widely used for the molar masses of atoms, molecular compounds, and ionic compounds, respectively. Informally, the word gram is often omitted.

If your instructor emphasizes terms other than those used in this book, you should, of course, use those terms. More important than the words are what you are able to do with them. What counts is the idea behind the words and the calculations that are based on them.

8.5 CONVERSION AMONG GRAMS, MOLES, AND NUMBER OF ATOMS, MOLECULES, OR FORMULA UNITS

In Chapter 10 you will learn that the number of moles of different chemicals are used to calculate the quantities that react with each other and the amount of product that results. There is no way to count or measure moles directly; but mass can be measured. The conversion factor from Equation 8.2,

$$\frac{\text{MM g X}}{\text{1 mol X}}$$

gives us a way to convert between mass and moles. In fact, you did this in Example 8.3, when you calculated the number of moles of aluminum in 75 grams. That conversion, g → mol, and its opposite, mol → g, are among the most frequent calculations in chemistry.

EXAMPLE 8.7: How many moles of calcium chloride are in 95.0 grams?

Begin this example by rewriting the above conversion factor specifically for calcium chloride as substance X. You will need to compute the molar mass of calcium chloride.

$40.08 \text{ g/mol} + (2)(35.45) \text{ g/mol} = 110.98 \text{ g } CaCl_2/\text{mol } CaCl_2$

Now you can use the above conversion factor between grams and moles to make the necessary calculation. What is the starting quantity? What is asked for? What is the unit path? Perform the calculation below.

GIVEN: 95.0 g $CaCl_2$ FIND: mol $CaCl_2$ FACTOR: 110.98 g $CaCl_2$/mol PATH: g → mol

$$95.0 \text{ g } CaCl_2 \times \frac{1 \text{ mol } CaCl_2}{110.98 \text{ g } CaCl_2} = 0.856 \text{ mol } CaCl_2$$

EXAMPLE 8.8: Find the mass of 3.52 moles of $Mg(NO_3)_2$.

First calculate the molar mass of the compound:

$Mg(NO_3)_2$:

$Mg(NO_3)_2$: 24.31 g/mol + (2)(14.01 g/mol) + (6)(16.00 g/mol) = 148.33 g/mol

You now know the mass of 1 mole of $Mg(NO_3)_2$. How will you find the mass of 3.52 moles of $Mg(NO_3)_2$? Complete the problem below.

GIVEN: 3.52 mol $Mg(NO_3)_2$ FIND: g $Mg(NO_3)_2$
FACTOR: 148.33 g $Mg(NO_3)_2$/mol PATH: mol → g

$$3.52 \text{ mol } Mg(NO_3)_2 \times \frac{148.33 \text{ g } Mg(NO_3)_2}{1 \text{ mol } Mg(NO_3)_2} = 522 \text{ g } Mg(NO_3)_2$$

Recall the conversion factor represented by Avogadro's number (from Equation 8.1):

$$\frac{6.022 \times 10^{23} \text{ units of X}}{1 \text{ mole of X}}$$

In Example 8.2 you used this to convert number of nitrogen atoms to moles of nitrogen atoms. X, in that case, was nitrogen atoms. X can be anything that is represented by a chemical formula, such as molecules or formula units of an ionic compound. What's more, using molar mass (Equation 8.2), the conversion can be carried one more step to the mass of the given number of molecules or formula units.

EXAMPLE 8.9: Find (a) the number of moles of methane, CH_4, in a sample containing 4.55×10^{22} molecules and (b) its mass.

A routine calculation with Avogadro's number solves the first problem.

GIVEN: 4.55×10^{22} CH_4 molecules FIND: mol CH_4
FACTOR: 6.022×10^{23} molecules/1 mol PATH: molecules → moles

$$4.55 \times 10^{22}\ CH_4\ \text{molecules} \times \frac{1\ \text{mol}\ CH_4}{6.022 \times 10^{23}\ CH_4\ \text{molecules}} = 0.0756\ \text{mol}\ CH_4$$

To solve part (b), you must change moles of methane to grams. That requires the molar mass of methane. Calculate that first.

CH_4: 12.01 g/mol + (4)(1.008 g/mol) = 16.04 g CH_4/mol

You are now ready for the setup to convert molecules to grams. Even though you have already finished the first step, complete the analysis all the way from molecules to grams to be sure you see it overall. Then set up and solve the problem.

GIVEN: FIND:

FACTORS:

PATH:

GIVEN: 4.55×10^{22} molecules CH_4 FIND: g CH_4
FACTORS: Avogadro's number, MM of CH_4, 16.04 g CH_4/mol
PATH: molecules → moles → grams

$$4.55 \times 10^{22}\ CH_4\ \text{molecules} \times \frac{1\ \text{mole}\ CH_4}{6.022 \times 10^{23}\ CH_4\ \text{molecules}} \times \frac{16.04\ \text{g}\ CH_4}{1\ \text{mole}\ CH_4} = 1.21\ \text{g}\ CH_4$$

EXAMPLE 8.10: Calculate the number of oxygen atoms in 65.0 grams of magnesium nitrate, $Mg(NO_3)_2$.

This time you must develop a conversion factor from the formula of the compound. How many oxygen atoms are in each formula unit of $Mg(NO_3)_2$? How many moles of oxygen atoms are in 1 mole of $Mg(NO_3)_2$? What conversion factor will take you between moles of oxygen atoms and moles of magnesium nitrate? Once you find that conversion factor, can you use it with other conversion factors that will connect grams of $Mg(NO_3)_2$ to the number of oxygen atoms? First analyze (GIVEN-FIND-FACTOR-PATH) the problem without the calculation setup.

GIVEN: FIND:

FACTORS:

PATH:

GIVEN: 65.0 g $Mg(NO_3)_2$ FIND: number oxygen atoms

FACTORS: MM of $Mg(NO_3)_2$; Avogadro's number; $\frac{6 \text{ mol O atoms}}{1 \text{ mol } Mg(NO_3)_2}$

There are two nitrate ions in each formula unit, and three oxygen atoms in each nitrate ion, giving six oxygen atoms per formula unit. Therefore there are six moles of oxygen atoms per mole of magnesium nitrate formula units.

We deliberately held back on the unit path in the answer until you could be sure of the last conversion factor. If you didn't have the unit path before, you probably can write it now. Also set up and solve the problem. Remember the above conversion factor in planning your unit path.

PATH: g $Mg(NO_3)_2$ → mol $Mg(NO_3)_2$ → mol O atoms → O atoms

$$65.0 \text{ g } Mg(NO_3)_2 \times \frac{1 \text{ mol } Mg(NO_3)_2}{148.33 \text{ g } Mg(NO_3)_2} \times \frac{6 \text{ mol O atoms}}{1 \text{ mol } Mg(NO_3)_2}$$

$$\times \frac{6.022 \times 10^{23} \text{ O atoms}}{1 \text{ mol O atoms}} = 1.58 \times 10^{24} \text{ O atoms}$$

8.6 PERCENTAGE COMPOSITION

The percentage composition of a compound is the percent by mass of each element in the compound. (See Section 2.6 for a discussion of percentage.) Percentage composition is found from the same data that are used to determine molar mass. Using numbers for Example 8.6, you found that 261.3 grams of barium nitrate contain 137.3 grams of barium. The percentage barium in the compound is calculated from the defining equation for any percentage,

$$\text{percentage} = \frac{\text{part quantity}}{\text{total quantity}} \times 100 \quad \text{(See Section 2.6)}$$

$$\% \text{ barium} = \frac{\text{grams barium}}{\text{grams barium nitrate}} \times 100 = \frac{137.3 \text{ g Ba}}{261.3 \text{ g } Ba(NO_3)_2} \times 100 = 52.54\% \text{ Ba}$$

In a similar manner the percent nitrogen and percent oxygen are, respectively,

$$\frac{(2)(14.01 \text{ g N})}{261.3 \text{ g } Ba(NO_3)_2} \times 100 = 10.72\% \text{ N}; \quad \frac{(6)(16.00 \text{ g O})}{261.3 \text{ g } Ba(NO_3)_2} \times 100 = 36.74\% \text{ O}$$

If all percentages have been calculated correctly they should total 100:

$$52.54\% \text{ Ba} + 10.72\% \text{ N} + 36.74\% \text{ O} = 100.00\%$$

EXAMPLE 8.11: Calculate the percentage of each element in $(NH_4)_2SO_4$.

Begin by finding the molar mass of the compound in the space below.

$(NH_4)_2SO_4$:

$(NH_4)_2SO_4$: (2)(14.01 g/mol) + (8)(1.008 g/mol) + 32.06 g/mol + (4)(16.00 g/mol) = 132.14 g/mol

Now calculate the percent nitrogen, using information from the molar mass setup.

GIVEN: 2 N at 14.01 g/mol in $(NH_4)_2SO_4$ at 132.14 g/mol FIND: % N

$$\text{EQN: } \% = \frac{\text{g N}}{\text{g } (NH_4)_2SO_4} \times 100 = \frac{(2)(14.01)\text{ g N}}{132.14\text{ g } (NH_4)_2SO_4} \times 100 = 21.20\ \%\ \text{N}$$

The procedure is the same for hydrogen, sulfur, and oxygen. Complete those percentage calculations in the space below. When you have finished, add them up to see if they check out to 100.

$$\frac{(8)(1.008\text{ g H})}{132.14\text{ g } (NH_4)_2SO_4} \times 100 = 6.103\%\ \text{H} \qquad \frac{32.06\text{ g S}}{132.14\text{ g } (NH_4)_2SO_4} \times 100 = 24.26\%\ \text{S}$$

$$\frac{(4)(16.00\text{ g O})}{132.14\text{ g } (NH_4)_2SO_4} \times 100 = 48.43\%\ \text{O}$$

21.20% N + 6.103% H + 24.26% S + 48.43% O = 99.99%

The 99.99% total, rather than 100.00%, is the result of rounding off the individual calculations; it is perfectly acceptable.

8.7 THE EMPIRICAL (SIMPLEST) FORMULA OF A COMPOUND

If a chemist wishes to determine the chemical formula of an unknown compound, he or she may begin by conducting an experiment to determine the mass of each element in a weighed sample of the compound. These mass relationships are often expressed as the percentage composition of the compound. As such, the percentage of each element represents the number of grams of that element in 100 grams of the compound. The essential information is the mass of each element in the sample, however it may be expressed.

This mass information can be used to determine the **empirical formula**, or **simplest formula**, of the substance. ("Empirical" means something that has been observed or determined by experiment.) An empirical formula gives the smallest whole-number ratio of atoms of each element in a compound. Thus NO_2 is an empirical formula showing that nitrogen and oxygen atoms are present in a 1:2 ratio. N_2O_4 is not an empirical formula. It shows a 2:4 ratio of nitrogen atoms to oxygen atoms. The 2:4 ratio has a common divisor, 2, and can therefore be "reduced" to 1:2, just as you reduce a fraction to lowest terms. Therefore, the empirical formula of N_2O_4 is NO_2.

NO_2 and N_2O_4 are both real compounds. They both have the same empirical formula, NO_2. They are more specifically described by their **molecular formulas, which show the actual number of atoms of each element in a molecule of the substance.** One molecule of NO_2 is made up of one nitrogen atom and two oxygen atoms; one molecule of N_2O_4 has two nitrogen atoms and four oxygen atoms.

Not all empirical formulas represent real compounds. For example, the molecular formula for hydrogen peroxide is H_2O_2, which gives HO as its empirical formula. There is no known stable substance with the formula HO.

The distinction between empirical and molecular formulas may be summarized as follows:

> If the subscripts of a formula can be divided evenly by a whole number, it is a molecular formula. If the subscripts of a formula cannot be reduced further, it is an empirical formula. It may also be a molecular formula.

Notice that *1 mole* of NO_2 contains *1 mole* of nitrogen atoms and *2 moles* of oxygen atoms. *The ratio of moles of atoms of the elements in a sample of a compound is the same as the ratio of atoms of those elements in a molecule.* In the calculations that follow you will actually find the ratio of *moles* of atoms, but interpret it as a ratio of atoms in the empirical formula.

In this section we will concentrate on how to determine empirical formulas from experimental data. In the next section we will show how the empirical formula is used in finding the molecular formula of a compound.

Suppose you are to find the empirical formula of a compound that is 13.6% carbon and 86.4% fluorine. That means than in a 100-gram sample of the compound there are 13.6 grams of carbon (13.6% of 100) and 86.4 grams of fluorine (86.4% of 100).

In an empirical formula problem it helps to organize your work in a table with the following headings:

ELEMENT	GRAMS	MOLES OF ATOMS	RATIO OF MOLES OF ATOMS	FORMULA RATIO
C	13.6			
F	86.4			

The grams of each element have already been entered into the table.

The next step is to convert grams of each element to moles of atoms by dividing by the molar mass of atoms. Notice: of *atoms*, not molecules. This distinction is important for fluorine. In other examples it is important for chlorine, bromine, iodine, hydrogen, oxygen, and nitrogen—all the elements that form diatomic molecules as *elements*, uncombined with other elements in chemical compounds. Empirical formulas are written for compounds, not elements. With fluorine you divide by 19.00 g F/1 mol F; you do not divide 38.00 g F_2/1 mol F_2. For carbon and fluorine in this example you get

$$13.6 \text{ g C} \times \frac{1 \text{ mol C}}{12.01 \text{ g C}} = 1.13 \text{ mol C}; \qquad 86.4 \text{ g F} \times \frac{1 \text{ mol F}}{19.00 \text{ g F}} = 4.55 \text{ mol F}$$

Please enter 1.13 for the moles of carbon atoms and 4.55 for the moles of fluorine atoms in the above table.

The moles of atoms are now in the same ratio as they appear in the empirical formula, which could be written $C_{1.13}F_{4.55}$. But chemical formula subscripts are written as whole numbers, not decimal fractions. The easiest way to get equivalent whole numbers, or numbers that can be changed to integers easily, is to divide each number of moles by the smallest number of moles. In this case the smallest number of moles is 1.13 for carbon. Thus

$$\text{C:} \quad \frac{1.13}{1.13} = 1.00 \qquad\qquad \text{F:} \quad \frac{4.55}{1.13} = 4.03$$

Enter 1.00 and 4.03 in the RATIO OF MOLES OF ATOMS column for carbon and fluorine, respectively. Using these numbers would give $C_{1.00}F_{4.03}$ for the empirical formula of the compound.

The last column in the table gives you a place to adjust the RATIO OF MOLES numbers to integers. In this case the 1.00 is rounded off to 1, and 4.03 is rounded off to 4. Enter these numbers. Minor roundoffs are to be expected in numbers that come from experimental measurements. The empirical formula of the compound can now be written as CF_4, which also happens to be the molecular formula of carbon tetrafluoride. The final table follows:

ELEMENT	GRAMS	MOLES OF ATOMS	RATIO OF MOLES OF ATOMS	FORMULA RATIO
C	13.6	1.13	1.00	1
F	86.4	4.55	4.03	4

The above procedure may be summarized as follows:

1. Prepare a table with the following headings: ELEMENT; GRAMS; MOLES OF ATOMS; RATIO OF MOLES OF ATOMS; FORMULA RATIO.
2. For each ELEMENT, enter the symbol and the number of GRAMS in the sample.
3. For each element, calculate the number of moles by dividing grams by the molar mass of atoms (atomic mass), and enter that value into the MOLES OF ATOMS column.
4. Divide each number of moles of atoms by the smallest number of moles of atoms. Enter the quotient into the RATIO OF MOLES OF ATOMS column in the table.
5. Express the RATIO OF MOLES OF ATOMS values as a ratio of integers. Enter the result in the FORMULA RATIO column. Write the empirical formula of the compound.

EXAMPLE 8.12: In the laboratory a 1.332-gram sample of an unknown compound is found to contain 0.360 gram of sodium, 0.220 gram of nitrogen, and the balance oxygen. Find the empirical formula of the compound.

The problem gives you the mass of sodium and nitrogen in the compound and data from which the mass of oxygen can be calculated. Do you see how to find the number of grams of oxygen in the 1.332-gram sample? If so, calculate that quantity.

$$\begin{array}{r} 1.332 \text{ g Na + N + O} \\ -\ 0.360 \text{ g Na} \\ \underline{-\ 0.220 \text{ g N}} \\ 0.752 \text{ g O} \end{array}$$

By subtracting the mass of sodium and the mass of nitrogen from the total mass of the compound, you get the mass of the third element, oxygen.

Once you have the mass of each element, complete steps 1 and 2 of the procedure in the spaces provided. Proceed with step 3 by calculating the moles of atoms of each element and entering the results in the table. Remember you want moles of *atoms*, not molecules.

ELEMENT	GRAMS	MOLES OF ATOMS	RATIO OF MOLES OF ATOMS	FORMULA RATIO
Na				
N				
O				

ELEMENT	GRAMS	MOLES OF ATOMS	RATIO OF MOLES OF ATOMS	FORMULA RATIO
Na	0.360	$\frac{0.360 \text{ g Na}}{22.99 \text{ g/mol}} = 0.0157$		
N	0.220	$\frac{0.220 \text{ g N}}{14.01 \text{ g/mol}} = 0.0157$		
O	0.752	$\frac{0.752 \text{ g O}}{16.00 \text{ g/mol}} = 0.0470$		

In finding the moles of nitrogen and oxygen atoms, you must divide by the number that corresponds to the *atomic* mass of the element, the molar mass of atoms.

To find the ratio of moles of atoms (step 4), divide each number of moles by the smallest number. Complete the calculations and fill in the RATIO OF MOLES OF ATOMS column in the above table.

ELEMENT	GRAMS	MOLES OF ATOMS	RATIO OF MOLES OF ATOMS	FORMULA RATIO
Na	0.360	$\frac{0.360 \text{ g Na}}{22.99 \text{ g/mol}} = 0.0157$	$\frac{0.0157}{0.0157} = 1.00$	
N	0.220	$\frac{0.220 \text{ g N}}{14.01 \text{ g/mol}} = 0.0157$	$\frac{0.0157}{0.0157} = 1.00$	
O	0.752	$\frac{0.752 \text{ g O}}{16.00 \text{ g/mol}} = 0.0470$	$\frac{0.0470}{0.0157} = 2.99$	

At this point you must change 1.00 and 2.99 to the closest whole numbers. Do so, entering the numbers into the FORMULA RATIO column of the table (step 5). Also, from those numbers write the empirical formula of the compound.

Empirical formula:

ELEMENT	GRAMS	MOLES OF ATOMS	RATIO OF MOLES OF ATOMS	FORMULA RATIO
Na	0.360	$\frac{0.360 \text{ g Na}}{22.99 \text{ g/mol}} = 0.0157$	$\frac{0.0157}{0.0157} = 1.00$	1
N	0.220	$\frac{0.220 \text{ g N}}{14.01 \text{ g/mol}} = 0.0157$	$\frac{0.0157}{0.0157} = 1.00$	1
O	0.752	$\frac{0.752 \text{ g O}}{16.00 \text{ g/mol}} = 0.0470$	$\frac{0.0470}{0.0157} = 2.99$	3

Empirical formula: $NaNO_3$

The next example shows what to do when the RATIO OF MOLES numbers do not round off to integers.

EXAMPLE 8.13: By analysis, a compound is found to be 54.0% iron and 46.0% sulfur. Find the empirical formula of the compound.

Remember that percentage composition of a compound is interpreted to be the grams of each element in a 100-gram sample. With that in mind, complete the first four steps of the procedure, through RATIO OF MOLES OF ATOMS.

ELEMENT	GRAMS	MOLES OF ATOMS	RATIO OF MOLES OF ATOMS	FORMULA RATIO
Fe				
S				

ELEMENT	GRAMS	MOLES OF ATOMS	RATIO OF MOLES OF ATOMS	FORMULA RATIO
Fe	54.0	$\frac{54.0 \text{ g Fe}}{55.85 \text{ g/mol}} = 0.967$	$\frac{0.967}{0.967} = 1.00$	
S	46.0	$\frac{46.0 \text{ g S}}{32.06 \text{ g/mol}} = 1.43$	$\frac{1.43}{0.967} = 1.48$	
			Empirical formula:	

The numbers 1.00 and 1.48 do not round off to a ratio of integers. The 1.48 is close to 1.5, which suggests an empirical formula $FeS_{1.5}$. But we need integers, not decimal fractions, in the formula. Do you see a pair of integers that would express the same 1/1.5 ratio between moles of iron atoms and sulfur atoms, giving an acceptable formula? If so, fill in the FORMULA RATIO column of the table and write the empirical formula above.

ELEMENT	GRAMS	MOLES OF ATOMS	RATIO OF MOLES OF ATOMS	FORMULA RATIO
Fe	54.0	$\frac{54.0 \text{ g Fe}}{55.85 \text{ g/mol}} = 0.967$	$\frac{0.967}{0.967} = 1.00$	2
S	46.0	$\frac{46.0 \text{ g S}}{32.06 \text{ g/mol}} = 1.43$	$\frac{1.43}{0.967} = 1.48$	3
			Empirical formula:	Fe_2S_3

By doubling both 1 and 1.5 you come up with 2 and 3, two whole numbers having the same ratio as 1 and 1.5.

Usually a mole ratio that is not a ratio of whole numbers can be made into one by multiplying all ratio numbers by the same small integer. In the above example, 2 was the number: $2 \times 1 = 2$ and $2 \times 1.5 = 3$. If 2 doesn't work, try 3. If that fails, try 4, and, if necessary 5. Usually 2 or 3 will succeed; but occasionally a larger integer is needed.

8.8 MOLECULAR FORMULAS FROM EMPIRICAL FORMULAS

In Section 8.7, NO_2 and N_2O_4 were used to show the difference between a molecular formula and an empirical formula. In that section you learned how laboratory data can be used to find the empirical formula of a compound. Once the empirical formula is known, how is the true molecular formula to be found? In other words, how do we know that dinitrogen tetroxide is N_2O_4 and not simply NO_2?

The answer to this question must be found in the laboratory. There are ways to determine the molar mass of a compound. You will learn one method in Chapter 12, and others will be considered in Chapter 17.

For the moment, assume that one of these methods has established that the molar mass of an "unknown" oxide of nitrogen is 92.0 g/mol. Assume also that its empirical formula has been found to be NO_2. If NO_2 is also the molecular formula of the compound, its molar mass is $14.01 + 2(16.00) = 46.01$ g/mol.

This is one-half of the real molar mass. It follows that there must be two empirical formula units in the molecular formula.

One way to think of this problem is to consider the molecular formula of the unknown compound as $(NO_2)_n$, where n is an integer that is the number of moles of empirical formula units in 1 mole of molecules. To find n, calculate the molar mass of the empirical formula unit and divide it into the molar mass of the compound:

$$n = \frac{\text{g/mol (molecular)}}{\text{g/mol (empirical)}} = \frac{92.0 \text{ g/mol}}{46.01 \text{ g/mol}} = 2 \qquad (8.3)$$

$$\text{Molecular formula} = n \times \text{empirical formula} = (NO_2)_2 = N_2O_4 \qquad (8.4)$$

EXAMPLE 8.14: The empirical formula of a compound has been found to be CH_3. Find the molecular formula of the substance if its molar mass is 30.0 g/mol.

Begin by finding the molar mass of the empirical formula unit.

CH_3:

CH_3: 12.01 g/mol + (3)(1.008 g/mol) = 15.03 g/mol

Considering the molecular formula to be $(CH_3)_n$, you can now find n, the number of empirical formula units in a mole, by substituting into Equation 8.3. Complete this calculation, and then write the molecular formula.

EQ: $$n = \frac{\text{g/mol (molecular)}}{\text{g/mol (empirical)}} = \frac{30.0 \text{ g/mol}}{15.03 \text{ g/mol}} = 2$$

$$\text{Molecular formula} = n \times \text{empirical formula} = (CH_3)_2 = C_2H_6$$

The molecular formula could be written $(CH_3)_2$, where the subscript 2 shows there are two CH_3 units in the molecule. Multiplying the subscripts within the parentheses by 2 gives the conventional formula with integral subscripts.

CHAPTER 8 PROBLEMS

1) Determine the molar masses of carbon dioxide, CO_2, and propyl alcohol, C_3H_7OH.

2) Calculate the molar mass of each of the following substances:

a) F_2 b) He c) K_2CO_3 d) H_2SO_4 e) Fe f) SO_2

3) Find the molar mass of: a) LiF b) K_2S c) $NaBrO_3$
d) $Al_2(Cr_2O_7)_3$ e) $C_9H_{11}NO_4$ f) $HOOC(CH_2)COOH$

4) Calculate the percentage of each element in aluminum phosphate, $AlPO_4$.

5) A component of smog, peroxyacetyl nitrate, commonly called PAN, causes the reddish-brown haze seen in the sky. Its molecular formula is $C_2H_3NO_5$. Find the percentage of each element in the compound.

6) Calculate the percentage composition of chromium sulfate, $Cr_2(SO_4)_3$.

7) How many molecules are in 2.18 moles of hydrogen gas, H_2?

8) How many moles of potassium are in 8.71×10^{22} atoms?

9) What is the mass of 0.358 moles of ammonium chloride, NH_4Cl?

10) Find the mass of 3.79×10^{22} molecules of tetraphosphorus decoxide, P_4O_{10}.

11) How many moles are contained in 16.4 grams of Fe_2O_3?

12) Calculate the number of molecules in 17.9 grams of CH_3OH.

13) How many moles of sodium nitrate, $NaNO_3$, are in 4.55×10^{25} formula units?

14) How many moles are in 124 grams of barium sulfate, $BaSO_4$?

15) Calculate the mass of 8.09×10^{21} formula units of $Mg(ClO_3)_2$.

16) Find the mass of 0.512 moles of C_3H_8.

17) How many formula units are in 0.169 grams of Na_2SO_4?

18) How many moles are in 1.19×10^{23} molecules of SO_3? What is the mass of this number of molecules?

19) How many oxygen molecules are in 80.0 grams of O_2? How many oxygen atoms are in 80.0 grams of O_2?

20) On analysis, a compound is found to be 60.7% nickel and 39.3% fluorine. Find the empirical formula of the compound.

21) A 5.83 gram sample of a compound is found to contain 0.292 grams of hydrogen, 3.50 grams of oxygen, and the balance is nitrogen. Calculate the empirical formula of the compound.

22) Nicotine, a stimulant that can be poisonous in large dosages, contains 74.1% carbon, 8.64% hydrogen, and 17.3% nitrogen. What is the empirical formula for nicotine?

23) What are the empirical and molecular formulas of a compound if it is 14.3% hydrogen and 85.7% carbon and its molar mass is 42.0 g/mol?

24) A 29.8 gram sample of a compound is found to contain 10.5 grams of carbon and 16.7 grams of fluorine. The rest is hydrogen. The molar mass of the compound is 34 g/mol. Find the empirical and molecular formulas of the compound.

25) A compound analyzes at 49.3% carbon, 2.1% hydrogen, and 48.6% chlorine. Its molar mass is 292 g/mol. Find the empirical and molecular formulas of the compound.

The next four problems are beyond the chemical skills listed at the beginning of the chapter, but they are logical extensions of the ideas already considered. First solve Problem 26 and check the answer, and then work Problems 27 to 29. Problem 26 is the most direct application of the method that underlies the other three.

26) How many grams of chlorine are in 59.7 grams of calcium chloride?

27) Nickel sulfate forms a hexahydrate, $NiSO_4 \cdot 6H_2O$, in which the crystal has six water molecules with each formula unit of the salt. The molar mass of the hydrate includes the water. Answer the following questions about the hydrate:

a) How many grams of sulfur are in 162 grams of the hydrate?
b) What percent of the hydrate is water?
c) How many moles of sulfate ion are in 840 grams of the hydrate?
d) How many grams of the hydrate must be used to obtain 52.8 grams of nickel?
e) If 45.1 grams of the hydrate are dissolved and reacted with sodium hydroxide, all of the nickel is changed to nickel hydroxide, $Ni(OH)_2$. Calculate the mass of the nickel hydroxide.

28) Dolomite, the white rock used in many gardens, is a "double salt" with the formula $CaCO_3 \cdot MgCO_3$. Answer the following questions about dolomite:

a) How many grams of calcium can be recovered from 665 grams of $CaCO_3 \cdot MgCO_3$?
b) If you wish to obtain 45.0 grams of magnesium from $CaCO_3 \cdot MgCO_3$, how many grams of dolomite should be used?
c) Calculate the calcium-to-magnesium ratio, g Ca/g Mg, in dolomite.

29) Copper sulfate pentahydrate, $CuSO_4 \cdot 5H_2O$, and copper nitrate hexahydrate, $Cu(NO_3)_2 \cdot 6H_2O$, are sold in containers that hold the same number of pounds. If the price of a container of the nitrate is $505, and the price of a container of the sulfate is $544, which is the better buy from the standpoint of pounds of copper received? In other words, for both hydrates, calculate the price of copper in dollars per unit mass.

++++++++++++++++++++++

30) Calculate the molar mass of each of the following:

a) LiCl b) SiO_2 c) CH_3COOH d) $Ca(OH)_2$ e) $(NH_4)_2CO_3$ f) $Ba_3(PO_4)_2$

31) a) How many molecules are in 2.89 mol NF_3?

b) Calculate the mass of 4.24×10^{19} atoms of argon.

32) a) What is the mass of 9.44×10^{24} molecules of C_2H_5OH?

b) How many atoms (repeat, atoms) are in 6.40 grams of I_2?

33) a) Find the mass of 3.16×10^{23} formula units of $Ca(NO_3)_2$.

b) How many formula units of $Ni_3(PO_4)_2$ are in 6.54 grams?

34) Calculate the percentage composition of:

a) $FeCl_3$ b) $C_2H_4Cl_2$ c) $Ca(OH)_2$ d) $Ba(ClO_3)_2$

35) Find the number of moles in a) 32.2 g $PbCl_2$ and b) 508 g $Ba(IO_4)_2$.

36) Calculate the mass of a) 2.42 mol $AgNO_3$ and b) 0.795 mol $Al_2(CO_3)_3$.

37) An unidentified compound is 11.1% nitrogen, 3.2% hydrogen, 41.3% chromium, and 44.4% oxygen. What is the empirical formula of the compound?

38) A 16.60 gram sample of an unknown ionic compound contains 5.70 grams of sodium and 2.98 grams of carbon, and the balance is oxygen. The molar mass of the compound is 134 g/mol. Find the empirical and molecular formulas of the compound.

ADDITIONAL PROBLEMS

39) An unknown hydrate of chromium(III) nitrate is found to have a molar mass of 400 g/mol. What is its formula?

40) A 25.0 pg sample of an unknown element, known to be tetratomic, contains 50.0 billion molecules. What is the molar mass of the molecules and the atomic mass of the element?

41) A 1239 mg sample of an organic compound of formula $C_xH_yO_z$, when burned in excess oxygen, yields 2723 mg CO_2 and 892 mg H_2O.
   a) How many mg C were present? How many mg H were present?
   b) How many mg O were in the $C_xH_yO_z$ sample?
   c) What is the empirical formula of $C_xH_yO_z$?

42) Each of the following are used as dietary sources of calcium: $Ca_3(C_6H_5O_7)_2$, $CaCO_3$, and $CaHPO_4$. How many mg of each are needed to obtain the RDA (recommended daily allotment) of 1000 mg Ca?

CHAPTER TEST

1) Find the molar mass of a) OF_2 b) C_2H_5OH c) $(NH_4)_3PO_4$.

2) How many molecules are in 44.1 grams of ozone, O_3?

3) What is the mass of 3.04×10^{24} formula units of $Cu(NO_3)_2$?

4) Find the percentage composition of a) C_4H_9COOH and b) Na_2CO_3.

5) How many moles are in 80.5 grams of oxalic acid, HOOCCOOH?

6) What is the mass of 0.272 mol $Al(NO_3)_3$?

7) The molar mass of a compound has been determined to be 124 grams per mole. Find (a) the empirical formula and (b) the chemical formula of the compound if it is 22.6% nitrogen, 6.45% hydrogen, 19.4% carbon, and 51.6% oxygen. (The compound is ionic. Its "chemical" formula corresponds to the molecular formula of a molecular compound.)

[illegible]

17. Find the [illegible]

[illegible] (CO) [illegible]

(a) [illegible] carbon [illegible]

18. [illegible] sample [illegible] gm [illegible] oxygen [illegible] of [illegible] 98 [illegible] that [illegible] 204

ADDITIONAL PROBLEMS

19. [illegible] of [illegible] iron [illegible]

(c) [illegible] molecular [illegible] (CH₄) [illegible]

(a) [illegible] (b) [illegible] molecular [illegible]

[illegible]

CHAPTER [illegible]

[illegible]

[illegible] equation [illegible]

[illegible]

[illegible]

The [illegible] formula and [illegible]

9. CHEMICAL EQUATIONS

PREREQUISITES

1. Section 6.6 presented the formulas and physical states of the elements which must be written when they appear in a chemical equation.

2. Chapter 7 covered names and formulas so that, if you are given the name of a compound in a reaction, you will be able to write its formula.

CHEMICAL SKILLS

You may refer to an approved periodic table while practicing this skill.

1. Given information from which you can write the formulas for all reactants and all products for each of the following types of reactions, write balanced chemical equations for the reactions:

Combination (synthesis)

Decomposition

Complete oxidation (burning) of a compound of carbon and hydrogen, or of carbon, hydrogen, and oxygen

Oxidation-reduction ("single replacement" equations only)

Precipitation and acid-base reactions (those which follow "double replacement" equations only)

Other reactions with identifiable reactants and products

9.1 THE MAKEUP OF A CHEMICAL EQUATION

If hydrochloric acid is poured onto a piece of solid calcium carbonate, gaseous carbon dioxide, liquid water, and a solution of calcium chloride are produced. The equation below describes this chemical change.

$$CaCO_3(s) + 2\ HCl(aq) \rightarrow CaCl_2(aq) + CO_2(g) + H_2O(\ell)$$

A chemical equation shows the formulas of the substances that enter into a chemical change, called the reactants ($CaCO_3$ and HCl), on the left side of the equation. The formulas of the substances produced, called the **products** ($CaCl_2$, CO_2, and H_2O), appear on the right. The reactants and products are separated by an arrow, →, indicating the direction of the change. In reading the equation the arrow may be read "yields," or any equivalent word.

Coefficients, numbers written in front of the chemical formulas, are used as necessary to **balance** the number of atoms of each element on the two sides of the equation. As in an algebraic equation, if the coefficient is 1, it usually is not written; if there is no coefficient in an equation, it is understood to be 1. The coefficients and "terms" in a chemical equation may be operated on just as they are in an algebraic equation. You can add something to or subtract something from both sides of a chemical equation, or you may multiply or divide both sides of an equation by the same number.

Chemical equations should also show the physical state of substances in the reaction. The symbol (g) is written immediately after the formula of a gas. Similarly (ℓ) identifies a liquid, and (s) a solid. If a substance is in aqueous (water) solution, the symbol (aq) is used.

Occasionally, the arrow is lengthened and a word, formula, temperature, or symbol is written above the arrow, and sometimes above and below, to describe reaction conditions.

EXAMPLE 9.1 There is a mistake in each of the following equations. What is wrong with each equation?

$H_2(g) + O_2(g) \rightarrow H_2O(\ell)$ ____________________

$Zn(s) + 2\ HCl(aq) \rightarrow ZnCl_2 + H_2(g)$ ____________________

$N(g) + 3\ H(g) \rightarrow NH_3(g)$ ____________________

The first equation is not balanced. It has two oxygen atoms on the left and one on the right. $2\ H_2(g) + O_2(g) \rightarrow 2\ H_2O(\ell)$ is the correct equation.

The state of $ZnCl_2$ is not shown in the second equation. The equation is $Zn(s) + 2\ HCl(aq) \rightarrow ZnCl_2(aq) + H_2(g)$.

The formulas of nitrogen and hydrogen are incorrect in the third equation. The correct equation is $N_2(g) + 3\ H_2(g) \rightarrow 2\ NH_3(g)$.

9.2 HOW TO WRITE CHEMICAL EQUATIONS

There are three separate steps in writing a chemical equation:

1. Write the correct chemical formula of each reactant on the left side of an arrow and the correct chemical formula of each product on the right. Separate two or more reactants and two or more products with plus signs (+).

2. Identify the state (gas, liquid, solid, aqueous solution) of the reactants and products. For elements in their standard states, see Table 6.3. For the present, the states of compounds will be given.
3. Balance the number of atoms of each element on each side of the equation by placing coefficients in front of the formulas of reactants and/or products.

It is particularly important to recognize that these are completely independent steps. Specifically,

DO NOT, in Step 1 "adjust" a formula on one side so it has the same number of atoms of some element that is on the other side.

DO NOT, in Step 3, change a correct chemical formula to balance an element.

DO NOT, in Step 3, place a coefficient inside a formula, to apply to only part of that formula, in order to balance an element.

DO NOT, in Step 3, add to either side of the equation the formula of some species that has nothing to do with the reaction, that is not present, or that may not even exist.

Quite often the word description of a chemical reaction does not identify all of the species in the equation. Your knowledge of chemistry must supply the additional information. Specifically, you should be familiar with the kinds of reactions listed at the beginning of this chapter. You will then be able to identify and write formulas for any species not mentioned in the description of the reaction.

9.3 COMBINATION REACTIONS

A **combination reaction**, sometimes called a **synthesis reaction**, occurs when two or more substances combine to form a single product. The reactants may be elements or compounds, perhaps one or more of each.

Quite often the description of the reaction will give the name of the product only. In such cases you will need to determine the formula, using the skills developed in Chapter 7.

The final step is to balance the equation. A general method for balancing equations is summarized below. This method will be illustrated in the examples that follow.

1. Put a coefficient of "1" in front of the most complicated formula. The most complicated formula is usually the one with the largest number of elements and/or the largest number of atoms.

2. Balance the elements one at a time in the following order:
   a. Start with elements in the most complicated formula.
   b. Save for last:
      1) elements appearing in more than two formulas, and
      2) elements in their uncombined forms.
3. If fractions were needed in balancing the elements in part 2, clear them by multiplying by an integer so that all coefficients are in simplest whole-number form.
4. DON'T go back and change a coefficient once written, except to clear fractions, as in step 3. NEVER change or add a formula to balance the equation.

EXAMPLE 9.2: Write the equation showing that solid sodium sulfide is formed from its elements.

The statement says solid sodium sulfide is formed. It is therefore the product of the reaction. The words "its elements" refer to the two elements, sodium and sulfur, in sodium sulfide. These are the reactants. An equation in word form could be written for the reaction:

sodium + sulfur → sodium sulfide

What are the formulas and physical states? Recall that both elements are solids. Their formulas are Na(s) and S(s). Using the methods developed in Section 7.8, the charges on the ions, Na^+ and S^{2-}, yield $Na_2S(s)$ as the formula of the compound. Rewrite the above equation, using chemical formulas in place of words.

$$Na(s) + \quad S(s) \rightarrow \quad Na_2S(s)$$

Notice that space has been left in front of each formula. This is for a coefficient if one is necessary.

The next step is to balance the atoms of each element on the two sides of the equation. First, put a "1" in front of the most complicated formula.

$$Na(s) + \quad S(s) \rightarrow 1\ Na_2S(s)$$

Na_2S is the most complicated: it has two different elements and three atoms.

Now balance Na and S.

$$2\ Na(s) + 1\ S(s) \rightarrow 1\ Na_2S(s)$$

The equation is now balanced. As noted earlier, the coefficient 1 is not usually written in an equation, so the equation is $2\ Na(s) + S(s) \rightarrow Na_2S(s)$.

EXAMPLE 9.3: Write the equation for the direct combination of aluminum and oxygen to form solid aluminum oxide.

The procedure is the same as in Example 9.2. Write the equation as far as the formulas of the reactants and products and a coefficient of 1 in front of the most complicated formula.

$Al(s) + \quad O_2(g) \rightarrow 1\ Al_2O_3(s)$

With two elements and a total of five atoms, Al_2O_3 is the most complicated formula.

Now balance aluminum and oxygen. Aluminum will be simple enough, but oxygen may not. Notice there are three oxygen atoms on the right side of the equation. They must come from oxygen molecules. Each oxygen molecule can be thought of as a package that contains two oxygen atoms. How many packages do you need to get a total of three atoms? Write the answer—the coefficient—as a fraction.

$2\ Al(s) + \frac{3}{2}\ O_2(g) \rightarrow 1\ Al_2O_3(s)$

To get three oxygens on the reactant side, the coefficient is $\frac{3}{2}$: $\frac{3}{2} \times 2 = 3$.

Now clear the fraction to get the simplest whole-number coefficients.

$4\ Al(s) + 3\ O_2(g) \rightarrow 2\ Al_2O_3(s)$

Multiplying all coefficients by two gives the final answer.

EXAMPLE 9.4: Write the equation for the formation of solid iron(III) hydroxide from its elements.

Complete the first step by writing the formulas of the reactants on the left and the formula of the product on the right.

$Fe(s) + \quad O_2(g) + \quad H_2(g) \rightarrow \quad Fe(OH)_3(s)$

Put a "1" in front of the most complicated formula.

$Fe(s) + \quad O_2(g) + \quad H_2(g) \rightarrow 1\ Fe(OH)_3(s)$

Now balance the elements, using fractions as needed.

$Fe(s) + \frac{3}{2}\ O_2(g) + \frac{3}{2}\ H_2(g) \rightarrow 1\ Fe(OH)_3(s)$

Finally, rewrite the equation without fractions.

$2\ Fe(s) +\ 3\ O_2(g) +\ 3\ H_2(g) \rightarrow 2\ Fe(OH)_3(s)$

Combination reactions can occur between two elements, between an element and a compound, and also between two compounds. In fact, the formation of $Fe(OH)_3$, the approximate formula of common rust, occurs in two steps. First, iron reacts with oxygen and water in the air to form iron(II) hydroxide. Then further reaction with oxygen and water vapor forms iron(III) hydroxide. Both of these combinations involve compounds as reactants. We will leave the equations to you.

EXAMPLE 9.5: Write the equation for the conversion of iron to iron(II) hydroxide upon exposure to oxygen and water vapor in the air. (Note: Water "vapor" is gaseous water.)

in words:	iron + oxygen + water → iron(II) hydroxide
equation:	$Fe(s) + O_2(g) + H_2O(g) \rightarrow Fe(OH)_2(s)$
most complicated:	$Fe(s) + O_2(g) + H_2O(g) \rightarrow 1\ Fe(OH)_2(s)$
balance H:	$Fe(s) + O_2(g) + 1\ H_2O(g) \rightarrow 1\ Fe(OH)_2(s)$
balance Fe:	$1\ Fe(s) + O_2(g) + 1\ H_2O(g) \rightarrow 1\ Fe(OH)_2(s)$
balance O:	$1\ Fe(s) + \frac{1}{2}\ O_2(g) + 1\ H_2O(g) \rightarrow 1\ Fe(OH)_2(s)$
clear fractions:	$2\ Fe(s) + O_2(g) + 2\ H_2O(g) \rightarrow 2\ Fe(OH)_2(s)$

Notice that oxygen is left for last for two reasons: First, it appears in three formulas. Second, it is an uncombined element.

EXAMPLE 9.6: Now complete the reaction for the formation of rust by writing the equation for the conversion to $Fe(OH)_3$ by further reaction of $Fe(OH)_2$ with water and oxygen.

in words:	iron(II) hydroxide + oxygen + water → iron(III) hydroxide
equation:	$Fe(OH)_2(s) + O_2(g) + H_2O(g) \rightarrow Fe(OH)_3(s)$
most complicated:	$Fe(OH)_2(s) + O_2(g) + H_2O(g) \rightarrow 1\ Fe(OH)_3(s)$
balance Fe:	$1\ Fe(OH)_2(s) + O_2(g) + H_2O(g) \rightarrow 1\ Fe(OH)_3(s)$
balance H:	$1\ Fe(OH)_2(s) + O_2(g) + \frac{1}{2}\ H_2O(g) \rightarrow 1\ Fe(OH)_3(s)$
balance O:	$1\ Fe(OH)_2(s) + \frac{1}{4}\ O_2(g) + \frac{1}{2}\ H_2O(g) \rightarrow 1\ Fe(OH)_3(s)$
clear fractions:	$4\ Fe(OH)_2(s) + O_2(g) + 2\ H_2O(g) \rightarrow 4\ Fe(OH)_3(s)$

9.4 DECOMPOSITION REACTIONS

A **decomposition reaction** is one in which a single reactant breaks down into two or more products, which may be elements or compounds. A decomposition reaction is exactly the opposite of a combination reaction.

EXAMPLE 9.7: Write the equation for the decomposition of solid mercury(II) oxide into its elements.

First write the formulas and states of the reactants and products.

$$HgO(s) \rightarrow Hg(\ell) + O_2(g)$$

Balancing the equation involves exactly the same thought processes you used in the first three examples. Proceed with the equation above.

$$2\ HgO(s) \rightarrow 2\ Hg(\ell) + O_2(g)$$

EXAMPLE 9.8: When heated, solid magnesium hydroxide decomposes into solid magnesium oxide and water vapor. Write the equation.

First write the formulas and states of the reactants and products.

$$Mg(OH)_2(s) \rightarrow MgO(s) + H_2O(g)$$

Now balance the atoms of each element, using coefficients only.

$$Mg(OH)_2(s) \rightarrow MgO(s) + H_2O(g)$$

This one was easy, as all coefficients turned out to be 1.

9.5 COMPLETE OXIDATION (BURNING) REACTIONS

When something burns in air, it reacts chemically with the abundant supply of oxygen in the atmosphere. Oxygen is therefore the unidentified second reactant in any burning reaction. If the compound that is burned contains only carbon and hydrogen—or carbon, hydrogen, and oxygen—and if the substance is burned completely, the end products are always carbon dioxide gas and water. The water that leaves a burning reaction is normally in the vapor state, $H_2O(g)$. Nevertheless, we usually write $H_2O(\ell)$, indicating its liquid state at room temperature and pressure. Because the products of this kind of burning reaction are always the same, they usually are not identified.

Essentially the same kind of reaction can occur under conditions other than burning. Compounds of carbon, hydrogen, and oxygen react with oxygen in living systems, yielding the same products, carbon dioxide and water—this time definitely a liquid. In such cases it is said that the compound has been completely oxidized.

From the above information you should be able to write formulas for all reactants and products for **burning** (combustion) or **complete oxidation reactions**, given only the formula of the substance burned or oxidized. The second reactant is oxygen, and the two products are always carbon dioxide and water.

EXAMPLE 9.9: Hexane, $C_6H_{14}(\ell)$, burns in air. Write the equation.

First write the formulas for all reactants and products on their proper sides of the equation.

$$C_6H_{14}(\ell) + \quad O_2(g) \rightarrow \quad CO_2(g) + \quad H_2O(\ell)$$

Strategy for balancing: C_6H_{14} is the most complicated substance, so it should be assigned a coefficient of 1. Oxygen appears in three places, once as an elemental substance. Therefore, it should be left to last. If you can get the carbon and hydrogen balanced, and then count up the number of oxygens on the right, you can balance the oxygen by inserting the coefficient that is necessary in front of O_2 on the left. Give it a try. Balance both carbon and hydrogen above as the next step, but leave oxygen unbalanced.

$$1\ C_6H_{14}(\ell) + \quad O_2(g) \rightarrow 6\ CO_2(g) + 7\ H_2O(\ell)$$

Now for the oxygen. Count up the oxygen atoms on the right and use whatever coefficient is necessary on the left to balance the equation. You may have to use a fraction. If so, clear it in the final step, written below.

$$1\ C_6H_{14}(\ell) + \frac{19}{2}\ O_2(g) \rightarrow 6\ CO_2(g) + 7\ H_2O(\ell)$$

$$2\ C_6H_{14}(\ell) + 19\ O_2(g) \rightarrow 12\ CO_2(g) + 14\ H_2O(\ell)$$

There are 19 oxygens on the right, to come from O_2 molecules with 2 oxygens each: $\frac{19}{2} \times 2 = 19$.

EXAMPLE 9.10: Write the equation for the complete oxidation of acetic acid, $CH_3COOH(\ell)$.

Begin by writing the unbalanced equation.

$$CH_3COOH(\ell) + \quad O_2(g) \rightarrow \quad CO_2(g) + \quad H_2O(\ell)$$

The formula for acetic acid has been written as an organic chemist would write it. This sometimes leads to careless errors in the last two parts of this example. With this warning, see if you can avoid them in balancing carbon and hydrogen, leaving oxygen unbalanced.

$$1\ CH_3COOH(\ell) + \quad O_2(g) \rightarrow 2\ CO_2(g) + 2\ H_2O(\ell)$$

Both carbon and hydrogen appear twice in the formula CH_3COOH, a fact sometimes overlooked when working with organic compounds.

Be careful as you complete the equation by balancing oxygen.

$$CH_3COOH(\ell) + 2\ O_2(g) \rightarrow 2\ CO_2(g) + 2\ H_2O(\ell)$$

The oxygen count on the right side of $CH_3COOH(\ell) + O_2(g) \rightarrow 2\ CO_2(g) + 2\ H_2O(\ell)$ is 6. This time they don't all come from O_2. Two of them come from CH_3COOH, the reactant being oxidized. The remaining four come from O_2, calling for a coefficient of 2. Always be sure to account for the oxygens present in the organic compound being oxidized as you select your coefficient for O_2.

9.6 OXIDATION-REDUCTION REACTIONS (SINGLE REPLACEMENT EQUATIONS)

The words *oxidize* and *oxidation* have a meaning in chemistry beyond "reaction with oxygen," as they have been used so far. If a substance loses electrons in a chemical change, it is said to have been **oxidized**. If electrons are lost by one substance, there must be another substance present to gain the electrons. That substance is said to have been **reduced**. An electron transfer reaction is classified as an **oxidation-reduction reaction**, frequently called a **redox reaction**.

The equation for one kind of redox reaction looks as if an element that reacts with a compound replaces one of the elements in the compound. The equation is therefore called a **single replacement** or **single displacement** equation. This section will be limited to redox reactions that may be described by single replacement equations.

EXAMPLE 9.11: Gaseous hydrogen is released and a solution of zinc chloride remains when zinc reacts with hydrochloric acid. Write the equation for the reaction.

For the first time in this chapter you must examine the description of the reaction very carefully to distinguish between reactants and products. See if you can get the formulas of the reactants and products on the correct sides of the unbalanced equation.

$$Zn(s) + HCl(aq) \rightarrow H_2(g) + ZnCl_2(aq)$$

The words *released* and *remains* in the description of the reaction suggest that hydrogen and zinc chloride are products of the reaction, and *reacts* identifies zinc and hydrochloric acid as reactants. You know from Chapter 7 that zinc commonly forms an ion with a +2 charge, so the formula of zinc chloride is $ZnCl_2$.

Balancing the equation is straightforward.

$$Zn(s) + 2\ HCl(aq) \rightarrow H_2(g) + ZnCl_2(aq)$$

The single replacement character of the above equation appears as you recognize that hydrogen is combined with chlorine as a compound on the left side of the equation. On the right side zinc has replaced hydrogen as the element combined with chlorine.

If you recognize the possibility of a single replacement equation between two given reactants, you can write the equation and predict the products of the redox reaction that may occur. The reactants are always an element and a compound; the products are one of the elements in the reacting compound and a new compound formed by the other elements. The equation has one of two general forms:

$$A + BX \rightarrow B + AX \qquad \text{or} \qquad Y + MZ \rightarrow Z + MY$$

In the first instance, element A has replaced element B in compound BX; in the second, Y has replaced Z in MZ. Both changes are redox reactions.

EXAMPLE 9.12: Lead reacts with a solution of copper(II) nitrate. Write the equation.

First write the formulas of the reactants only on the left side of the arrow; leave the right side of the equation blank.

$Pb(s) + Cu(NO_3)_2(aq) \rightarrow$

In considering what to do next, regard the nitrate ion as a unit, just as the chloride ion was a unit in Example 9.11. Now, if this reaction will be described by a single replacement equation, the lead must either replace the copper and form a compound with the nitrate ion, or it must replace the nitrate ion and form a compound with copper. Which do you predict? Write the formulas for the expected products on the right side of the arrow. Remember, lead commonly forms Pb^{2+}, not Pb^{4+}.

$Pb(s) + Cu(NO_3)_2(aq) \rightarrow Cu(s) + Pb(NO_3)_2(aq)$

The Pb^{2+} ion forms an ionic compound with the negatively charged nitrate ion, leaving metallic copper behind. Positively charged lead ions could hardly form a compound with positively charged copper ions.

Now balance the equation.

$Pb(s) + Cu(NO_3)_2(aq) \rightarrow Cu(s) + Pb(NO_3)_2(aq)$

Balancing the equation in Example 9.12 was easy, with all coefficients being 1. There is something to be gained in the thought process, though, so let's examine it closely.

You quickly see that there is only one lead atom on the left side of the equation and one on the right, so lead is balanced. The same applies to copper. Now, rather than balancing nitrogen and oxygen separately, notice that these elements are present on both sides of the equation as nitrate ions. Whenever a polyatomic ion appears in exactly the same form on both sides of the equation, without any substance present that comes from the decomposition of the ion, the ion may be balanced as a unit. In this example you count two nitrate ions on the left and two on the right, so the entire ion is balanced at once. Other examples of this approach will arise shortly.

EXAMPLE 9.13: Write the equation for the reaction of chlorine gas with a solution of sodium bromide.

Write the unbalanced equation, with the formulas of the identified reactants on the left and the formulas of your predicted products on the right. This example is a little bit different, but the idea is the same.

$Cl_2(g) + NaBr(aq) \rightarrow Br_2(\ell) + NaCl(aq)$

The reacting element forms a negatively charged ion this time, and therefore replaces the negatively charged ion in the reacting compound.

Balance the equation.

$Cl_2(g) + 2\ NaBr(aq) \rightarrow Br_2(\ell) + 2\ NaCl(aq)$

9.7 PRECIPITATION AND ACID-BASE REACTIONS (DOUBLE REPLACEMENT EQUATIONS)

Precipitation Many times when two compounds are brought together in water solutions, an insoluble (solid) ionic product is formed. The insoluble compound is called a **precipitate**, and the process is a **precipitation reaction**. Equations for precipitation reactions look as if ions present in the reactants are "changing partners." They are commonly called **double replacement** or **double displacement** equations. The precipitation of silver chloride when solutions of silver nitrate and sodium chloride are combined is a good example:

$$AgNO_3(aq) + NaCl(aq) \rightarrow AgCl(s) + NaNO_3(aq)$$

The positively charged silver ion from one reactant combines with the negatively charged chloride ion from the other reactant to form the silver chloride precipitate, indicated by the (s) following the formula. At the same time, the negatively charged ion from the first reactant appears to combine with the positively charged ion from the second reactant. The sodium nitrate from the second combination remains in solution, as indicated by the (aq) following its formula.

If you recognize the "changing partners" character of a precipitation reaction, you can predict the identity of the products from the formulas of the reactants. The key is a knowledge of the charges on the ions. In the above case, Ag^+ combines with Cl^-, so the formula of the first product is AgCl. Also, Na^+ combines with NO_3^-, thus the second product is $NaNO_3$. Try it in the following example.

EXAMPLE 9.14: When aqueous sodium hydroxide is added to a solution of copper(II) nitrate, copper(II) hydroxide precipitates. Write the equation.

You are given the names of both reactants and one product. Write the unbalanced equation, including the formula of the unidentified product.

$$Cu(NO_3)_2(aq) + NaOH(aq) \rightarrow Cu(OH)_2(s) + NaNO_3(aq)$$

When Cu^{2+} from the first reactant combines with OH^- from the second reactant, that leaves Na^+ from the second reactant to join NO_3^- from the first. $NaNO_3$, sodium nitrate, is the second product.

Now balance the equation. Remember what was said about treating polyatomic ions as units when they are not broken up in the reaction. Neither the nitrate ion nor the hydroxide ion changes form in this reaction, so they may be balanced as units. It is much easier than trying to balance oxygen, appearing, as it does, in all four compounds.

$Cu(NO_3)_2(aq) + 2\ NaOH(aq) \rightarrow Cu(OH)_2(s) + 2\ NaNO_3(aq)$

Acid-Base Reactions: Neutralization As we learned in Chapter 7, an acid is a source of H^+ ions, and H is the first element in its formula. Compounds containing the OH^- ion are bases. When an acid of the general formula HX (where X^- is some negative ion) reacts with a base of the general formula MOH (where M^+ is some metal ion), the H^+ from the acid joins the OH^- from the base to form H_2O, water. The result is called a **neutralization reaction.**

Equations for neutralization reactions follow the double replacement pattern:

$$HX(aq) + MOH(aq) \rightarrow MX(aq) + H_2O(\ell)$$
$$\text{acid} + \text{base} \rightarrow \text{salt} + \text{water}$$

The (aq) after the other product, the salt MX, indicates that it is soluble. It is possible to form an insoluble salt as a product in a neutralization reaction, in which case the result is both an acid-base reaction *and* a precipitation reaction.

EXAMPLE 9.15: Write the equation for the neutralization reaction between solutions of sulfuric acid and potassium hydroxide.

Begin with the unbalanced equation.

$$H_2SO_4(aq) + \quad KOH(aq) \rightarrow \quad K_2SO_4(aq) + \quad H_2O(\ell)$$

The salt is formed from the K^+ ion from the base and the SO_4^{2-} ion from the acid. The charges on the ions dictate the formula of the salt: K_2SO_4.

Now balance the equation.

$$H_2SO_4(aq) + 2\ KOH(aq) \rightarrow K_2SO_4(aq) + 2\ H_2O(\ell)$$

Acid-Base Reactions: Formation of a Weak Acid Some acids are stronger than others. You may know that battery acid, which is sulfuric acid, H_2SO_4, is strong enough to cause severe burns if spilled on your skin. However, household vinegar, which contains acetic acid, $HC_2H_3O_2$, is swallowed with gusto on salads quite regularly!

There is a natural tendency for strong acids to react to form weak acids. Most reactions of this type fit the double replacement form. An example is:

$$HCl(aq) + NaC_2H_3O_2(aq) \rightarrow HC_2H_3O_2(aq) + NaCl(aq)$$

This reaction occurs because the strong acid, hydrochloric acid, forms the weak acid, acetic acid.

We will learn more about how to recognize weak and strong acids in later chapters. For now, we are interested in how to write the products of such reactions and then balance them.

EXAMPLE 9.16: Write the equation for the reaction between solutions of sulfuric acid and potassium formate, $KCHO_2$.

The unbalanced equation begins with $H_2SO_4(aq) + KCHO_2(aq) \rightarrow$

Can you write the formulas of the products in this double replacement equation? Perhaps the formula of the not-too-familiar formate ion, including its charge, may cause you to hesitate. What is the charge on the potassium ion? If that ion is separated from neutral potassium formate, $KCHO_2$, what is the formula—including charge—of what remains? Now go ahead; complete the right side of the above equation.

$$H_2SO_4(aq) + KCHO_2(aq) \rightarrow K_2SO_4(aq) + HCHO_2(aq)$$

If a potassium ion, K^+, is separated from electrically neutral $KCHO_2$, the negative charge on the formate ion that remains must be equal in magnitude to the +1 charge on the potassium ion. Therefore the charge on the formate ion is -1, and the ion formula is CHO_2^-. This ion combines one-to-one with hydrogen ion, H^+, to form a molecule of formic acid, $HCHO_2$. The other ions, K^+ and SO_4^{2-}, yield K_2SO_4 to complete the equation.

Now balance the equation.

$$H_2SO_4(aq) + 2\ KCHO_2(aq) \rightarrow K_2SO_4(aq) + 2\ HCHO_2(aq)$$

9.8 OTHER REACTIONS WITH IDENTIFIABLE REACTANTS AND PRODUCTS

There are many kinds of chemical reactions that do not fit into one of the classifications considered here. Without instruction beyond the scope of this chapter, you cannot be expected to predict the identities and formulas of reaction products for these chemical changes. If given the formulas of all species in an unclassified reaction, however, writing the equation becomes nothing more than an exercise in balancing atoms.

EQUATION BALANCING AND WRITING EXERCISES

Equation Balancing: Balance the following equations using the simplest whole number coefficients.

1) $H_2S(s) + SO_2(g) \rightarrow S(s) + H_2O(g)$

2) $H_2S(g) + O_2(g) \rightarrow S(s) + H_2O(g)$

3) $H_2S(g) + HNO_3(aq) \rightarrow NO(g) + H_2O(\ell) + S(s)$

4) $SO_2(g) + O_2(g) + CaO(s) \rightarrow CaSO_4(s)$

Equation Writing: Write the balanced chemical equation for each reaction described. Formulas are given only for compounds that are not likely to be familiar to you or are not included in Chapter 7. If you have difficulty with any formula, check Chapter 7 for the method by which it may be written. Tables 7.2 and 7.3 will be helpful.

Classify each equation as combination, decomposition, combustion, single replacement (redox), or double replacement. For double replacement, specify if the reaction is a precipitation, neutralization, or formation of a weak acid.

Physical states, (g), (ℓ), (s), or (aq) should be included in all equations. Table 7.2 should be helpful for the elements. Compounds may be assumed to be in aqueous solution unless otherwise stated.

5) Solid tetraphosphorus hexoxide is formed by direct combination of its elements.

6) Aluminum replaces hydrogen when the metal is placed in hydrochloric acid.

7) Carbonic acid is unstable, decomposing to carbon dioxide and liquid water.

8) Barium hydroxide solution is neutralized with sulfuric acid. A precipitate forms.

9) Barium carbonate precipitates when solutions of barium chloride and ammonium carbonate are combined.

10) Ammonium sulfate is the only product of the reaction between solutions of ammonia and sulfuric acid.

11) Propane, C_3H_8, burns in air.

12) Sodium fluoride reacts with nitric acid.

13) Potassium chlorate decomposes into potassium chloride and oxygen.

14) Zinc replaces lead when it reacts with lead(II) nitrate solution.

15) Hydrobromic acid reacts with potassium hydroxide.

16) Elemental chlorine reacts with a solution of potassium iodide.

17) Precipitation occurs when silver nitrate solution is treated with dihydrogen sulfide gas (commonly known simply as hydrogen sulfide gas).

18) Solid phosphorus triiodide is formed from its elements.

19) Sugar, $C_{12}H_{22}O_{11}$, is burned in air.

20) Copper(II) chloride and water are the products of a reaction between copper(II) oxide and hydrochloric acid.

21) Solid barium peroxide, BaO_2, decomposes into oxygen and solid barium oxide.

22) Butanal, C_3H_7CHO, is oxidized completely.

23) Ammonia is prepared from its elements.

24) Carbon dioxide and liquid water are two of the three products of a reaction between solid magnesium carbonate and hydrochloric acid.

25) Sodium formate, $NaCHO_2$, reacts with sulfuric acid.

26) Zinc reacts with hydrochloric acid.

27) Pentanol, $C_5H_{11}OH$, is burned in air.

28) Precipitation occurs in the reaction of sodium iodate with silver nitrate.

29) Sulfuric acid neutralizes nickel hydroxide.

30) Potassium nitrite reacts with sulfuric acid.

31) Mercury(II) bromide reacts with ammonium sulfide.

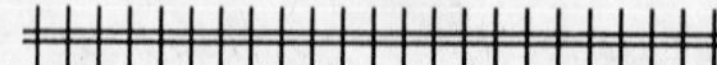

Equation Balancing: Balance the following equations using the simplest whole number coefficients.

32) $SO_2(g) + H_2O(\ell) + CaS(aq) \rightarrow S(s) + Ca(OH)_2(aq)$

33) $SO_2(g) + O_2(g) + CaO(s) \rightarrow CaSO_4(s)$

34) $KMnO_4(aq) + KOH(aq) \rightarrow K_2MnO_4(aq) + O_2(g) + H_2O(\ell)$

35) $Fe_3O_4(s) + Al(s) \rightarrow Fe(s) + Al_2O_3(s)$

Equation Writing: Write the balanced chemical equation for each reaction described. Formulas are given only for compounds that are not likely to be familiar to you or are not included in Chapter 7. If you have difficulty with any formula, check Chapter 7 for the method by which it may be written. Tables 7.2 and 7.3 will be helpful.

36) Solid aluminum chloride may be formed by direct combination of its elements.

37) Solid ammonium nitrite may be decomposed into nitrogen and liquid water.

38) Magnesium hydroxide precipitates when solutions of magnesium sulfate and sodium hydroxide react.

39) CH_3CHO is completely oxidized.

40) Sulfuric acid reacts with potassium acetate, $KC_2H_3O_2$.

41) Iron(II) chloride and sodium phosphate solutions react to form a solid. (Which of the two products do you suspect remains in solution?)

42) Sugar, $C_{12}H_{22}O_{11}$, decomposes into carbon and liquid water when heated.

43) Zinc hydroxide reacts with sulfuric acid.

44) Solid lithium oxide reacts with water to form lithium hydroxide solution.

45) Aluminum releases hydrogen from phosphoric acid. The other product is a solid.

46) Sodium oxalate, $Na_2C_2O_4$, reacts with hydrochloric acid.

47) When solid calcium sulfate dihydrate, $CaSO_4 \cdot 2\ H_2O$, is heated, water vapor and solid calcium sulfate are the products.

48) Cobalt reacts with a solution of copper(II) nitrate. The product is a cobalt(II) compound

49) Magnesium nitride is formed from its elements.

50) Sulfuric acid reacts with sodium hydrogen carbonate, releasing carbon dioxide and water as two of three products.

51) Liquid $C_7H_{15}COOH$ is completely oxidized.

52) A double precipitation occurs when iron(III) sulfate reacts with barium hydroxide.

ADDITIONAL PROBLEMS

53) An unknown metal reacts with water to form hydrogen gas and a slightly soluble hydroxide. Identify the metal and write the equation for the reaction, showing the hydroxide as a precipitate.

54) When heated gently, solid ammonium oxalate decomposes to ammonia gas, water vapor, and a mixture of carbon monoxide and carbon dioxide gases. When heated strongly, the products are the same except that solid elemental carbon forms instead of carbon monoxide gas. Write equations for both reactions.

55) Batteries produce electrical energy from chemical reactions. The lead-acid battery reacts lead metal and lead(IV) oxide solids in contact with aqueous sulfuric acid to form water and solid lead(II) sulfate. The alkaline dry cell reacts solid manganese(IV) oxide and zinc metal in contact with water to form solid manganese(III) hydroxide and zinc hydroxide. Write equations for both reactions.

56) A common metallurgical process is to "roast" sulfide ores by heating them in air. Balance these roasting equations.

a) $Cu_2S(s) + O_2(g) \rightarrow Cu(s) + SO_2(g)$
b) $PbS(s) + O_2(g) \rightarrow PbSO_4(s)$
c) $MoS_2(s) + O_2(g) \rightarrow MoO_3(s) + SO_2(g)$

CHAPTER TEST

For each reaction below, write the balanced chemical equation. Classify each equation. You may refer to an approved periodic table.

1) Solid sodium chloride is formed from its elements.

2) Solid iron(III) sulfide is decomposed into its elements.

3) Solid aluminum hydroxide is decomposed into solid aluminum oxide and liquid water.

4) Ethane, C_2H_6, is burned completely in air.

5) Liquid propanoic acid, C_2H_5COOH, is oxidized completely.

6) Aluminum displaces nickel ion from a solution of nickel nitrate.

7) Lead(II) chloride precipitates when solutions of lead(II) nitrate and sodium chloride are combined.

8) Potassium cyanide, KCN, reacts with hydrochloric acid.

9) Sodium carbonate reacts with barium chloride.

10) Nitric acid is neutralized with potassium hydroxide.

10. STOICHIOMETRY

PREREQUISITES

1. Dimensional analysis (Section 2.4) is the standard method for solving stoichiometry problems.

2. The section on percentage yield includes the kind of percentage calculations discussed in Section 2.6.

3. Most of the reactions in this chapter are described in words only, using chemical names rather than formulas. It is therefore necessary to change the names to formulas, as you learned in Chapter 7.

4. Conversion factors from moles of one substance in a chemical equation to moles of another are based on the coefficients of balanced chemical equations (Chapter 9).

5. Gram → mole and mole → gram conversions, introduced in Section 8.5, are essential skills in solving stoichiometry problems.

CHEMICAL SKILLS

1. The stoichiometry 3-step. Given a chemical reaction for which the equation may be written and given the quantity of any substance in the reaction, calculate the quantity of any other substance in the reaction. Either quantity may be in moles or grams of any substance.

2. Percentage yield. Given two of the group (a) theoretical yield, (b) actual yield, or (c) percentage yield, or information from which they may be obtained, calculate the third.

3. Percentage yield. Given the equation for a reaction and the percentage yield, calculate the amount of reactant required to produce a specified actual yield.

4. Limiting reactant. Given the equation for a reaction and quantities of any two reactants, identify the limiting reactant and calculate the theoretical yield of any product. Also calculate the quantity of any non-limiting (excess) reactants that are left at the end of the reaction.

5. Analysis of mixtures. Given two of the group (a) amount of a mixture analyzed, (b) percentage of a chemical substance in that mixture, and (c) the amount of a product obtained by a reaction of that chemical substance, calculate the third.

10.1 THE QUANTITATIVE MEANING OF A CHEMICAL EQUATION

Stoichiometry is the name given to the relationships between quantities of reactants and products in a chemical change. These relationships are based on the number of moles of different species expressed by the coefficients in the *balanced* equation for the reaction. It is impossible to overstate the importance of writing a balanced equation *before* starting the calculations. *Everything depends on the balanced equation.*

The equation for the reaction between carbon and oxygen to produce carbon monoxide is

$$2\ C(s) \ + \ O_2(g) \ \rightarrow \ 2\ CO(g) \qquad (10.1)$$

The equation says that two atoms of carbon react with one molecule of oxygen to produce two molecules of carbon monoxide. It also says two moles of carbon react with one mole of oxygen to produce two moles of carbon monoxide.

$2\ C(s)$	+	$O_2(g)$	→	$2\ CO(g)$
2 molecules		1 molecule		2 molecules
2 moles		1 mole		2 moles

Now consider how many moles of carbon are required to completely react with three moles of oxygen. This is three times the amount of oxygen shown in the equation. It follows that three times as much carbon is needed: $3 \times 2 = 6$ moles of carbon.

This same result can be reached by dimensional analysis. A chemical equation provides conversion factors between any pair of substances in the reaction. For Equation 10.1, these conversion factors are

between C and O_2: $\dfrac{2\ \text{moles C}}{1\ \text{mole}\ O_2}$ or $\dfrac{1\ \text{mole}\ O_2}{2\ \text{moles C}}$

between C and CO: $\dfrac{2\ \text{moles C}}{2\ \text{moles CO}}$ or $\dfrac{2\ \text{moles CO}}{2\ \text{moles C}}$

between O_2 and CO: $\dfrac{1\ \text{mole}\ O_2}{2\ \text{moles CO}}$ or $\dfrac{2\ \text{moles CO}}{1\ \text{mole}\ O_2}$

Think only in terms of *moles* of substances when you set up conversion factors to solve stoichiometry problems. The coefficients—the numbers in front of the chemical formulas in the equation—represent the number of moles of each species. They provide conversion relationships between any two species in the reaction. The species can both be reactants, both products, or one a reactant and one a product.

Using only the relationship between carbon and oxygen, the unit path for finding the number of moles of carbon that react with 3 moles of oxygen is mol O_2 → mol C. We must choose the proper conversion factor to accomplish this unit change. The correct calculation setup is

$$3\ \text{mol}\ O_2 \times \frac{2\ \text{mol C}}{1\ \text{mol}\ O_2} = 6\ \text{mol C}$$

The units of the starting quantity (mol O_2) cancel, and the units of the wanted quantity (mol C) appear in the answer.

This step, from moles of the *given* substance to moles of the *wanted* substance, is present in every stoichiometry problem.

EXAMPLE 10.1: How many moles of oxygen are required to produce 4.62 moles of carbon monoxide that is formed by direct combination of its elements?

First you need the equation for the reaction. Write it here.

$2\ C(s)\ +\ O_2(g)\ \rightarrow\ 2\ CO(g)$

Identify the quantity GIVEN, the quantity you are asked to FIND, the conversion FACTOR, and the unit PATH. Solve the problem here.

GIVEN: FIND:

FACTOR: PATH:

GIVEN: 4.62 mol CO FIND: mol O_2
FACTOR: 2 mol CO/1 mol O_2 PATH: mol CO → mol O_2

$$4.62\ \text{mol CO} \times \frac{1\ \text{mol } O_2}{2\ \text{mol CO}} = 2.31\ \text{mol } O_2$$

Did you get 9.24 mol O_2 instead of 2.31 mol O_2? Be sure you attach the formulas to the mole ratios and that your units cancel correctly.

Note that the conversion factor between two substances depends on the reaction. For a different reaction between the same substances, there may be a different relationship.

Carbon and oxygen also produce carbon dioxide when they react. The equation and the resulting conversion factors between C and O_2 are

$$C(s) + O_2(g) \rightarrow CO_2(g) \qquad \frac{1\ \text{mol C}}{1\ \text{mol } O_2} \quad \text{or} \quad \frac{1\ \text{mol } O_2}{1\ \text{mol C}}$$

These ratios are not the same as those based on Equation 10.1.

Mole ratios for a reaction are always given by the coefficients from the balanced equation for that particular reaction. This is why a balanced equation is so important when solving stoichiometry problems.

EXAMPLE 10.2: How many moles of gaseous ammonia will be produced by the complete reaction of 1.92 moles of hydrogen gas with excess nitrogen gas?

Begin by writing the equation for the reaction. Then identify the starting quantity, the wanted quantity, the conversion factor, and the unit path.

Equation:

GIVEN: FIND:

FACTOR: PATH:

Equation: $N_2(g) + 3\ H_2(g) \rightarrow 2\ NH_3(g)$

GIVEN: 1.92 mol H_2 FIND: mol NH_3 FACTOR: $\frac{2 \text{ mol } NH_3}{3 \text{ mol } H_2}$ PATH: mol H_2 → mol NH_3

Now set up and solve the problem.

$$1.92 \text{ mol } H_2 \times \frac{2 \text{ mol } NH_3}{3 \text{ mol } H_2} = 1.28 \text{ mol } NH_3$$

10.2 THE "STOICHIOMETRY 3-STEP"

Once the equation for a reaction is known—and therefore the mole ratios between species in the reaction—there are three steps by which you can solve any stoichiometry problem. These three steps will be developed by three examples, and then combined into a single setup for solving the whole problem. The reaction used will be the complete burning of liquid heptane, $C_7H_{16}(\ell)$, which is described by the equation

$$C_7H_{16}(\ell) + 11\ O_2(g) \rightarrow 7\ CO_2(g) + 8\ H_2O(\ell) \qquad (10.2)$$

The overall problem is: How many grams of oxygen are required to react with 501 grams of heptane by Equation 10.2? Now for the first step:

EXAMPLE 10.3: How many moles of heptane are in 501 grams of heptane?

This is straightforward gram → mole conversion, using molar mass. The molar mass of heptane is 100.20 g/mol. Set up the problem and solve.

GIVEN: 501 g C_7H_{16} FIND: mol C_7H_{16} FACTOR: MM C_7H_{16} PATH: g C_7H_{16} → mol C_7H_{16}

$$501 \text{ g } C_7H_{16} \times \frac{1 \text{ mol } C_7H_{16}}{100.20 \text{ g } C_7H_{16}} = 5.00 \text{ mol } C_7H_{16}$$

EXAMPLE 10.4: How many moles of oxygen are required to react with 5.00 moles of heptane in Equation 10.2?

Here is the conversion from moles of one species to moles of another that is present in every stoichiometry problem: mol C_7H_{16} → mol O_2. The procedure is the same as in Examples 10.1 and 10.2. Solve the problem.

GIVEN: 5.00 mol C_7H_{16} FIND: mol O_2
FACTOR: 11 mol O_2/1 mol C_7H_{16} PATH: mol C_7H_{16} → mol O_2

$$5.00 \text{ mol } C_7H_{16} \times \frac{11 \text{ mol } O_2}{1 \text{ mol } C_7H_{16}} = 55.0 \text{ mol } O_2$$

The conversion factor between heptane and oxygen is: 11 mol O_2/1 mol C_7H_{16} according to Equation 10.2. The numbers correspond with the coefficients in the equation.

EXAMPLE 10.5: What is the mass of 55.0 moles of oxygen gas?

A mole → gram conversion is needed this time. Solve the problem.

GIVEN: 55.0 mol O_2 FIND: g O_2 FACTOR: MM of O_2 PATH: mol O_2 → g O_2

$$55.0 \text{ mol } O_2 \times \frac{32.00 \text{ g } O_2}{1 \text{ mol } O_2} = 1.76 \times 10^3 \text{ g } O_2$$

Did you use 16.00 g/mol? Remember oxygen gas is O_2, not O.

Now let us review what has been done. The overall objective was to find the number of grams of oxygen needed to react with 501 grams of heptane.

In Example 10.3, the unit path was: grams heptane → moles heptane;

In Example 10.4, the unit path was: moles heptane → moles oxygen;

In Example 10.5, the unit path was: moles oxygen → grams oxygen.

If these unit paths are combined—taken in sequence—the result goes all the way from the starting quantity to the wanted quantity:

GIVEN: 501 g C_7H_{16} FIND: g O_2
FACTOR: MM of C_7H_{16}, mole ratio from equation, MM of O_2
PATH: g C_7H_{16} → mol C_7H_{16} → mol O_2 → g O_2

Using this path, the entire problem can be written in a single setup:

$$501 \text{ g } C_7H_{16} \times \frac{1 \text{ mol } C_7H_{16}}{100.20 \text{ g } C_7H_{16}} \times \frac{11 \text{ mol } O_2}{1 \text{ mol } C_7H_{16}} \times \frac{32.00 \text{ g } O_2}{1 \text{ mol } O_2} = 1.76 \times 10^3 \text{ g } O_2$$

If the purpose of the problem is to find the grams of oxygen required, the intermediate answers to Examples 10.3 and 10.4 are not necessary.

Given a reaction equation, the procedure for solving a stoichiometry problem may be summarized in three steps:

1. Convert given quantity to moles: g → mol using molar mass.
2. Convert moles of given substance to moles of wanted substance: mol → mol using mole ratio from balanced equation.
3. Convert moles of wanted substance to quantity units required: mol → g using molar mass.

The unit path for the general solution of a stoichiometry problem is

given quantity → moles given → moles wanted → wanted quantity

We will call the above procedure the "stoichiometry 3-step." In stating the stoichiometry 3-step, the word *quantity* is used rather than grams because mass is only one of several quantity units that appear in stoichiometry problems. Gas quantity, for example, is usually measured in volume.

EXAMPLE 10.6: Calculate the mass of hydrogen required to produce 39.8 grams of ammonia in the reaction $3\ H_2(g) + N_2(g) \rightarrow 2\ NH_3(g)$.

To help you see clearly where you are starting and where you are going, set up the GIVEN-FIND-FACTOR-PATH analysis of the problem.

GIVEN: FIND:

FACTOR:

PATH:

GIVEN: 39.8 g NH_3 FIND: g H_2

FACTOR: MM NH_3, 3 mol H_2/2 mol NH_3, MM of H_2

PATH: g NH_3 → mol NH_3 → mol H_2 → g H_2

The first step in the stoichiometry 3-step is to change the 39.8 grams of ammonia (MM = 17.03 g/mol) to moles. Set up this first step, but do not calculate the intermediate answer.

$$39.8 \text{ g } NH_3 \times \frac{1 \text{ mol } NH_3}{17.03 \text{ g } NH_3} \times \underline{\hspace{3cm}}$$

Now that the first step in the stoichiometry 3-step is completed, go to Step 2. Extend the setup for the conversion from moles of ammonia to moles of hydrogen. Again, do not calculate the intermediate answer.

$$39.8 \text{ g } NH_3 \times \frac{1 \text{ mol } NH_3}{17.03 \text{ g } NH_3} \times \frac{3 \text{ mol } H_2}{2 \text{ mol } NH_3} \times \underline{\hspace{4cm}} =$$

The conversion factor from moles of ammonia to moles of hydrogen matches the coefficients of these substances in the equation. This is the inverse of the ratio in Example 10.2.

The third part of the stoichiometry 3-step is to change moles of hydrogen to grams at 2.016 grams per mole. Complete the setup, and this time calculate the answer.

$$39.8 \text{ g } NH_3 \times \frac{1 \text{ mol } NH_3}{17.03 \text{ g } NH_3} \times \frac{3 \text{ mol } H_2}{2 \text{ mol } NH_3} \times \frac{2.016 \text{ g } H_2}{1 \text{ mol } H_2} = 7.07 \text{ g } H_2$$

The conversion factors between grams and moles of given and wanted species and from moles of given species to moles of wanted species are routine for all stoichiometry problems. We will therefore omit the FACTOR statement from the examples ahead unless there is some special reason for calling your attention to them.

EXAMPLE 10.7: How many grams of calcium phosphate will precipitate if excess calcium nitrate solution is added to a solution containing 3.21 grams of sodium phosphate?

As before, begin with the equation for the reaction. Follow with the starting quantity, the wanted quantity, and the unit path.

Equation:

GIVEN: FIND: PATH:

$3 \; Ca(NO_3)_2(aq) + 2 \; Na_3PO_4(aq) \rightarrow Ca_3(PO_4)_2(s) + 6 \; NaNO_3(aq)$

GIVEN: 3.21 g Na_3PO_4 FIND: g $Ca_3(PO_4)_2$

PATH: g Na_3PO_4 → mol Na_3PO_4 → mol $Ca_3(PO_4)_2$ → g $Ca_3(PO_4)_2$

Follow the same 3-step procedure used in the above example to write the complete setup, and calculate the answer.

$$3.21 \text{ g } Na_3PO_4 \times \frac{1 \text{ mol } Na_3PO_4}{163.94 \text{ g } Na_3PO_4} \times \frac{1 \text{ mol } Ca_3(PO_4)_2}{2 \text{ mol } Na_3PO_4} \times \frac{310.18 \text{ g } Ca_3(PO_4)_2}{1 \text{ mol } Ca_3(PO_4)_2} = 3.04 \text{ g } Ca_3(PO_4)_2$$

The given and wanted masses are expressed in kilograms in the next example. Technically you must convert kilograms to grams in order to apply the mole-mole conversion in the middle of the stoichiometry 3-step. The result in grams must be converted to kilograms at the end. The unit path thus becomes

kg given → g given → mol given → mol wanted → g wanted → kg wanted

The first and last steps are inverse operations. You multiply by 1000 in changing kilograms to grams, and divide by 1000 in changing grams back to kilograms. In terms of numbers, the first and last conversions may be omitted—but not in terms of units.

To justify this omission in terms of units, we introduce the **kilomole**. Just as a kilogram is 1000 grams, a kilomole is 1000 moles; just as the symbol for a kilogram is kg, the symbol for a kilomole is kmol. Extending this thinking into the middle conversion in the stoichiometry 3-step, just as molar mass is grams per mole, "kilomolar" mass is kilograms per kilomole.

Numerically, molar mass and kilomolar mass are identical. The molar mass of water, for example, is 18.02 g/mol. Then

$$\frac{18.02 \text{ g}}{1 \text{ mol}} = \frac{1000}{1000} \times \frac{18.02 \text{ g}}{1 \text{ mol}} = \frac{(1000)(18.02) \text{ g}}{1000 \text{ mol}} = \frac{18.02 \text{ kg}}{1 \text{ kmol}}$$

What this means is that a stoichiometry problem may be solved in kilograms and kilomoles in exactly the same way as it is solved in grams and moles. In such a case, the unit path becomes

kg given → kmol given → kmol wanted → kg wanted

In the laboratory and in industry the same reasoning leads to working stoichiometry problems in milligrams (mg) and millimoles (mmol), pounds and pound-moles, or even tons and ton-moles!

EXAMPLE 10.8: Calculate the number of kilograms of coke (assume it to be pure carbon, C) needed to completely remove the iron from ore that contains 55 kilograms of Fe_2O_3. The equation is:

$$2\ Fe_2O_3(s) + 3\ C(s) \rightarrow 3\ CO_2(g) + 4\ Fe(s)$$

Aside from the kilounits, the procedure is identical to that in the previous examples. Set up the problem completely and solve.

GIVEN: 55 kg Fe_2O_3 FIND: kg C

PATH: kg Fe_2O_3 → kmol Fe_2O_3 → kmol C → kg C

$$55 \text{ kg } Fe_2O_3 \times \frac{1 \text{ kmol } Fe_2O_3}{159.70 \text{ kg } Fe_2O_3} \times \frac{3 \text{ kmol C}}{2 \text{ kmol } Fe_2O_3} \times \frac{12.01 \text{ kg C}}{1 \text{ kmol C}} = 6.2 \text{ kg C}$$

10.3 PERCENTAGE YIELD

A stoichiometry problem gives the impression that a reaction will yield the full amount of a product calculated. This is called the **theoretical yield**. It makes no allowance for side reactions, impurities, unavoidable loss of material, or any number of other things that can cause the **actual yield** to be less than the theoretical yield. This leads to the **percentage yield**, in which the actual yield is stated as a percentage of the theoretical yield. Mathematically,

$$\text{percentage yield} = \frac{\text{actual product yield}}{\text{theoretical product yield}} \times 100 \qquad (10.3)$$

Notice that percentage yield relates to a *product* of the reaction, not a reactant. In the problem setups that follow, we will use "act" as an abbreviation for actual product yield and "theo" to represent theoretical product yield.

Equation 10.3 is used to calculate percentage yield whenever you are given or can find both the actual and theoretical yields.

EXAMPLE 10.9: In the reaction described in Example 10.7,

$$3\ Ca(NO_3)_2(aq) + 2\ Na_3PO_4(aq) \rightarrow Ca_3(PO_4)_2(s) + 6\ NaNO_3(aq)$$

the theoretical yield was 3.04 g $Ca_3(PO_4)_2$ from 3.21 g Na_3PO_4. If only 2.84 grams are actually recovered, what is the percentage yield?

This is a straightforward substitution of given values into Equation 10.3.

GIVEN: 2.84 g $Ca_3(PO_4)_2$ act; 3.04 g $Ca_3(PO_4)_3$ theo FIND: % yield

EQN: $$\% \text{ yield} = \frac{\text{act yield}}{\text{theo yield}} \times 100$$

$$= \frac{2.84 \text{ g } Ca_3(PO_4)_2 \text{ actual}}{3.04 \text{ g } Ca_3(PO_4)_2 \text{ theoretical}} \times 100 = 93.4\% \text{ yield}$$

The "parts-per-hundred" nature of percentage was discussed in Section 2.6. By this approach,

$$\text{X percentage yield} = \frac{\text{X grams actual product yield}}{\text{100 grams theoretical product yield}} \qquad (10.4)$$

The ratio in Equation 10.4 is a conversion factor with which we can change in either direction between grams of actual product and grams of theoretical product. Whenever percentage yield is *given* in a problem, it is more convenient to solve the problem with the conversion factor based on Equation 10.4 than to solve algebraically with Equation 10.3.

Suppose the reaction in Examples 10.7 and 10.9 operates consistently at 93.4% yield, based on the starting quantity of sodium phosphate. This fact, stated as the conversion factor given in Equation 10.4, can be used to predict the actual yield for any given quantity of sodium phosphate. Assume the reaction begins with 5.08 g Na_3PO_4 and you want to find the grams of calcium phosphate you can expect to recover. Begin by calculating the theoretical yield exactly as in Example 10.7. Then convert to actual yield.

$$\text{g } Na_3PO_4 \rightarrow \text{mol } Na_3PO_4 \rightarrow \text{mol } Ca_3(PO_4)_2 \rightarrow \text{g } Ca_3(PO_4)_2 \text{ theo} \rightarrow \text{g } Ca_3(PO_4)_2 \text{ act}$$

$$5.08 \text{ g } Na_3PO_4 \times \frac{1 \text{ mol } Na_3PO_4}{163.94 \text{ g } Na_3PO_4} \times \frac{1 \text{ mol } Ca_3(PO_4)_2}{2 \text{ mol } Na_3PO_4} \times \frac{310.18 \text{ g } Ca_3(PO_4)_2 \text{ theo}}{1 \text{ mol } Ca_3(PO_4)_2}$$

$$\times \frac{93.4 \text{ g } Ca_3(PO_4)_2 \text{ act}}{100 \text{ g } Ca_3(PO_4)_2 \text{ theo}} = 4.49 \text{ g } Ca_3(PO_4)_2 \text{ actual}$$

The first line above is the setup for the calculation of theoretical yield, as in Example 10.7. The next line—the last conversion factor—changes theoretical yield to actual yield.

EXAMPLE 10.10: Solid sodium nitrate can be decomposed to solid sodium nitrite and oxygen gas. If the percentage yield for the reaction is 80.4%, calculate the grams of sodium nitrite that will result from the decomposition of 955 grams of sodium nitrate.

As usual, you must begin with the equation.

$$2\ NaNO_3(s) \rightarrow 2\ NaNO_2(s) + O_2(g)$$

Write the starting and wanted quantities, the unit path for the entire problem, and set up the stoichiometry 3-step to the theoretical yield, but do not calculate that intermediate answer.

GIVEN: 955 g $NaNO_3$ FIND: g $NaNO_2$ act

PATH: g $NaNO_3$ → mol $NaNO_3$ → mol $NaNO_2$ → g $NaNO_2$ theo → g $NaNO_2$ act

$$955 \text{ g } NaNO_3 \times \frac{1 \text{ mol } NaNO_3}{85.00 \text{ g } NaNO_3} \times \frac{2 \text{ mol } NaNO_2}{2 \text{ mol } NaNO_3} \times \frac{69.00 \text{ g } NaNO_2 \text{ theo}}{1 \text{ mol } NaNO_2} \times \underline{\qquad\qquad} =$$

Now extend the setup by the conversion factor from Equation 10.4 that changes theoretical yield to actual yield. Calculate the answer.

$$955 \text{ g } NaNO_3 \times \frac{1 \text{ mol } NaNO_3}{85.00 \text{ g } NaNO_3} \times \frac{2 \text{ mol } NaNO_2}{2 \text{ mol } NaNO_3} \times \frac{69.00 \text{ g } NaNO_2 \text{ theo}}{1 \text{ mol } NaNO_2}$$

$$\times \frac{80.4 \text{ g } NaNO_2 \text{ act}}{100 \text{ g } NaNO_2 \text{ theo}} = 623 \text{ g } NaNO_2 \text{ actual}$$

A manufacturer of chemicals must know the percentage yield of each process in order to determine how much raw material is required for a certain amount of product. If, for example, 50.0 kg $NaNO_2$ are to be produced—the actual yield is to be 50.0 kg—by the reaction in Example 10.10, how many kilograms of $NaNO_3$ are needed with a percentage yield of 80.4%? The starting quantity this time is a product, and that is where the percentage yield is used. You begin by converting the intended actual yield to the LARGER theoretical yield on which production must be planned.

$$50.0 \text{ kg } NaNO_2 \text{ act} \times \frac{100 \text{ kg } NaNO_2 \text{ theo}}{80.4 \text{ kg } NaNO_2 \text{ act}} \times \underline{\qquad\qquad}$$

From here the stoichiometry 3-step may be applied. The complete setup is

GIVEN: 50.0 kg $NaNO_2$ act FIND: kg $NaNO_3$ needed
PATH: kg $NaNO_2$ act → kg $NaNO_2$ theo → kmol $NaNO_2$ → kmol $NaNO_3$ → kg $NaNO_3$

$$50.0 \text{ kg } NaNO_2 \text{ act} \times \frac{100 \text{ kg } NaNO_2 \text{ theo}}{80.4 \text{ kg } NaNO_2 \text{ act}} \times \frac{1 \text{ kmol } NaNO_2}{69.00 \text{ kg } NaNO_2}$$

$$\times \frac{2 \text{ kmol } NaNO_3}{2 \text{ kmol } NaNO_2} \times \frac{85.00 \text{ kg } NaNO_3}{1 \text{ kmol } NaNO_3} = 76.6 \text{ kg } NaNO_3 \text{ needed}$$

Notice that the stoichiometry conversion is applied to the theoretical product yield, not to the actual yield.

EXAMPLE 10.11: A method for preparing sodium sulfate is summarized in the equation below. The percentage yield is 77.2%, based on the starting mass of NaOH. Calculate the grams of sodium hydroxide needed to produce 15.0 grams of sodium sulfate.

$$2\ S(s) + 3\ O_2(g) + 4\ NaOH(aq) \rightarrow 2\ Na_2SO_4(aq) + 2\ H_2O(\ell).$$

First write the setup that changes actual mass of sodium sulfate to theoretical mass. Then complete the problem with the stoichiometry 3-step.

GIVEN: 15.0 g Na_2SO_4 FIND: g NaOH needed
PATH: g Na_2SO_4 act → g Na_2SO_4 theo → mol Na_2SO_4 → mol NaOH → g NaOH

$$15.0 \text{ g } Na_2SO_4 \text{ act} \times \frac{100 \text{ g } Na_2SO_4 \text{ theo}}{77.2 \text{ g } Na_2SO_4 \text{ act}} \times \frac{1 \text{ mol } Na_2SO_4}{142.04 \text{ g } Na_2SO_4}$$

$$\times \frac{4 \text{ mol NaOH}}{2 \text{ mol } Na_2SO_4} \times \frac{40.00 \text{ g NaOH}}{1 \text{ mol NaOH}} = 10.9 \text{ g NaOH needed}$$

10.4 LIMITING REACTANT PROBLEMS

Suppose you operate a small bicycle shop and build customized bicycles. At one particular time you have in your shop eight wheels, six handlebars, and a large supply of everything else you need to build bicycles. How many bicycles can you build from that inventory? Eight? Six? Four? Three? The answer, of course, is four. That's when you run out of wheels. Even though you have, at the beginning, more wheels than handlebars, the limiting item is wheels because you use them two at a time.

After you have built four bicycles, how many handlebars will be left? You have six to begin with, and you use one on each of the four bicycles, a total of four. The number left will be 6 - 4 = 2.

Chemical reactions between two or more reactants work in exactly the same way. The reaction takes place until one of the reactants, called the **limiting reactant**, is used up. Some of the other reactant—the **excess reactant**—will be left over.

Consider the reaction between zinc and sulfur, which takes place according to the equation: Zn(s) + S(s) → ZnS(s). If you begin with 8 moles of zinc and 6 moles of sulfur, you will run out of sulfur after 6 moles of zinc sulfide have formed. Because the reaction consumes zinc and sulfur on a one-to-one basis, only 6 moles of zinc will be consumed. Thus 8 - 6 = 2 moles of zinc will remain. Sulfur, the reactant completely used up, is the limiting reactant: it limits the amount of zinc sulfide which can form. Zinc, the reactant left over, is the excess reactant. Using the equation formulas as column heads in a table, this analysis is summarized as follows:

	Zn(s)	+	S(s)	→	ZnS(s)
Moles at start	8		6		0
Moles used (-), produced (+)	-6		-6		+6
Moles at end	2		0		6

The idea behind limiting reactants is simple enough, but calculations become a bit more complicated when grams are introduced. Preparing a table beneath the reaction equation, as above, sometimes helps. Consider this problem:

> How many grams of zinc sulfide can be produced by 8.50 grams of zinc and 6.84 grams of sulfur according to the equation Zn(s) + S(s) → ZnS(s)? Also, how many grams of which element will remain unreacted?

We begin by rewriting the equation as headings for a table of information given and wanted. This time the given quantities are in grams. The molar mass of each species is included in the table.

	Zn(s)	+	S(s)	→	ZnS(s)
Grams at start	8.50		6.84		0
Molar mass	65.38		32.06		97.44
Moles at start	______		______		______

The moles of both reactants are needed to determine the limiting reactant. The grams (first line) divided by molar mass (second line) gives moles:

$$8.50 \text{ g Zn} \times \frac{1 \text{ mol Zn}}{65.38 \text{ g Zn}} = 0.130 \text{ mol Zn}; \qquad 6.84 \text{ g S} \times \frac{1 \text{ mol S}}{32.06 \text{ g S}} = 0.213 \text{ mol S}$$

	$Zn(s)$ +	$S(s)$ →	$ZnS(s)$
Grams at start	8.50	6.84	0
Molar mass	65.38	32.06	97.44
Moles at start	0.130	0.213	0
Moles used, produced	___	___	___

The reaction proceeds until the limiting reactant is used up. Which will it be, zinc or sulfur? In this case the elements combine in a 1:1 mole ratio, so the limiting reactant is simply the smaller number of moles, 0.130 moles of zinc. The same number of moles of sulfur will be used. How many moles of product will there be? The equation shows 1 mole of ZnS for each mole of Zn, so there will be 0.130 mole of ZnS (assuming 100% yield).

	$Zn(s)$ +	$S(s)$ →	$ZnS(s)$
Grams at start	8.50	6.84	0
Molar mass	65.38	32.06	97.44
Moles at start	0.130	0.213	0
Moles used, produced	-0.130	-0.130	+0.130
Moles at end	___	___	___

A minus sign has been included to show where a substance was used, and a plus sign shows what has been produced. Subtraction of moles used from moles at the start gives moles at the end; addition of moles produced to moles at the start gives moles at the end. This is simply the algebraic addition of the last two number lines:

	$Zn(s)$ +	$S(s)$ →	$ZnS(s)$
Grams at start	8.50	6.84	0
Molar mass	65.38	32.06	97.44
Moles at start	0.130	0.213	0
Moles used, produced	-0.130	-0.130	+0.130
Moles at end	0	0.083	0.130
Grams at end	___	___	___

The moles of the limiting reactant is zero at the end. Finally, multiplying moles of zinc sulfide and sulfur by their respective molar masses gives the mass of product and mass of excess reactant that remains:

$$0.130 \text{ mol ZnS} \times \frac{97.44 \text{ g ZnS}}{1 \text{ mol ZnS}} = 12.7 \text{ g ZnS}; \quad 0.083 \text{ mol S} \times \frac{32.06 \text{ g S}}{1 \text{ mol S}} = 2.7 \text{ g S}$$

	$Zn(s)$ +	$S(s)$ →	$ZnS(s)$
Grams at start	8.50	6.84	0
Molar mass	65.38	32.06	97.44
Moles at start	0.130	0.213	0
Moles used, produced	-0.130	-0.130	+0.130
Moles at end	0	0.083	0.130
Grams at end	0	2.7	12.7

As a check, note that the total mass left at the end of the reaction (2.7 g + 12.7 g = 15.4 g) is the same as the mass at the beginning of the reaction (8.50 g + 6.84 g = 15.34 g), within ±1 of the last significant figure. Remember, mass is neither created nor destroyed in chemical reactions; it is merely changed from one substance into another. (This conservation does not apply to moles, which may or may not remain the same in a chemical change.)

You might wonder what would happen if you selected the wrong reactant as the limiting reactant. The answer appears if we examine the three "moles" lines of the table if sulfur was used as the limiting reactant:

	Zn(s)	+ S(s)	→ ZnS(s)
Moles at start	0.130	0.213	0
Moles used, produced	-0.213	-0.213	+0.213
Moles at end	-0.083	0	0.213

The subtraction of 0.213 moles of zinc used from 0.130 moles present gives a negative number of moles—a negative quantity of matter. This is impossible. It signals that the wrong reactant has been used as the limiting reactant.

The table in the above sample problem is a detailed analysis of a simple limiting reactant problem—simple because the reactants combine in a 1:1 mole ratio. Nevertheless, the problem illustrates the whole procedure, which can be summarized in these steps:

1. Calculate the number of moles of each reactant at the start.
2. Estimate (or calculate, if necessary) the mole ratio in which the reactants are present. Compare this with the ratio in which they react, as indicated by the coefficients in the equation. From this, identify the limiting reactant.
3. From the moles of limiting reactant, calculate the number of moles of product formed. If the problem also asks for the mass of excess reactant that remains, find the number of moles used and subtract from the moles at the start to get the moles that remain.
4. Convert the moles from Step 3 to grams.

EXAMPLE 10.12: A solution containing 43.5 g $Ca(NO_3)_2$ is added to a solution that contains 39.5 g NaF. Calcium fluoride precipitates completely, as shown by the equation below.

(a) How many grams of CaF_2 precipitate?
(b) How many grams of which reactant are in excess?

$$Ca(NO_3)_2(aq) + 2\ NaF(aq) \rightarrow CaF_2(s) + 2\ NaNO_3(aq)$$

Grams at start	______	______	______
Molar mass	______	______	______
Moles at start	______	______	______

The tabulation of given information has been started. The grams of reactants are given. Place them in the table, as well as the molar masses of the reactants. The first step in the procedure is to calculate the moles of each reactant. Do this, and enter the results into the table. (The sodium nitrate that remains in solution is of no interest in this problem.)

	$Ca(NO_3)_2(aq)$ +	2 $NaF(aq)$ →	$CaF_2(s)$ + 2 $NaNO_3(aq)$
Grams at start	43.5	39.5	0
Molar mass	164.10	41.99	78.08
Moles at start	0.265	0.941	0
Moles used, produced	______	______	______

The next step is to identify the limiting reactant by comparing the mole ratio of reactants according to the equation to mole ratio actually present.

$$\text{From equation} = \frac{2 \text{ mol NaF}}{1 \text{ mol } Ca(NO_3)_2} = \frac{2}{1}; \quad \text{present} = \frac{0.941 \text{ mol NaF}}{0.265 \text{ mol } Ca(NO_2)_3} > \frac{3}{1}$$

From these figures, can you predict which reactant is the limiting reactant? Try to reason it out without calculating, but don't delay too long. Then actually calculate the moles of both reactants that will be used until the limiting reactant is gone, as well as the moles of CaF_2 formed. Fill in the blanks in the table, and see if your prediction is confirmed.

	$Ca(NO_3)_2(aq)$ +	2 $NaF(aq)$ →	$CaF_2(s)$ + 2 $NaNO_3(aq)$
Grams at start	43.5	39.5	0
Molar mass	164.10	41.99	78.08
Moles at start	0.265	0.941	0
Moles used, produced	-0.265	-0.530	+0.265
Moles at end	______	______	______

If there are over three times as many moles of NaF as $Ca(NO_3)_2$, and you need only twice as many, there is more NaF than can be used by all the $Ca(NO_3)_2$. $Ca(NO_3)_2$ is the limiting reactant. This appears in the numbers:

$$0.265 \text{ mol } Ca(NO_3)_2 \times \frac{2 \text{ mol NaF}}{1 \text{ mol } Ca(NO_3)_2} = 0.530 \text{ mol NaF required}$$

0.941 mol NaF—more than enough—is available, so $Ca(NO_3)_2$ is the limiting reactant. The same conclusion is reached from the other side:

$$0.941 \text{ mol NaF} \times \frac{1 \text{ mol } Ca(NO_3)_2}{2 \text{ mol NaF}} = 0.470 \text{ mol } Ca(NO_3)_2 \text{ required.}$$

Only 0.265 mol $Ca(NO_3)_2$ is available, not enough to react with all of the NaF. We again conclude that $Ca(NO_3)_2$ is the limiting reactant.

Calculate the moles of NaF that remain unreacted, and list again the moles of CaF_2 produced (moles at end).

	$Ca(NO_3)_2(aq)$ +	2 $NaF(aq)$ →	$CaF_2(s)$ + 2 $NaNO_3(aq)$
Grams at start	43.5	39.5	0
Molar mass	164.10	41.99	78.06
Moles at start	0.265	0.941	0
Moles used, produced	-0.265	-0.530	+0.265
Moles at end	0	0.411	0.265
Grams at end	______	______	______

Finding the grams of NaF that remain and the grams of CaF_2 produced are straightforward mole → gram conversions. Complete the problem.

	$Ca(NO_3)_2(aq)$ +	2 $NaF(aq)$ →	$CaF_2(s)$ + 2 $NaNO_3(aq)$
Grams at start	43.5	39.5	0
Molar mass	164.10	41.99	78.06
Moles at start	0.265	0.941	0
Moles used, produced	-0.265	-0.530	+0.265
Moles at end	0	0.411	0.265
Grams at end	0	17.3	20.7

$$0.411 \text{ mol NaF} \times \frac{41.99 \text{ g NaF}}{1 \text{ mol NaF}} = 17.3 \text{ g NaF}$$

$$0.265 \text{ mol } CaF_2 \times \frac{78.08 \text{ g } CaF_2}{1 \text{ mol } CaF_2} = 20.7 \text{ g } CaF_2$$

Do not be concerned that the total number of grams at the start (43.5 + 39.5 = 83.0) is more than the total number of grams at the end (17.3 + 20.7 = 38.0). The missing grams are the mass of the other product, $NaNO_3$.

EXAMPLE 10.13: Calculate the mass of PbI_2 that will precipitate if 252 grams of $Pb(NO_3)_2$ and 165 grams of NaI are dissolved in the same container of water. Also identify the reactant that is in excess, and find the number of grams that will remain unreacted. The equation is

	2 $NaI(aq)$ +	$Pb(NO_3)_2(aq)$ →	$PbI_2(s)$ + 2 $NaNO_3(aq)$
Grams at start	______	______	______
Molar mass	______	______	______
Moles at start	______	______	______
Moles used, produced	______	______	______
Moles at end	______	______	______
Grams at end	______	______	______

	2 $NaI(aq)$ +	$Pb(NO_3)_2(aq)$ →	$PbI_2(s)$ + 2 $NaNO_3(aq)$
Grams at start	165	252	0
Molar mass	149.9	331.2	461.0
Moles at start	1.10	0.761	0
Moles used, produced	-1.10	-0.550	+0.550
Moles at end	0	0.211	0.550
Grams at end	0	69.9	254

10.5 ANALYSIS OF MIXTURES

One of the most common uses of mole-to-mole calculations based on chemical equations is to determine how much of a certain chemical is present in a sample containing several chemicals mixed together. The experiment must be designed so that only the chemical of interest reacts. In the following problems you will be told what substance is being analyzed and what reaction is being used.

As an example, you might be interested in the sulfuric acid content of battery acid. A good reaction to use to test for sulfuric acid is to add sodium hydroxide.

$$H_2SO_4(aq) \quad + \quad 2\ NaOH(aq) \quad \rightarrow \quad Na_2SO_4(aq) \quad + \quad 2\ H_2O(\ell)$$

By measuring the amount of sodium hydroxide needed to react completely with all the sulfuric acid, you can calculate how many grams of sulfuric acid were originally present in the sample you tested. Suppose that you found that a 45.0-gram sample of battery acid required 18.0 grams of sodium hydroxide for complete neutralization. How many grams of sulfuric acid were present? The method for this calculation is, as you probably expected, the stoichiometry 3-step:

$$\text{g NaOH} \rightarrow \text{mol NaOH} \rightarrow \text{mol } H_2SO_4 \rightarrow \text{g } H_2SO_4$$

$$18.0 \text{ g NaOH} \times \frac{1 \text{ mol NaOH}}{40.00 \text{ g NaOH}} \times \frac{1 \text{ mol } H_2SO_4}{2 \text{ mol NaOH}} \times \frac{98.08 \text{ g } H_2SO_4}{1 \text{ mol } H_2SO_4} = 22.1 \text{ g } H_2SO_4$$

The results of the analysis of a mixture are usually expressed as a percentage. As was the case with percentage yield, the mathematical form to use depends on whether the percentage is wanted or given:

% WANTED: use equation

$$\text{percentage of A in a mixture} = \frac{\text{grams of A}}{\text{grams of mixture}} \times 100 \qquad (10.5)$$

% GIVEN: use conversion factor

$$\text{percentage of A in a mixture} = \frac{\text{grams of A}}{100 \text{ grams of mixture}} \qquad (10.6)$$

For the battery acid analyzed above the percentage sulfuric acid is:

$$\text{percentage } H_2SO_4 = \frac{22.1 \text{ g } H_2SO_4}{45.0 \text{ g sample}} \times 100 = 49.1\%\ H_2SO_4$$

EXAMPLE 10.14: What is the percentage of salt in seawater if a 75.0-gram sample precipitates 6.50 grams of silver chloride when treated with excess silver nitrate solution?

Let's plan our strategy before starting work on this problem. We are going to calculate percentage. This means we will use Equation 10.5, which includes the mass of the sample and the mass of salt (NaCl) as its variables. The mass of the sample is given, but nothing is said about the mass of salt in the solution. We can find that, though, by calculating the amount of NaCl needed to precipitate 6.50 g AgCl. So we have two problems. First, we must find the mass of NaCl by the stoichiometry 3-step. Second, we use that value in Equation 10.5 to calculate percentage.

As usual, begin by writing the balanced equation for the reaction. Identify the given and find quantities, plan your unit path, set up the problem, and calculate the mass of NaCl that reacted to produce 6.50 g AgCl.

$NaCl(aq) + AgNO_3(aq) \rightarrow AgCl(s) + NaNO_3(aq)$

GIVEN: 6.50 g AgCl FIND: g NaCl PATH: g AgCl → mol AgCl → mol NaCl → g NaCl

$$6.50 \text{ g AgCl} \times \frac{1 \text{ mol AgCl}}{143.4 \text{ g AgCl}} \times \frac{1 \text{ mol NaCl}}{1 \text{ mol AgCl}} \times \frac{58.44 \text{ g NaCl}}{1 \text{ mol NaCl}} = 2.65 \text{ g NaCl}$$

Now that you know the mass of NaCl in the 75.0-gram sample of seawater, you can express it as a percentage. Complete the problem.

GIVEN: 75.0 g sample, 2.65 g NaCl FIND: % NaCl

$$\text{EQN: \% NaCl} = \frac{\text{mass of NaCl}}{\text{mass of sample}} \times 100 = \frac{2.65 \text{ g NaCl}}{75.0 \text{ g sample}} \times 100 = 3.53\% \text{ NaCl}$$

EXAMPLE 10.15: If an ore assays as 35.0% aluminum oxide, how many kilograms of aluminum can be recovered from each ton (908 kg) according to the equation $2\ Al_2O_3(s) + 3\ C(s) \rightarrow 4\ Al(s) + 3\ CO_2(g)$?

Now you have been *given* the percentage of the substance in the mixture. Use this as a conversion factor, as written in Equation 10.6, *before* the stoichiometry 3-step. To be sure you see the way, we suggest that you write the unit path first.

GIVEN: 1 ton (908 kg) ore FIND: kg Al
FACTOR: 35.0 kg Al_2O_3/100 kg ore, plus three 3-step conversion factors
PATH: kg ore → kg Al_2O_3 → kmol Al_2O_3 → kmol Al → kg Al

Now go for the calculation setup. Complete the problem.

$$908 \text{ kg ore} \times \frac{35.0 \text{ kg } Al_2O_3}{100 \text{ kg ore}} \times \frac{1 \text{ kmol } Al_2O_3}{101.96 \text{ kg } Al_2O_3} \times \frac{4 \text{ kmol Al}}{2 \text{ kmol } Al_2O_3} \times \frac{26.98 \text{ kg Al}}{1 \text{ kmol Al}} = 168 \text{ kg Al}$$

10.6 EPILOGUE

In this chapter you have been introduced to a pattern for solving stoichiometry problems. It has been called the stoichiometry 3-step. The choice of that unattractive name was deliberate. It is intended to impress you again and again with the three steps in the process. The stoichiometry 3-step is a direct and logical method that always works. You have seen it applied to quantity measurements in grams of any pure substance. In Chapter 12 you will see this stoichiometry method again, applied to reactions of gases. In the chapter on solutions you will use the method again, expanded to quantity measured in volume of solution. In the chapter on thermochemistry the same reasoning is used. In electrochemical reactions, not included in this book, you will find that the same three steps can be applied to such unexpected units as amperes (electric current) and hours, minutes, or seconds.

Stoichiometry problems sometimes become quite long, but they are all basically the same. They all fit the pattern of the stoichiometry 3-step. Learn that sequence of steps well, and any lingering mystery of stoichiometry calculations will disappear.

CHAPTER 10 PROBLEMS

1) Aluminum combines with oxygen to produce solid aluminum oxide. How many moles of aluminum are required to react with 7.37 moles of oxygen? How many moles of aluminum oxide are produced by this reaction?

2) How many moles of oxygen are produced by the decomposition of 6.09 grams of solid potassium chlorate into solid potassium chloride and oxygen?

3) In the reaction $2\ PbS(s) + 3\ O_2(g) \rightarrow 2\ PbO(s) + 2\ SO_2(g)$, how many moles of lead(II) sulfide are required to produce 39.6 grams of lead(II) oxide?

4) Silver bromide, a photographic chemical, may be precipitated by combining solutions of sodium bromide and silver nitrate. How many grams of silver nitrate are needed to produce 2.50×10^2 grams of silver bromide?

5) $2\ NaCl(s) + H_2SO_4(aq) \rightarrow Na_2SO_4(aq) + 2\ HCl(g)$ is the equation for a reaction by which hydrogen chloride gas is produced. How many kilograms of Na_2SO_4 by-product can be recovered from the reaction of 2.7 kg of NaCl?

6) How many grams of lithium hydride are required to produce 43.9 g $LiAlH_4$ in the reaction $4\ LiH(s) + AlCl_3(s) \rightarrow LiAlH_4(s) + 3\ LiCl(s)$?

7) Calculate the mass of NO_2 required to react with 11.7 g H_2 in the reaction $7\ H_2(g) + 2\ NO_2(g) \rightarrow 2\ NH_3(g) + 4\ H_2O(\ell)$.

8) Calculate the grams of carbon dioxide that can be produced by the complete combustion of 816 grams of butane, $C_4H_{10}(g)$.

9) Calculate the number of grams of aluminum that must react according to the equation $8\ Al(s) + 3\ Fe_3O_4(s) \rightarrow 4\ Al_2O_3(s) + 9\ Fe(g)$ to yield 31.5 grams of iron.

10) Acetylene gas, C_2H_2, is made commercially by adding water to calcium carbide, CaC_2, according to $CaC_2(s) + 2\ H_2O(\ell) \rightarrow C_2H_2(g) + Ca(OH)_2(aq)$. If an acetylene manufacturer recovers 9.13×10^3 kilograms of acetylene by reacting 3.00×10^4 kilograms of CaC_2 with an excess of water, what is the percent yield for the reaction?

11) A common laboratory experiment is the production of aspirin, $C_9H_8O_4$, by the reaction of salicylic acid, $C_7H_6O_3$, with acetic anhydride, $C_4H_6O_3$. The equation is $C_7H_6O_3(s) + C_4H_6O_3(\ell) \rightarrow C_9H_8O_4(s) + HC_2H_3O_2(\ell)$. If the yield for this reaction is 84.6% and a student begins with 75.0 grams of salicylic acid and excess acetic anhydride, how much aspirin will be produced?

12) When aluminum reacts with phosphoric acid to produce hydrogen gas, the yield of solid aluminum phosphate is known to be 62.7%, based on the starting mass of phosphoric acid. If a student's actual yield was 73.9 grams of aluminum phosphate, how many grams of acid were used?

13) 316 grams of ammonia, NH_3, produce 1210 grams of NH_4Br by reacting with excess bromine:

$$3\ Br_2(\ell) + 8\ NH_3(g) \rightarrow 6\ NH_4Br(g) + N_2(g).$$

Calculate the percentage yield for this reaction.

14) The following reaction shows how nitric acid can be made from nitrogen dioxide by the reaction $3\ NO_2(g) + H_2O(\ell) \rightarrow 2\ HNO_3(aq) + NO(g)$. If the percentage yield for this reaction is 92.3%, calculate the number of grams of nitrogen dioxide needed to produce 46.8 grams of nitric acid.

15) The yield of carbon dioxide is known to be 79.4% in the complete burning of ethane, $C_2H_6(g)$. Calculate the mass of carbon dioxide that will be released by the burning of 23.6 grams of ethane.

16) A solution of 155 grams of KI is added to a solution of 175 grams of HNO_3:

$$6\ KI(aq) + 8\ HNO_3(aq) \rightarrow 6\ KNO_3(aq) + 2\ NO(g) + 3\ I_2(s) + 4\ H_2O(\ell).$$

How many grams of NO will be produced? Which reactant is present in excess, and how many grams will remain unreacted?

17) 50.0 grams of hydrogen and 50.0 grams of oxygen are ignited to form water. Calculate the mass of water formed, and the mass of oxygen or hydrogen that remains unreacted.

18) If 54.9 grams of solid aluminum hydroxide are neutralized with 91.0 grams of sulfuric acid, how many grams of aluminum sulfate will be produced? Which reactant is in excess, and how many grams are left?

Problems 19 to 25 are stoichiometry problems that are a bit more complicated. Some involve mixtures, others impure substances, and still others more complex limiting reactant problems. All can be solved with the stoichiometry 3-step, but each step may require a series of conversions.

19) Calculate the grams of lime, CaO, that can be recovered by decomposing 98.6 grams of limestone that is 88.4% $CaCO_3$. The reaction is

$$CaCO_3(s) \rightarrow CaO(s) + CO_2(g).$$

20) Find the mass of 79.2% H_2SO_4 sulfuric acid solution that is required to prepare 6.5 kilograms of HCl by the reaction

$$2\ NaCl(s) + H_2SO_4(aq) \rightarrow Na_2SO_4(aq) + 2\ HCl(g).$$

21) $4\ C_2H_5Cl(g) + 4\ NaPb(s) \rightarrow (C_2H_5)_4Pb(g) + 4\ NaCl(s) + 3\ Pb(s)$ is the equation by which the gasoline additive tetraethyl lead is made. The density of the product is 1.66 g/mL. 0.85 mL of this product are added to every liter of gasoline treated. Calculate the mass of ethyl chloride, C_2H_5Cl, needed to prepare the tetraethyl lead for a full 62-liter tank of gasoline.

22)
$$4\ FeS_2(s) + 11\ O_2(g) \rightarrow 2\ Fe_2O_3(s) + 8\ SO_2(g)$$
$$2\ SO_2(g) + O_2(g) \rightarrow 2\ SO_3(g)$$
$$SO_3(g) + H_2O(\ell) \rightarrow H_2SO_4(aq)$$

The three equations above summarize the chemistry by which iron pyrites, FeS_2, are used in preparing sulfuric acid. How many kilograms of rock that is 21.6% FeS_2 must be processed to obtain 25 kilograms of sulfuric acid?

23) Oleum is a solution of SO_3 in H_2SO_4 that is formed in one step of the commercial preparation of sulfuric acid. When oleum is diluted with water, the SO_3 becomes more sulfuric acid: $SO_3(g) + H_2O(\ell) \rightarrow H_2SO_4(aq)$. Calculate the percentage SO_3 in the oleum if 108 grams of pure H_2SO_4 are produced from 100 grams of oleum by adding water.

24) Potassium chlorate decomposes to potassium chloride and oxygen on heating. A student is given a 4.09-gram mixture of potassium chloride and potassium chlorate and asked to determine the percent $KClO_3$ in the mixture. She heats it until the $KClO_3$ is completely decomposed and collects the oxygen produced. If 0.895 grams of oxygen are produced, what percentage potassium chlorate should she report?

25) Barium carbonate can be precipitated in the following reaction:

$$NaHC_2O_4 + Cl_2 + 5\ NaOH + 2\ Ba(NO_3)_2 \rightarrow 2\ NaCl + 2\ BaCO_3 + 3\ H_2O + 4\ NaNO_3$$

Calculate the grams of barium carbonate that can be produced by adding excess barium nitrate solution to a solution containing 0.10 mol $NaHC_2O_4$, 0.12 mol NaOH, and 0.040 mol Cl_2.

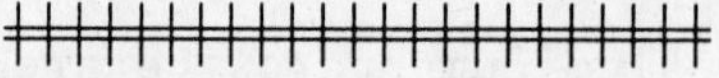

26) Answer the following questions about the complete oxidation of $C_{10}H_{22}$:

a) How many moles of $C_{10}H_{22}$ are needed to react with 41.3 moles of O_2?

b) What mass of CO_2 results from the burning of 0.169 moles of $C_{10}H_{22}$?

c) How many grams of water result when 84.9 grams of CO_2 are produced?

27) The reaction $3\ HNO_2(aq) \rightarrow 2\ NO(g) + HNO_3(aq) + H_2O(\ell)$ occurs in preparing nitric acid.

a) Calculate the grams of HNO_2 that are required to produce 274 grams of HNO_3.

b) If the reaction produces 9.78 g H_2O, what mass of NO will be formed?

28) A laundry bleach reacts as follows:

$$NaClO(aq) + NaCl(aq) + H_2O(\ell) \rightarrow Cl_2(aq) + 2\ NaOH(aq).$$

a) How many grams of NaClO must react to produce 81.6 grams of chlorine?

b) What mass of NaOH will result from the reaction of 75.8 g NaCl?

29) $Cr_2O_3(s) + 3\ C(s) \rightarrow 2\ Cr(s) + 3\ CO(g)$ is the reaction by which chromium is obtained from an oxide ore.

a) What mass of carbon monoxide results from the reaction of 93.7 grams of chromium(III) oxide?

b) How many grams of carbon are required to produce 434 grams of chromium?

c) How many kilograms of chromium can be obtained by processing one ton (907 kilograms) of ore that is 4.37% Cr_2O_3?

30) The medical compound $C_3H_5O_9N_3$ can be prepared by the reaction between glycerine and nitric acid: $C_3H_8O_3(aq) + 3\ HNO_3(aq) \rightarrow C_3H_5O_9N_3(s) + 3\ H_2O(\ell)$.

a) How many grams of nitric acid are used to make 163 grams of $C_3H_5O_9N_3$?

b) How many grams of $C_3H_5O_9N_3$ will be recovered from the reaction of 79.7 grams of $C_3H_8O_3$ if the percentage yield is 91.2%?

31) $2\ Ca_3(PO_4)_2(s) + 6\ SiO_2(s) + 10\ C(s) \rightarrow 6\ CaSiO_3(s) + 10\ CO(g) + P_4(s)$ is the reaction by which phosphorus is prepared from calcium phosphate.

a) How many kilograms of carbon are needed to produce 4.60 kg P_4?

b) 4.02 kilograms of phosphorus are recovered in the reaction described in part a). Calculate the percentage yield.

c) The percentage yield is known to be 84.3%, based on the starting quantity of silicon dioxide. Calculate the kilograms of silicon dioxide required to produce 5.4 kilograms of phosphorus.

d) 425 grams of silicon dioxide are mixed with 596 grams of calcium phosphate and excess carbon. The reaction proceeds until the limiting reactant is exhausted. How many grams of $CaSiO_3$ will result? Also, calculate the mass of the excess reactant that remains.

32) $P_4S_3(s) + 8\ O_2(g) \rightarrow P_4O_{10}(s) + 3\ SO_2(g)$ is the reaction for the burning of one kind of match tip.

a) If a 31-milligram match tip produces 31 milligrams of P_4O_{10}, what is the percentage yield for the reaction?

b) If the reaction has a 74.9% yield, calculate the mass of SO_2 formed by a match tip containing 43 milligrams of P_4S_3.

33) Chlorine can be prepared by the reaction:

$$MnO_2(s) + 4\ HCl(aq) \rightarrow MnCl_2(aq) + 2\ H_2O(\ell) + Cl_2(g).$$

a) What mass of $MnCl_2$ will result from the reaction of 52.9 grams of MnO_2 if the percent yield is 92.4%?

b) How many grams of MnO_2 must react to produce 100.0 grams of chlorine if the percent yield is 92.4%?

c) If 0.394 moles of MnO_2 and 61.8 grams of HCl are available for the reaction, which proceeds until the limiting reactant is consumed, how many grams of Cl_2 will be produced? How many grams of which reactant will remain?

34) A 92.0-gram sample of coal is found to yield 8.79 grams of SO_2 upon complete combustion. What is the percentage of sulfur in the sample?

35) Chlorine may be produced by the electrolysis of salt solutions according to the equation 2 NaCl(aq) + 2 $H_2O(\ell)$ | Cl2(g) + H2(g) + 2 NaOH(aq). How many grams of a 15.0% salt solution would be required to produce 225 grams of chlorine?

ADDITIONAL PROBLEMS

36) How many grams of bicarbonate of soda ($NaHCO_3$) are needed to neutralize the stomach acid in 1.0 pint of 0.010 M HCl?

37) A 2.37-gram sample of a chloride of formula MCl_3 yields 4.61 grams of AgCl when treated with excess $AgNO_3$. What is the atomic mass of M?

38) A 1.96-gram sample known to consist of a mixture of Na_2SO_4, MM = 142.01 g/mol, and K_2SO_4, MM = 174.26 g/mol, when treated with excess $BaCl_2$, form 2.82 grams of $BaSO_4$, MM = 233.4 g/mol. What is the %Na_2SO_4 by mass? [Hints: Let x = g Na_2SO_4, then g K_2SO_4 = 1.96 - x; so moles $SO_4{}^{2-}$ from Na_2SO_4 = x/142.04 and from K_2SO_4 = (1.96 - x)/174.26; multiply total moles $SO_4{}^{2-}$ by 233.4 to get 2.82 g $BaSO_4$; solve for x.]

39) Phosphoric acid may be made commercially by treating calcium phosphate with hydrofluoric acid. When 35 kg of $Ca_3(PO_4)_2$, MM = 310.18 g/mol, is mixed with 17 kg of HF, MM = 20.01 g/mol, how many kg of H_3PO_4 are produced if the process yield is 74%?

CHAPTER TEST

1) How many grams of C_3H_7OH must be completely oxidized to produce 35.7 grams of carbon dioxide?

2) Calculate the mass of oxygen that will react with 6.34 grams of C_3H_7OH in the same reaction as in Question 1.

3) $4\ HNO_3(aq) + 3\ Na(s) \rightarrow 2\ H_2O(\ell) + NO(g) + 3\ NaNO_3(aq)$. Calculate the percent yield if 12.5 grams of NO are recovered from the reaction of 31.6 grams of sodium according to the above equations.

4) The percent yield of SO_2 is 88.9% when carbon disulfide is burned according to the equation $CS_2(\ell) + 3\ O_2(g) \rightarrow 2\ SO_2(g) + CO_2(g)$. How many grams of CS_2 must be burned to produce an actual yield of 119 grams of SO_2?

5) How many grams of magnesium hydroxide will precipitate when a solution containing 7.50 grams of magnesium nitrate is added to a solution containing 5.36 grams of sodium hydroxide? Assume the reaction proceeds until the limiting reactant is completely used. How many grams of which original compound will remain unreacted?

6) Solutions of iodine may be analyzed by the reaction:

$$As_2O_3(s) + 2\ I_2(aq) + 5\ H_2O(\ell) \rightarrow 2\ H_3AsO_4(aq) + 4\ HI(aq)$$

If 225 grams of solution react completely with 4.40 grams of As_2O_3, what is the percentage of iodine in the solution?

11. ATOMIC STRUCTURE

PREREQUISITE

1. Section 6.3 presented the names, symbols, and atomic numbers of the common elements whose atomic structures will be discussed.

CHEMICAL SKILLS

1. Calculate frequency ν and wavelength λ given the other and the speed of light.

2. Calculate the energy E or frequency ν of light given the other and Planck's constant h.

3. State the bands of the electromagnetic spectrum, including the visible colors, in order of wavelength, frequency, and energy.

4. For the hydrogen spectrum, state the n_i values for each band, UV, visible, IR, and the n_o values for the visible line colors.

5. Calculate any one of the values of ν, n_i, and n_f for hydrogen given the values of the other two and the Rydgerg constant R_H.

6. Sketch the Bohr orbits and energy level diagram for hydrogen. Show the electron energy level change for any line in the emission spectrum.

7. State the notation for any subshell and draw the shape of an electron wave within that subshell.

8. Using only a periodic table, identify the valence electrons and give the shorthand notation and noble gas notation for the ground state electron structure of any s, p, or d block element.

9. Give the possible n, ℓ, m, and s values for any electron in the ground state of an atom.

10. Draw the shape, including axis designations, for any s, p, or d orbital.

11. For any given electron structure, identify it as ground state, excited state, or impossible (violation of the Pauli exclusion principle). For excited states, idnetify which effect, n, ℓ, m, or s, has been violated.

11.1 QUESTIONS ARISING FROM THE NUCLEAR ATOM

In Chapter 6 we learned that atoms are made up of protons, neutrons, and electrons arranged in the **nuclear atom**. The protons and the neutrons are found in the nucleus while the electrons are outside the nucleus, making up the rest of the volume of the atom.

One of the questions that puzzled scientists about the nuclear atom was the location of the electron. The electron is negative, and the nucleus is positive. Negative charges are attracted to positive charges. So why does the electron stay outside the nucleus? It would seem that the attractive +/- force would pull the electron right into the nucleus! Also, just where *is* the electron within the relatively large volume of the atom?

During the years following the discovery of the nuclear atom, much work was done on figuring out the energy and location of the electron. The results of that work have greatly improved our understanding of the structure of the atom. In this chapter we will present the modern view of the atom.

11.2 LIGHT AND ATOMIC SPECTRA

Understanding the modern view of the atom requires some knowledge of the nature of light. Light may be thought of as **waves** moving through space. Figure 11.1 shows a light wave.

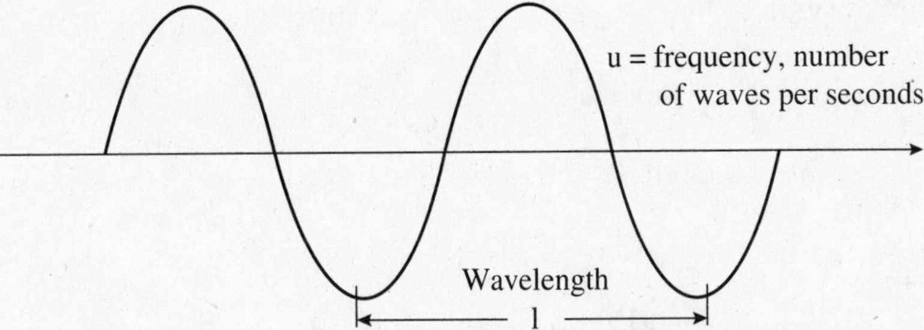

Figure 11-1 Light as waves.

Several terms and symbols are used to describe waves. Waves have **wavelength**, λ (the Greek letter lambda). Wavelength is the distance between peaks of the waves. (It can be the distance between *any* two identical points on neighboring waves). Wavelength is usually measured in meters (m.) Waves have **speed**, c. The speed of light is 2.998×10^8 meters/second (m/s.) Waves have **frequency**, ν (the Greek letter nu). Frequency is the number of waves that go by each second. A complete wave is called a **cycle**. The unit of frequency is "number of waves"/second. This can be written without a term for "number of waves" as 1/second (read "per second"). Or, more commonly, by using the term "cycle" for a wave, the unit is cycles/second. The metric unit for cycles per second is the hertz, which we will not use. Speed, wavelength, and frequency are related by the equation:

$$c = \lambda\nu, \qquad c = 2.998 \times 10^8 \text{ m/s} \tag{11.1}$$

The units for each are:

$$\frac{\text{m}}{\text{s}} = \text{m} \cdot \frac{1}{\text{s}}$$

Equation 11.1 tells us that the product of frequency and wavelength is a constant. Thus λ and ν are *inversely proportional*. As one gets bigger, the other gets smaller. Light of short wavelength has a high frequency. (If the waves are short, a lot of them go by in a second.) Light of long wavelength has a low frequency. (Fewer long waves would go by in a second.)

Light also has **energy**, E. Energy has the unit joule, J (pronounced like "jewel"). The energy carried by a beam of light depends on its frequency, ν. As its frequency increases, so does the energy of a light wave.

$$E = h\nu, \quad h = 6.626 \times 10^{-34} \text{ J}\cdot\text{s} \tag{11.2}$$

The constant h is called Planck's constant. With energy in joules and frequency in 1/seconds, h is equal to 6.626×10^{-34} joule•seconds.

Equation 11.2 gives us the energy of a single wave of light. This means that each wave, like the one shown in Figure 11.1, carries a definite amount or "packet" of energy. The amount of energy carried by a light wave is called a **quantum**. (The plural of "quantum" is "quanta.")

The size of each quantum of energy depends on the frequency of the light. A short, high frequency wave carries a lot of energy. Long, low frequency waves carry much less.

EXAMPLE 11.1: What is the frequency and energy of green light with a wavelength of 515 nm? Set up the problem in the usual way. (Remember that 1 nm = 10^{-9} m.)

GIVEN: FIND: EQN:

GIVEN: $\lambda = 515$ nm FIND: ν, E EQN: $c = \lambda\nu$, $E = h\nu$
$c = 2.998 \times 10^{8}$ m/s
$h = 6.626 \times 10^{-34}$ J•s

$$\nu = \frac{c}{\lambda} = \frac{2.998 \times 10^{8} \text{ m/s}}{515 \text{ nm} \times \dfrac{10^{-9} \text{ m}}{1 \text{ nm}}} = 5.82 \times 10^{14} \text{ 1/s (cycles/second)}$$

$$E = h\nu = (6.626 \times 10^{-34} \text{ J}\cdot\text{s})(5.82 \times 10^{14} \text{ 1/s}) = 3.86 \times 10^{-19} \text{ J}$$

Light can have a variety of frequencies and wavelengths. Taken together, all types of light make up the **electromagnetic spectrum**. This range of frequencies and wavelengths is divided into bands. Figure 11.2 shows the bands of the electromagnetic spectrum.

The spectrum begins at the left with radio and television waves. These waves are the longest and have the lowest frequency and energy. Moving from left to right in Figure 11.2, frequency and energy increase and wavelength decreases. The highest energy, shortest waves are gamma rays.

Visible light occurs in the middle of the spectrum. Visible light is divided into colors. A common way to remember the order of the colors is the "name" ROY G BIV. Each letter is the first of the word for its color: <u>r</u>ed, <u>o</u>range, <u>y</u>ellow, <u>g</u>reen, <u>b</u>lue, <u>i</u>ndigo, and <u>v</u>iolet. Red light has the longest and lowest energy waves. Violet light is shortest and most energetic.

The locations of the bands nearest visible light are easy to remember. Ultraviolet (UV) light is above ("ultra") violet. Infrared (IR) is below ("infra") red.

long λ low ν low E	radio TV	micro waves	infrared (IR)	visible	ultraviolet (UV)	x rays	gamma rays	short λ high ν high E

Visible Colors: Red Orange Yellow Green Blue Indigo Violet

FIGURE 11.2. THE ELECTROMAGNETIC SPECTRUM.

You have probably seen the light emitted from neon signs (red), sodium street lights and auto fog lights (yellow), and mercury street lights (blue). These are all examples of light emitted by atoms. Figure 11.3 shows the spectra of the light emitted by these atoms. Because the black-and-white figure cannot show colors, they are written below each spectrum.

The amazing fact about the spectra of the atoms is that they consist of *lines*. Almost all other light sources give off a continuous spectrum of colors. The hydrogen spectrum is particularly interesting. It has only four visible lines. How does hydrogen emit only those few frequencies of light?

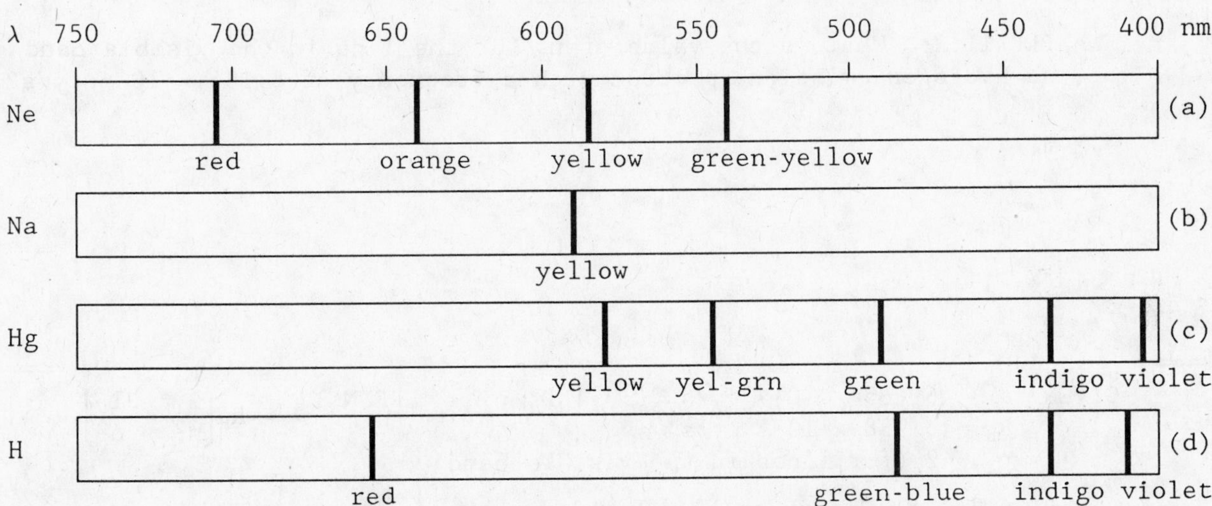

FIGURE 11.3. THE LINE EMISSION SPECTRA OF NEON, SODIUM, MERCURY, AND HYDROGEN.

After the discovery of the visible emission lines of hydrogen, two additional sets of lines were found. One set of lines was found in the UV, the other in the IR. In addition, the violet line seen in Figure 11.3d was discovered to be many, many closely spaced lines. The more complete line spectrum of hydrogen is shown in Figure 11.4.

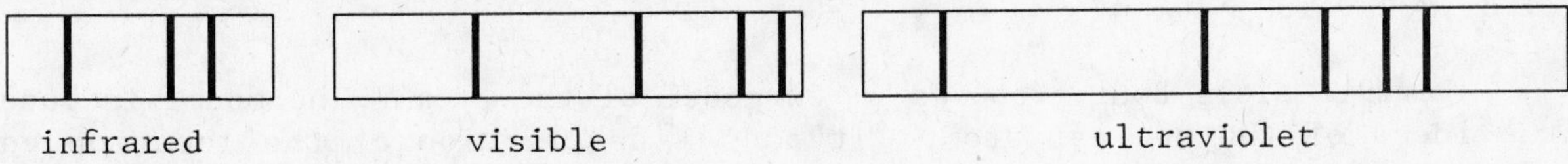

FIGURE 11.4. LINE EMISSION SPECTRUM OF HYDROGEN: UV, VISIBLE, AND IR.

Atomic spectra are another example of quantization in atoms. These atoms are giving off just certain definite frequencies of light. One feature of quantized values, such as frequency here, is that they depend on a whole number called a **quantum number**. In hydrogen the frequencies of all the hydrogen emission lines are shown to depend on the quantum number n in a relationship known as the **Rydberg equation**:

$$\nu = R_H\left(\frac{1}{n_f^2} - \frac{1}{n_i^2}\right) \qquad \text{with } n_i > n_f \qquad (11.3)$$

Light of frequency ν is emitted. R_H, the Rydberg constant, has the value 3.288×10^{15} cycles/second. The quantum number n has simple whole number values: 1, 2, 3, etc. The subscript f in n_f stands for "final." The subscript i in n_i means "initial." When a hydrogen atom gives off light, it changes from its "initial state" before the light is emitted to the "final state" after the light is emitted. Note that $n_{initial}$ must be greater than n_{final}.

The lines in the visible band of the spectrum are produced when $n_{final} = 2$ and $n_{initial} = 3$ (the red line), 4 (the green-blue line), 5 (the blue-indigo line), 6, etc. (the violet lines). The family of UV lines come from $n_{final} = 1$ and $n_{initial} = 2, 3, 4$, etc. Those in the IR have $n_f = 3$ and $n_i = 4, 5, 6$, etc.

The Rydberg equation gives a mathematical description of the hydrogen line spectrum. However, it does not describe what is going on inside the atom.

EXAMPLE 11.2: What is the value of n_i for the line in the visible band of the hydrogen emission spectrum with a frequency of 6.165×10^{14} cy/s?

GIVEN: FIND: EQN:

GIVEN: $\nu = 6.165 \times 10^{14}$ cy/s FIND: n_i EQN: $\nu = R_H\left(\frac{1}{n_f^2} - \frac{1}{n_i^2}\right)$

$R_H = 3.288 \times 10^{15}$ cy/s

$n_f = 2$ (for lines in the visible band)

$$6.165 \times 10^{14} \text{ cy/s} = 3.288 \times 10^{15} \text{ cy/s} \left(\frac{1}{(2)^2} - \frac{1}{n_i^2}\right)$$

$$0.1875 = \frac{1}{4} - \frac{1}{n_i^2}, \quad -0.6250 = -\frac{1}{n_i^2}, \quad \frac{1}{0.0650} = n_i^2$$

$n_i = 4$ (This is the blue-green line.)

11.3 BOHR ATOM

In 1913 Niels Bohr proposed a new model of the atom. The model includes the ideas of the nuclear atom. It adds a description of the location and energy of the electron. According to Bohr, the electron travels in circular orbits around the nucleus. The movement of the electron around the nucleus is similar to that of a planet around the sun. If fact, the Bohr atom is known as the "planetary model." The Bohr model is shown in Figure 11.5.

The Bohr model explains why the electron is not pulled into the nucleus by the attraction between e^- and p^+. The stability of the electron in its orbit can be understood by thinking about the motion of a planet around the sun. The planet is in constant motion. This motion gives the planet **inertia** which tends to keep the planet moving in a straight line. However, the planet is attracted to the sun by gravity. This attraction bends the path of the planet into a circular orbit.

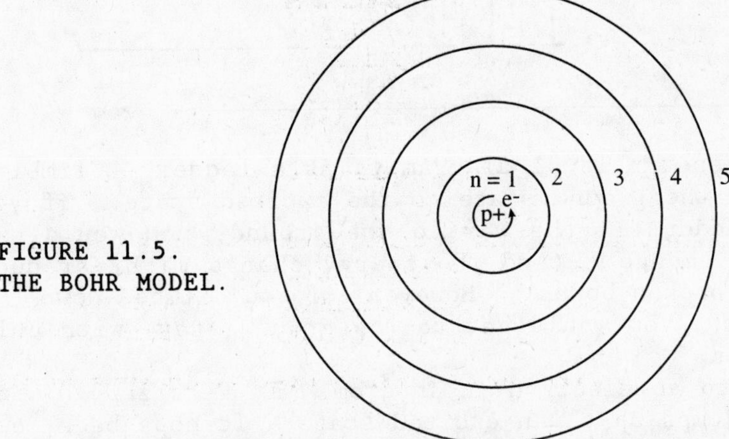

FIGURE 11.5.
THE BOHR MODEL.

The Bohr model also explains the quantized frequencies of light emitted by atoms. The electron can travel only in certain orbits. The radius of each allowed orbit depends on the quantum number n. This is the same quantum number which appears in the Rydberg equation. Using calculations based on well-known laws of physics, Bohr calculated the value of the radius, r, of each orbit.

$$r = a_o n^2 \qquad a_o = 0.0529 \text{ nm} \tag{11.4}$$

The constant a_o is called the **Bohr radius**. It is the radius of the first orbit, having the value of n = 1 as shown in Figure 11.5. Its value, 0.0529 nm, agrees very well with the known radius of the hydrogen atom.

When the electron is in its lowest energy state, known as the **ground state**, it is in the orbit closest to the nucleus. This orbit has n = 1. When the atom gains energy, it goes to a higher energy state called an **excited state**. The absorbed energy pushes the electron to an orbit farther from the nucleus. The excited states of the electron are the orbits with n = 2, 3, 4, etc. *Orbits with larger n values have higher energies*. When the atom returns to the ground state, the electron falls from the higher orbit back to the n = 1 orbit. The difference in energy is given off as light. The frequency of the light follows the Planck equation, $E = h\nu$.

Figure 11.6 shows how the energy levels of the Bohr model explain the line emission spectrum of the hydrogen atom. When the excited state electron falls to the n = 1 ground state, UV light is given off. In the Rydberg equation this corresponds to a fall from $n_{initial}$ = 2, 3, 4, etc. to n_{final} = 1. Note that UV light is higher in energy than either visible or IR. From Figure 11.6 you can see that the electron loses the most energy when it falls to n = 1.

The electron does not always fall directly to the ground state. It may stop at any energy level on the way down. Again looking at Figure 11.6, if the electron falls to n = 2, visible light is emitted. In the Rydberg equation this corresponds to a fall from $n_{initial}$ = 3, 4, 5, etc. to n_{final} = 2. When the electron drops from n_i = 4, 5, 6, etc. to n_f = 3, infrared light is emitted.

Figure 11.6 ENERGY LEVEL DIAGRAM FOR HYDROGEN. Changes in energy levels produce lines in the emission spectrum.

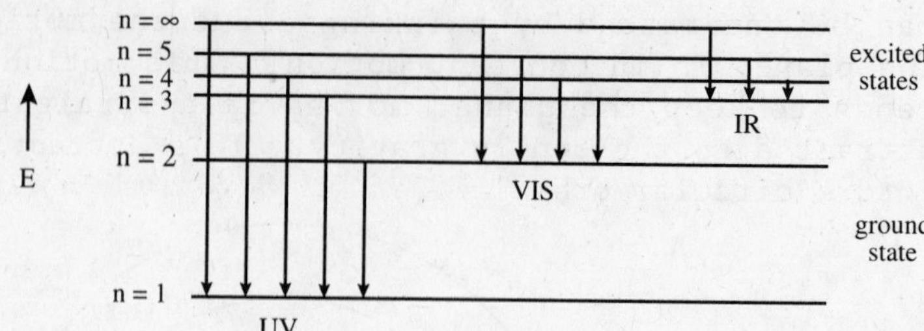

One way to think of the energy level diagram is as a ladder. Climbing the ladder is like going from the ground state to the excited state. If you jumped off the top of the ladder all the way to the ground, you would experience the greatest energy change. (And the energy change might produce some physical changes such as broken bones!) However, if you climbed down the ladder a rung or two at a time, you would get to the ground state with only small energy changes in between.

A similar thing happens to an electron in an atom. It gets zapped with energy from the outside. It jumps up to an excited state. It hops back down by giving off energy as light. It can get down in a big leap and give off high energy UV light. Or it can go down in smaller steps and give off lower energy IR and then visible light.

EXAMPLE 11.3: Suppose that an excited electron in a hydrogen atom gives off green-blue light and then UV light. Draw the energy level diagram of Figure 11.6 and the Bohr orbits of Figure 11.5. On each drawing place arrows to show the movement of the electron as it first emits green-blue light and then emits UV light.

The green-blue line is the *second* line in the hydrogen spectrum: red, then green-blue, then indigo, then violet. All the visible lines have n_f = 2. The n_i values are 2 (red), 3 (green-blue), etc. So the green-blue light emitted is caused by the electron moving from n = 4 to n = 2. The UV lines all have n_f = 1, so the final jump down is from n = 2 to n = 1.

11.4 WAVE MODEL OF THE ATOM

Both light and the electon in the hydrogen atom are quantized. This led scientists to compare the behavior of light and electrons. In 1923 Louis deBroglie proposed that the electron behaves as a wave. This was soon confirmed in the laboratory. Shortly after , Erwin Schroedinger applied "wave equations" to electrons and developed the quantum mechanical model of the atom.

A discussion of the Schroedinger equation is beyond the scope of this book. It is a complex mathematical relationship that is the basis of our present understanding of the atom. One of the major outcomes of the quantum mechanical model and the Schroedinger equation are the **quantum numbers**. Four such numbers are required to describe the behavior of the electron in the atom. That description is the focus of the remainder of this chapter.

Quantum numbers are not an idea that was made up to create a theoretical model of the atom. They are a fundamental part of nature, just like the law of gravity or the existance of + and - charges.

11.5 QUANTUM NUMBERS: SHELLS AND SUBSHELLS

The Bohr model required just one quantum number, n, to describe the hydrogen atom. The four quantum numbers of the wave model describe all atoms, not just hydrogen. Taken together, they tell us the positions and energies of electrons in atoms. Each of them describes a different property of the electron wave, called an **orbital**. In this section we will describe the first two quantum numbers.

The quantum number n is called the **principal quantum number**. It tells us the *distance* of the electron from the nucleus. The values of n are positive whole numbers: 1, 2, 3, on up to infinity. As n increases, the electron is farther and farther from the nucleus. This means that the attractive +/- force between the electron and the nucleus becomes less and less. (It reaches zero at n = infinity.) Electrons with the same value of n are said to be in the same **shell**.

Next is ℓ, the **secondary quantum number**. It influences the energy of the electron by determining its angular momentum. (The idea of angular momentum is beyond the scope of our discussion and is not required to understand the influence of quantum numbers on the structure of atoms.) Electrons with the same n and ℓ are in the same **subshell**. The values of ℓ are positive whole numbers, beginning with 0 and ending with n - 1. The use of letter symbols for ℓ avoids confusion with the numbers used for n. The symbols used for ℓ values and for subshells are:

n value	ℓ value	letter	subshell	
1	0	s	1s	first shell
2	0	s	2s	second shell
2	1	p	2p	
3	0	s	3s	third shell
3	1	p	3p	
3	2	d	3d	
4	0	s	4s	fourth shell
4	1	p	4p	
4	2	d	4d	
4	3	f	4f	

Beyond $\ell = 3$, the letters are alphabetical: $\ell = 4$ is g, 5 is h, etc. Notice that each shell contains one more subshell than the one before it.

The value of ℓ determines the *shape* of the electron wave (orbital). Those in s subshells are spheres. The p subshell waves are dumbbell-shaped. The d subshells contain cloverleaf-shaped waves. Drawings of these electron waves appear in Figure 11.7. Orbitals in the f subshells and beyond are too complex to draw easily. We will not be concerned with them.

You should be able to write the symbols for the shells and subshells. You should know their quantum number values (n and ℓ). And you should be able to draw the shapes of the electron waves in s, p, and d subshells.

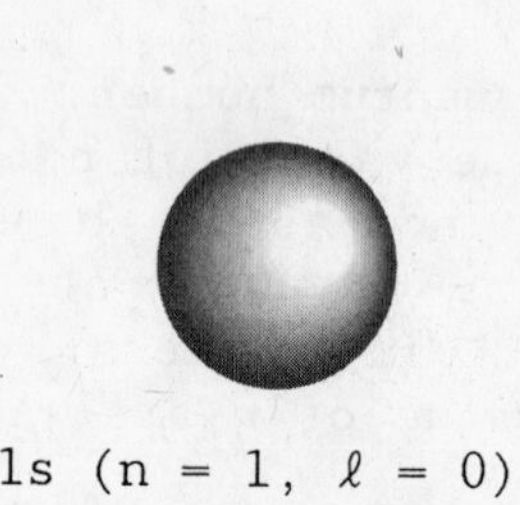

1s ($n = 1$, $\ell = 0$)

2s ($n = 2$, $\ell = 0$)

a. The first subshell in each shell is always an s orbital. All s orbitals are spheres. They increase in size (average distance from the nucleus) as n increases. ℓ is always zero for an s orbital.

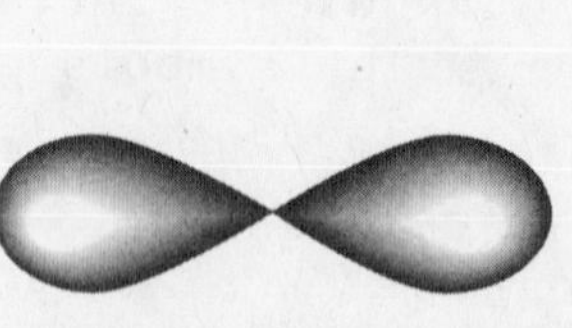

2p ($n = 2$, $\ell = 1$)

b. All p orbitals are dumbbell shaped. They are the second subshell in each shell, beginning with the second shell.

3d ($n = 3$, $\ell = 2$)

c. d orbitals are cloverleafs. The first two shells have no d orbitals.

FIGURE 11.7. SHAPES AND SIZES OF ORBITALS IN s, p, AND d SUBSHELLS.

EXAMPLE 11.4: How would you write the symbol for the subshell with $n = 5$ and $\ell = 1$? What shape are the orbitals in this subshell? Draw the shape.

This is the 5p subshell. It is dumbbell-shaped.

The Bohr model explained atomic spectra using quantized energy levels in the hydrogen atom which depend on the value of n. (Review Figure 11.6.) Bohr could not explain the spectra of atoms other than hydrogen. One reason for this is that each principal energy level is divided into sublevels, one for each subshell. The subshell energies within each shell are in the order s < p < d < f and so on. This is illustrated in Figure 11.8.

Notice that the energies of the subshells "fan out" from the shell levels. One result of this is that the 4s subshell is lower in energy than the 3d subshell. In the next two sections we will see that this "overlap" of subshell energies has a profound effect on the organization of the periodic table.

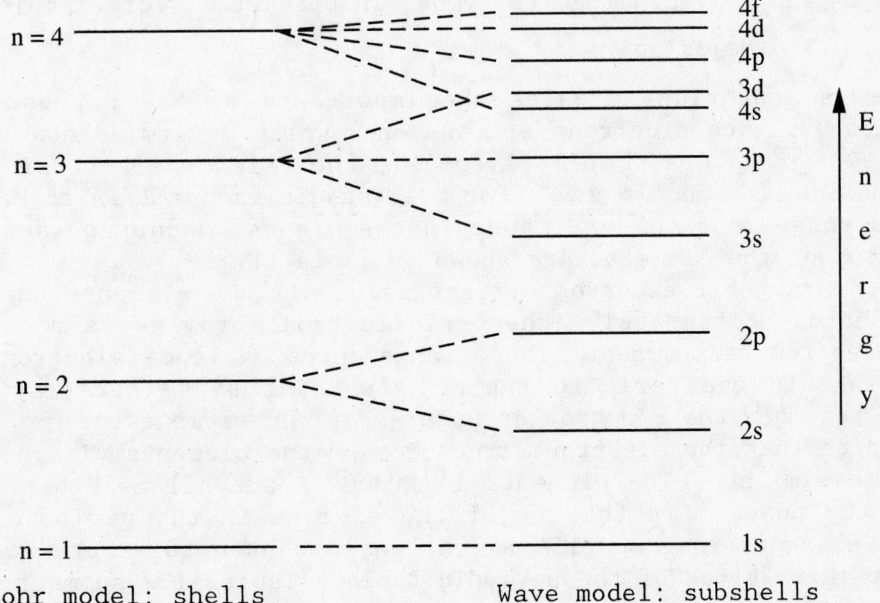

FIGURE 11.8. ENERGY LEVEL DIAGRAMS SHOWING SHELLS SPLIT INTO SUBSHELLS. Only the first four shells are shown.

11.6 PERIODIC ARRANGEMENT OF SHELLS AND SUBSHELLS

There are over 100 elements in the periodic table. Each neutral element has a different number of electrons. As we learned in Chapter 6, the number of protons (and thus electrons) in a neutral atom is given by its atomic number, Z. For example, sulfur, Z = 16, has 16 protons and thus 16 electrons.

The electrons in an atom are arranged in shells and subshells called the **electron structure** of the atom. Each shell and subshell has a maximum number of electrons it can hold. The s subshells hold 2 electrons, p subshells hold 6, d subshells 10, and f subshells 14. The maximum number of electrons in the subshells in the first four shells is thus:

	1s: 2 e^-	2s: 2 e^-	3s: 2 e^-	4s: 2 e^-
		2p: 6 e^-	3p: 6 e^-	4p: 6 e^-
			3d: 10 e^-	4d: 10 e^-
				4f: 14 e^-
Total:	2	8	18	32

This information does not need to be memorized. As we shall see shortly, the periodic table contains this information!

In the ground state of each atom, the electrons fill in the lower subshells first. The filling of subshells from lower to higher energy is called the **Aufbau process**. ("Aufbau" means "building up" in German.) For example, sulfur, with 16 e^-, puts 2 e^- in the 1s, 2 e^- in the 2s, and 6 e^- in the 2p. This leaves 16 − 2 − 2 − 6 = 6 e^- to begin filling the third shell. After putting 2 of these electrons in the 3s subshell, just four are left for the 3p subshell. Notice that this does not completely fill the 3p subshell.

A common way to write electron structures is to use **shorthand notation**. In this method the number of electrons in each subshell is listed as a superscript. For example:

shorthand: S: $1s^2 2s^2 2p^6 3s^2 3p^4$

It is also common to show only the electrons beyond the last noble gas. The noble gas core is shown by placing its symbol in brackets. For example:

noble gas: S: $[Ne]3s^2 3p^4$

The noble gas notation stresses the importance of the highest energy electrons. These **valence electrons** are the ones which determine how the element reacts. *For groups IA to IIB (1 to 12), the valence electrons are the electrons beyond the last noble gas. For elements in groups IIIA to VIIIA (13 to 18), the valence electrons are those in the highest occupied shell.* The valence electrons of the elements are shown in Table 11.1.

Trying to learn their electron structures of all the elements one by one would be very hard. Fortunately, the periodic table provides a much easier way to learn electron structures. The arrangement of valence electrons falls into patterns within the periodic table. *Learning these patterns is the easiest way to remember the electron structures of the elements.*

First, notice that the electron structures of the elements within a given family are quite similar. The elements in group IA are all ns^1 where n, the principal quantum number, is the row of the element in the periodic table. Similarly, the elements in group IIA are all ns^2. Taken together, groups IA and IIA make up the **s block** of the periodic table. Table 11.2 summarizes this relationship.

The next set of families, groups IIIB to IIB (3 to 12), are filling d subshells. These elements are in the **d block**. Generally, these elements have

TABLE 11.1: VALENCE ELECTRON STRUCTURES OF THE ELEMENTS

1 IA	2 IIA	3 IIIB	4 IVB	5 VB	6 VIB	7 VIIB	8 ←VIIIB	9 VIIIB	10 VIIIB→	11 IB	12 IIB	13 IIIA	14 IVA	15 VA	16 VIA	17 VIIA	18 VIIIA(0)
H $1s^1$																	He $1s^2$
Li $2s^1$	Be $2s^2$											B $2s^22p^1$	C $2s^22p^2$	N $2s^22p^3$	O $2s^22p^4$	F $2s^22p^5$	Ne $2s^22p^6$
Na $3s^1$	Mg $3s^2$											Al $3s^23p^1$	Si $3s^23p^2$	P $3s^23p^3$	S $3s^23p^4$	Cl $3s^23p^5$	Ar $3s^23p^6$
K $4s^1$	Ca $4s^2$	Sc $4s^23d^1$	Ti $4s^23d^2$	V $4s^23d^3$	Cr $4s^13d^5$	Mn $4s^23d^5$	Fe $4s^23d^6$	Co $4s^23d^7$	Ni $4s^23d^8$	Cu $4s^13d^{10}$	Zn $4s^23d^{10}$	Ga $4s^24p^1$	Ge $4s^24p^2$	As $4s^24p^3$	Se $4s^24p^4$	Br $4s^24p^5$	Kr $4s^24p^6$
Rb $5s^1$	Sr $5s^2$	Y $5s^24d^1$	Zr $5s^24d^2$	Nb $5s^14d^4$	Mo $5s^14d^5$	Tc $5s^14d^6$	Ru $5s^14d^7$	Rh $5s^14d^8$	Pd $5s^14d^9$	Ag $5s^14d^{10}$	Cd $5s^24d^{10}$	In $5s^25p^1$	Sn $5s^25p^2$	Sb $5s^25p^3$	Te $5s^25p^4$	I $5s^25p^5$	Xe $5s^25p^6$
Cs $6s^1$	Ba $6s^2$	La* $6s^25d^1$	Hf $6s^25d^2$	Ta $6s^25d^3$	W $6s^15d^5$	Re $6s^25d^5$	Os $6s^25d^6$	Ir $6s^25d^7$	Pt $6s^15d^9$	Au $6s^15d^{10}$	Hg $6s^25d^{10}$	Tl $6s^26p^1$	Pb $6s^26p^2$	Bi $6s^26p^3$	Po $6s^26p^4$	At $6s^26p^5$	Rn $6s^26p^6$
Fr $7s^1$	Ra $7s^2$	Ac# $7s^26d^1$															

*Lanthanides	Ce $6s^24f^2$	Pr $6s^24f^3$	Nd $6s^24f^4$	Pm $6s^24f^5$	Sm $6s^24f^6$	Eu $6s^24f^7$	Gd $6s^24f^75d^1$	Tb $6s^24f^9$	Dy $6s^24f^{10}$	Ho $6s^24f^{11}$	Er $6s^24f^{12}$	Tm $6s^24f^{13}$	Yb $6s^24f^{14}$	Lu $6s^24f^{14}5d^1$
#Actinides	Th $7s^26d^2$	Pa $7s^25f^26d^1$	U $7s^25f^46d^1$	Np $7s^25f^46d^1$	Pu $7s^25f^6$	Am $7s^25f^7$	Cm $7s^25f^76d^1$	Bk $7s^25f^9$	Cf $7s^25f^{10}$	Es $7s^25f^{11}$	Fm $7s^25f^{12}$	Md $7s^25f^{13}$	No $7s^25f^{14}$	Lr $7s^25f^{14}6d^1$

just filled the "ns" subshell and are adding electrons to the "(n−1)d" subshell. Their electron structures are $ns^2(n-1)d^x$. The value of x can be obtained by counting the columns of the d block. Be sure to start with group IIIB (3), the scandium family, with x = 1: $4s^2 3d^1$ for Sc. Also note that x is the group number minus 2 (the two electrons in the s subshell): 3 − 2 = 1.

In each row, the d subshell which is filling has an n value *one less* than the s subshell which was just filled. The order, summarized in Table 11.2, is: 4s then 3d in the first row of the d block, 5s then 4d in the second row, and 6s followed by 5d in the last row. (There are, however, several "unusual" structures in the d block. Note, for example, Cr and Cu in the first row of the d block. The reasons for these variations are complex, and we won't discuss them here.)

Groups IIIA to VIIIA (13 to 18) have the electron structure ns^2np^x. Again, n is the row number. These elements are filling the **p block**. The value of x is the column within the p block. Note that the first column of the p block is group IIIA (13). The second is group IVA (14), and so on. For example, oxygen is in the 2nd row and the 4th column of the p block. Its valence electron structure is thus $2s^2\ 2p^4$. (The one exception in the p block is He. Helium has the structure $1s^2$. It has no p electrons.)

The f block elements are also shown in Tables 11.1 and 11.2. We will not cover their electron structures in the examples and problems in this text. These elements are rare. Also, their electron structures show several exceptions to the patterns mentioned above.

TABLE 11.2: THE PERIODIC ARRANGEMENT OF SHELLS AND SUBSHELLS

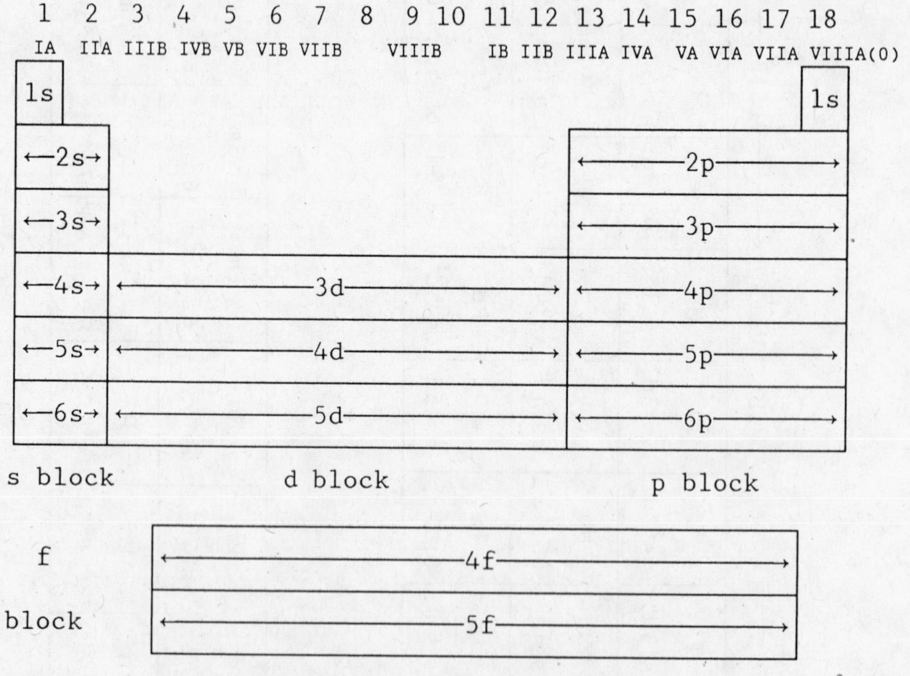

EXAMPLE 11.5: Write the valence electron structures of the following elements. Use only the periodic table inside the back cover.

(a) P

(b) Fe

(c) Br

s block: ns^x, d block: $ns^2(n-1)d^x$, p block: ns^2np^x

(a) P: 3rd row, 3rd column of p block: $3s^23p^3$

(b) Fe: 1st row of d block, 6th column: $4s^23d^6$

(c) Br: 4th row, 5th column of p block: $4s^24p^5$

Example 11.5 shows that the noble gas notation sometimes lists more than just the valence electrons. The notation starts with the last noble gas, in brackets. Then *all* the subshells beyond that are shown. In many cases, the electrons beyond the last noble gas *are* the valence electrons. This is true of P and Fe in the previous example:

	valence electrons	noble gas notation
P:	$3s^23p^3$	$[Ne]3s^23p^3$
Fe:	$4s^23d^6$	$[Ar]4s^23d^6$

For bromine, however, there is a difference:

	valence electrons	noble gas notation
Br:	$4s^24p^5$	$[Ar]4s^23d^{10}4p^5$

The 3d electrons in bromine are not valence electrons. They do not participate in the reactions of bromine.

By locating the element in the periodic table (column and row) we can identify the valence electrons and write the electron structure in noble gas notation. If you are asked for the full shorthand notation, just replace the noble gas in brackets with its full subshell listing. Remember, you do not have to memorize the order in which the subshells fill. The periodic table is your guide. Start at 1s (H, He), then 2s (Li, Be), then 2p (B to Ne), and so on. Let the rows and columns of the periodic table lead you through the subshells, in order! For phosphorus, iron, and bromine:

	valence electrons	noble gas notation	shorthand notation
${}_{15}P$:	$3s^23p^3$	$[{}_{10}Ne]3s^23p^3$	$1s^22s^22p_63s^23p^3$
${}_{26}Fe$:	$4s^23d^6$	$[{}_{18}Ar]4s^23d^6$	$1s^22s^22p^63s^23p^64s^23d^6$
${}_{35}Br$:	$4s^24p^5$	$[{}_{18}Ar]4s^23d^{10}4p^5$	$1s^22s^22p^63s^23p^64s^23d^{10}4p^5$

There is a quick way to check that you have not left out some electrons. Notice in the above summary that the atomic numbers of the elements have been included. For example, $_{15}P$ indicates that the atomic number of phosphorus is 15. This means there are 15 protons in a phosphorus atom.. It also means thare are 15 electrons in the neutral atom. In general, *the sum of the electrons in the shorthand notation must equal the atomic number of the element.* This may be done by adding all the superscripts in the complete shorthand notation. For the noble gas notation, this is checked by adding the atomic number of the noble gas and all the superscripts:

$_{15}P$:	10 + 2 + 5 = 15	and	2 + 2 + 6 + 2 + 3 = 15
$_{26}Fe$:	18 + 2 + 6 = 26	and	2 + 2 + 6 + 2 + 6 + 2 + 6 = 26
$_{35}Br$:	18 + 2 + 10 + 5 = 35	and	2 + 2 + 6 + 2 + 6 + 2 + 10 + 5 = 35

EXAMPLE 11.6: For the following elements give the valence electrons, the noble gas notation and the shorthand notation.

	valence electrons	noble gas notation	shorthand notation
Ca			
Sn			
Hg			

	valence	noble gas notation	shorthand notation
$_{20}Ca$	$4s^2$	$[_{18}Ar]4s^2$	$1s^22s^22p^63s^23p^64s^2$
		18 + 2 = 20	2 +2 +6 +2 +6 +2 = 20
$_{50}Sn$	$5s^25p^2$	$[_{36}Kr]5s^24d^{10}5p^2$	$1s^22s^22p^63s^23p^64s^23d^{10}4p^65s^24d^{10}5p^2$
		36+2+10+2 = 50	2 +2 +6 +2 +6 +2 +10 +6 +2 +10 +2 = 50
$_{80}Hg$	$6s^25d^{10}$	$[_{54}Xe]6s^24f^{14}5d^{10}$	$1s^22s^22p^63s^23p^64s^23d^{10}4p^65s^24d^{10}5p^66s^24f^{14}5d^{10}$
		54+2+14+10 = 80	2+2+6+2+6+2+10+6+2+10+6+2+14+10 = 80

11.7 MAGNETIC AND SPIN QUANTUM NUMBERS

The previous two sections have presented the first two quantum numbers and the shell and subshell structure of atoms. This knowledge alone is enough to understand most of the differences in properties of atoms. The next two sections present the last two quantum numbers and give a more detailed discussion of the properties of orbitals.

The **magnetic quantum number** has the symbol m. Electrons in the same subshell which have the same value of m are in the same **orbital**. The values of m are the positive and negative whole numbers, beginning with zero and ending with ℓ: $m = 0, \pm 1, \pm 2 \ldots \pm \ell$. Each value of m indicates a different *direction in space* toward which the orbital is pointing. It is common to identify these orbitals by their position on a set of x, y, z axes. Figure 11.9 shows the orbitals within the first three shells. The symbols for these orbitals are:

subshell	m values	number of orbitals	symbols
1s	m = 0	1	1s
2s	m = 0	1	2s
2p	m = +1,0,-1	3	$2p_x$, $2p_y$, $2p_z$
3s	m = 0	1	3s
3p	m = +1,0,-1	3	$3p_x$, $3p_y$, $3p_z$
3d	m = +2,+1,0,-1,-2	5	$3d_{xy}$, $3d_{xz}$, $3d_{yz}$, $3d_{x^2-y^2}$, $3d_{z^2}$

Notice that each subshell has 2 more orbitals than the one before it in the same shell.

The use of the x-, y-, z-axes subscripts helps in drawing the orbitals. Begin by drawing the x-, y-, z-axes as three straight lines at right angles to each other. The s orbitals can be drawn as a circle and then shaded in to indicate a sphere. Remember that the spheres get larger as n increases. 1s is smaller than 2s, and 2s is smaller than 3s. Review Figure 11.9a.

The p orbitals can be drawn as figure eights. There are always just three p orbitals. The p_x orbital is along the x-axis. Then p_y is along the y-axis, and p_z is along the z-axis. The p orbitals are at right angles to each other just like the axes are. Take a look again at Figure 11.9b.

There are five d orbitals. Four of them are cloverleaves, which can be drawn like two figure eights. Three of the d orbitals are *between the axes*: d_{xy}, d_{xz}, and d_{yz}. The d_{xy} is between the x-axis and the y-axis. It is in the xy plane. Similarly, d_{xz} is between the x- and z-axes, and d_{yz} is between the y- and z-axes. These are shown in Figure 11.9c.

The other two d orbitals are *on the axes*. The $d_{x^2-y^2}$ has its cloverleaf on the x- and y-axes. The d_{z^2} is not a cloverleaf. It has a dumbbell on the z-axis and a doughnut-shaped surface in the xy plane (Figure 11.9c).

There is no correlation between the value of m and the axis of the orbital. That is, we cannot say that the p orbital with m = 0 is definitely on the x-axis. The reason for this is that the axis labels are artificial. They don't really exist. (If you could look at an atom, you wouldn't see little lines sticking out labeled "x," "y," "z"!) The axis labels just help us to draw the orbitals in the right directions compared to each other.

The m stands for "magnetic" quantum number. The motion of charged particles, such as electrons, produces a magnetic field. Each value of m represents a specific direction of the magnetic field. As you can tell from Figure 11.9, only certain directions are allowed.

You should be able to write the symbols for the shells, subshells, and orbitals. You should know their quantum number values (n, ℓ, and m). And you should be able to draw the orbitals from 1s to 3d, as shown in Figure 11.9.

The last quantum number is s, the **spin quantum number**. Like m, s is a magnetic quantum number. The magnetic field comes from the spin of the electron. Within an orbital an electron can have one of only two values of s: $+\frac{1}{2}$ or $-\frac{1}{2}$. We usually say that the direction of the magnetic field can be either "up" or "down." Arrows are used to show these directions. Spin up is "↑" and spin down is "↓." Each orbital can hold only two electrons, one with spin up and one with spin down: ↑↓. These electrons are said to be "spin paired." Of course, within an atom there are no up and down directions. This means that spin up could be either $s = +\frac{1}{2}$ or $s = -\frac{1}{2}$. All we know for certain is that if two electrons are in the same orbital, one has $s = +\frac{1}{2}$ and the other has $s = -\frac{1}{2}$.

The four quantum numbers and their properties are summarized in Table 11.3.

a. All s orbitals are spheres. They increase in size (average distance from the nucleus) as n increases. ℓ and m are both zero. s is $+\frac{1}{2}$ or $-\frac{1}{2}$.

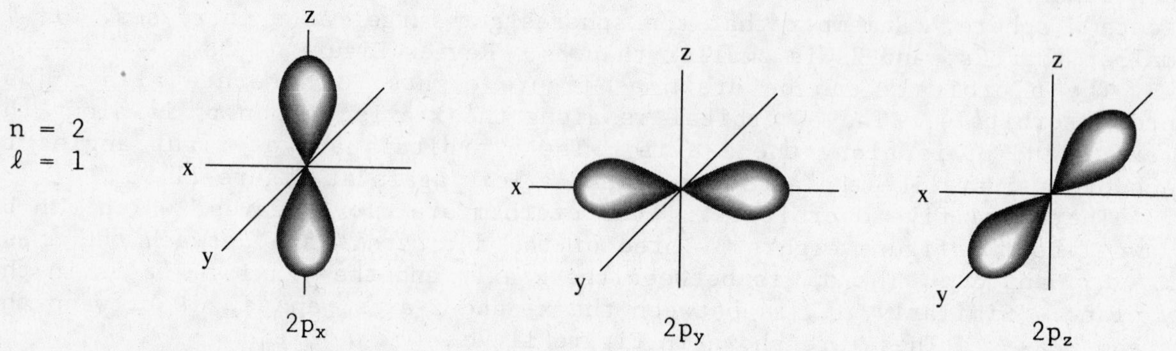

b. In p orbitals (here $2p_x$, $2p_y$, and $2p_z$), $\ell = 1$ and m = 0, +1, or −1. All are dumbbell in shape. The direction is given by m (as x, y, and z).

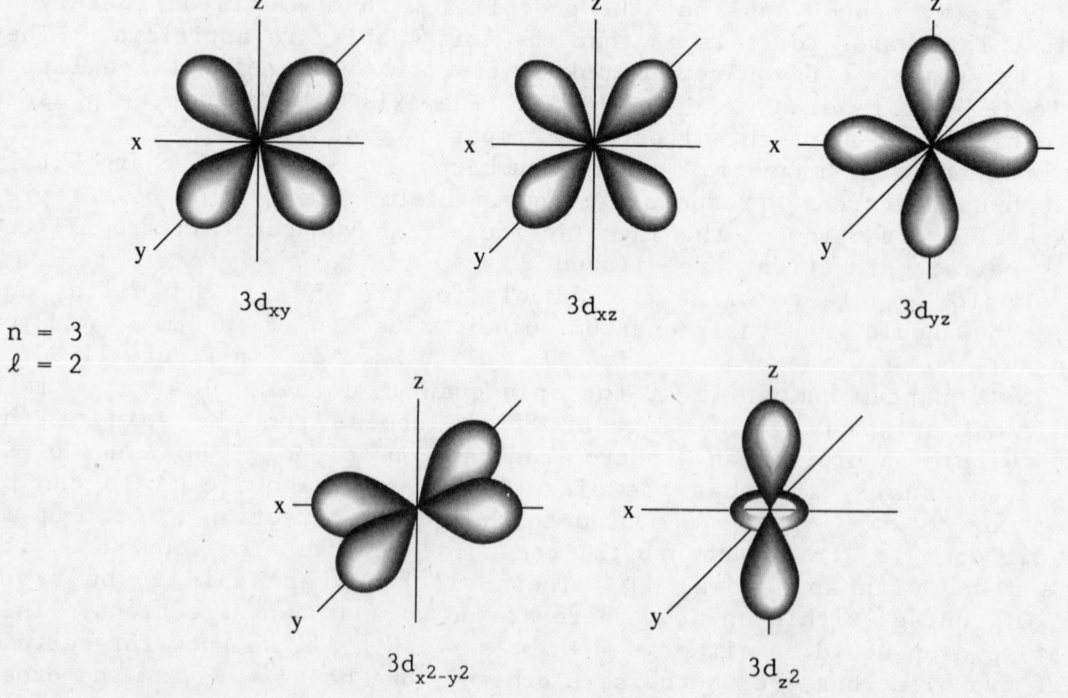

c. In d orbitals (here 3d) $\ell = 2$, and there are five possible orientations (m = 0, +2, −2, +1, or −1) called d_{xy}, d_{xz}, d_{yz}, $d_{x^2-y^2}$, and d_{z^2}.

FIGURE 11.9. SHAPES AND DIRECTIONS OF ORBITALS.

TABLE 11.3: THE QUANTUM NUMBERS

Symbol	Name	Values	Electron Position	Electron Energy	Identifies
n	principal	1, 2, 3...∞	distance	+/- attraction	shell
ℓ	secondary	0, 1, 2...n-1	shape	angular momentum	subshell
m	magnetic	0, ±1, ±2...±ℓ	orbital direction	orbital magnetic	orbital
s	spin	+½, -½	spin direction	spin magnetic	electron

EXAMPLE 11.7: (a) What are the possible values of n, ℓ, m, and s for an electron in a $3d_{xy}$ orbital? (b) What is the symbol for an orbital which has n = 4, ℓ = 1, and is on the z-axis? (c) Draw these two orbitals.

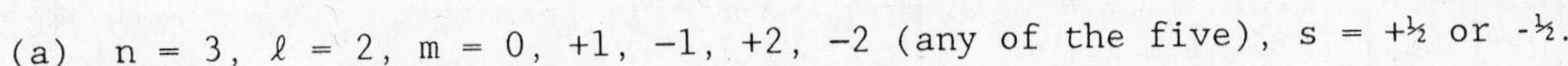

(a) n = 3, ℓ = 2, m = 0, +1, -1, +2, -2 (any of the five), s = +½ or -½.
(b) $4p_z$
(c)

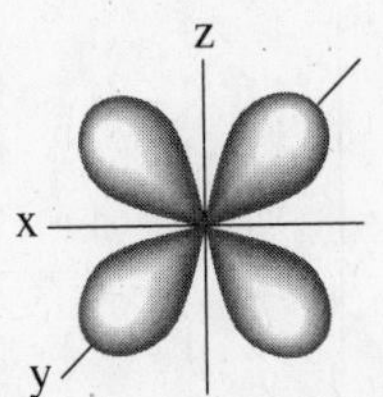

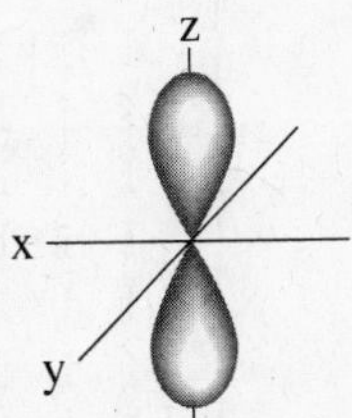

11.8 ELECTRON STRUCTURES OF THE ELEMENTS

Electrons have a definite, lowest energy arrangement in most atoms. This ground state electron structure the first 36 elements are shown in Table 11.4. There are several patterns that occur in the table.

First and most important, *no two electrons in an atom have the same set of four quantum numbers*. This is known as the **Pauli exclusion principle**. For example, the two electrons in helium are both in the 1s orbital.

He: [↑↓] (1s) ground state

This means that they both have n = 1, ℓ = 0, and m = 0. But they have different spins, represented as ↑↓. Therefore, one has s = +½ and the other has s = -½. We say that such an orbital is "full." (No more than two electrons

TABLE 11.4: ELECTRON STRUCTURES OF THE FIRST 36 ELEMENTS

Z	Element	1s	2s	2p			3s	3p			3d					4s	4p		
				x	y	z		x	y	z	xy	xz	yz	z^2	x^2-y^2		x	y	z
1	H	↑																	
2	He	↑↓																	
3	Li	↑↓	↑																
4	Be	↑↓	↑↓																
5	B	↑↓	↑↓	↑															
6	C	↑↓	↑↓	↑	↑														
7	N	↑↓	↑↓	↑	↑	↑													
8	O	↑↓	↑↓	↑↓	↑	↑													
9	F	↑↓	↑↓	↑↓	↑↓	↑													
10	Ne	↑↓	↑↓	↑↓	↑↓	↑↓													
11	Na	filled					↑												
12	Mg	filled					↑↓												
13	Al	filled					↑↓	↑											
14	Si	filled					↑↓	↑	↑										
15	P	filled					↑↓	↑	↑	↑									
16	S	filled					↑↓	↑↓	↑	↑									
17	Cl	filled					↑↓	↑↓	↑↓	↑									
18	Ar	filled					↑↓	↑↓	↑↓	↑↓									
19	K	filled					filled									↑			
20	Ca	filled					filled									↑↓			
21	Sc	filled					filled				↑					↑↓			
22	Ti	filled					filled				↑	↑				↑↓			
23	V	filled					filled				↑	↑	↑			↑↓			
24	Cr	filled					filled				↑	↑	↑	↑	↑	↑			
25	Mn	filled					filled				↑	↑	↑	↑	↑	↑↓			
26	Fe	filled					filled				↑↓	↑	↑	↑	↑	↑↓			
27	Co	filled					filled				↑↓	↑↓	↑	↑	↑	↑↓			
28	Ni	filled					filled				↑↓	↑↓	↑↓	↑	↑	↑↓			
29	Cu	filled					filled				↑↓	↑↓	↑↓	↑↓	↑↓	↑			
30	Zn	filled					filled				↑↓	↑↓	↑↓	↑↓	↑↓	↑↓			
31	Ga	filled					filled				filled					↑↓	↑		
32	Ge	filled					filled				filled					↑↓	↑	↑	
33	As	filled					filled				filled					↑↓	↑	↑	↑
34	Se	filled					filled				filled					↑↓	↑↓	↑	↑
35	Br	filled					filled				filled					↑↓	↑↓	↑↓	↑
36	Kr	filled					filled				filled					↑↓	↑↓	↑↓	↑↓

can occupy a given orbital.) Every orbital which contains two electrons must have them spin paired. If they appeared as ↑↑ or ↓↓, they would both have the same values of n, ℓ, m and s. This would be a violation of the Pauli exclusion principle. No such violation has ever been known to occur.

	1s	
He:	↑↑	not possible--violates Pauli exclusion principle

Electrons generally fill the lowest shells (lowest n values) first. We will call this the **n effect**. The first two elements, H and He, have their electrons in the 1s orbital, n = 1. This is the lowest energy state for the electrons because they are closest to the nucleus. They therefore feel the greatest +/- attraction. When the n = 1 shell is filled, electrons go into the n = 2 shell next. (For atoms with a large number of electrons, not all the electrons go into the lower shells first. In a moment we will give an example of this and explain why it happens.)

	1s	2s	
He:	↑	↑	excited state--violates n effect

Within a shell, electrons fill the subshells in the order: s, p, d, f (lowest ℓ value first). We will call this the **ℓ effect**. Consider the electron structure of Li, which has three electrons.

	1s	2s	2p			
Li:	↑↓	↑				ground state

The first two electrons fill the n = 1 shell. The third electron goes into the n = 2 shell (the n effect). There are two subshells in the n = 2 shell: 2s and 2p. The electron occupies the 2s subshell. 2s has ℓ = 0 while 2p has ℓ = 1. The 2s subshell is lower in energy than the 2p subshell. The reasons for the differences in energy of subshells are complex. We will not discuss them. (As we shall see, there are some exceptions to the ℓ effect also.)

	1s	2s	2p			
Li:	↑↓		↑			excited state--violates ℓ effect

Within a subshell electrons 1) occupy DIFFERENT ORBITALS (have different m values) but 2) have the SAME SPIN (same s value). This is known as **Hund's rule**. We will call the first part the **m effect** and the second part the **s effect**. Take a look at the electron structure of carbon.

	1s	2s	$2p_x$	$2p_y$	$2p_z$	
C:	↑↓	↑↓	↑	↑		ground state

The first two electrons fill the n = 1 shell (n effect). The next two electrons fill the 2s subshell (ℓ effect). The two electrons in the 2p subshell are in different orbitals: $2p_x$ and $2p_y$ (the m effect). And they have the same spin: both up (the s effect). Each of the two parts of Hund's rule can be understood by looking at the behavior of electrons.

First, electrons occupy different orbitals. If they were both in the same orbital, they would be in the same region of space. (Both might be along the x-axis, for example.) Their negative charges would repel each other strongly. This is known as **electron pair repulsion**. The repulsion is less when the two electrons are not as close together. (One might be on the x-axis while the other is on the y-axis, for example.)

	1s	2s	$2p_x$	$2p_y$	$2p_z$	
C:	↑↓	↑↓	↑↓			excited state--violates Hund's rule (m effect)

Second, electrons have the same spin. The electron's spin produces a magnetic field, either up or down. When electrons have the same spin, their magnetic fields are aligned, the lowest energy state.

	1s	2s	$2p_x$	$2p_y$	$2p_z$	
C:	↑↓	↑↓	↑	↓		excited state--violates Hund's rule (s effect)

The next element, nitrogen, has three electrons in the 2p subshell. All of them are in different orbitals. Following Hund's rule, all three have the same spin.

	1s	2s	$2p_x$	$2p_y$	$2p_z$	
N:	↑↓	↑↓	↑	↑	↑	ground state

The next element, oxygen, has four electrons in the 2p subshell. Because there are only three orbitals, not all of the four electrons can have their own orbital. The fourth electron must pair up with one of the first three.

	1s	2s	$2p_x$	$2p_y$	$2p_z$	
O:	↑↓	↑↓	↑↓	↑	↑	ground state

This pattern continues until the 2p subshell is filled at neon. The 3s and 3p subshells fill in the same way as did the 2s and 2p subshells. At argon, element 18, we have the electron structure:

	1s	2s	$2p_x$	$2p_y$	$2p_z$	3s	$3p_x$	$3p_y$	$3p_z$	
Ar:	↑↓	↑↓	↑↓	↑↓	↑↓	↑↓	↑↓	↑↓	↑↓	ground state

The next electron, added for the element potassium, K, does not go into the 3d subshell. It occupies the 4s orbital instead. The 4s orbital of potassium is lower in energy than the 3d orbitals. The electron structure of K is:

	1s	2s	$2p_x$	$2p_y$	$2p_z$	3s	$3p_x$	$3p_y$	$3p_z$	4s	3d	
K:	↑↓	↑↓	↑↓	↑↓	↑↓	↑↓	↑↓	↑↓	↑↓	↑		ground state

TABLE 11.5: SUMMARY OF ORBITAL FILLING EFFECTS

Effect		Pattern	Reason
Pauli		no two e^- have same n, ℓ, m, and s	basic principle of nature
n		fill lower n first	e^- gets closer to nucleus for better +/− attraction
ℓ		within the same shell: fill lower ℓ first (s < p < d < f)	reasons are complex
Hund's Rule	m	within the same subshell: fill different m first (different orbitals)	electron pair repulsion
Hund's Rule	s	within the same subshell: e^-'s have same s (same spin)	magnetic fields aligned

conflict here. The n effect says that 3d (n = 3) should be lower than 4s (n = 4). The ℓ effect says that 4s (ℓ = 0) should be lower than 3d (ℓ = 2). In all the cases we have seen up to now, the n effect "wins." That is, the electron goes into the lower shell. Beginning here, and several more times for the higher elements, the ℓ effect becomes more important than the n effect. We will show you examples of this in the next section. Within the first 36 elements, this occurs just once: 4s fills before 3d.

	1s	2s	$2p_x$ $2p_y$ $2p_z$	3s	$3p_x$ $3p_y$ $3p_z$	4s	3d	
K:	↑↓	↑↓	↑↓ ↑↓ ↑↓	↑↓	↑↓ ↑↓ ↑↓		↑	excited state violates ℓ effect

EXAMPLE 11.8: Identify which of the following electron structures are not for the ground state of the element. Point out which effect makes the electron structure not that of the ground state. Then write the correct ground state for the element.

(a)

	1s	2s	2p	3s	3p
Na:	↑↓	↑↓	↑↓ ↑↓ ↑↓		↑

(b)

	1s	2s	2p	3s	3p
Si:	↑↓	↑↓	↑↓ ↑↓ ↑↓	↑↓	↑ ↓

(c)

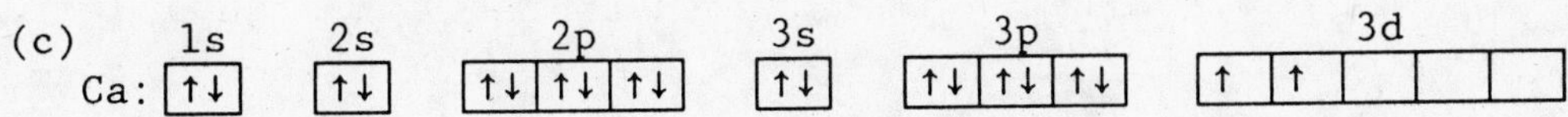

	1s	2s	2p	3s	3p	3d
Ca:	↑↓	↑↓	↑↓ ↑↓ ↑↓	↑↓	↑↓ ↑↓ ↑↓	↑ ↑

(a) Not the ground state--violates the ℓ effect. Ground state:

	1s	2s	2p	3s	3p
Na:	↑↓	↑↓	↑↓ ↑↓ ↑↓	↑	

(b) Not the ground state--violates Hund's rule (s effect). Ground state:

	1s	2s	2p	3s	3p
Si:	↑↓	↑↓	↑↓ ↑↓ ↑↓	↑↓	↑ ↑

(c) Not the ground state--violates the ℓ effect. Ground state:

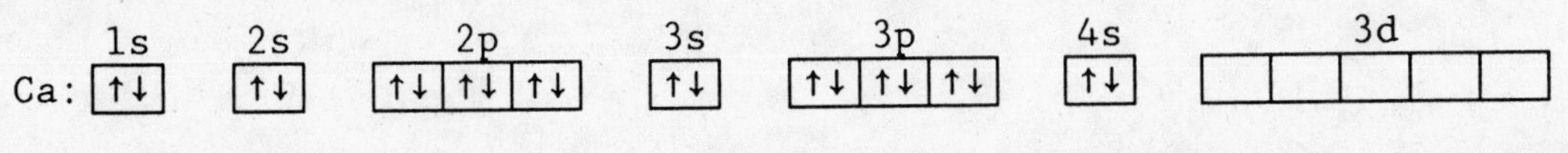

	1s	2s	2p	3s	3p	4s	3d
Ca:	↑↓	↑↓	↑↓ ↑↓ ↑↓	↑↓	↑↓ ↑↓ ↑↓	↑↓	

When we began writing electron structures, we wrote boxes for each orbital and arrows for each electron. These detailed electron structures can be written for the elements, again without memorization. To do this, we begin with the complete shorthand listing of subshells. Then Hund's rule is applied to show the arrangement of electrons in the orbitals of each subshell.

As an example, phosphorus is $1s^2\ 2s^2\ 2p^6\ 3s^2\ 3p^3$. The s subshells have just one orbital. The p subshells have three. Shown from lowest to highest energy, the box diagram is:

3p	☐	☐	☐
3s		☐	
2p	☐	☐	☐
2s		☐	
1s		☐	

This vertical arrangement of energy levels emphasizes that the orbitals within each subshell have the same energy.

The boxes are filled in with electrons following Hund's rule. Electrons in the same subshell go in different orbitals. Their spins are the same as far as possible. If two electrons must go in the same orbital, their spins will be opposite. Following these rules, we obtain the detailed electron structure of phosphorus.

$3p^3$	↑	↑	↑
$3s^2$		↑↓	
$2p^6$	↑↓	↑↓	↑↓
$2s^2$		↑↓	
$1s^2$		↑↓	

EXAMPLE 11-9. Draw the detailed electron structure (the box diagram) for iron.

$_{26}Fe$: $1s^2\ 2s^2\ 2p^6\ 3s^2\ 3p^6\ 4s^2\ 3d^6$

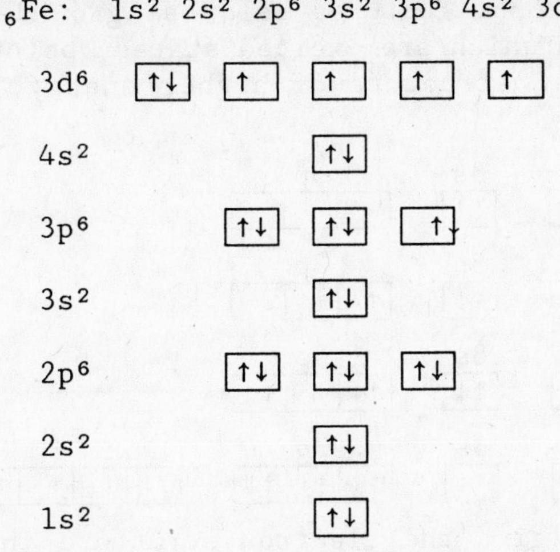

CHAPTER 11 PROBLEMS

1) What is the wavelength in nanometers of light with a frequency of $3.75 \times 10^{14}\ s^{-1}$?

2) What is the wavelength in meters and nanometers of light with an energy of 2.25×10^{-18} joule?

3) Place these bands of light in order from highest frequency on the left to lowest frequency on the right: green, infrared, orange, microwave, and x-ray.

4) What frequency of light is emitted when an electron in a hydrogen atom falls from $n = 5$ to $n = 3$? Is this light in the Lyman, Balmer, Paschen, or Brackett series?

5) Sketch the Rydberg energy level diagram and the Bohr planetary model showing the change in the previous problem.

6) Give the n and ℓ values for these subshells: a) 4p, b) 6s, c) 3f, d) 5d.

7) For each of these elements in the ground state a) give the electron structure in shorthand notation, b) in noble gas form and c) identify the valence electrons: Li, P, Se, Mo, and Hg.

8) Give the possible sets of $\{n, \ell, m, s\}$ for the electrons in a) $4p^2$, b) $6s^2$, c) $4d^5$, and d) $3p^4$.

9) Draw boundary surface diagrams, with axis designations, for all the orbitals in these subshells: a) 4p, b) 6s, c) 3f, and d) 5d.

10) In each of the following cases, state how many ground state electrons are present with each combination of quantum numbers.

a) How many electrons in magnesium have $s = +\frac{1}{2}$?

b) How many electrons in manganese have $n = 3$ and $\ell = +1$?

c) How many electrons in sulfur have $m = 0$ and $s = +\frac{1}{2}$?

d) How many total electrons can have $n = 2$?

e) How many total electrons can have $n = 3$ and $\ell = +2$?

11) Identify each of the electron structures below as ground state, excited state, or impossible. For those which are excited states, point out which effect (n, ℓ, m, or s) makes that structure of higher energy, and write its proper ground state.

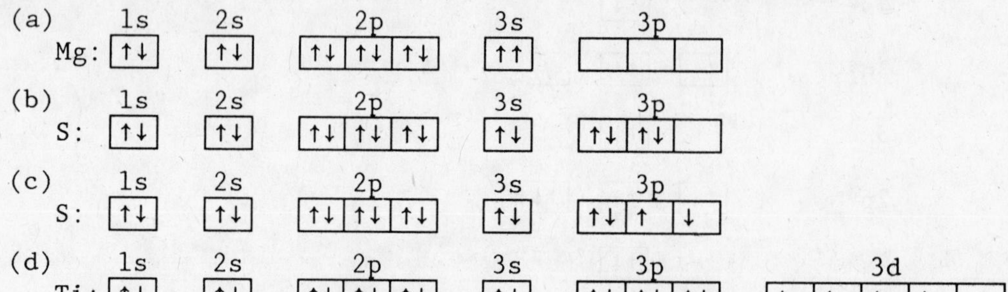

12) For each of these elements give the electron structure in "box" format: a) O, b) Al, c) Ni, d) Sr, and e) Bi.

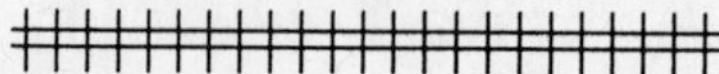

13) What is the frequency of light with a wavelength of 23.4 mm?

14) What is the frequency and energy of light with a wavelength of 1.22 km?

15) Place these bands of light in order from highest energy on the left to lowest energy on the right: ultraviolet, blue, microwaves, yellow, radio waves.

16) What is the n_o value for the Paschen line with a wavelength of 1282 nm?

17) Sketch the Rydberg energy level diagram and the Bohr planetary model showing the change in the previous problem.

18) Give the n and ℓ values for these subshells: a) 5d, b) 2d, c) 4s, d) 3p.

19) For each of these elements in the ground state a) give the electron structure in shorthand notation and b) in noble gas form and c) identify the valence electrons: Ne, Br, As, Fe, and K.

20) Give the possible sets of {n, ℓ, m, s} for the electrons in the highest energy subshell of these elements: a) Be, b) Ar, c) Mn, and d) Cl.

21) Draw boundary surface diagrams, with axis designations, for all the orbitals in these subshells: a) 5d, b) 2d, c) 4s, d) 3p.

22) In each of the following cases, state how many ground state electrons are present with each combination of quantum numbers.

a) How many electrons in neon have n = 2 and s = -½?

b) How many electrons in sodium have ℓ = 0?

c) How many electrons in phosphorus have ℓ = 1 and m = +1?

d) How many total electrons have n = 3?

e) How many total electrons have n = 4 and ℓ = 3?

23) Identify each of the electron structures below as ground state, excited state, or impossible. For those which are excited states, point out which effect (n, ℓ, m, or s) makes that structure of higher energy, and write its proper ground state.

(a)

	1s	2s	2p			3s	3p		
Al:	↑↓	↑↓	↑↓	↑↓	↑↓	↑↓		↓	

(b)

	1s	2s	2p			3s	3p		
P:	↑↓	↑↓	↑↓	↑↓	↑↓	↑	↑↓	↑	↑

(c)

	1s	2s	2p			3s	3p		
S:	↑↓	↑↓	↑↓	↑↓	↑↓	↑↓	↑↑	↑	↑

(d)

	1s	2s	2p			3s	3p			4s	3d				
Fe:	↑↓	↑↓	↑↓	↑↓	↑↓	↑↓	↑↓	↑↓	↑↓	↑↓	↑↓	↑↓	↑↓		

24) For each of these elements give the electron structure in "box" format: a) F, b) Ca, c) V, d) Kr, and e) Hg.

ADDITIONAL PROBLEMS

25) One of the lines in the hydrogen emission spectrum occurs at 434.2 nm. Find the values of n_i and n_o and show this electronic transition both on the Rydberg energy level diagram and on the Bohr planetary model.

26) Why doesn't the Bohr model work for atoms with more than one electron?

27) What is Hund's rule? Give an example of an excited state which does not follow the m effect of Hund's rule. Give an example of a violation of the s effect of Hund's rule. Explain these two effects in energy terms.

28) As you may have noticed in Table 11.1, chromium has the ground state electron structure $[Ar]4s^13d^5$ rather than the expected $[Ar]4s^23d^4$. This means that the $4s^13d^5$ arrangement is lower in energy than is $4s^23d^4$. Draw the box diagram for these two cases and explain their energy difference in terms of n, ℓ, m, and s effects.

CHAPTER TEST

1) Calculate the energy of the light emitted when the hydrogen electron falls from n = 4 to n = 1.

2) What are the n values for the second lowest energy line in the Balmer series? Show this electron change on both the Rydberg energy level diagram and the Bohr planetary model.

3) Put these bands of light in order from longest wavelength on the left to shortest wavelength on the right: indigo, green, ultraviolet, infrared, and gamma radiation.

4) Give the electron structure of antimony in shorthand notation, in noble gas notation, and as a box diagram. Identify the valence electrons.

5) Is the electron structure shown below for oxygen the ground state, an excited state, or impossible? If it is an excited state, is this due to the n, ℓ, m, or s effects?

	1s	2s	2p		
Al:	↑↓	↑↓	↑↓	↑↓	

12. THE GAS LAWS

PREREQUISITES

1. The ideal gas law requires that algebraic equations be solved for an unknown variable, using methods developed in Section 1.10.

2. Section 2.2 describes direct and inverse proportionalities that are the basis of the reasoning approach to solving gas law problems.

3. You will use the dimensional analysis method (Section 2.4) to solve gas stoichiometry problems.

4. "Per" relationships, introduced in Section 2.5, are the sources of several conversion factors.

5. The concept of density (Section 3.12) appears in gas problems.

6. Section 10.2 introduced the "stoichiometry 3-step," which is applied to reactions of gases in this chapter.

CHEMICAL SKILLS

1. Pressure units. Given a gas pressure expressed in atmospheres, torr, or kilopascals, express that pressure in the other units.

2. Charles' law; Boyle's law; Avogadro's law. Given both initial values of variables in any of the above laws and the final value of one variable, calculate the final value of the other variable.

3. Combined gas laws. Given the initial values of pressure, volume, temperature, and number of moles of a gas and final values of three of those variables, calculate the remaining final value.

4. Ideal gas law. Given values for all but one of the variables in the ideal gas equation (P, V, n, T), calculate the value of the remaining variable.

5. Standard temperature and pressure (STP). Given the volume (or number of moles) of any gas at STP, calculate the number of moles (or volume).

6. Density and molar mass. Given the density of a gas at a particular pressure and temperature, or data from which the density may be determined, calculate the molar mass of the gas; given the identity or molar mass of a gas, calculate its density at specified temperature and pressure.

7. Gas stoichiometry. Given a chemical reaction for which an equation may be written and the quantity of any substance in the reaction, calculate the quantity of any other substance in the reaction. Either quantity may be in volume of gas at specified temperature and pressure or in grams.

8. Dalton's law of partial pressures. Given all but one of the following, calculate the missing quantity: partial pressure of each gas in a mixture and the total pressure of the mixture.

9. Dalton's law of partial pressures. Given the molar composition of a mixture of gases, or information from which it may be found, and the total pressure, calculate the partial pressure of any component.

10. Graham's law of effusion. Given the relative rates or times for a known gas and an unknown gas to escape from an effusion apparatus, both gases at the same starting conditions, calculate the molar mass of the unknown gas.

12.1 MEASUREMENTS WITH GASES

The four measurements that are commonly associated with gases are:

Volume. A gas fills the container that holds it. The volume (V) of a gas is therefore the same as the volume of its container. Volume is usually expressed in liters (L), and sometimes milliliters (mL).

Temperature. The temperature of a gas is measured with a thermometer and expressed in degrees Celsius (°C). In its relationship with other variables, temperature must be expressed as absolute temperature in kelvins (K). Degrees Celsius may be converted to kelvins simply by adding 273.15, an exact (by definition) relationship:

$$K = °C + 273.15 \qquad (12.1)$$

In this book we will use Celsius temperatures to a precision of ±1°C, so that the Kelvin temperature may be obtained by adding just 273.

Quantity: Mass or Moles. Though the techniques are a bit more involved than with liquids or solids, the mass (m) of a gas may be determined by weighing. In its relationship to the other variables, however, the amount of gas present is best expressed in number of moles (n). Grams of a substance may be converted to moles by dividing by molar mass (MM) in grams per mole.

Pressure. By definition, pressure is force per unit area. The SI unit of pressure that corresponds to the definition is newtons per square meter. One newton per square meter is called a **pascal** (Pa). A pascal is so small, it is usually converted to the unit 1000 times larger, the kilopascal (kPa).

Air pressure is normally measured with a **barometer**. The pressure of a gas in a closed container is often measured with a **manometer**. These instruments usually measure the height of a column of mercury, which is proportional to pressure when corrected for thermal expansion. The **millimeter of mercury** is the resulting pressure unit. This unit is commonly called the torr, the height in millimeters of a column of mercury at 0°C.

Very high pressures are often given in atmospheres (atm). One atmosphere is the "normal" atmospheric pressure at sea level. It is defined as exactly 760 torr.

The relationships between the three major pressure units are

$$101.3 \text{ kPa} = 760 \text{ torr} \equiv 1 \text{ atm (exactly)} \qquad (12.2)$$

Changing from one pressure unit to another uses one of the conversion factors derived directly from Equation 12.2:

760 torr/atm   101.3 kPa/atm   760 torr/101.3 kPa

In this book we will generally express pressures only in atm and torr.

EXAMPLE 12.1: What pressure in torr and in kilopascals corresponds to 0.882 atmospheres?

GIVEN: 0.882 atm   FIND: torr
FACTOR: 760 torr/atm   PATH: atm → torr

$$0.882 \text{ atm} \times \frac{760 \text{ torr}}{1 \text{ atm}} = 6.7 \times 10^2 \text{ torr}$$

GIVEN: 0.882 atm   FIND: kPa
FACTOR: 101.3 kPa/atm   PATH: atm → kPa

$$0.882 \text{ atm} \times \frac{101.3 \text{ kPa}}{1 \text{ atm}} = 89.3 \text{ kPa}$$

12.2 THE GAS LAWS: VOLUME, TEMPERATURE, PRESSURE, AND MOLES

Several relationships between the measurable properties of a gas are known as the "gas laws." Three of them, Charles' law, Boyle's law, and Avogadro's law, are introduced at this time. They are considered separately in this section and combined in the next.

Charles' Law It is an experimental fact that the volume occupied by a fixed number of moles of gas at constant pressure is directly proportional to the absolute temperature (note: ABSOLUTE temperature, that is, in kelvins):

$$V \propto T \quad \text{(moles and pressure constant, T in kelvin)} \qquad (12.3)$$

This relationship is known as Charles' law. The meaning of the law is that if you increase the temperature of a gas sample at constant pressure, the volume increases in the same proportion. If you reduce the temperature, the volume is reduced accordingly. In a direct proportion the variables change in the same direction.

By introducing a proportionality constant, C, Charles' law may be expressed as an equation:

$$V = C \times T \tag{12.4}$$

Solving for C,

$$\frac{V}{T} = C \tag{12.5}$$

This shows that if the pressure of a fixed amount of gas is constant, the ratio V/T is constant.

The most common application of Charles' law is to calculate the final volume, V_2, of a fixed amount of gas at a final temperature, T_2, if its initial volume was V_1 at a starting temperature of T_1, both volumes measured at the same pressure.

Algebra Method. The problem may be solved in either of two ways. The algebra method is based on the mathematical relationship between the four variables, which comes from Equation 12.5:

$$\frac{V_2}{T_2} = C = \frac{V_1}{T_1} \qquad\qquad \frac{V_2}{T_2} = \frac{V_1}{T_1} \tag{12.6}$$

If given values for three of the four variables in Equation 12.6, the equation can be solved algebraically for the missing value, the given quantities substituted, and the remaining value calculated.

Suppose that a sample of gas occupies 5.00 L at 275 K. What volume will it occupy at 325 K? The problem may be summarized as:

	Volume	Temperature	Pressure	Amount
Initial value (1)	$V_1 = 5.00$ L	$T_1 = 275$ K	P_1 = constant	n_1 = constant
Final value (2)	V_2 = ?	$T_2 = 325$ K	P_2 = constant	n_2 = constant

Solving Equation 12.6 for V_2 and substituting the given values:

$$V_2 = \frac{V_1 T_2}{T_1} = \frac{5.00 \text{ L} \times 325 \text{ K}}{275 \text{ K}} = 5.91 \text{ L}$$

By rearranging the above setup slightly, we find that the new volume can be calculated by multiplying the initial volume by a ratio of absolute temperatures. We will call this the *factor method*. Here is how it works:

When the temperature of the 5.00 L sample of gas increases from 275 K to 325 K, will its volume increase or decrease? Will it be more than 5.00 L or less? If volume is directly proportional to temperature, volume and temperature must increase together. (This is logical. Most things expand when they are heated.)

If the final volume is reached by multiplying 5.00 L by a ratio of temperatures, and the final volume will be greater than 5.00 L, the ratio must be greater than 1. The larger temperature must be in the numerator, and the smaller temperature must be in the denominator:

$$V_2 = V_1 \times (\text{T factor, >1}) = V_1 \times \frac{T_2}{T_1} = 5.00 \text{ L} \times \frac{325 \text{ K}}{275 \text{ K}} = 5.91 \text{ L}$$

The setup derived by proportional reasoning is the same as that obtained by solving Equation 12.6 for V_2. The methods are equivalent, and you are free to choose the one that appeals most to you—unless, of course, your instructor directs you to use one or the other. Either way, we suggest that you concentrate on learning one method and disregard the other. If you are choosing for yourself, compare methods through the first few examples—you might want to glance at the introduction to Section 12.4 too—and then drop the one that appeals to you the least. We will, as a matter of routine, show both methods for all examples.

EXAMPLE 12.2: The volume of a gas is 3.90 liters at 72°C. What will the volume become if the temperature is changed to −18°C at constant pressure?

To help you visualize the problem, begin by filling in the blanks in a table like the one used earlier:

Volume	Temperature	Pressure	Amount
V_1 =	T_1 =	P_1 =	n_1 =
V_2 =	T_2 =	P_2 =	n_2 =

Volume	Temperature	Pressure	Amount
V_1 = 3.90 L	T_1 = 72°C = 345 K	P_1 = constant	n_1 = constant
V_2 = ?	T_2 = −18°C = 255 K	P_2 = constant	n_2 = constant

Notice that temperature, though given in degrees Celsius, is also entered in kelvins, reached by adding 273 to the given Celsius temperature. THIS IS VERY IMPORTANT. *All equations involving gases require that temperature be in kelvins, not degrees Celsius.* We suggest that you change Celsius degrees to kelvins as soon as you start to work on a problem.

Factor Method. This time the temperature comes down. What will happen to the volume when temperature decreases? Will volume increase (___) or decrease (___)? Will the temperature ratio by which the initial volume must be multiplied be more than 1 (___) or less than 1 (___)? Check the correct answers.

In a direct proportion the variables move in the same direction, so a drop in temperature will cause a decrease in volume. This is logical. If you cool a balloon, it will shrivel up. The initial volume must be multiplied by a temperature ratio less than 1.

Having reached the conclusions above, set up the problem and solve. (Note: The table summary of the problem includes and substitutes for the GIVEN-FIND-FACTOR-PATH format we normally use.)

$$V_2 = V_1 \times (\text{T factor}, <1) = 3.90 \text{ L} \times \frac{255 \text{ K}}{345 \text{ K}} = 2.88 \text{ L}$$

Algebra Method. Solve Equation 12.6 for V_2, substitute known values, and calculate the answer.

Equation 12.6: $\frac{V_2}{T_2} = \frac{V_1}{T_1}$ $\qquad V_2 =$

$$V_2 = \frac{V_1T_2}{T_1} = \frac{3.90 \text{ L} \times 255 \text{ K}}{345 \text{ K}} = 2.88 \text{ L}$$

Boyle's Law It can be shown experimentally that at constant temperature, the volume of a fixed quantity of gas is inversely proportional to pressure.

$$V \propto \frac{1}{P} \quad \text{(moles and temperature constant)} \tag{12.7}$$

This is Boyle's law. In an inverse proportion, one variable increases as the other decreases; they move in opposite directions. Inserting the proportionality constant B and solving the resulting equation for B gives

$$V = B \times \frac{1}{P} \tag{12.8}$$

$$PV = B \tag{12.9}$$

In an inverse proportionality it is the product of the variables that is constant. Again using subscripts to indicate initial and final conditions and equating the two products:

$$P_1V_1 = P_2V_2 \tag{12.10}$$

If given any three of the above four variables, the fourth may be found by means of Equation 12.10.

As before, it is possible to solve this kind of a problem by the factor method. If a gas has a volume of 24.0 liters at 3.00 atmospheres, and the pressure is reduced to 2.00 atmospheres, what will happen to the volume? Will it increase or decrease? Remember that the pressure-volume relationship is an inverse proportion, and that as one goes up, the other goes down. The drop in pressure therefore causes an increase in volume. The pressure ratio must be greater than 1, with the larger pressure in the numerator. The setup is

$$V_2 = 24.0 \text{ L} \times \frac{3.00 \text{ atm}}{2.00 \text{ atm}} = 36.0 \text{ L}$$

The setup is identical to what would have been obtained by solving Equation 12.10 for V_2 and substituting the given values.

EXAMPLE 12.3: A gas fills a space of 1.64 liters at 755 torr. What volume will it occupy if pressure is increased to 793 torr at the same temperature? Enter the given values below.

$V_1 =$	$T_1 =$	$P_1 =$	$n_1 =$
$V_2 =$	$T_2 =$	$P_2 =$	$n_2 =$

$V_1 = 1.64$ L	$T_1 =$ constant	$P_1 = 755$ torr	$n_1 =$ constant
$V_2 = ?$	$T_2 =$ constant	$P_2 = 793$ torr	$n_2 =$ constant

Factor Method. If the pressure changes from 755 torr to 793 torr, do you expect the volume to increase (___) or decrease (___)? Check the right answer.

If pressure increases, volume will *decrease*. Volume and pressure are inversely proportional; they move in opposite directions. This is reasonable; if you push on the outside of a balloon, the volume goes down.

Now select the pressure ratio by which the initial volume must be multiplied to get the final volume. Set up and solve the problem.

$$V_2 = V_1 \times (\text{P factor, } <1) = 1.64 \text{ L} \times \frac{755 \text{ torr}}{793 \text{ torr}} = 1.56 \text{ L}$$

If the volume is to decrease, the pressure ratio must be less than 1, with the lower pressure on top.

Algebra Method. Solve Equation 12.10 for V_2, substitute values, and calculate the answer.

Equation 12.10: $P_1V_1 = P_2V_2 \quad V_2 =$

$$V_2 = \frac{V_1P_1}{P_2} = \frac{1.64 \text{ L} \times 755 \text{ torr}}{793 \text{ torr}} = 1.56 \text{ L}$$

Avogadro's Law The volume of a gas at a given temperature and pressure also depends on the number of moles of gas present. Almost 200 years ago, Avogadro proposed that equal volumes of gases contained equal numbers of molecules, if both are at the same pressure and temperature. It follows that at constant P and T the volume of a gas is directly proportional to the number of moles present:

$$V \propto n \quad \text{(pressure and temperature constant)} \tag{12.11}$$

Inserting the proportionality constant A gives:

$$V = A \times n \tag{12.12}$$

For a change from V_1 and n_1 to V_2 and n_2 which is done at constant pressure and temperature:

$$\frac{V_1}{n_1} = A = \frac{V_2}{n_2} \qquad \frac{V_1}{n_1} = \frac{V_2}{n_2} \tag{12.13}$$

Problems involving changes in volume and moles may be solved by using Equation 12.13 or by proportional reasoning.

EXAMPLE 12.4: How many moles of helium must be added to a balloon currently holding 6.00 moles in order to increase its volume from 125 liters to 175 liters? Temperature and pressure are constant.

V_1 =	T_1 =	P_1 =	n_1 =
V_2 =	T_2 =	P_2 =	n_2 =

V_1 = 125 L	T_1 = constant	P_1 = constant	n_1 = 6.00 moles
V_2 = 175 L	T_2 = constant	P_2 = constant	n_2 = ?

Factor Method. The problem says moles of helium will be added. If moles of gas increases, what will happen to the volume, increase or decrease? Set up and solve. Be careful! The problem asks for the moles to be added.

$$n_2 = n_1 \times (\text{V factor}, >1) = 6.00 \text{ mol} \times \frac{175 \text{ L}}{125 \text{ L}} = 8.40 \text{ mol}$$

8.40 mol at end − 6.00 mol at start = 2.40 mol added

V and n are directly proportional. Therefore, to increase V, n must also be increased. The gas expands as more moles are pumped in. The starting number of moles must be multiplied by a volume ratio that is greater than 1 to get the final number of moles. The difference between the final quantity and the starting quantity is the number of moles added.

Algebra Method. Use Equation 12.13 to calculate n_2. Then find the number of moles added.

Equation 12.13: $\frac{V_1}{n_1} = \frac{V_2}{n_2}$ $\quad n_2 =$

Equation 12.13: $\frac{V_1}{n_1} = \frac{V_2}{n_2}$ $\quad n_2 = \frac{n_1 V_2}{V_1} = \frac{6.00 \text{ mol} \times 175 \text{ L}}{125 \text{ L}} = 8.40 \text{ mol}$

8.40 mol at end − 6.00 mol at start = 2.40 mol added

The difference between the final quantity and the starting quantity is the number of moles added.

12.3 THE IDEAL GAS LAW

So far we have held constant two of the four variables, pressure, temperature, volume, and moles. We will now combine all four into a single relationship. We have already pointed out that volume is directly proportional to absolute temperature (Proportionality 12.3), inversely proportional to pressure (Proportionality 12.7), and directly proportional to moles (Proportionality 12.11). If volume is proportional to each of three variables, it is proportional to their product (Section 2.2):

$$V \propto \frac{n \times T}{P} \tag{12.14}$$

Inserting a proportionality constant, R, known as the universal gas constant, and rearranging gives us

$$V = \frac{nRT}{P} \qquad \text{so} \qquad PV = nRT \tag{12.15}$$

PV = nRT is known as the **ideal gas law** or **ideal gas equation**. It is valid for any gas that behaves as an "ideal" gas, which most gases do at ordinary conditions of temperature and pressure. We will assume that this equation is valid for all gases considered in this book.

The value of the gas constant, R, depends on the units used for P, V, n, and T. T is always in kelvins, and n is always in moles. V is usually in liters; if given in milliliters, it is readily changed to liters. Chemists most often use atmospheres or torr for pressure measurements. The values of R for these two pressure units are

$$R = 0.08205 \text{ L·atm/mol·K} = 62.36 \text{ L·torr/mol·K}$$

These values are used constantly; they should be memorized.

Don't be concerned about the "strange" units for R. Proportionality constants frequently have complicated units when three or more proportionalities are combined. Units serve a valuable purpose in checking the correctness of algebraic solutions in problem solving, as you will soon see.

In most applications of the ideal gas law, you are given all but one of the variables, or information from which all but one can be found, and asked to calculate the remaining variable. This is a straightforward problem in algebra. If you use the units of R correctly, you are guaranteed that your algebraic operations have been performed correctly. The following example illustrates the procedure.

EXAMPLE 12.5: Find the temperature at which 0.667 moles of helium in a 4.89 liter vessel will exert a pressure of 3.25 atmospheres.

This is clearly a problem that is to be solved algebraically, not by dimensional analysis. List the given quantitles below. Use "?" to identify the quantity you are asked to find. Then solve the ideal gas equation for the unknown variable. Do not calculate the answer at this time.

n = P = Ideal Gas Eqn:

T = V = Solved:

---

n = 0.667 mol P = 3.25 atm $PV = nRT$

T = ? V = ? so $T = \frac{PV}{nR}$

The ideal gas equation, $PV = nRT$, is solved for the wanted quantity, T, by dividing by nR. (If you have difficulty with algebraic manipulations such as this, you may wish to review this material in Section 1.10.)

One more decision must be made before you are ready to substitute into the equation: Which value of R will you use? Look at the given information again: 0.465 mole, 15°C and 735 torr. Now select the R value you will use. Check your choice: 0.08205 L•atm/mol•K (___) or 62.36 L•torr/mol•K (___)

0.08205 L•atm/mol•K. Pressure is given in atm, so you use the form of R that includes atm.

Now you are ready to substitute information from the problem into the equation. Important: Include units! You will see that units give you a check on the correctness of your algebra as well as the substitutions themselves. Extend the $T = PV/nR$ equation, which is repeated below, make your substitutions, and solve for the answer.

$T = \frac{PV}{nR} =$

$$T = \frac{PV}{nR} = \frac{3.25 \text{ atm} \times 4.89 \text{ L}}{0.667 \text{ mol} \times \dfrac{0.08205 \text{ L}\cdot\text{atm}}{\text{mol}\cdot\text{K}}} = 2.90 \times 10^2 \text{ K} = 17°\text{C}$$

If your algebraic solution of the ideal gas equation and substitutions of all given values have been correct, all unwanted units will cancel, leaving only K, the unit of the quantity you were asked to find.

$$\frac{\text{atm} \times \text{L}}{\text{mol} \times \dfrac{\text{L}\cdot\text{atm}}{\text{mol}\cdot\text{K}}} = \frac{1}{\dfrac{1}{\text{K}}} = \text{K}$$

The ideal gas law requires that the amount of gas be in units of moles. However, in the laboratory, we often determine the the amount as mass in grams. It is a simple one step conversion from grams to moles using the molar mass of the gas. The next example illustrates this conversion.

EXAMPLE 12.6: What will be the volume of 18.6 grams of argon at 15°C and 735 torr?

Analyze the problem. The ideal gas law relates P, V, n, and T. Which is the unknown? Have you been given the values of the others? If not, can you calculate them? Fill in the given values below. Use "?" for the unknown.

P = n =

V = T =

$P = 735 \text{ torr}$ $n = 18.6 \text{ g Ar} \times \dfrac{1 \text{ mol Ar}}{39.95 \text{ g Ar}} = 0.466 \text{ mol He}$

$V = ?$ $T = 15°\text{C} = 288 \text{ K}$

You were not given moles directly, so you had to calculate them from the given grams.

Substitute the above values into the ideal gas equation solved for V. Which value of R should you use? Calculate the answer.

P is in torr so use R = 62.36 L•torr/mol•K.

$$V = \frac{nRT}{P} = \frac{0.466 \text{ mol Ar} \times \dfrac{62.36 \text{ L}\cdot\text{torr}}{\text{mol}\cdot\text{K}} \times 288 \text{ K}}{735 \text{ torr}} = 11.4 \text{ L}$$

In particular, notice the temperature factor. If you had forgotten to change Celsius degrees to kelvins and used 15°C, the nonsense units of L•°C/K would have alerted you to the mistake.

EXAMPLE 12.7: How many moles of chlorine are in 4.22 liters at 24°C and 745 torr?

$V = 4.22$ L, $T = 24°C = 297$ K, $P = 745$ torr, $n = ?$

$$n = \frac{PV}{RT} = \frac{745 \text{ torr} \times 4.22 \text{ L}}{\dfrac{62.36 \text{ L}\cdot\text{torr}}{\text{mol}\cdot\text{K}} \times 297 \text{ K}} = 0.170 \text{ mol}$$

12.4 CHANGES OF PRESSURE, TEMPERATURE, VOLUME, AND/OR MOLES

When the gas laws were introduced, only P-V, T-V, and n-V changes were discussed. Several other types of changes are possible. P and T may change with n and V constant. Or three variables may change with only one constant: P, V, and T may vary while n stays the same. Methods for solving such problems are discussed next.

In Section 12.2 we saw that, when the other two variables are held constant, V is directly proportional to T. Notice where the factors V and T are located in the ideal gas equation, $PV = nRT$. V is on the left side of the equation and T is on the right. We have also seen that V is directly proportional to n. Again the variables are on opposite sides of the equation. But P and V appear as factors on the same side of the equation, and they are indirectly proportional to each other. These and other observations lead to the following generalization:

> *Variable factors on opposite sides of an equation are directly proportional (either both go up or both go down); variable factors on the same side of an equation are inversely proportional (as one goes up, the other goes down).*

This idea is helpful when solving problems by the factor method.

If the ideal gas equation, $PV = nRT$, is expressed for initial state (subscript 1) and final state (subscript 2), the result is

$$P_1V_1 = n_1RT_1 \quad \text{and} \quad P_2V_2 = n_2RT_2 \tag{12.16}$$

The ratio of the final state to initial state values gives an equation useful for solving problems by the algebra method.

$$\frac{P_2V_2}{P_1V_1} = \frac{n_2RT_2}{n_1RT_1} \quad \text{or} \quad \frac{P_2V_2}{P_1V_1} = \frac{n_2T_2}{n_1T_1} \tag{12.17}$$

Equation 12.17 is the basis for solving gas law problems algebraically.

Suppose that the temperature of a fixed amount of gas in a rigid tank (constant volume) is increased from 25°C to 125°C. At the lower temperature the pressure was 275 torr. You are to find the pressure at the higher temperature. Here is the thought process for both methods:

Factor Method. How are pressure and temperature related? They are on opposite sides of the ideal gas equation, PV = nRT. Therefore, they are directly proportional. They move in the same direction. If temperature goes up, pressure goes up. (This is logical; if you heat up a gas, the pressure builds up.) This means that the initial pressure must be multiplied by a temperature ratio greater than 1:

$$P_2 = P_1 \times (\text{T factor}, >1) = 275 \text{ torr} \times \frac{398 \text{ K}}{298 \text{ K}} = 367 \text{ torr}$$

Remember that all temperatures in gas law problems must be in K, not °C.

Algebra Method. Refer to Equation 12.17. At constant volume, $V_1 = V_2$; and at a constant quantity of gas, $n_1 = n_2$. Equation 12.17 may therefore be simplified:

$$\frac{P_2V_2}{P_1V_1} = \frac{n_2T_2}{n_1T_1} \quad \text{so} \quad \frac{P_2}{P_1} = \frac{T_2}{T_1}$$

This equation may now be solved for P_2, the known values substituted, and the answer calculated:

$$P_2 = \frac{P_1T_2}{T_1} = \frac{275 \text{ torr} \times 398 \text{ K}}{298 \text{ K}} = 367 \text{ torr}$$

Both methods (factor and algebra) reach the same calculation setup.

EXAMPLE 12.8: An air compressor is set to switch on when the pressure in a storage tank falls to 7.2 atm. It turns off when the pressure reaches 8.0 atm. At 7.2 atm the tank holds 17.3 moles of compressed gas. If temperature and volume remain constant, how many moles are present at 8.0 atm? Fill in the data table below, then solve by the factor method first.

P_1 =	V_1 =	n_1 =	T_1 =
P_2 =	V_2 =	n_2 =	T_2 =

P_1 = 7.2 atm	V_1 = constant	n_1 = 17.3 mol	T_1 = constant
P_2 = 8.0 atm	V_2 = constant	n_2 = ?	T_2 = constant

Factor Method. From PV = nRT, how are P and n related, directly (__) or indirectly (__)? Do P and n change in the same direction (__) or in opposite directions (__)? Must the initial quantity be multiplied by a ratio greater than 1 (__) or less than 1 (__)?

P and n are on opposite sides of the equation, so they are directly proportional and they move in the *same direction*. Pressure increases, so n must also increase; the ratio must be greater than 1. This is logical; if you add more gas to a container, pressure will build up.)

Set up and solve the problem.

$$n_2 = n_1 \times (\text{P factor}, >1) = 17.3 \text{ mol} \times \frac{8.0 \text{ atm}}{7.2 \text{ atm}} = 19 \text{ mol}$$

Algebra Method. Simplify Equation 12.17, $\frac{P_2V_2}{P_1V_1} = \frac{n_2T_2}{n_1T_1}$, by crossing out the factors that are constant, solve for the wanted variable, substitute known values, and calculate the answer.

With T and V constant, Equation 12.17 may be simplified to

$$\frac{P_2}{P_1} = \frac{n_2}{n_1}. \quad \text{Therefore} \quad n_2 = \frac{n_1P_2}{P_1} = \frac{17.3 \text{ mol} \times 8.0 \text{ atm}}{7.2 \text{ atm}} = 19 \text{ mol}$$

Combined Gas Laws For a fixed amount of a given gas at different conditions of pressure, volume and temperature, $n_1 = n_2$. In this case only n drops out of Equation 12.17, leaving

$$\frac{P_2V_2}{P_1V_1} = \frac{T_2}{T_1} \quad \text{or} \quad \frac{P_1V_1}{T_1} = \frac{P_2V_2}{T_2} \qquad (12.18)$$

Equation 12.18 is sometimes called the **combined gas laws**. It shows the relationship between pressure, temperature, and volume of a fixed amount of gas. If we know all three initial conditions and change two of them, we can calculate the value to which the third will adjust.

EXAMPLE 12.9: A gas has a volume of 4.18 liters at 23°C and 712 torr. What volume will it occupy if the temperature is changed to 16°C and the pressure adjusted to 686 torr?

Begin by assembling a data table.

P_1 = 712 torr	V_1 = 4.18 L	n_1 = constant	T_1 = 23°C = 296 K
P_2 = 686 torr	V_2 = ?	n_2 = constant	T_2 = 16°C = 289 K

Factor Method. Consider what will happen to the volume as a result of the temperature change alone. How are volume and temperature related, directly or indirectly? If the pressure remains constant at 712 torr, will the volume increase (__) or decrease (__) as the temperature goes from 23°C to 16°C? Check the correct answer.

Volume will *decrease* with a decrease in temperature. From PV = nRT, V and T are on opposite sides and so are directly proportional; they move in the same direction. (This is logical; if you cool a gas balloon, it shrivels up—this is also Charles' law.)

Begin the setup of the problem by writing the initial volume multiplied by the proper temperature ratio. Do not calculate the answer.

$$V_2 = V_1 \times (\text{T factor}, <1) \times (\text{P factor}, \_\_) = 4.18\ \text{L} \times \frac{289\ \text{K}}{296\ \text{K}} \times \frac{\qquad\qquad}{} =$$

Both temperatures must be expressed in kelvins. Because volume is less at the reduced temperature, the smaller temperature is on top.

Calculation of the setup so far would give the volume the gas would occupy at the initial pressure, 712 torr, and the final temperature, 289 K. Now consider the pressure change. What will happen to the volume if the pressure is reduced from 712 torr to 686 torr? Will the volume increase (___) or decrease (___)?

Volume will *increase* as pressure decreases. From PV = nRT, V and P are on the same side, so are inversely proportional; they move in opposite directions. (This logical; if you let up on the pressure, the gas will expand—this is also Boyle's law.)

Now select the proper pressure ratio and complete the fraction in the calculation setup above. Calculate the answer.

$$V_2 = V_1 \times (\text{T factor, } <1) \times (\text{P factor, } > 1) = 4.18 \text{ L} \times \frac{289 \text{ K}}{296 \text{ K}} \times \frac{712 \text{ torr}}{686 \text{ torr}} = 4.24 \text{ L}$$

If volume is to increase as a result of the pressure change, the larger pressure must be in the numerator of the pressure ratio.

Algebra Method. Solve Equation 12.18 for V_2, substitute values from the table above, and calculate the answer.

Equation 12.18: $\frac{P_1V_1}{T_1} = \frac{P_2V_2}{T_2}$ $V_2 =$

$$V_2 = \frac{V_1T_2P_1}{T_1P_2} = \frac{4.18 \text{ L} \times 289 \text{ K} \times 712 \text{ torr}}{296 \text{ K} \times 686 \text{ torr}} = 4.24 \text{ L}$$

12.5 STANDARD TEMPERATURE AND PRESSURE

Because gas volume depends on pressure and temperature, volume is not a suitable measurement of quantity. It may be used, however, if both pressure and temperature are specified. Standard temperature and pressure, STP, is often used for this purpose. Standard temperature is 0°C, or 273.15 K (exactly). Standard pressure is 1 atmosphere, or 760 torr (exactly). Combined gas law problems frequently call for the conversion of a gas volume at nonstandard conditions to the volume at STP, or vice versa. The procedure is the same as that developed in the previous section.

EXAMPLE 12.10: A gas occupies 5.42 liters at −32°C and 0.859 atmosphere. What will be the volume of the same gas at STP?

From a starting volume of 5.42 liters, the temperature change is from −32°C to 0°C, and the pressure change is from 0.859 atm to 1.00 atm. Set up the problem and calculate the final volume.

$P_1 = 0.859$ atm	$V_1 = 5.42$ L	n_1 = constant	$T_1 = -32°C = 241$ K
$P_2 = 1.00$ atm	$V_2 = ?$	n_2 = constant	$T_2 = 0°C = 273$ K

$$V_2 = V_1 \times (\text{T factor } >1) \times (\text{P factor } <1) = 5.42 \text{ L} \times \frac{273 \text{ K}}{241 \text{ K}} \times \frac{0.859 \text{ atm}}{1.00 \text{ atm}} = 5.27 \text{ L}$$

Factor Method. Temperature increases from −32°C to 0°C. Volume is directly proportional to temperature, so it must also increase. The temperature ratio is greater than 1. Pressure increases from 0.859 atm to 1.00 atm. Pressure and volume are inversely proportional, so volume decreases. The pressure ratio is less than 1.

Algebra Method. Using Equation 12.18,

$$V_2 = \frac{V_1 T_2 P_1}{T_1 P_2} = \frac{5.42 \text{ L} \times 273 \text{ K} \times 0.859 \text{ atm}}{241 \text{ K} \times 1.00 \text{ atm}} = 5.27 \text{ L}$$

12.6 GAS DENSITY

In Section 3.12, density was defined as mass per unit volume:

$$\text{Density} \equiv \frac{\text{mass}}{\text{volume}} = \frac{m}{V} = \frac{\text{grams}}{\text{liter}} = \frac{g}{L} \qquad (12.19)$$

The usual units of density are grams per cubic centimeter or grams per milliliter. However, the densities of gases are so low that they are generally expressed in grams per liter, g/L.

Because the volume of a fixed quantity (mass) of a gas depends on temperature and pressure, the density of a gas also depends on those two variables. Therefore, if we are to speak meaningfully of the density of a gas, temperature and pressure must be specified. Two types of conditions will be considered in this section: those at STP and those at some other specified temperature and pressure.

Gas Density at STP; Molar Volume A useful relationship in working with gases is the link between the number of moles in a gas sample and the volume it occupies. The molar volume of a gas is the volume occupied by one mole. We will use the symbol $\overline{V}$ to represent molar mass. Mathematically,

$$\overline{V} = \text{Molar volume} \equiv \frac{\text{volume}}{\text{moles}} = \frac{V}{n} = \frac{\text{liters}}{\text{mole}} = \frac{L}{mol} \qquad (12.20)$$

Like gas density, molar volume depends on temperature and pressure. Molar volume is usually considered at STP. Its value for any gas may be calculated by substituting 273.15 K and 1.000 atm into the ideal gas equation, solved for V/n, liters per mole:

$$\frac{V}{n} = \frac{RT}{P} = \frac{0.08205 \text{ L·atm/mol·K} \times 273.15 \text{ K}}{1.000 \text{ atm}} = 22.41 \text{ L/mol} \qquad (12.21)$$

IMPORTANT: Notice the restriction on 22.41 L/mol. It applies only to gases AT STANDARD TEMPERATURE AND PRESSURE. It may not be used for a gas at other conditions of temperature and pressure, and it must NEVER be used for a liquid or a solid.

Molar volume is quite similar to molar mass (MM). Their units are liters per mole, L/mol, for molar volume, and grams per mole, g/mol, for molar mass:

$$\text{Molar mass} \equiv \frac{\text{mass}}{\text{moles}} = \frac{m}{n} = \frac{\text{grams}}{\text{mole}} = \frac{g}{\text{mol}} \qquad (12.22)$$

You will soon see that molar volume and molar mass are used in similar ways. But there is an important difference between them. Each individual substance has its own molar mass. It is always the same, at any temperature and pressure, whether the substance be a solid, a liquid or a gas. But all ideal gases have the same molar volume at any given temperature and pressure; and at STP that molar volume is 22.41 L/mol.

In summary, we have identified three quantities, density, molar volume, and molar mass. Their definitions and customary units for gases are

$$\text{Density (d)} \equiv \frac{\text{mass}}{\text{volume}} = \frac{m}{V} = \frac{\text{grams}}{\text{liter}} = \frac{g}{L} \qquad (12.19)$$

$$\text{Molar volume } (\overline{V}) \equiv \frac{\text{volume}}{\text{moles}} = \frac{V}{n} = \frac{\text{liters}}{\text{mole}} = \frac{L}{\text{mol}} = \begin{matrix}22.41 \text{ L/mol} \\ \text{(gas at STP)}\end{matrix} \qquad (12.20)$$

$$\text{Molar mass (MM)} \equiv \frac{\text{mass}}{\text{moles}} = \frac{m}{n} = \frac{\text{grams}}{\text{mole}} = \frac{g}{\text{mol}} \qquad (12.22)$$

In practice, the above relationships are used as conversion factors for one- or two-step dimensional analysis conversions between the mass, number of moles, and volume of a gas at STP.

EXAMPLE 12.11: How many moles are in 1.26 liters of a gas at STP?

Identify the given and find quantities. Then select the needed conversion factor, develop the unit path, and solve the problem.

GIVEN: 1.26 L FIND: mol FACTOR: 22.41 L/mol PATH: L → mol

$$1.26 \text{ L} \times \frac{1 \text{ mol}}{22.41 \text{ L}} = 0.0562 \text{ mol}$$

EXAMPLE 12.12: What volume will be occupied by 2.85 grams of sulfur dioxide, SO_2, at STP?

We have no information by which to make a direct conversion from the mass of SO_2 to its volume, but we can get from mass to moles. It will take two steps to solve this problem. Go to it.

GIVEN: 2.85 g SO_2 FIND: L SO_2

FACTOR: MM of SO_2, 22.41 L/mol PATH: g → mol → L

$$2.85 \text{ g } SO_2 \times \frac{1 \text{ mol } SO_2}{64.06 \text{ g } SO_2} \times \frac{22.41 \text{ L } SO_2}{1 \text{ mol } SO_2} = 0.997 \text{ L } SO_2$$

The molar mass, density, and molar volume of a gas are related by the following equation:

$$MM = d \times \overline{V} \qquad \frac{g}{mol} = \frac{g}{L} \times \frac{L}{mol} \qquad (12.23)$$

If any two of the three ratios in Equation 12.23 is known, the third can be calculated.

EXAMPLE 12.13: Calculate the molar mass of a gas whose density is 1.25 g/L at STP.

The problem appears to have only one given quantity; actually, there are two. Be sure to include units as you substitute into Equation 12.23.

GIVEN: $d = 1.25$ g/L, $\overline{V} = 22.41$ L/mol FIND: MM (g/mol)

EQN: $MM = d \times \overline{V} = \frac{1.25 \text{ g}}{L} \times \frac{22.41 \text{ L}}{mol} = 28.0 \text{ g/mol}$

As long as STP is specified, 22.41 L/mol may be regarded as a given quantity.

EXAMPLE 12.14: What is the density of methane, CH_4, at STP?

GIVEN: MM of CH_4, 22.41 L/mol FIND: d (g CH_4/L)

EQN: $d = \frac{MM}{V} = \frac{16.04 \text{ g/mol}}{22.41 \text{ L/mol}} = 0.716 \text{ g/L}$

The molar mass of methane was calculated from its formula. Equation 12.23 was solved for density. The molar mass and the molar volume at STP were substituted and the density was calculated.

Gas Density and Molar Mass at Non-STP Conditions Standard temperature, 0°C (32°F) is not a very comfortable temperature at which to work. Nor is atmospheric pressure easily adjusted to 760 torr. As a consequence, most gas measurements are made at some other temperature and pressure. The ideal gas law may be used to determine the molar mass of a gas at these conditions too.

Suppose that you find that the density of an unknown gas is 3.35 g/L at 735 torr and 37°C. You would like to calculate its molar mass in g/mol. We can treat the density, d = m/V, as two given values: a mass of 8.35 g and a volume of 1.00 L. The given values are:

$m = 8.35 \text{ g}, \quad V = 1.00 \text{ L}, \quad P = 735 \text{ torr}, \quad T = 37°C = 310 \text{ K}$

Solving for the molar mass is a two step process. Begin by solving PV = nRT for n, and calculating its value from the given P, V, and T. Then calculate MM = m/n by dividing the given mass by your calculated moles. These two steps are:

$$n = \frac{PV}{RT} = \frac{735 \text{ torr} \times 1.00 \text{ L}}{62.36 \text{ L·torr/mol·K} \times 310 \text{ K}} = 0.0380\underline{2} \text{ moles}$$

$$MM = \frac{m}{n} = \frac{3.35 \text{ g}}{0.0380\underline{2} \text{ mol}} = 88.1 \text{ g/mol}$$

Calculator tip: The n value in your calculator display is 0.0380207. Don't clear it yet. For the next step press "÷" then 3.35, then "=." (The display is 0.0113494.) You have calculated n/m. To get MM = m/n, hit "1/x." The display is now 88.110014; round to 88.1 g/mol.

EXAMPLE 12.15: 2.94 grams of a gas exert a pressure of 0.284 atmospheres in a 3.60-liter container at 265°C. Calculate the molar mass of the substance.

The procedure is the same as in the worked-out example above. Complete the problem.

GIVEN: m = 2.94 g, P = 0.284 atm, V = 3.60 L, T = 265°C = 538 K

FIND: MM (g/mol)

EQN: PV = nRT, MM = m/n

$$n = \frac{PV}{RT} = \frac{0.284 \text{ atm} \times 3.60 \text{ L}}{0.08205 \text{ L·atm/mol·K} \times 538 \text{ K}} = 0.0231\underline{6} \text{ mol}$$

$$MM = \frac{m}{n} = \frac{2.94 \text{ g}}{0.0231\underline{6} \text{ mol}} = 127 \text{ g/mol}$$

EXAMPLE 12.16: Calculate the molar mass of a gas whose density is 1.25 g/L at 760.0 torr and 273 K.

GIVEN: m = 1.25 g, V = 1.00 L, P = 760.0 torr, T = 273 K

FIND: MM (g/mol)

EQN: PV = nRT, MM = m/n

$$n = \frac{PV}{RT} = \frac{760.0 \text{ torr} \times 1.00 \text{ L}}{62.36 \text{ L·torr/mol·K} \times 273 \text{ K}} = 0.0446\underline{4} \text{ mol}$$

$$MM = \frac{m}{n} = \frac{1.25 \text{ g}}{0.0446\underline{4} \text{ mol}} = 28.0 \text{ g/mol}$$

This is the same problem as in Example 12.13, which was done at STP using 22.41 L/mol. As you can see, the ideal gas law works at STP too.

EXAMPLE 12.17: Find the density of methane, CH_4, at 22°C and 0.961 atm.

Unlike molar mass, density does not appear in either form of the ideal gas equation. But the quantities that make up density do. Equation 12.21 shows clearly that density is mass/volume (m/V). The mass of 1 mole is the molar mass, 16.04 g for 1.00 mole of CH_4. The volume of 1.00 mole of gas can be obtained from the ideal gas law, PV = nRT, solved for V. List the given quantities as requested below. Show the ideal gas and density equations needed to solve the problem.

GIVEN: m = , n = FIND:

T = , P = EQNS:

GIVEN: m = 16.04 g, n = 1.00 mol FIND: density, d (g/L)
T = 22°C = 295 K, P = 0.961 atm EQNS: PV = nRT, d = m/V

Now you are ready to substitute the given values into the solved equation and calculate the density of the gas.

$$V = \frac{nRT}{P} = \frac{1.00 \text{ mol} \times 0.08205 \text{ L·atm/mol·K} \times 295 \text{ K}}{0.961 \text{ atm}} = 25.1\underline{9} \text{ L}$$

$$d = \frac{m}{V} = \frac{16.04 \text{ g}}{25.1\underline{9} \text{ L}} = 0.637 \text{ g/L}$$

12.7 GAS STOICHIOMETRY

In Section 10.2 you learned a general procedure for solving stoichiometry problems that we called the stoichiometry 3-step. Its unit path is

given quantity → given moles → wanted moles → wanted quantity

The given and wanted quantities were usually in grams, and molar mass was the conversion factor between grams and moles in the first and third steps. The conversion factor for the given moles → wanted moles step came from the coefficients of the equation.

For gases, the usual measurement of quantity is volume at a specified pressure and temperature. To apply the stoichiometry 3-step, we must be able to convert back and forth between volume of gas and moles of gas. If the volume is measured at STP, the molar volume at those conditions, 22.41 L/mol, serves as a conversion factor. (See Examples 12.11 and 12.12.)

EXAMPLE 12.18: How many liters of oxygen, measured at STP, will be released by the decomposition of 2.58 grams of potassium chlorate into potassium chloride and oxygen?

The equation, please.

$$2\ KClO_3(s)\ \rightarrow\ 2\ KCl(s)\ +\ 3\ O_2(g)$$

Next write the starting quantity, wanted quantity, conversion factors, and unit path. Remember that volume is desired at STP, so 22.41 L/mol may be used as a conversion factor between moles and liters.

GIVEN: 2.58 g $KClO_3$ FIND: L O_2 at STP

FACTOR: MM of $KClO_3$, 22.41 L/mol PATH: g $KClO_3$ → mol $KClO_3$ → mol O_2 → L O_2

This is the stoichiometry 3-step (Section 10.2) with the third step using 22.41 L/mol to convert moles of oxygen to liters of oxygen. Set up the problem and solve for the answer.

$$2.58\ g\ KClO_3 \times \frac{1\ mol\ KClO_3}{122.55\ g\ KClO_3} \times \frac{3\ mol\ O_2}{2\ mol\ KClO_3} \times \frac{22.41\ L\ O_2}{1\ mol\ O_2} = 0.708\ L\ O_2$$

EXAMPLE 12.19: How many grams of zinc must react with hydrochloric acid in order to produce 4.35 liters of hydrogen, measured at STP?

Begin with the reaction equation.

$Zn(s) + 2\ HCl(aq) \rightarrow H_2(g) + ZnCl_2(aq)$

This time the starting quantity is 4.35 L H_2 at STP. You are looking for the mass of zinc. Molar volume at STP and molar mass are the conversion factors. Write the complete setup for the problem and calculate the answer.

PATH: L H_2 → mol H_2 → mol Zn → g Zn

$$4.35\ \text{L H}_2 \times \frac{1\ \text{mol H}_2}{22.41\ \text{L H}_2} \times \frac{1\ \text{mol Zn}}{1\ \text{mol H}_2} \times \frac{65.38\ \text{g Zn}}{1\ \text{mol Zn}} = 12.7\ \text{g Zn}$$

The next two examples show the ideal gas law approach to non-STP gas stoichiometry problems.

EXAMPLE 12.20: During the combustion of ethane, $C_2H_6(g)$, 40.4 grams of liquid water are collected. How many liters of ethane, measured at 3.55 atm and 33°C, were burned?

Begin with the equation for the reaction.

$2\ C_2H_6(g) + 7\ O_2(g) \rightarrow 4\ CO_2(g) + 6\ H_2O(\ell)$

The starting quantity is in grams this time. The first two steps of the stoichiometry 3-step are performed in the usual way. Set up the problem as far as moles of ethane, and calculate the result.

GIVEN: 40.4 g H_2O

FACTOR: MM H_2O, mole ratio

FIND: mol C_2H_6

PATH: g H_2O → mol H_2O → mol C_2H_6

$$40.4\ \text{g H}_2\text{O} \times \frac{1\ \text{mol H}_2\text{O}}{18.02\ \text{g H}_2\text{O}} \times \frac{2\ \text{mol C}_2\text{H}_6}{6\ \text{mol H}_2\text{O}} = 0.747\ \text{mol C}_2\text{H}_6$$

The next task is to convert moles of gas to volume at the specified temperature and pressure. Use PV = nRT, solved for V, to do this.

GIVEN: n = 0.747 mol C_2H_6, P = 3.55 atm, T = 33°C = 306 K FIND: L C_2H_6

EQN: $V = \frac{nRT}{P} = \frac{0.747 \text{ mol } C_2H_6 \times 0.08205 \text{ L·atm/mol·K} \times 306 \text{ K}}{3.55 \text{ atm}} = 5.28 \text{ L } C_2H_6$

EXAMPLE 12.21: How many grams of zinc must react with hydrochloric acid to produce 4.35 liters of hydrogen, measured at 760.0 torr and 273 K?

Begin with the equation.

$Zn(s) + 2\ HCl(aq) \rightarrow H_2(g) + ZnCl_2(aq)$

Complete the problem, calculating first the number of moles of gas to be produced, then the number of grams of zinc needed.

GIVEN: V = 4.35 L H_2, P = 760.0 torr, T = 273 K FIND: g Zn
FACTOR: mole ratio from equation, MM of Zn PATH: mol H_2 → mol Zn → g Zn

EQN: $n = \frac{PV}{RT} = \frac{760.0 \text{ torr} \times 4.35 \text{ L}}{62.36 \text{ L·torr/mol·K} \times 273 \text{ K}} = 0.194 \text{ mol } H_2$

$0.194 \text{ mol } H_2 \times \frac{1 \text{ mol Zn}}{1 \text{ mol } H_2} \times \frac{65.38 \text{ g Zn}}{1 \text{ mol Zn}} = 12.7 \text{ g Zn}$

This is the same problem as Example 12.19, but worked by the ideal gas law method rather than by molar volume at STP. Note that the ideal gas law method works at any given pressure and temperature, including STP.

All gas stoichiometry problems considered so far have had a gas volume at one end of the calculation setup and grams of some species at the other. What if both given and wanted quantities are expressed in gas volumes at specified temperatures and pressures? This kind of problem involves **Avogadro's law**, which says that **equal volumes of different gases, measured at the same temperature and pressure, contain the same number of molecules (or moles of molecules)**.

In Section 12.2 you learned that the volume of a gas at a given temperature and pressure is directly proportional to the number of moles. This was expressed in Equation 12.13, which may be rearranged to give

$$\frac{V_1}{V_2} = \frac{n_1}{n_2} \qquad (12.24)$$

This equation says that the ratio of volumes of gases measured at the same temperature and pressure is the same as the ratio of moles. We've been using the ratio of moles, given by the coefficients in the balanced equation, as the conversion factor for the middle part of the stoichiometry 3-step right along. Equation 12.24 gives us a shortcut for problems in which both the given and wanted quantities are gas volumes measured at the same temperature and pressure.

Consider the reaction between hydrogen and oxygen to form water. The equation is $2\ H_2(g) + O_2(g) \rightarrow 2\ H_2O(\ell)$. The mole ratios and volume ratios of hydrogen and oxygen are the same:

$$\frac{2 \text{ moles } H_2}{1 \text{ mole } O_2} = \frac{2 \text{ liters } H_2}{1 \text{ liter } O_2} \qquad \text{(at constant temperature and pressure)}$$

If we want to find the volume of hydrogen that will react with 2.34 liters of oxygen, both measured at the same temperature and pressure, the second ratio is a conversion factor that gives us a one-step unit path: $L\ O_2 \rightarrow L\ H_2$:

$$2.34 \text{ L } O_2 \times \frac{2 \text{ L } H_2}{1 \text{ L } O_2} = 4.68 \text{ L } H_2$$

EXAMPLE 12.22: What volume of ammonia will be produced by 6.33 liters of hydrogen in the reaction $N_2(g) + 3\ H_2(g) \rightarrow 2\ NH_3(g)$ if both volumes are measured at the same temperature and pressure?

Write below the unit path and the conversion factor you will use. Then solve the problem.

GIVEN: 6.33 L H_2 FIND: L NH_3 FACTOR: 2 L NH_3/3 L H_2 PATH: L H_2 → L NH_3

$$6.33 \text{ L } H_2 \times \frac{2 \text{ L } NH_3}{3 \text{ L } H_2} = 4.22 \text{ L } NH_3$$

The next question is, how do you solve a volume-to-volume problem when the volumes are measured at different temperatures and pressures? One way is to adjust the given volume to the temperature and pressure at which the wanted volume is to be measured. Then the problem becomes a volume-to-volume problem at the same temperature and pressure and you can proceed as in Example 12.22. The next example takes you through the steps:

EXAMPLE 12.23: What volume of nitrogen, measured at 18°C and 0.795 atm, is required to produce 4.22 liters of ammonia, measured at 85°C and 25.8 atm? The equation is $N_2(g) + 3\ H_2(g) \rightarrow 2\ NH_3(g)$

The first step is to change the given volume of ammonia at its starting conditions to the volume the same gas (ammonia) would occupy at the temperature and pressure at which the nitrogen volume is to be measured. A table, such as that used in Section 12.4, analyzes this part of the problem:

$P_1 =$ $V_1 =$ $n_1 =$ $T_1 =$

$P_2 =$ $V_2 =$ $n_2 =$ $T_2 =$

$P_1 = 25.8$ atm, $V_1 = 4.22$ L, $n_1 =$ constant, $T_1 = 85°C = 358$ K
$P_2 = 0.795$ atm, $V_2 = ?$ $n_2 =$ constant, $T_2 = 18°C = 291$ K

A problem of this type will always give the volume, temperature, and pressure of one of the gases. They make up the initial values. In this problem the values are given for ammonia. The final temperature and pressure values are those given for the second gas—nitrogen in this example. V_2 in this part of the problem is not for the second gas, nitrogen, but for the first gas, ammonia. We are looking for the volume the ammonia would occupy at the temperature and pressure at which the nitrogen is to be measured. If we get that, the problem is just like Example 12.22, converting one gas volume to the other, both measured at the same temperature and pressure.

Begin the calculation setup from 4.22 L NH_3 at the initial conditions to the volume of ammonia at the final conditions. Use either proportional reasoning or Equation 12.18, whichever method you used in Section 12.4. One more thing: We are not really interested in a calculated answer here; the setup alone will do.

$$4.22 \text{ L } NH_3 \times \frac{291 \text{ K}}{358 \text{ K}} \times \frac{25.8 \text{ atm}}{0.795 \text{ atm}} \times \underline{\qquad\qquad}$$

The setup so far represents the volume of ammonia at 18°C and 0.795 atm, the temperature and pressure at which the nitrogen volume is to be measured. At the same temperature and pressure the volume conversion factor from ammonia to nitrogen may be taken directly from the coefficients in the equation. Extend the setup and calculate the answer.

$$4.22 \text{ L } NH_3 \times \frac{291 \text{ K}}{358 \text{ K}} \times \frac{25.8 \text{ atm}}{0.795 \text{ atm}} \times \frac{1 \text{ L } N_2}{2 \text{ L } NH_3} = 55.7 \text{ L } N_2$$

EXAMPLE 12.24: What volume of oxygen, measured at 14°C and 0.202 atm, is needed to burn all of the carbon monoxide in a 2.91 L container at 4.65 atm and 26°C?

Begin with the reaction equation.

$$2\ CO(g) + O_2(g) \rightarrow 2\ CO_2(g)$$

There are 2.91 L CO at 4.65 atm and 26°C. You want the volume of oxygen at 14°C and 0.202 atm. Complete the problem.

P_1 = 4.65 atm, V_1 = 2.91 L, n_1 = constant, T_1 = 26°C = 299 K
P_2 = 0.202 atm, V_2 = ? n_2 = constant, T_2 = 14°C = 287 K

$$2.91 \text{ L CO} \times \frac{287 \text{ K}}{299 \text{ K}} \times \frac{4.65 \text{ atm}}{0.202 \text{ atm}} \times \frac{1 \text{ L } O_2}{2 \text{ L CO}} = 32.1 \text{ L } O_2$$

12.8 DALTON'S LAW OF PARTIAL PRESSURES

Particles in the gaseous state are widely separated from each other and there is little or no interaction between them. When two or more gases are mixed, each one behaves almost as if it alone occupies the container, conforming to the ideal gas law independently of the other gases present. Each gas exerts a partial pressure that depends on the number of moles of that gas in the mixture.

If n_A is the number of moles of gas A, its partial pressure, p_A, is

$$p_A = \frac{n_A RT}{V} \qquad (12.25)$$

If two or more gases are combined, the sum of their partial presssures, Σp, is

$$\Sigma p = p_A + p_B + \ldots = \frac{n_A RT}{V} + \frac{n_B RT}{V} + \ldots = (n_A + n_B + \ldots)\frac{RT}{V} = \frac{nRT}{V} \qquad (12.26)$$

The sum $n_A + n_B + \ldots$ represents the total number of moles of all gases in the mixture. This is n in the ideal gas equation, PV = nRT. If that equation is solved for P, the result is

$$P = \frac{nRT}{V} \qquad (12.27)$$

Combining Equations 12.26 and 12.27 gives a mathematical statement of Dalton's Law of Partial Pressure:

$$\Sigma p = P = p_A + p_B + \ldots \qquad (12.28)$$

In words, *the total pressure exerted by a gaseous mixture is equal to the sum of the partial pressures of the gases in the mixture.*

EXAMPLE 12.25: The partial pressure of helium is 442 torr, and the partial pressure of neon is 312 torr in a mixture of the two gases. Find the total pressure of the mixture.

This is a straight substitution into Equation 12.28.

GIVEN: p_{He} = 442 torr, p_{Ne} = 312 torr FIND: P

EQN: $P = p_{He} + p_{Ne}$ = 442 torr + 312 torr = 754 torr.

EXAMPLE 12.26: A gaseous mixture of oxygen, sulfur dioxide, and sulfur trioxide exerts a total pressure of 3.32 atmospheres. What is the partial pressure of sulfur dioxide if the partial pressures of oxygen and sulfur trioxide are 1.04 atmospheres and 1.29 atmospheres, respectively?

GIVEN: $P = 3.23$ atm, $p_{O_2} = 1.04$ atm, $p_{SO_3} = 1.29$ atm FIND: p_{SO_2}

EQN: $P = p_{SO_2} + p_{O_2} + p_{SO_3}$

$p_{SO_2} = P - p_{O_2} - p_{SO_3} = 3.32 \text{ atm} - 1.04 \text{ atm} - 1.29 \text{ atm} = 0.99 \text{ atm}$

At constant temperature and volume, the pressure exerted by a gas is directly proportional to the number of moles present. (Recall that variables on opposite sides of the equal sign in an equation are directly proportional, as pressure and moles in PV = nRT.) It follows that the fraction of the total pressure that is caused by one component in a mixture is the same as that component's fraction of total number of molecules in the mixture. If one-fourth of the molecules in a gaseous mixture is made up of oxygen molecules, oxygen is responsible for one-fourth of the pressure of the mixture. If X_A is the fraction of A molecules in a mixture, then

$$p_A = X_A \times P \tag{12.29}$$

The "fraction of molecules" has a name. It is **the mole fraction of A, the ratio of moles of A in a mixture to the total number of moles of all gases in the mixture**:

$$X_A = \frac{\text{moles of A}}{\text{total moles in mixture}} \tag{12.30}$$

EXAMPLE 12.27: A gaseous mixture contains 0.274 mole of methane, 0.072 mole of ethane, and 0.011 mole of propane. The mixture exerts a pressure of 1.60 atmospheres. Find the partial pressure of each component in the mixture.

To find the mole fraction of any component, you must know the total number of moles in the mixture. Find that first.

0.274 mol + 0.072 mol + 0.011 mol = 0.357 moles total

We will concentrate first on the methane. What is its mole fraction?

GIVEN: 0.274 mol methane, 0.357 total mol FIND: $X_{methane}$

EQN: $X_{methane} = \dfrac{0.274 \text{ moles of methane}}{0.357 \text{ total moles}} = 0.768$

Equation 12.29 can be used to find the partial pressure of methane.

GIVEN: $X_{methane} = 0.768$, $P = 1.60$ atm FIND: $p_{methane}$

EQN: $p_{methane} = X_{methane} \times P = 0.768 \times 1.60 \text{ atm} = 1.23 \text{ atm}$

Notice that it is not necessary to calculate the mole fraction of methane as an intermediate answer. The total pressure can just as easily be multiplied by the mole fraction in its fractional form:

$$\frac{0.274}{0.357} \times 1.60 \text{ atm} = 1.23 \text{ atm}$$

Complete the problem by calculating the partial pressures of the other two gases.

$$p_{ethane} = \frac{0.072}{0.357} \times 1.60 \text{ atm} = 0.32 \text{ atm}$$

$$p_{propane} = \frac{0.011}{0.357} \times 1.60 \text{ atm} = 0.049 \text{ atm}$$

If you have, indeed, found the partial pressures of all components of the mixture, their sum must be the total pressure:

$$1.23 \text{ atm} + 0.32 \text{ atm} + 0.049 \text{ atm} = 1.60 \text{ atm}.$$

At the beginning of this section it was indicated that each component of a mixture conforms to the ideal gas law as if it alone occupied the container. This means that the ideal gas equation may be applied to the mixture of gases as a whole, or *to each individual gas in the mixture*. Therefore, if you have all the information you need for one gas in a mixture, other gases may be ignored when solving a problem about the first gas.

EXAMPLE 12.28: 0.442 mole of carbon dioxide, 0.106 mole of carbon monoxide, and 0.230 mole of nitrogen occupy a 4.15-liter container at 33°C. Find the partial pressure of the nitrogen in atmospheres.

This problem can be solved by applying the ideal gas equation to nitrogen alone.

GIVEN: 0.230 mol N_2, 4.15 L, 33°C (306 K) FIND: p_{N_2} EQN: $p_{N_2}V = n_{N_2}RT$

$$p_{N_2} = \frac{n_{N_2}RT}{V} = \frac{(0.230 \text{ mol } N_2)(0.08205 \text{ L·atm/mol·K})(306 \text{ K})}{(4.15 \text{ L})} = 1.39 \text{ atm}$$

12.9 EFFUSION AND DIFFUSION

Gas under pressure escapes through a small opening. This process is called **effusion**. Once released, the gas distributes itself throughout the larger container, perhaps the atmosphere itself. This is an example of **diffusion**. The mechanics of these processes are different, but they both involve a rate at which a gas travels.

The relative rates of effusion and diffusion of two gases are inversely proportional to the square root of the ratio of the molar masses of the gases:

$$\frac{r_A}{r_B} = \sqrt{\frac{MM_B}{MM_A}} \qquad (12.31)$$

This relationship is known as **Graham's Law of Effusion**. Notice its inverse character. *The gas with the smaller molar mass effuses at a higher rate, and vice versa*. This fact, plus the fact that the rates are related to the square roots of the molar masses, makes it possible to solve problems by reason as well as substitution into a memorized equation. This is seen in the next example.

EXAMPLE 12.29: Compute the ratio of the effusion and diffusion rates of gaseous NH_3 to that of gaseous HCl.

The problem is a straightforward application of Equation 12.31.

Using Equation 12.31 with NH_3 as A and HCl as B gives

$$\frac{r_{NH_3}}{r_{HCl}} = \sqrt{\frac{MM_{HCl}}{MM_{NH_3}}} = \sqrt{\frac{36.46}{17.03}} = \sqrt{2.141} = 1.463$$

The reasoning approach to this problem is simply to recognize that the ratio of rates will be equal to the square root of the ratio of molar masses of the gases. Calculate that value, as above, without regard to the rate ratio at the left end of the line. The answer, 1.463, tells you that one gas effuses 1.463 times as fast as the other. Which is the faster? According to the italicized statement above, it is the gas with the lower molar mass. Thus the conclusion is drawn,

$$\frac{r_{NH_3}}{r_{HCl}} = 1.463$$

Even if you don't use the reasoning approach to solve effusion problems, you should test your answer by the same criterion: Does the answer show that the gas with the lower molar mass effuses faster? Light molecules zip around at higher velocities than heavy molecules, so they effuse more rapidly. In other words, they escape in less time—and measuring the time required for a given amount of gas to escape through an opening is the easiest way to measure rate of effusion.

Time is inversely proportional to rate; the faster something is done, the less time it takes to do it. Equation 12.31 can be extended to include a time ratio that is inversely related to the rate ratio:

$$\frac{r_A}{r_B} = \frac{t_B}{t_A} = \sqrt{\frac{MM_B}{MM_A}} \qquad (12.32)$$

If you choose to solve problems algebraically from equations, remember that the square root of the molar mass ratio is inversely related to rate, but directly related to time. And always use reason to verify your numerical answer.

EXAMPLE 12.30: Argon is placed in an effusion apparatus at a certain pressure and temperature. It is allowed to effuse until a certain quantity escapes. It takes 145 seconds. The system is flushed and an unknown gas is placed into the apparatus at the same initial temperature and pressure. The same quantity of this gas escapes in 131 seconds. Find the molar mass of the unknown gas.

Begin by setting up the proper ratios. From Graham's law, are time and molar mass directly (__) or indirectly (__) related? The given times may be expressed as the ratio t_{Ar}/t_X, which is equal to the square root of which molar mass ratio, MM_{Ar}/MM_X (__) or MM_X/MM_{Ar} (__)? Check the answers before going on.

Times and molar masses are directly related; time is directly proportional to the square root of molar mass. The correct proportion is therefore

$$\frac{t_{Ar}}{t_X} = \sqrt{\frac{MM_{Ar}}{MM_X}}$$

Now substitute the given values and solve the problem.

GIVEN: $t_{Ar} = 145$ s, $t_X = 131$ s, MM_{Ar} FIND: MM_X

EQN: $\frac{t_{Ar}}{t_X} = \sqrt{\frac{MM_{Ar}}{MM_X}}$

$$\frac{145 \text{ s}}{131 \text{ s}} = \sqrt{\frac{39.95 \text{ g/mol}}{MM_X}} \quad \text{so} \quad 1.23 = \frac{39.95 \text{ g/mol}}{MM_X}$$

$$MM_X = 32.5 \text{ g/mol}$$

The unknown gas, which effuses in less time than argon, has a lower molar mass than argon. This is reasonable.

EXAMPLE 12.31: Find the molar mass of an unknown gas if 94.0 seconds are required for a given amount to escape from an effusion apparatus, but only 54.0 seconds are required for the same quantity of neon to escape under similar starting conditions.

GIVEN: $t_{Ne} = 54.0$ s, $t_X = 94.0$ s, MM_{Ne} FIND: MM_X

EQN: $\frac{t_X}{t_{Ne}} = \sqrt{\frac{MM_X}{MM_{Ne}}}$

$$\frac{94.0 \text{ s}}{54.0 \text{ s}} = 1.74 = \sqrt{\frac{MM_X}{20.18 \text{ g/mol}}} \qquad MM_X = 61.1 \text{ g/mol}$$

The unknown requires more time to escape, so it must have the larger molar mass.

CHAPTER 12 PROBLEMS

1) Complete the indicated conversion:

a) 830 torr = ______________ atm = ______________ kPa

b) 2.31 atm = ______________ torr = ______________ kPa

2) A gas occupies 4.87 liters at 14°C. Calculate the volume it will occupy at 71°C and the same pressure.

3) 9.15 liters of a gas exert 462 torr at a certain temperature. What will the pressure be if the volume is reduced to 7.96 liters?

4) A gas fills a 56.0-cm^3 cylinder while exerting a pressure of 3.61 atm. At what volume will the same gas have a pressure of 0.972 atm if the temperature remains constant?

5) How many moles of helium will a 75.0-L balloon hold if a 27.5-L balloon holds 1.76 moles at the same temperature and pressure?

6) What pressure will be exerted by 1.42 moles of butane, C_4H_{10}, in a 10.0-liter cylinder at 75°C?

7) How many moles of an ideal gas will fill 7.07 liters at a pressure of 1.81 atmospheres and a temperature of 25°C?

8) A sample of oxygen is collected in a laboratory experiment. Find its mass if the volume is 618 mL at 790 torr and 15°C.

9) If 1.74 moles of carbon dioxide, CO_2, are in a steel cylinder at 2.90 atm and 743°C, calculate the volume of the cylinder.

10) At what temperature will 3.12 grams of methane, CH_4, exert a pressure of 994 torr in a 10.0-liter cylinder?

11) A fixed quantity of gas fills a volume of 10.6 liters at 638 torr and 83°C. Find the volume it will occupy if the pressure is 1076 torr and the temperature rises to 96°C.

12) If 392 cm^3 of helium at 32°C and 925 torr are cooled to −15°C and the pressure is reduced to 775 torr, calculate the new volume the gas will occupy.

13) A sample of gas has a volume of 5.79 liters at 25°C and 518 torr. What will be the volume of this gas at STP?

14) 10.0 liters of hydrogen at 44°C and 1.35 atm are contained in a cylinder with an adjustable piston. The piston is moved until the same gas occupies 2.32 liters, and the temperature is raised to 54°C. Find the pressure in the cylinder.

15) An air compressor reduced a sample of air originally at 25°C and 750 torr to 0.193 L at 68.1 atm and 80°C. Calculate the original volume of the air.

16) A 8.75-liter balloon holds 0.450 moles of gas at 785 torr. If 0.250 moles of gas are added, the pressure is observed to be 812 torr. What is the new volume of the balloon, assuming constant temperature?

17) Find the volume that would be occupied by 0.141 moles of oxygen at STP.

18) What is the volume of 9.22 grams of nitrogen at STP?

19) How many moles of methane, CH_4, the main constituent of natural gas, are contained in 846 milliliters of methane measured at STP?

20) What is the mass of helium in a balloon if its volume is 4.39 L at STP?

21) Calculate the molar mass of a gas if 905 mL at STP weigh 2.28 grams.

22) What volume would be occupied by 3.59 grams of hydrogen at standard temperature and pressure?

23) What is the density of propane, C_3H_8, at STP?

24) The STP density of an unidentified gas is 1.43 grams per liter. What is the molar mass of that gas?

25) Carbon monoxide, CO, is the air pollutant produced in the largest quantity in the United States. What is the density of CO at 18°C and 0.725 atm?

26) What is the molar mass of an unknown gas if 8.28 grams occupy 4.41 liters at 740 torr and 27°C?

27) How many liters of NO at 22°C and 8.00×10^2 torr will be produced by 42.6 grams of HNO_3 in the reaction

$$3\ Ag(s) + 4\ HNO_3(aq) \rightarrow 3\ AgNO_3(aq) + NO(g) + 2\ H_2O(\ell)?$$

28) Carbon dioxide combines with potassium oxide according to the following equation: $CO_2(g) + K_2O(s) \rightarrow K_2CO_3(s)$. If 9.54 liters of CO_2 at 2.30 atm and 75°C combine with excess K_2O, how many grams of K_2CO_3 will result?

29) Calculate the volume of chlorine at 31°C and 751 torr that will be produced by the reaction of 4.51 grams of potassium permanganate, $KMnO_4$, with excess HCl. The equation for the reaction is

$$2\ KMnO_4(aq) + 16\ HCl(aq) \rightarrow 2\ MnCl2(aq) + 5\ Cl_2(g) + 8\ H_2O(\ell) + 2\ KCl(aq)$$

30) Acetylene gas, C_2H_2, is made commercially by adding water to calcium carbide, CaC_2, according to: $CaC_2(s) + 2\ H_2O(\ell) \rightarrow C_2H_2(g) + Ca(OH)_2(aq)$. How many kilograms of calcium carbide must be used to make 125 liters of acetylene at 7.50 atm and 20°C?

31) The following equation describes the production of glucose, $C_6H_{12}O_6$, during photosynthesis: $6\ CO_2(g) + 6\ H_2O(\ell) \rightarrow C_6H_{12}O_6(aq) + 6\ O_2(g)$. How many liters of air at 23°C are needed to produce 15.0 milligrams of $C_6H_{12}O_6$? The partial pressure of CO_2 in air is 16 torr.

32) Gasoline—assume it to be octane, C_8H_{18}—undergoes combustion in automobile engines as follows: $2\ C_8H_{18}(\ell) + 25\ O_2(g) \rightarrow 16\ CO_2(g) + 18\ H_2O(\ell)$. How many liters of CO_2, measured at 21°C and 725 torr, are produced when sufficient gasoline reacts with 35.8 liters of oxygen, also measured at 21°C and 725 torr?

33) $C_3H_8(g) + 5\ O_2(g) \rightarrow 3\ CO_2(g) + 4\ H_2O(\ell)$ is the equation for the combustion of propane. How many liters of propane, measured at 18.5 atm and −23°C, are needed to yield 45.7 liters of CO_2, measured at 1.15 atm and 45°C?

34) Sulfur dioxide, which is a by-product of smeltering processes, can form "acid rain" in the atmosphere: $2\ SO_2(g) + O_2(g) + 2\ H_2O(\ell) \rightarrow 2\ H_2SO_4(aq)$. How many liters of oxygen at 740 torr and 25°C will be required to react with 3250 liters of SO_2 at 620 torr and 47°C?

35) What is the total pressure of a mixture of oxygen and nitrogen if their partial pressures are 351 torr and 581 torr respectively?

36) A student collects oxygen by bubbling it through water. The resulting gas is a mixture of oxygen and water vapor at a total pressure of 749 torr. Find the partial pressure of the oxygen if the water vapor has a partial pressure of 24.8 torr.

37) A mixture of noble gases contains 4.74 mol He, 0.72 mol Ne, and 0.13 mol Ar. Find the partial pressure of each gas if the total pressure is 5.49 atm.

38) 98.1 grams of methane, CH_4, are mixed with 165 grams of ethane, C_2H_6. The pressure of the mixture is 0.954 atm. Find the partial pressure of each.

39) 1.73 mol CO, 0.355 mol CO_2, and 1.85 mol C_2H_6 occupy 5.50 liters at 17°C. Find the partial pressure of CO in atmospheres.

40) 18.2 g NH_3, 26.9 g N_2, and 5.41 g H_2 fill a 6.14-liter reaction vessel at 41°C. Calculate the partial pressure of nitrogen.

41) Compute the ratio of the effusion rate of argon to that of krypton.

42) An unknown gas effuses from a container in 86 seconds. The same number of moles of oxygen requires 73 seconds for an identical quantity to escape from the same starting conditions. Find the molar mass of the unknown gas.

43) An effusion apparatus allows a certain quantity of butane, C_4H_{10}, to escape in 136 seconds. 106 seconds are required for the same number of moles of an unknown gas to be released. What is the molar mass of the second gas?

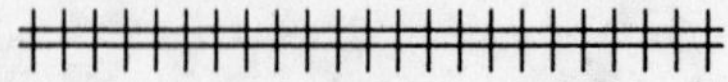

44) Express 0.856 atmospheres in torr and kilopascals.

45) 27.4 liters of a gas at 38°C are cooled to −8°C at constant pressure. What is the final volume of the gas?

46) The pressure of a gas is 771 torr when the volume is 25.9 cm^3. What will the pressure be if the volume is changed to 43.6 cm^3 at constant temperature?

47) How many moles of an ideal gas must be added to a container which currently holds 0.154 moles of gas at 4.95 L to increase its volume to 6.00 L?

48) 2.49 moles of argon exert a pressure of 1.21 atmospheres at 35°C. What is the volume of the container?

49) At what temperature will 1.59 moles of neon in a 34.2-liter container exert a pressure of 739 torr?

50) Approximately 2.0×10^{10} liters of ammonia, NH_3, measured at 2.90 atm and 743°C, were produced in the United States in a certain year. Find its mass.

51) How many moles of gas are in 2.72 liters at 1.33 atmospheres and 26°C?

52) A weather balloon is filled with 81.7 liters of helium at STP. When the balloon reached an altitude of 6.00 kilometers the atmospheric pressure was 0.426 atm and the temperature was −25°C. What was the volume of the balloon under these conditions?

53) What will be the volume of ammonia at 30°C and 805 torr if its volume at 5°C and 752 torr is 20.0 liters?

54) Helium occupies 946 ft^3 at 15°C and 763 torr. Calculate the volume it will occupy at −6°C and 420 torr.

55) A container originally holds 25.0 moles of gas in a 125 liter volume at 27°C. If 5.0 moles of gas escape and the temperature of the container drops to 14°C, what will be the volume of the container?

56) What is the STP volume of a balloon filled with 1.58 moles of hydrogen.

57) Calculate the STP volume of 48.0 grams of SO_2.

58) How many moles of gas at STP can be contained in a room with a volume of 4.9×10^4 liters?

59) What is the mass of 5.68 liters of fluorine at STP?

60) Calculate the molar mass of a gas if its STP density is 1.29 g/L.

61) The density of air at 1.00 atmosphere and 25°C is 1.19 g/L. If air were a pure gas, what would be its molar mass? Actually air is a mixture of about 20% oxygen and 80% nitrogen. What would you expect the "molar mass" of air to be, based on these percentages?

62) What is the molar mass of a gas if a 242 mL container at 23°C and 565 torr contains 1.93 grams of the gas?

63) Calculate the volume of H_2 at 24°C and 742 torr released by the reaction of 12.9 grams of aluminum in $2\ Al(s) + 2\ H_3PO_4(aq) \rightarrow 3\ H_2(g) + 2\ AlPO_4(s)$.

64) How many grams of $C_{12}H_{22}O_{11}$ are produced along with 30.4 liters of oxygen at 0.219 atm and 8°C in 11 $H_2O(\ell)$ + 12 $CO_2(g)$ → $C_{12}H_{22}O_{11}(s)$ + 12 $O_2(g)$?

65) 6 KI(aq) + 8 $HNO_3(aq)$ → 6 $KNO_3(aq)$ + 2 NO(g) + 3 $I_2(s)$ + 4 $H_2O(M)$. What mass of KI must be used to release 1.28 liters of NO at 1.12 atm and 36°C?

66) 7.54 liters of NO_2, measured at 6.49 atm and 17°C, react according to the equation 3 $NO_2(g)$ + $H_2O(\ell)$ → 2 $HNO_3(aq)$ + NO(g). How many liters of NO, measured at 0.942 atm and 30°C, are produced?

67) Calculate the volume of NH_3, measured at 18.3 atm and 12°C, that is required to react with 346 liters of bromine vapor, measured at 0.880 atm and 38°C. The equation is 3 $Br_2(g)$ + 8 $NH_3(g)$ → 6 NH4Br(s) + $N_2(g)$.

68) The partial pressures in a mixture of gases are: H_2, 0.73 atm; N_2, 1.06 atm; CH_4, 1.09 atm; CO_2, 1.48 atm. What is the total pressure of the mixture?

69) A sample of air exerts a pressure of 753 torr. If the partial pressures of N_2, O_2, and CO_2 are, respectively, 577 torr, 141 torr, and 13 torr, what pressure is exerted by water vapor plus all other residual gases in the air?

70) When 42.5 grams of N_2, 7.85 grams of NO, and 31.4 grams of NO_2 are mixed, they exert a total pressure of 3.21 atmospheres. What is the partial pressure of each gas?

71) 9.03 g NO_2 and 2.94 g N_2O_4 occupy a 3.16-liter vessel at 15°C. Calculate the partial pressure of NO_2.

72) Which gas effuses more rapidly, CO or CO_2? What is the ratio of the effusion rate of CO to CO_2?

73) It takes 93 seconds for a certain number of moles of CH_4 to pass through an orifice. The same quantity of a different gas requires 158 seconds from the same starting conditions. Find the molar mass of the second gas.

ADDITIONAL PROBLEMS

74) Aerosol cans should not be placed in a fire because of the danger of explosion. If a can with a pressure of 1.25 atm at 22°C explodes when heated to 975°C, what maximum pressure was reached?

75) In order for a helium balloon to rise it must have a total mass less than the mass of the air it displaces, given by the density of air times the volume of the balloon: $m_{tot} = d_{air} \times V_{balloon}$. To lift a balloon of total mass 225 kg at a pressure of 635 torr and a temperature of −7°C, what mass of helium is needed? The density of air at these conditions is 1.10 g/L.

76) A 963 mg sample of an unknown organic compound of general formula $C_xH_yO_z$, when burned in excess oxygen, forms 1225 mL of carbon dioxide gas collected at 743 torr and 19°C and 1002 mL of water vapor collected at 752 torr and 114°C. a) How many mg of carbon were present? of hydrogen? b) How many mg of oxygen were in the original $C_xH_yO_z$ sample? c) What is the empirical formula of the compound?

77) The distance which a gas diffuses also follows Graham's law. If an unknown diffuses 37.2 cm in the time it takes ammonia to diffuse 83.9 cm, what is the molar mass of the unknown? [Hint: Do light molecules travel a longer or shorter distance than heavy molecules in a given time? In other words, is distance directly or inversely proportional to molar mass?]

CHAPTER TEST

1) Express 25.6 torr in atmospheres and kilopascals.

2) What pressure is exerted by 2.45 moles of methane, CH_4, in a 4.12-liter container at 34°C? Answer in atmospheres.

3) An outdoor neon sign has a volume of 1.21 liters. It contains neon at 0.110 atmospheres and -4°C. If the neon could be transferred to a variable volume container, what volume would it occupy at inside room conditions of 23°C and 0.962 atmospheres?

4) A sample of hydrogen occupies 1.91 liters at STP. Find its volume at 1.36 atmospheres and 78°C.

5) What will be the volume of 79.8 grams of NH_3 at STP?

6) 58.9 grams of a gas fills a 16.8 liter container at STP. Find the molar mass of the gas.

7) The density of an unidentified gas is 1.47 g/L at 1.05 atm and 22°C. Calculate the molar mass of the gas.

8) How many liters of oxygen, measured at 0.304 atm and 16°C, are needed for the complete combustion of 124 grams of $C_5H_{12}(\ell)$?

9) Calculate the grams of $KClO_3$ required to produce 92.0 mL O_2 at 692 torr and 19°C in the reaction 2 $KClO_3(s) \rightarrow 2\ KCl(s) + 3\ O_2(g)$.

10) $CS_2(\ell) + 3\ O_2(g) \rightarrow CO_2(g) + 2\ SO_2(g)$. What volume of SO_2, at 1.73 atm and 112°C, are produced by 74.4 liters of oxygen, at 0.583 atm and 21°C?

11) 0.980 mol F_2, 0.662 mol Cl_2, and 0.796 mol Br_2 occupy 83.7 liters at 37°C. Find the partial pressure of chlorine in torr.

12) How does the effusion rate of SO_2 compare with that of SO_3?

13) 84 seconds and 61 seconds are the times for equal quantities of two gases to escape from an effusion apparatus under similar starting conditions. If the 84 second gas is SO_2, what is the molar mass of the other gas?

13. THERMOCHEMISTRY

PREREQUISITES

1. In Section 2.2 you learned that proportionalities can be changed to equations by introducing a proportionality constant. Specific heat, heat of fusion, and heat of vaporization are presented as proportionality constants.

2. Changes in heat content (enthalpy) are associated with chemical reactions and the equations (Chapter 9) that describe them.

3. The quantitative meaning of a chemical equation discussed in Section 10.1 is expanded to include changes in heat content, which are then used in stoichiometry problems.

CHEMICAL SKILLS

1. Heat change for a temperature change of a single phase. Given three of the following calculate the fourth: (a) mass of a sample; (b) temperature change, or initial and final temperatures; (c) specific heat; (d) heat change of the sample.

2. Heat transfer. Given the mass, specific heat, and initial temperature of both a hot object and a cold object, calculate the final temperature when heat flow between the two objects reaches equilibrium.

3. Heat change for a phase change at constant temperature. Given two of the following, calculate the third: (a) mass of a sample that changes phase; (b) heat of fusion or heat of vaporization; (c) heat change of the sample.

4. Heat change for both phase change and temperature change. Given the mass of a sample and its initial and final phases and temperatures, plus the necessary specific heats and heats of fusion and/or vaporization, calculate the heat change as the sample changes from the initial condition to the final condition.

5. Heat change for a reaction. Given two of the following, calculate the third: (a) ΔH for a reaction whose equation is known; (b) the quantity of one of the reactants or products; (c) net heat change for that quantity of reactant or product.

6. Calorimetry. Calculate the ΔH of a chemical reaction that occurs in a calorimeter if given (a) the mass and specific heat of a substance to absorb

or provide heat in a calorimeter (usually water); (b) the temperature change in the calorimeter, or initial and final temperatures; (c) mass of a reactant, and (d) the equation for its reaction in the calorimeter.

7. Hess' law. Given ΔH for two or more reactions whose equations may be added to produce a final equation, perform that addition and calculate ΔH for the resulting reaction.

8. ΔH from heats of formation. Given the heat of formation of each substance in a chemical equation, calculate ΔH for the reaction.

9. Heat of formation from ΔH. Given ΔH for a reaction and the heat of formation of each substance in the reaction except one, calculate the remaining heat of formation.

13.1 ENERGY: HEAT, WORK, AND ENERGY UNITS

Energy has many forms. Two of the most common are heat and work. When energy is produced, by a chemical reaction for example, that energy is often in the form of heat or work. The gasoline burned in a car produces energy by a chemical reaction. Some of the energy is used to move the car (work energy) and some is used to warm up the surroundings (heat energy). For this chapter, we will assume that all the energy produced by chemical reactions is in the form of heat—that no work is done at all. This is as if the gasoline used in the car was just spilled on the ground and ignited. The same amount of energy is produced this way, but all of it is in the form of heat.

All forms of energy are expressed in the same SI unit, the **joule**, (pronounced as if it were spelled "jool," as in school). A joule, symbol J, is a rather small unit for measuring energy changes for chemical reactions, so we usually use the **kilojoule, kJ**, which is 1000 times larger than the joule. (Until about 1980, the **calorie** was the common energy unit used in chemistry. One calorie is defined as 4.184 joules. The Calorie, spelled with a capital "C," used in nutrition is a kilocalorie, 1000 calories.)

13.2 SPECIFIC HEAT

It has been found experimentally that the amount of heat required to raise the temperature of any pure substance is proportional to (a) the mass of the sample and (b) the temperature change, ΔT. Using Q for amount of heat gained or lost, and m for mass, $Q \propto m$ and $Q \propto \Delta T$. Whenever one variable is proportional to two or more other variables, it is proportional to the product of those variables: $Q \propto m \times \Delta T$. Introducing a proportionality constant, c, changes the proportionality to an equation:

$$Q = mc\Delta T \qquad (13.1)$$

Solving Equation 13.1 for c gives joules per gram per degree Celsius as the units of the proportionality constant:

$$c = \frac{Q}{m\Delta T} = \frac{J}{g \cdot {}^\circ C} \qquad (13.2)$$

The proportionality constant, c, is called **specific heat**. It is a property of a pure substance. It is the amount of heat required to raise the temperature of one gram of the substance one degree Celsius. The specific heat of water under ordinary conditions is, to three significant figures, 4.18 J/g•°C. This number is used so often, it should be memorized. The specific heats of other substances may be found in handbooks.

A comment about ΔT: The Greek *delta*, Δ, is used in scientific writing to show change. It always means the final value of something minus the initial value: $\Delta X = X_{final} - X_{initial}$. If X_{final} is larger than $X_{initial}$, then ΔX is a positive quantity; if X_{final} is smaller than $X_{initial}$, then ΔX is negative. If temperature rises in Equation 13.1, ΔT is a positive quantity, so Q is also positive; if temperature drops, ΔT and Q are negative. A positive Q indicates heat flow into a substance, while a negative Q shows heat flows out of the sample.

13.3 CALCULATIONS WITH SPECIFIC HEAT

If a sample of matter does not change between the liquid phase and solid or gas phase, a specific heat problem can never be more complex than being given three of the four items in Equation 13.1 and being asked to find the fourth. The recommended procedure is to solve the equation algebraically for the wanted variable, substitute the given information *including units*, and then calculate the answer. If the problem has been set up correctly, the units will cancel to leave the proper unit for the required quantity.

EXAMPLE 13.1: How much heat energy is required to raise the temperature of 44.0 grams of lead from room temperature, 21°C, to its melting point, 327°C? The specific heat of lead is 0.13 J/g•°C.

This is straight substitution into Equation 13.1. Calculate the answer.

GIVEN: $m = 44.0$ g, $T_1 = 21°C$, $T_2 = 327°C$, $c = 0.13$ J/g•°C FIND: Q

EQN: $Q = mc\Delta T = 44.0 \text{ g} \times \frac{0.13 \text{ J}}{\text{g}\cdot°\text{C}} \times (327 - 21)°\text{C} = 1.8 \times 10^3 \text{ J}$

The answer is given in two significant figures because of the specific heat.

EXAMPLE 13.2: In a laboratory experiment, 58.1 grams of copper lose 1.6×10^3 joules in cooling from 99.8°C to 28.4°C. Calculate the specific heat of copper.

Notice that the word "lose" in the problem indicates heat is flowing from the copper, so Q is a negative quantity. Use the space at the top of the next page to solve Equation 13.1 for the wanted variable. Then substitute the given values, and calculate the answer.

GIVEN: $m = 58.1$ g, $Q = -1.6 \times 10^3$ J, $T_1 = 99.8°C$, $T_2 = 28.4°C$ FIND: c

EQN: $$c = \frac{Q}{m\Delta T} = \frac{-1.6 \times 10^3 \text{ J}}{(58.1 \text{ g})(28.4 - 99.8)°C} = 0.39 \text{ J/g}\cdot°C$$

Notice that both Q and ΔT are negative quantities. The division of one by the other yields a positive quotient. Specific heat is always positive because of the way it is defined.

EXAMPLE 13.3: To what temperature will 117 grams of nickel be heated if it starts at 18.3°C and absorbs 2.46×10^3 joules? The specific heat of nickel is 0.444 J/g•°C.

This time the wanted quantity, final temperature, does not appear in Equation 13.1. It is a part of ΔT, however. Use the equation to find ΔT, and then we will figure out how to get the final temperature.

GIVEN: $m = 117$ g, $T_1 = 18.3°C$, $Q = 2.46 \times 10^3$ J, $c = 0.444$ J/g•°C FIND: T_2

EQN: $$\Delta T = \frac{Q}{mc} = \frac{2.46 \times 10^3 \text{ J}}{117 \text{ g}} \times \frac{\text{g}\cdot°C}{0.444 \text{ J}} = 47.4°C$$

Now a little reasoning: The starting temperature was 18.3°C. The temperature went up 47.4°C. What was the final temperature?

$$T_2 = T_1 + \Delta T = 18.3°C + 47.4°C = 65.7°C$$

When two substances that are at different temperatures are placed in contact with each other, heat will flow from the warmer one to the colder one. Heat flow will stop when they are both at the same final temperature, a condition called *thermal equilibrium*. In cases such as this, the heat gained by the cold substance is equal to the heat lost by the warm object, but the heat changes are opposite in sign. Remember that heat gained is positive, and heat lost is negative. Expressed as an equation, heat change of cold substance = -heat change of warm substance

$$mc\Delta T_{cold} = -mc\Delta T_{warm} \qquad (13.3)$$

This equation can be used to solve for the final temperature, T_2, which is reached by both substances.

EXAMPLE 13.4: A 25-gram block of iron ($c = 0.452$ J/g•°C) is initially at 225°C. It is dropped into 75 grams of water ($c = 4.18$ J/g•°C), initially at 25°C. What is the final temperature?

First identify the given quantities, the quantity to find, and the equation to be used.

GIVEN: For H_2O: $m = 75$ g, $c = 4.18$ J/g•°C, $T_1 = 25$°C FIND: T_2
For Fe: $m = 25$ g, $c = 0.452$ J/g•°C, $T_1 = 225$°C

EQN: $mc\Delta T_{H_2O} = -mc\Delta T_{Fe}$

Notice that ΔT appears on both sides of the equation—but they are not the same ΔT's. For this problem, write ΔT as $T_2 - T_1$. The values of T_1 are given for both water and iron. The value of T_2 is the same for both, and it is the quantity you are asked to find. Because T_2 appears in two places in the equation, it is easier to substitute directly into the original equation rather than first solving for the unknown, as we usually recommend. Be sure to use the mass, specific heat, and temperatures of water on one side of the equation and the values for iron on the other. Make the substitutions, but don't solve for T_2 yet.

$$75 \text{ g} \times \frac{4.18 \text{ J}}{\text{g•°C}} \times (T_2 - 25)\text{°C} = -25 \text{ g} \times \frac{0.452 \text{ J}}{\text{g•°C}} \times (T_2 - 225)\text{°C}$$

This equation looks complicated! It can be simplified by canceling the units that appear on both sides of the equation. Do this, and then by algebra solve for T_2 in °C. Be careful: the minus signs are important.

$$75 \times 4.18 \times (T_2 - 25) = -25 \times 0.452 \times (T_2 - 225)$$
$$313.5(T_2 - 25) = -11.3(T_2 - 225)$$
$$313.5\ T_2 - 7837.5 = -11.3\ T_2 + 2542.5$$
$$324.8\ T_2 = 10{,}380$$
$$T_2 = 32°C$$

The result is not rounded for significant figures until the final answer.

13.4 HEAT OF FUSION; HEAT OF VAPORIZATION

When the temperature of a pure substance has been raised to its melting point, any additional heat that is absorbed is used to break down the crystal structure rather than to increase the temperature further. Melting is a constant temperature process. Boiling also occurs at constant temperature. After a liquid has absorbed enough heat to get to the boiling point, any additional energy is used to separate the liquid particles from each other and change the substance to a gas.

The heat absorbed in a change of phase is directly proportional to the mass of the sample:

$$Q_{fus} \propto m \qquad\qquad Q_{vap} \propto m \tag{13.4}$$

The subscript "fus" refers to fusion, or melting, and "vap" to vaporization, or boiling. The proportionalities are changed to equations by proportionality constants that are properties of the substance called **heat of fusion**, ΔH_{fus}, and **heat of vaporization**, ΔH_{vap}, respectively:

$$Q_{fus} = m \times \Delta H_{fus} \qquad\qquad Q_{vap} = m \times \Delta H_{vap} \tag{13.5}$$

Solved for the proportionality constants, heats of fusion and vaporization become conversion factors for dimensional analysis setups. Their units are energy units per mass unit. Heats of fusion are usually given in joules per gram, but, because heats of vaporization are so much larger, they usually appear in kilojoules per gram:

$$\Delta H_{fus} = \frac{Q_{fus}}{m} = \frac{J}{g} \qquad\qquad \Delta H_{vap} = \frac{Q_{vap}}{m} = \frac{kJ}{g} \tag{13.6}$$

Heats of fusion and heats of vaporization are readily available in handbooks.

Change of phase calculations are direct applications of the above equations.

EXAMPLE 13.5: Calculate the energy required to melt 75.0 grams of ice if its heat of fusion is 335 joules per gram.

GIVEN: 75.0 g, ΔH_{fus} = 335 J/g FIND: Q

EQN: $Q = m \times \Delta H_{fus} = 75.0\ g \times \frac{335\ J}{g} = 2.51 \times 10^4\ J = 25.1\ kJ$

EXAMPLE 13.6: 86.1 kilojoules are required to vaporize 58.2 grams of a liquid. Calculate its heat of vaporization in kilojoules per gram.

GIVEN: Q = 86.1 kJ, m = 58.2 g FIND: ΔH_{fus}

EQN: $\Delta H_{fus} = \frac{Q}{m} = \frac{86.1\ kJ}{58.2\ g} = 1.48\ kJ/g$

13.5 COMBINED TEMPERATURE CHANGE AND CHANGE OF PHASE PROBLEMS

The energy required to heat 10.0 grams of water at 20.0°C to its boiling point, 100.0°C, and then to change that water to steam at that temperature is the sum of two calculations. First is the energy to raise the temperature:

$$Q_1 = mc\Delta T = 10.0\ g \times \frac{4.18\ J}{g\cdot °C} \times (100.0 - 20.0)°C = 3.34 \times 10^3\ J = 3.34\ kJ$$

The heat of vaporization of water is 2.26 kJ/g. Therefore, the energy to change the water to steam at the boiling point is

$$Q_2 = m\Delta H_{vap} = 10.0\ g \times \frac{2.26\ kJ}{g} = 22.6\ kJ$$

Adding the two energies gives the total requirement:

Energy to heat water to boiling + energy to boil = total energy

3.34 kJ + 22.6 kJ = 25.9 kJ

Figure 13.1 is a heating curve that is typical of any pure substance. The graph is not to scale. The sloped portions of the graph trace the changes that occur in the solid, liquid, and gaseous phases, all of which are calculated with Equation 13.1, using specific heat. The horizontal regions are the constant temperature changes of phase that are calculated by Equation 13.5. In the example just completed, 3.34 kJ is the energy in passing from point A to point B, and 22.6 kJ is the energy from B to C. Both energies would be measured along the horizontal axis of the graph.

It is helpful to sketch a curve like Figure 13.1 when solving combined temperature and change of state problems. It shows clearly what calculations must be made. They can then be handled individually and totaled at the end.

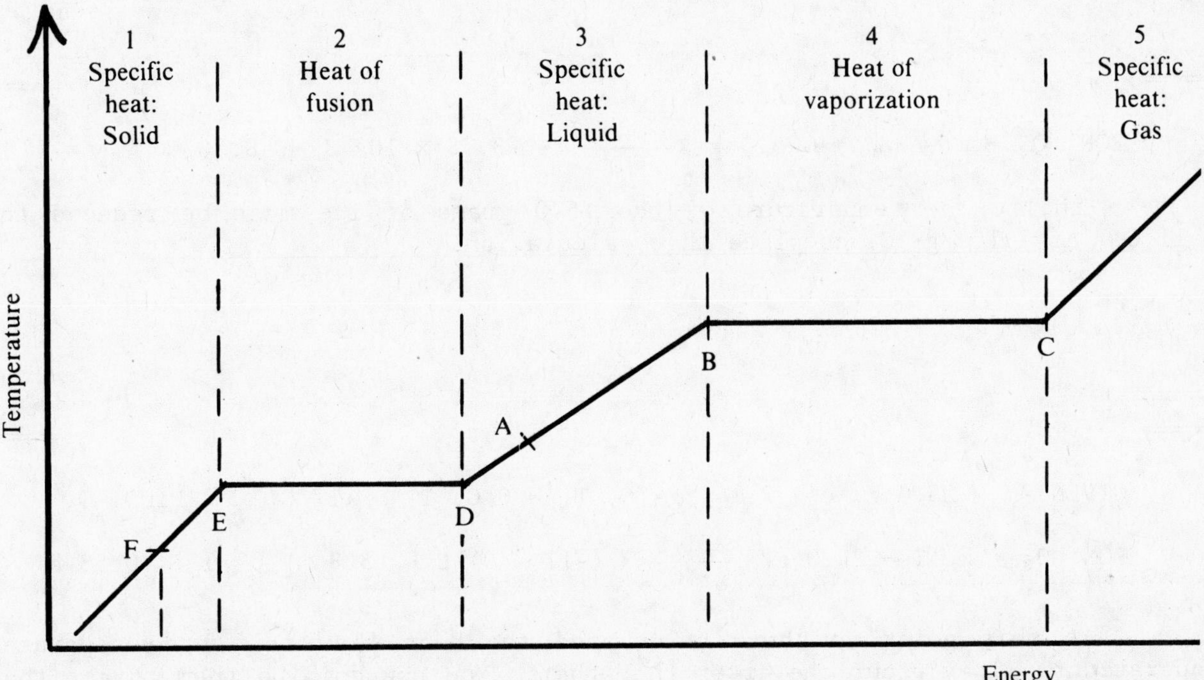

Figure 13.1

EXAMPLE 13.7: Calculate the energy that must be removed from 25.0 grams of water at 21°C as it is converted into ice at -11°C. The heat of fusion of water is 335 J/g, and the specific heat of ice is 2.1 J/g•°C.

There are three steps in this problem. First, the liquid water must be cooled from 21°C to the freezing point, 0°C (A to D in Figure 13.1). Second, the water must be frozen at 0°C (D to E). Third, the temperature of the ice at 0°C must be reduced to -11°C (E to F). Begin by calculating the heat change as the liquid water cools to the freezing point.

GIVEN: $m = 25.0\ g$, $c = 4.18\ J/g{\cdot}°C$, $T_1 = 21°C$, $T_2 = 0°C$ FIND: Q_1

EQN: $Q_1 = mc\Delta T = 25.0\ g \times \frac{4.18\ J}{g{\cdot}°C} \times (0 - 21)°C = -2.2 \times 10^3\ J = -2.2\ kJ$

Next, how much energy is removed as the water freezes? Freezing is the opposite of melting. The heat change is the same as it is in melting, except it is a flow from the water instead of into it. ΔH is therefore negative, and so is Q. Calculate Q for the freezing process. (The required data are repeated at the top of the next page.)

GIVEN: $m = 25.0\ g$, $\Delta H_{fus} = 335\ J/g$ FIND: Q_2

$$EQN:\ Q_2 = m(-\Delta H_{fus}) = 25.0\ g \times \frac{-335\ J}{g} = -8.38 \times 10^3\ J = -8.38\ kJ$$

Finally the temperature of the 25.0 grams of ice must be reduced to -11°C. At 2.1 J/g•°C, complete that calculation.

GIVEN: $m = 25.0\ g$, $c = 2.1\ J/g{\cdot}°C$, $T_1 = 0°C$, $T_2 = -11°C$ FIND: Q_3

$$EQN:\ Q_3 = mc\Delta T = 25.0\ g \times \frac{2.1\ J}{g{\cdot}°C} \times (-11 - 0)°C = -5.8 \times 10^2\ J = -0.58\ kJ$$

At this point you have calculated the heat flow for three separate operations. They occur one after the other. You now combine them to get the total heat change for the entire process. Calculate the total amount of heat taken from the water to produce ice at -11°C.

$$Q = Q_1 + Q_2 + Q_3 = (-2.2\ kJ) + (-8.38\ kJ) + (-0.58\ kJ) = -11.2\ kJ$$

13.6 HEAT CHANGES IN CHEMICAL REACTIONS

A measurable amount of energy is released or absorbed in nearly all chemical reactions. This energy is formally referred to as **enthalpy** and identified by the symbol H. It is sometimes called **heat content**. The enthalpy change for a chemical reaction is ΔH.

ΔH_{fus} and ΔH_{vap} are measured for a definite quantity of a substance undergoing a physical change. In the same way, ΔH for a chemical change is measured for the number of moles of each species in the chemical equation that accompanies it. ΔH may be written either as a separate item next to the chemical equation or within the equation as if it were a reactant or product. For example, the formation of water may be written:

$$2\ H_2(g) + O_2(g) \rightarrow 2\ H_2O(\ell) + 571.5\ kJ \quad (13.7)$$

$$2\ H_2(g) + O_2(g) \rightarrow 2\ H_2O(\ell) \qquad \Delta H = -571.5\ kJ \quad (13.8)$$

Equations such as 13.7 and 13.8 which show the change in heat content are colled **thermochemical equations**.

The negative sign for ΔH indicates that the system is losing heat to the surroundings. This is suggested in Equation 13.7 by showing energy as a product of the reaction. A reaction in which energy is produced, or released to the surroundings, is an **exothermic reaction**. ΔH is always negative for an exothermic reaction.

If the reactions are turned around—if we decompose water into its elements—the ΔH becomes positive:

$$2\ H_2O(\ell) + 571.5\ kJ \rightarrow 2\ H_2(g) + O_2(g) \tag{13.9}$$

$$2\ H_2O(\ell) \rightarrow 2\ H_2(g) + O_2(g) \qquad \Delta H = +571.5\ kJ \tag{13.10}$$

This time energy of the system increases, as indicated by the addition of energy in Equation 13.9. Reactions that absorb energy—that have a positive ΔH—are **endothermic reactions**.

Phase designations are essential when writing thermochemical equations. This is because the same reaction will have a different ΔH if the state of any substance is other than that for which the measurement was made. For example, ΔH = -483.7 kJ if water vapor, $H_2O(g)$, is the product in Equation 13.8. The difference is the heat of vaporization of water.

The conversion factors that may be drawn from a chemical equation may be expanded to include the energy term. Thus, from any of the four equations 13.7 to 13.10, these conversion factors are

$$\Delta H = \frac{571.5\ kJ}{2\ mol\ H_2(g)} = \frac{571.5\ kJ}{1\ mol\ O_2(g)} = \frac{571.5\ kJ}{2\ mol\ H_2O(\ell)} \tag{13.11}$$

What Equation 13.11 is saying is that the energy change in a chemical reaction is proportional to the quantity of chemical used. If hydrogen is looked upon as a fuel in Equations 13.7 and 13.8, it is logical to expect that if you use half the amount of fuel, you should get half as much energy.

$$H_2(g) + \tfrac{1}{2}\ O_2(g) \rightarrow H_2O(\ell) + 285.8\ kJ$$

Note that the value of ΔH is tied to a specific equation. The standard form of an equation is balanced with simplest whole-number coefficients.

EXAMPLE 13.8: Calculate the energy released if you burn 15.0 grams of hydrogen according to $2\ H_2(g) + O_2(g) \rightarrow 2\ H_2O(\ell) + 571.5\ kJ$.

The key conversion factor is ΔH = -571.5 kJ/2 mol H_2 (Equation 13.11). This makes possible the unit conversion: mol H_2 → kJ. Can you get to moles of H_2 from the information given? Certainly. The first step is the same as in the stoichiometry 3-step. In the second step you can go directly from the moles of given substance to energy. Set up and solve.

GIVEN: 15.0 g H_2 FIND: Q

FACTOR: MM of H_2, ΔH = -571.5 kJ/2 mol H_2 PATH: g H_2 → mol H_2 → kJ

$$15.0\ g\ H_2 \times \frac{1\ mol\ H_2}{2.016\ g\ H_2} \times \frac{-571.5\ kJ}{2\ mol\ H_2} = -2.13 \times 10^3\ kJ$$

EXAMPLE 13.9: ΔH is -8389 kJ for the burning of hexane:
$2\ C_6H_{14}(\ell) + 19\ O_2(g) \rightarrow 12\ CO_2(g) + 14\ H_2O(\ell) + 8389\ kJ$. How many grams of hexane must be burned to release 2.00×10^4 kJ of heat?

Set up and solve the problem.

GIVEN: -2.00×10^4 kJ FIND: g C_6H_{14}
FACTOR: -8389 kJ/2 mol C_6H_{14}, MM C_6H_{14} PATH: kJ → mol C_6H_{14} → g C_6H_{14}

$$-2.00 \times 10^4\ kJ \times \frac{2\ mol\ C_6H_{14}}{-8389\ kJ} \times \frac{86.17\ g\ C_6H_{14}}{1\ mol\ C_6H_{14}} = 4.11 \times 10^2\ g\ C_6H_{14}$$

13.7 CALORIMETRY

The most common laboratory device for measuring heat change in a chemical reaction is a **calorimeter**. An ideal calorimeter is a thermally isolated part of the universe that allows no transfer of heat into it or out of it. Any heat released within the calorimeter is absorbed by something inside the calorimeter. In other words, the sum of all heat exchanges in a calorimeter is zero:

$$Q_1 + Q_2 + Q_3 + \ldots = \Sigma Q = 0 \qquad (13.12)$$

(The Greek sigma, Σ, is used to represent the sum of two or more values.) In a typical experiment, a measured mass, m, of liquid water is placed into the calorimeter. A thermometer measures the temperature change caused by the chemical reaction. Knowing the specific heat of water, 4.18 J/g•°C, the heat gained or lost by the water, Q_w, is found by Equation 13.1: $Q_w = mc\Delta T$.

If a heat change occurs within a calorimeter, some heat is transferred to or taken from the calorimeter itself. In a sensitive experiment, this must be taken into account. In a typical introductory chemistry experiment, the heat transfer to or from the calorimeter is very small. We will regard it as being negligible in all problems in this book.

If the heat transfer to or from the calorimeter is disregarded, there are only two heat changes in a calorimeter experiment, the heat of reaction, Q_r, and the heat change of the water, Q_w. Equation 13.12 thus becomes

$$Q_r + Q_w = 0 \qquad (13.13)$$

The essence of Equation 13.13 is that all of the heat change in the reaction is transferred to or from the water. Solving for Q_r,

$$Q_r = -Q_w \qquad (13.14)$$

To get the ΔH of a reaction from Q_r, the heat change must be found in terms of joules per mole (n) of some species in the reaction:

$$\Delta H \text{ per mole} = \frac{Q_r}{n} \qquad (13.15)$$

Combining Equations 13.14 and 13.15,

$$\Delta H \text{ per mole} = \frac{-Q_w}{n} \qquad (13.16)$$

EXAMPLE 13.10: A 2.00-gram sample of sodium hydroxide is reacted with excess hydrochloric acid in a styrofoam-cup calorimeter containing 50.0 grams of water. The temperature of the water changes from 20.1°C to 37.3°C. The reaction equation is $NaOH(s) + HCl(aq) \rightarrow NaCl(aq) + H_2O(\ell)$. Calculate ΔH for the reaction. (Assume all heat is absorbed by the water.)

First calculate the heat absorbed by the water, using Equation 13.1.

GIVEN: m = 50.0 g, c = 4.18 J/g•°C, T_1 = 20.1°C, T_2 = 37.3°C FIND: Q_w

EQN: $Q_w = 50.0 \text{ g} \times \frac{4.18 \text{ J}}{\text{g}\cdot{}^\circ\text{C}} \times (37.3 - 20.1)^\circ\text{C} = 3.59 \times 10^3 \text{ J} = 3.59 \text{ kJ}$

Now calculate the number of moles of sodium hydroxide reacting.

GIVEN: 2.00 g NaOH FIND: mol NaOH FACTOR: MM of NaOH PATH: g → mol

$$2.00 \text{ g NaOH} \times \frac{1 \text{ mol NaOH}}{40.00 \text{ g NaOH}} = 0.0500 \text{ mol NaOH}$$

Now use Equation 13.16 to calculate ΔH/mol NaOH.

GIVEN: Q_w = 3.59 kJ, n = 0.0500 mol NaOH FIND: ΔH/mol NaOH

EQN: $\Delta H/\text{mol NaOH} = \frac{-Q_w}{n} = \frac{-3.59 \text{ kJ}}{0.0500 \text{ mol NaOH}} = -71.8 \text{ kJ/mol NaOH}$

The equation has only 1 mole NaOH, so ΔH = -71.8 kJ for the reaction $NaOH(s) + HCl(aq) \rightarrow NaCl(aq) + H_2O(\ell)$.

EXAMPLE 13.11: If the experiment described in Example 13.10 is repeated with sulfuric acid instead of hydrochloric acid, 2.00 g NaOH again produce a ΔT of 17.2°C in 50.0 g H_2O. This time the reaction equation is 2 NaOH(aq) + H_2SO_4(aq) → Na_2SO_4(aq) + 2 $H_2O(\ell)$. Calculate ΔH for this reaction.

Begin as before, finding ΔH per mole of NaOH.

As before, ΔH/mol NaOH = -71.8 kJ/mol NaOH. All the data are the same, and they yield the same result.

Remember, ΔH is tied to a specific equation. You have calculated the ΔH for the reaction of *1 mole* of NaOH. Write the chemical equation for the reaction, keeping 1 for the coefficient of NaOH. *Caution!* Fractions will be needed. Include the value of ΔH in the equation.

NaOH(aq) + ½ H_2SO_4(aq) → ½ Na_2SO_4(aq) + $H_2O(\ell)$ + 71.8 kJ

Notice that this is not the standard form of an equation because it does not use simplest whole-number coefficients. Rewrite the equation clearing the fractions. What happens to the value of ΔH?

2 NaOH(aq) + H_2SO_4(aq) → Na_2SO_4(aq) + 2 $H_2O(\ell)$ + 143.6 kJ

Multiplying the equation by 2 doubles ΔH: 2 × (-71.8 kJ) = -143.6 kJ.

13.8 ENTHALPY SUMMATION: HESS' LAW

The enthalpy change for a specific chemical reaction has a definite value. We often describe a chemical change as if it occurs in a series of steps, leading to the final result represented by the equation. The steps may be real or imaginary. Furthermore, there are often several different reaction paths that lead to the same final reaction equation. This notwithstanding, ΔH for the reaction is always the same.

Because ΔH is independent of the path of a reaction, it can be calculated if the equation can be written as the sum of a series of equations for which each individual ΔH is known. For example:

$$C(s) + \tfrac{1}{2} O_2(g) \rightarrow CO(g) \qquad \Delta H = -110.5 \text{ kJ} \qquad (13.17)$$

$$CO(g) + \tfrac{1}{2} O_2(g) \rightarrow CO_2(g) \qquad \Delta H = -283.0 \text{ kJ} \qquad (13.18)$$

$$C(s) + O_2(g) \rightarrow CO_2(g) \qquad \Delta H = -393.5 \text{ kJ} \qquad (13.19)$$

This process is called **enthalpy summation,** and it is an example of what is known as **Hess' law**.

Enthalpy summation can be used to find the ΔH of a reaction when it cannot be measured directly. Burning carbon to carbon monoxide, Equation 13.17, for example, cannot be controlled to produce only carbon monoxide. Some of the carbon burns to carbon dioxide before another portion has even been changed to carbon monoxide. It is possible, however, to burn carbon all the way to carbon dioxide and measure the -393.5 kJ/mol of carbon burned. It is also possible to burn carbon monoxide to carbon dioxide, for which ΔH is -283.0 kJ/mol. By reversing the second measured reaction and adding it to the first, the ΔH for the carbon to carbon monoxide reaction is produced:

$$C(s) + O_2(g) \rightarrow CO_2(g) \qquad \Delta H = -393.5 \text{ kJ}$$
$$CO_2(g) \rightarrow CO(g) + \tfrac{1}{2} O_2(g) \qquad \Delta H = +283.0 \text{ kJ}$$
$$C(s) + \tfrac{1}{2} O_2(g) \rightarrow CO(g) \qquad \Delta H = -110.5 \text{ kJ}$$

Notice that a reaction that is exothermic in one direction is endothermic in the opposite direction. This means that the sign of ΔH must be changed when the equation is reversed. This has been done in the $CO_2 \rightarrow CO$ equation.

To find the ΔH of a reaction by enthalpy summation, you must know the ΔH of all other reactions in the summation. You then proceed as follows:

1. Draw an addition line. Beneath it, write the equation whose ΔH is to be found by adding other equations.
2. Above the addition line, write each equation that is to be added, showing ΔH at the right.
3. Reverse any equation above the line that must be reversed in order to make a species appear on the correct side of the final equation. Be sure to change the sign of ΔH.
4. Multiply the equations above the line by whatever numbers are necessary to produce the correct coefficients when the equations are added, or to eliminate substances not present in the final equation. Be sure to multiply ΔH values.
5. Add the equations and ΔH values.

Combustion is one kind of reaction for which the ΔH is readily measured. The **heat of combustion of a substance, ΔH_c, is the ΔH of the reaction in which 1 mole of that substance reacts with oxygen and is completely burned.** If the substance contains only carbon and/or hydrogen, and possibly oxygen, the products are carbon dioxide gas and/or liquid water (see Section 9.5). Combustion data make it possible to calculate the ΔH of an "imaginary" reaction.

Note that the specific equation for which ΔH_c is defined must have a coefficient of 1 for the substance undergoing combustion. This often makes it necessary to use a fractional coefficient for oxygen.

EXAMPLE 13.12: Use the heats of combustion of methane, $CH_4(g)$, carbon, and hydrogen to calculate ΔH for the imaginary reaction in which 1 mole of methane is formed from its elements:

$$C(s) + 2\ H_2(g) \rightarrow CH_4(g)$$

Before starting on the above procedure, let's be sure we understand the idea of heat of combustion. The definition above says that *1 mole* of the substance is burned completely. ΔH_c = -890.2 kJ/mol for methane. Write the equation for the heat of combustion of methane, $CH_4(g)$. Place ΔH for this reaction at the right.

$CH_4(g) + 2\ O_2(g) \rightarrow CO_2(g) + 2\ H_2O(\ell)$ ΔH = -890.2 kJ

$C(s) + 2\ H_2(g) \rightarrow CH_4(g)$

The first step in the procedure calls for the final equation to be written below an addition line. This has been done. Above the line will be the three heat of combustion equations, starting with methane. Space has been left for the heat of combustion equations for C(s) and $H_2(g)$, which have values of -393.5 kJ/mol and -285.8 kJ/mol, respectively. Complete Step 2 of the procedure by writing the two remaining equations. You will have to make a decision when you come to hydrogen.

$CH_4(g) + 2\ O_2(g) \rightarrow CO_2(g) + 2\ H_2O(\ell)$ ΔH = -890.2 kJ

$C(s) + O_2(g) \rightarrow CO_2(g)$ ΔH = -393.5 kJ

$H_2(g) + \frac{1}{2}\ O_2(g) \rightarrow H_2O(\ell)$ ΔH = -285.8 kJ

$C(s) + 2\ H_2(g) \rightarrow CH_4(g)$

The heat of combustion equation is for *1* mole of a substance. All other coefficients must be adjusted to balance with that 1 mole. In the combustion of hydrogen, therefore, the coefficient for oxygen must be ½.

Now check the compounds above the addition line to be sure they are on the same side of the equation as they are below the line. Reverse any that are not as you rewrite the three equations above the line.

$C(s) + 2\ H_2(g) \rightarrow CH_4(g)$

$$\begin{array}{rl} CO_2(g) + 2\ H_2O(\ell) \rightarrow CH_4(g) + 2\ O_2(g) & \Delta H = -(-890.2) = +890.2\ kJ \\ C(s) + O_2(g) \rightarrow CO_2(g) & \Delta H = -393.5\ kJ \\ H_2(g) + \frac{1}{2}\ O_2(g) \rightarrow H_2O(\ell) & \Delta H = -285.8\ kJ \\ \hline C(s) + 2\ H_2(g) \rightarrow CH_4(g) & \end{array}$$

$CH_4(g)$ is on the right in the final equation, but it was on the left in the combustion equation. The first equation therefore had to be reversed. The sign of ΔH was also changed.

If the equations were to be added in their present form, they would not produce the equation below the line. Notice that the carbon dioxide on the left side of the first equation is canceled by the carbon dioxide on the right side of the second equation. This is well, as there is no carbon dioxide in the final equation.

Water molecules do not cancel. There are two water molecules on the left side of the first equation, but only one on the right side of the third equation. But there is no water in the final equation, so it must cancel. If the third equation is multiplied by 2, there will be two water molecules on both sides of the arrows and they will cancel. This is Step 4 in the procedure. Double the third equation, and write it in the space left for it. Will you double the ΔH value too? When you have decided, add the ΔH's to get the ΔH of the reaction (Step 5).

$$\begin{array}{rl} CO_2(g) + 2\ H_2O(\ell) \rightarrow CH_4(g) + 2\ O_2(g) & \Delta H = -(-890.2) = +\ 890.2\ kJ \\ C(s) + O_2(g) \rightarrow CO_2(g) & \Delta H = -393.5\ kJ \\ & \Delta H = \\ \hline C(s) + 2\ H_2(g) \rightarrow CH_4(g) & \Delta H = \end{array}$$

$$\begin{array}{rl} CO_2(g) + 2\ H_2O(\ell) \rightarrow CH_4(g) + 2\ O_2(g) & \Delta H = -(-890.2) = +890.2\ kJ \\ C(s) + O_2(g) \rightarrow CO_2(g) & \Delta H = -393.5\ kJ \\ 2\ H_2(g) + O_2(g) \rightarrow 2\ H_2O(\ell) & \Delta H = 2(-285.8) = -571.6\ kJ \\ \hline C(s) + 2\ H_2(g) \rightarrow CH_4(g) & \Delta H = -\ 74.9\ kJ \end{array}$$

In doubling the third equation, the ΔH must also be doubled. If you burn twice as much hydrogen, you will surely produce twice as much energy. Now not only the water and carbon dioxide cancel, but the oxygen too. The enthalpy summation is complete.

You probably wonder why anybody would want to find the ΔH of an imaginary reaction like that in Example 13.12. Actually there is a very good reason. The answer, -74.9 kJ/mol, is the **heat of formation** of methane. The symbol for heat of formation is ΔH_f°. If the heat of formation of a compound is known, it may be used in many other calculations. The heat of formation of a substance is **the enthalpy change of a reaction in which one mole of the substance is produced from its elements**. The elements must be in their natural states at 25°C and 1 atmosphere pressure. $\Delta H_f^\circ = 0$ for all elements in their standard states. There is no chemical change when "one mole of the (element) is produced from its elements."

Equation 13.17 for the reaction between carbon and oxygen to produce 1 mole of carbon monoxide is the heat of formation equation for carbon monoxide.

Similarly, Equation 13.19 is the heat of formation equation for carbon dioxide. In each case, the heat of formation equation has on its right side exactly 1 mole of the compound and nothing else. On the left side it has only the elements that appear in the compound in whatever number of moles may be necessary to form 1 mole of compound. Sometimes, as in Equation 13.17, this requires a fractional coefficient for an element.

EXAMPLE 13.13: Write the heat of formation equation for each of the following compounds: $HCl(g)$; $Fe_2O_3(s)$; $H_2SO_4(\ell)$; $HIO_3(s)$.

$$\frac{1}{2} H_2(g) + \frac{1}{2} Cl_2(g) \rightarrow HCl(g) \qquad 2 Fe(s) + \frac{3}{2} O_2(g) \rightarrow Fe_2O_3(s)$$

$$H_2(g) + S(s) + 2 O_2(g) \rightarrow H_2SO_4(\ell)$$

$$\frac{1}{2} H_2(g) + \frac{1}{2} I_2(g) + \frac{3}{2} O_2(g) \rightarrow HIO_3(s)$$

EXAMPLE 13.14: ΔH_c = -1367 kJ/mol for ethanol, $C_2H_5OH(\ell)$. Calculate its heat of formation. ΔH_f° = -393.5 kJ/mol for $CO_2(g)$ and -285.8 kJ/mol for $H_2O(\ell)$.

As in Example 13.12, three equations and the heat of formation equation for the product are needed. Write all four equations. Include the three known ΔH values.

$$\begin{array}{rll} C_2H_5OH(\ell) + 3 O_2(g) \rightarrow 2 CO_2(g) + 3 H_2O(\ell) & \Delta H = -1367 & kJ \\ C(s) + O_2(g) \rightarrow CO_2(g) & \Delta H = -393.5 & kJ \\ H_2(g) + \frac{1}{2} O_2(g) \rightarrow H_2O(\ell) & \Delta H = -285.8 & kJ \\ \hline 2 C(s) + 3 H_2(g) + \frac{1}{2} O_2(g) \rightarrow C_2H_5OH(\ell) & & \end{array}$$

The combustion of ethanol yields water and carbon dioxide as products. One mole of ethanol is the only product in the heat of formation equation, and the necessary number of moles of each element are the reactants.

As in Example 13.12, an equation must be reversed, and this time two equations must be multiplied to make the equations add up to the heat of formation equation. Perform those operations and total the resulting ΔH's for the answer.

Equation	ΔH
$3\ H_2O(\ell) + 2\ CO_2(g) \rightarrow C_2H_5OH(\ell) + 3\ O_2(g)$	ΔH = -(-1367) = +1367 kJ
$2\ C(s) + 2\ O_2(g) \rightarrow 2\ CO_2(g)$	ΔH = 2(-393.5) = - 787.0 kJ
$3\ H_2(g) + \frac{3}{2}\ O_2(g) \rightarrow 3\ H_2O(\ell)$	ΔH = 3(-285.8) = - 857.4 kJ
$2\ C(s) + 3\ H_2(g) + \frac{1}{2}\ O_2(g) \rightarrow C_2H_5OH(\ell)$	ΔH = - 277.4 kJ

If the heat of formation of an organic compound is known, its heat of combustion can be calculated.

EXAMPLE 13.15: ΔH_f° = -229.0 kJ/mol for nonane, $C_9H_{20}(\ell)$. Again, ΔH_f° = -393.5 kJ/mol for $CO_2(g)$ and -285.8 kJ/mol for $H_2O(\ell)$. Calculate the heat of combustion of nonane.

Begin by writing the heat of formation equations for all three compounds. Then write the equation for the combustion of $C_9H_{20}(\ell)$ beneath the line where it must appear as the result of adding the other equations.

Equation	ΔH
$9\ C(s) + 10\ H_2(g) \rightarrow C_9H_{20}(\ell)$	ΔH = -229.0 kJ
$C(s) + O_2(g) \rightarrow CO_2(g)$	ΔH = -393.5 kJ
$H_2(g) + \frac{1}{2}\ O_2(g) \rightarrow H_2O(\ell)$	ΔH = -285.8 kJ
$C_9H_{20}(\ell) + 14\ O_2(g) \rightarrow 9\ CO_2(g) + 10\ H_2O(\ell)$	

This problem is a bit different, but the procedure of reversing, multiplying, and adding equations is the same. Complete the problem.

$C_9H_{20}(\ell) \rightarrow 9\ C(s) + 10\ H_2(g)$	$\Delta H = -(-229.0) = +\ 229.0$ kJ
$9\ C(s) + 9\ O_2(g) \rightarrow 9\ CO_2(g)$	$\Delta H = 9(-393.5) = -3542$ kJ
$10\ H_2(g) + 5\ O_2(g) \rightarrow 10\ H_2O(\ell)$	$\Delta H = 10(-285.8) = -2858$ kJ
$C_9H_{20}(\ell) + 14\ O_2(g) \rightarrow 9\ CO_2(g) + 10\ H_2O(\ell)$	$\Delta H = -6171$ kJ

Example 13.15 can be used to illustrate an easier way to find the ΔH of a reaction if ΔH_f° is known for each substance in the reaction. Taking the four ΔH values and arranging them in a line, and in the order that suits the end purpose, the enthalpy summation can be written

$$-3542 + (-2858) + 229.0 = -6171 \text{ kJ} \qquad (13.20)$$

Now note where each of these terms came from:

-3542 = 9(-393.5), which is 9 times the ΔH_f° of $CO_2(g)$;

-2858 = 10(-285.8), which is 10 times the ΔH_f° of $H_2O(\ell)$;

229.0 = -(-229.0), which is the opposite of the ΔH_f° of $C_9H_{20}(\ell)$.

Substituting these values into Equation 13.20,

$$9(-393.5) \quad + \quad 10(-285.5) \quad - \quad (-229.0) \quad = -6171 \text{ kJ} \qquad (13.21)$$

$$9\ \Delta H_f^\circ \text{ of } CO_2(g) + 10\ \Delta H_f^\circ \text{ of } H_2O(\ell) \quad - \quad \Delta H_f^\circ \text{ of } C_9H_{20}(\ell) = \Delta H \qquad (13.22)$$

The first two terms in the two equations above represent the sum of the heats of formation of the products of the reaction, each ΔH_f° multiplied by the number of moles in the reaction equation. From this is subtracted the sum of the heats of formation of the reactants, each ΔH_f° multiplied by the number of moles in the reaction equation. In this example, $C_9H_{20}(\ell)$ and $O_2(g)$ are the reactants. $\Delta H_f^\circ = 0$ for an element, and there is only 1 mole of $C_9H_{20}(\ell)$, so the indicated sum of the ΔH_f° of the reactants is the ΔH_f° of $C_9H_{20}(\ell)$. Putting these words into an equation gives the simplification promised:

$$\Delta H = \Sigma\ \Delta H_{f\ \text{products}}^\circ - \Sigma\ \Delta H_{f\ \text{reactants}}^\circ \qquad (13.23)$$

To use Equation 13.23,

1. Write the equation for the reaction.
2. Beneath each formula write the ΔH_f° for the substance, multiplied by the number of moles of that substance in the equation.
3. Add the ΔH's on the right side of the equation and subtract the sum of the ΔH's on the left (sum of products - sum of reactants). Set equal to the ΔH of the reaction and solve for the unknown.

For Example 13.15 these three steps are

$$C_9H_{20}(\ell) + 14\ O_2(g) \rightarrow 9\ CO_2(g) + 10\ H_2O(\ell)$$
$$-229.0 \qquad 0 \qquad 9(-393.5) \qquad 10(-285.8)$$

$$\Delta H = 9(-393.5) + 10(-285.8) - (-229.0) = -3542 - 2858 + 229.0 = -6171\ kJ$$

This is the same as Equation 13.21.

The above procedure can be used to find the ΔH of any reaction for which the equation and the heats of formation of each compound are known. It can also be used to find the heat of formation of a compound in the equation if the ΔH of the reaction and all other heats of formation are known.

EXAMPLE 13.16: Calculate the heat of formation of $SO_2(g)$ if the heat of formation of $SO_3(g)$ is -395.2 kJ and ΔH is -196.6 kJ for the reaction $2\ SO_2(g) + O_2(g) \rightarrow 2\ SO_3(g)$.

$$2\ SO_2(g) + O_2(g) \rightarrow 2\ SO_3(g)$$
$$2(\Delta H_f^\circ) \qquad 0 \qquad 2(-395.2)$$

$$2(-395.2) - 2(\Delta H_f^\circ) = -196.6$$

$$\Delta H_f^\circ = \frac{-196.6 - 2(-395.2)}{-2} = -296.9\ kJ/mol$$

EXAMPLE 13.17: $Fe_2O_3(s) + 3\ CO(g) \rightarrow 2\ Fe(s) + 3\ CO_2(g)$. Find the enthalpy change for this reaction from the following heats of formation: Fe_2O_3, -822.2 kJ/mol; CO, -110.5 kJ/mol; CO_2, -393.5 kJ/mol.

$$Fe_2O_3(s) + 3\ CO(g) \rightarrow 2\ Fe(s) + 3\ CO_2(g)$$
$$-822.2 \qquad 3(-110.5) \qquad 0 \qquad 3(-393.5)$$

$$\Delta H = 3(-393.5) - (-822.2) - 3(-110.5) = -1180.5 + 822.2 + 331.5 = -26.8\ kJ$$

CHAPTER 13 PROBLEMS

1) How much energy will be lost by a 195-gram bar of lead as it cools from 90.0°C to 25.0°C? The specific heat of lead is 0.13 J/g•°C.

2) The tungston filament in a photoflood lamp rises from 20°C to 3600°C when in use. If a 1.5-gram filament absorbs 760 joules in that temperature change, what is the average specific heat of tungston over that range?

3) The iron burner on an electric stove weighs 925 grams. If the burner absorbs 2.51×10^5 joules while heating from 18°C, what is the final temperature of the burner? c = 0.473 J/g•°C for iron.

4) A piece of tin absorbed 2.67×10^3 joules upon heating from 20.0°C to 75.0°C. The specific heat of tin is 0.213 J/g•°C. What is the mass of the tin?

5) How much energy is absorbed by 3.50 kg of titanium in going from 25.0°C to 82.0°C if the specific heat of titanium is 0.594 J/g•°C?

6) If a piece of silicon, specific heat 0.703 J/g•°C, loses 3.15 kilojoules while changing from 54.8°C to 9.8°C, what is the mass of the sample?

7) Both 95 grams of copper (specific heat = 0.38 J/g•°C) and 95 grams of aluminum (specific heat = 0.88 J/g•°C) absorb 3.5 kJ of heat. a) Without calculating, predict which will reach the higher temperature. b) Confirm your prediction by calculating the temperature increase of each.

8) What is the final temperature when 7.5 grams of aluminum (c = 0.88 J/g•°C) at 175°C are added to 125 grams of water (c = 4.18 J/g•°C) at 17°C?

9) When a 25-gram block of a metal alloy at 215°C is dropped into 85 grams of water (c = 4.18 J/g•°C) at 22°C, the final temperature is 37°C. What is the specific heat of the alloy?

10) The heat of fusion of sodium chloride is 519 J/g. How much energy is required to melt 85.4 grams of sodium chloride?

11) How much heat is released by one pound (453.6 grams) of steam as it condenses? The heat of vaporization of water is 2.26 kJ/g.

12) A sample of an alloy is allowed to freeze in a device such that the heat released is absorbed by water. 925 grams of water rise from 23.4°C to 28.8°C as 773 grams of the metal freeze. Assuming all of the heat lost by the metal went to the water, what is the heat of fusion of the alloy?

13) How much energy is required to convert 141 grams of water at 62°C to superheated steam at 114°C? The specific heat of steam is 2.0 J/g•°C, and the heat of vaporization of water is 2.26 kJ/g.

14) The specific heat of Cu(s) is 0.38 J/g•°C, and of Cu(ℓ), 0.49 J/g•°C. Calculate the energy released by 56 grams of liquid copper initially at 1420°C if it cools to room temperature, 24°C. The freezing point of copper is 1083°C, and its heat of fusion is 205 J/g.

15) A jeweler melts 24.9 grams of pure gold in preparing an alloy for use in making settings for precious stones. How much energy is required to heat the liquid from room temperature, 26°C, to 1140°C? The melting point of gold is 1063°C. Its specific heat as a solid is 0.13 J/g•°C, and as a liquid, 0.15 J/g•°C. The heat of fusion is 64 J/g.

16) The oxidation of glucose takes place according to the equation

$$C_6H_{12}O_6(s) + 6\ O_2(g) \rightarrow 6\ H_2O(\ell) + 6\ CO_2(g) \qquad \Delta H = -2820\ kJ$$

How much energy is obtained by the oxidation of 325 grams of glucose?

17) The decomposition of potassium chlorate is as follows:

$$2\ KClO_3(s) \rightarrow 2\ KCl(s) + 3\ O_2(g) \qquad \Delta H = +89.5\ kJ$$

If 649 kJ are absorbed in decomposing $KClO_3$, how many liters of oxygen, measured at STP, will be produced?

18) The following equation shows the heat of solution for HCl(g):

$$HCl(g) + H_2O(\ell) \rightarrow H_3O^+(aq) + Cl^-(aq) \qquad \Delta H = -73.3\ kJ$$

What is the heat flow when 7.38 L of HCl are dissolved at STP?

19) NaCl(s)+ 3.89 kJ → $Na^+(aq)$ + $Cl^-(aq)$ is the equation for dissolving NaCl. How many grams of NaCl will dissolve if 5.08 kJ are consumed in the change?

20) A 5.00-gram sample of potassium chlorate is dissolved in a calorimeter containing 50.0 grams of water at 23.50°C. The temperature falls to 15.42°C. What is the heat of solution of $KClO_3$ in kJ/mol? (Consider only the heat absorbed by the water.)

21) A 2.00-gram sample of potassium hydroxide is dissolved in excess sulfuric acid in a calorimeter containing 45.0 grams of water at 17.5°C. The temperature rises to 43.2°C. What is the heat of reaction in kJ? (Consider only the heat absorbed by the water.)

22) A 1.50-gram sample of calcium hydroxide is dissolved in excess hydrochloric acid in a calorimeter containing 75.0 grams of water at 21.7°C. If the heat of reaction is -128 kJ, what is the final temperature of the water? (Consider only the heat absorbed by the water.)

23) Calculate ΔH for $F_2(g) + 2\ HCl(g) \rightarrow Cl_2(g) + 2\ HF(g)$. The heats of formation for HCl(g) and HF(g) are, respectively, -167.4 and -1269.0 kJ/mol.

24) $Ca(OH)_2(s) \rightarrow CaO(s) + H_2O(g)$ is the equation for the reaction that occurs on heating slaked lime. Calculate ΔH if ΔHf° is -986.6 kJ/mol for $Ca(OH)_2(s)$, -635.5 kJ/mol for $CaO(s)$, and -241.8 kJ/mol for $H_2O(g)$.

25) Calculate the heat of formation of $POCl_3(g)$ from the thermochemical equation below. The heats of formation of the other compounds in the reaction are -399.2 kJ/mol for PCl_5, -241.8 kJ/mol for $H_2O(g)$, and -92.5 kJ/mol for $HCl(g)$.

$$PCl_5(g) + H_2O(g) \rightarrow POCl_3(g) + 2\ HCl(g) \qquad \Delta H = -136.0\ kJ$$

26) $O_2(g) + 2\ SO_2(g) + 2\ H_2O(\ell) \rightarrow 2\ H_2SO_4(\ell)$ is one step in the process by which sulfuric acid is prepared. ΔH for this reaction is -457.2 kJ. Calculate the standard heat of formation for $H_2SO_4(\ell)$, given these other heats of formation: $SO_2(g)$, -296.9 kJ/mol; $H_2O(\ell)$, -285.8 kJ/mol.

27) Calculate the heat of formation of phenol, $C_6H_5OH(s)$, if its heat of combustion is -3053.5 kJ/mol. ΔH_f° is -393.5 kJ/mol for $CO_2(g)$ and -285.8 kJ/mol for $H_2O(\ell)$.

28) An experiment is performed in which the enthalpy changes of two acid-base neutralizations are measured. Expressed in net ionic form, the thermochemical equations are

$$H^+(aq) + OH^-(aq) \rightarrow H_2O(\ell) \qquad \Delta H = -55.9\ kJ$$

$$HCN(aq) + OH^-(aq) \rightarrow H_2O(\ell) + CN^-(aq) \qquad \Delta H = -12.1\ kJ$$

Calculate the ΔH for the reaction $HCN(aq) \rightarrow H^+(aq) + CN^-(aq)$ from these data.

29) How much energy will be released on the complete combustion of 66.4 grams of butane, $C_4H_{10}(g)$ (ΔH_f° = -124.7 kJ/mol)? The heat of formation of $CO_2(g)$ is -393.5 kJ/mol; of $H_2O(\ell)$, -285.8 kJ/mol.

++++++++++++++++++++++

30) Calculate the energy required to heat 395 grams of aluminum from 32°C to 450°C. The average specific heat of aluminum over that range is 0.899 J/g•°C.

31) 151 grams of an unknown metal loses 3.54×10^3 joules as it cools from 210°C to 35°C. What is the specific heat of the metal?

32) A copper wire weighing 135 grams loses 4.42×10^3 joules in cooling from a starting temperature of 125°C. Calculate the final temperature of the wire. The specific heat of copper is 0.385 J/g•°C.

33) A 12.0-gram lead bullet is fired into a brick wall. Some of the kinetic energy of the bullet is transferred to the wall, and some is retained by the bullet, which becomes hot as it is smashed and drops to the ground. How much energy is absorbed if the temperature increases from 19.0°C to 32.9°C? The specific heat of lead is 0.16 J/g•°C.

34) What is the final temperature when 225 grams of titanium (c = 0.594 J/g•°C) at 155°C are added to 775 grams of water (c = 4.18 J/g•°C) at 27°C?

35) If molten iron at 1840°C from a blast furnace is cast into 28-kilogram ingots, how much heat is lost as the ingot cools to room temperature, 27°C? The iron solidifies at 1530°C, and its heat of fusion is 267 J/g. The specific heat of iron is 0.44 J/g•°C as a solid, and 0.45 J/g•°C as a liquid.

36) Cadmium is used in preparing low-melting alloys. How much energy is required to melt 1500 grams of cadmium, initially at 23°C, and heat the liquid to 390°C? The specific heat of the solid is 0.23 J/g•°C, and of the liquid, 0.27 J/g•°C. The heat of fusion is 54 J/g at the melting point, 321°C.

37) Propane burns according to the equation

$$C_3H_8(g) + 5\ O_2(g) \rightarrow 3\ CO_2(g) + 4\ H_2O(\ell) \qquad \Delta H = -2050\ kJ$$

If 26.7 kJ are produced by this reaction, how many grams of propane burned?

38) Calculate the energy (heat) released by the reaction of 50.0 mL 1.50 M $CuCl_2$ in the reaction $Zn(s) + CuCl_2(aq) \rightarrow Cu(s) + ZnCl_2(aq) + 215\ kJ$. (See Chapter 14 for a discussion of M, molarity.)

39) In the reaction $4\ NH_3(g) + 5\ O_2(g) \rightarrow 4\ NO(g) + 6\ H_2O(g) + 904.7\ kJ$, how many liters of NO at STP will accompany the release of 207 kJ?

40) When 2.50 grams of aluminum are added to excess hydrochloric acid in 275 grams of water, the temperature increases from 20.5°C to 62.8°C. What is ΔH for the reaction?

41) When 50.0 mL of 2.00 M NH_3 and 50.0 mL of 1.00 M H_2SO_4 are mixed, the temperature of the 100.0 grams of water solution (c = 4.18 J/g•°C) increases by 12.4°C. What is ΔH for the reaction? (See Chapter 14 for a discussion of M, molarity.)

42) The Ostwald process for making nitric acid takes place in three steps:

$$4\ NH_3(g) + 5\ O_2(g) \rightarrow 4\ NO(g) + 6\ H_2O(\ell) \qquad \Delta H = -1168.8\ kJ$$

$$2\ NO(g) + O_2(g) \rightarrow 2\ NO_2(g) \qquad \Delta H = -\ 113.1\ kJ$$

$$3\ NO_2(g) + H_2O(\ell) \rightarrow 2\ HNO_3(aq) + NO(g) \qquad \Delta H = -\ 138.4\ kJ$$

What is ΔH for the overall reaction:

$$NH_3(g) + 2\ O_2(g) \rightarrow HNO_3(aq) + H_2O(\ell) \qquad \Delta H = ?$$

43) Calculate ΔH for the reaction $CaO(s) + H_2O(\ell) \rightarrow Ca(OH)_2(s)$ from the following heats of formation:

$CaO(s)$, -635.5 kJ/mol; $Ca(OH)_2(s)$, -986.59 kJ/mol; $H_2O(\ell)$, -285.8 kJ/mol.

44) For $KClO_3(s)$, ΔH_f° = -391.2 kJ/mol; for $KCl(s)$, ΔH_f° = -435.9 kJ/mol. Use these data to find ΔH for $2\ KClO_3(s) \rightarrow 2\ KCl(s) + 3\ O_2(g)$.

45) Find the heat of formation of acetaldehyde if ΔH = -2584.4 for the reaction $2\ CH_3CHO(g) + 5\ O_2(g) \rightarrow 4\ CO_2(g) + 4\ H_2O(\ell)$. ΔH_f° = -285.8 kJ/mol for $H_2O(\ell)$; ΔH_f° = -393.5 kJ/mol for $CO_2(g)$.

ADDITIONAL PROBLEMS

46) When 4500 joules of heat (to 4 significant figures) are added to 2.50 grams of ice at -15°C, what is the final temperature and physical state of the H_2O? ΔH_{fus} = 335 J/g, ΔH_{vap} = 2.26 kJ/g, c_s = 2.1 J/g•°C, c_l = 4.18 J/g•°C, and c_g = 2.0 J/g•°C.

47) In heat transfer problems in this chapter we have ignored the heat absorbed by the container. Suppose that 50.0 g of hot water at 73.0°C are added to 50.0 g of cold water at 17.0°C in a styrofocam cup. If the final temperature is 43.7°C, how much heat was absorbed by the styrofoam cup? c = 4.18 J/g•°C for water. [Hint: Heat lost equals heat gained so $Q_{cold} + Q_{styro} = -Q_{hot}$. Find Q_{styro}.]

48) When 0.112 grams of aluminum are dissolved in excess hydrochloric acid in 75.0 grams of water, the water temperature increases from 21.5°C to 32.9°C. What is ΔH in kJ/mol Al and in kJ? c = 4.18 J/g•°C for H_2O.

49) What is ΔH for the "hydroformylation" reaction below?

$$CH_2{=}CH_2(g) \quad + \quad CO(g) \quad + \quad H_2(g) \quad \rightarrow \quad CH_3\text{-}CH_2\text{-}CHO(g)$$

The heats of combustion of $CH_2{=}CH_2$ and $CH_3\text{-}CH_2\text{-}CHO$ are -1411 kJ/mol and -1806 kJ/mol, respectively. The needed heats of formation are -110.5 kJ/mol CO(g), -393.5 kJ/mol$CO_2(g)$, and -285.8 kJ/mol $H_2O(\ell)$.

CHAPTER TEST

1) 67.6 grams of cobalt are heated from 12.4°C to 47.5°C. How much energy does the cobalt absorb if its specific heat is 0.46 J/g•°C?

2) Calculate the mass of graphite that can be heated from 21°C to 235#C by 215 kJ if specific heat of graphite is 0.71 J/g•°C.

3) What is the specific heat of sulfur if 46.2 grams lose 23.6 kJ in cooling from 730°C to 31.4°C?

4) How many grams of nickel can be melted by 25 kJ of energy if its heat of fusion is 310 J/g?

5) It takes 3.97×10^4 joules to vaporize 16.2 grams of a certain liquid. Calculate its heat of vaporization.

6) What is the final temperature when 4.65 grams of iron, c = 0.473 J/g•°C, at 175°C are added to 27 grams of water, c = 4.18 J/g•°C, at 27°C?

7) An alloy of lead, tin, and antimony is used for typemetal. As a solid, its specific heat is 0.14 J/g•°C; as a liquid, 0.17 J/g•°C. Its heat of fusion is 25 J/g at its melting point, 297°C. Calculate the energy required to heat a 350-gram slug of this metal from room temperature, 23°C, to its casting temperature, 345°C.

8) When 2.00 grams of solid sodium hydroxide are added to 50.0 grams of water, the temperature rises from 22.5°C to 33.1°C. What is ΔH? (Consider only the heat absorbed by the water.)

9) Energy for photosynthesis comes from the sun and causes the following reaction in plants: $6\ CO_2(g) + 6\ H_2O(\ell) \rightarrow C_6H_{12}O_6(s) + 6\ O_2(g)$. How much energy must a plant absorb to produce 355 g $C_6H_{12}O_6(s)$ if $\Delta H = 2820$ kJ?

10) Calculate ΔH_f° of $Al_2O_3(s)$ if $\Delta H = 2157$ kJ for the reaction

$$2\ Al_2O_3(s) + 3\ C(s) \rightarrow 4\ Al(s) + 3\ CO_2(g).$$

$$\Delta H_f^\circ = -393.5\ \text{kJ/mol for } CO_2(g).$$

11) Calculate ΔH_c for $C_3H_6(g)$ from the following heats of formation: $C_3H_6(g)$, +20.4 kJ/mol; $CO_2(g)$, -393.5 kJ/mol; $H_2O(\ell)$, -285.8 kJ/mol.

PART III. PHYSICAL AND CHEMICAL CHANGES

14. SOLUTIONS

PREREQUISITES

1. You will use percentage calculations (Section 2.6) to express percentage concentration of a solution.

2. Name-and-formula skills developed in Chapter 7 are used throughout this chapter.

3. You will continue to write chemical equations, as you learned in Chapter 9.

4. The conversion between grams and moles (Section 8.6) is performed in the examples without explanation or comment.

5. Stoichiometry, introduced in Section 10.2 and expanded to gases in Section 12.7, is further expanded to include reactions that occur in solution.

6. Some solution properties depend on the mole fraction (Section 12.8) of the solute or solvent.

CHEMICAL SKILLS

1. Percentage composition. Given two of the following, calculate the third: (a) mass of solute; (b) mass of solvent, or mass of solution (or density data from which mass of solution can be found); (c) percentage concentration.

2. Mole fraction. Given two of the following, calculate the third: (a) moles of solute, or mass of a known solute; (b) moles of solvent, or mass of a known solvent; (c) mole fraction.

3. Molality. Given two of the following, calculate the third: (a) moles of solute, or mass of a known solute; (b) mass of solvent; (c) molality.

4. Molarity. Given two of the following, calculate the third: (a) moles of solute, or mass of a known solute; (b) volume of solution; (c) molarity.

5. Unit conversions. Given any of the following, or information from which any of the following may be determined, calculate all others: (a) percentage concentration; (b) mole fraction; (c) molality; (d) molarity (given the density).

6. Dilution. Given any three of the following, calculate the fourth: (a) initial molarity; (b) initial volume; (c) final molarity; (d) final volume.

7. Solution stoichiometry. For a reaction whose chemical equation is known, given the (a) mass, of any species, (b) volume of any gaseous species at designated temperature and pressure, or (c) volume of any solution at stated molarity, calculate any similar quantity of any other species.

8. Titration: standardization. For a reaction whose chemical equation is known, calculate the molarity of a solution from the mass of a primary standard and titration data.

9. Titration: unknowns. For a reaction whose chemical equation is known, given the volumes of two solutions that react with each other and the molarity of the standard, calculate the molarity of the unknown.

14.1 SOLUTION CONCENTRATION: PERCENTAGE

A solution is a mixture in which one substance called the **solute** is uniformly distributed in another substance called the **solvent**. Solutions can be identified for all possible combinations of solids, liquids, and gases, but this chapter will be limited primarily to water solutions of solutes commonly found in an introductory chemistry course.

One common way to express the concentration of a solution is by the percentage of solute. Percentages are usually given in terms of mass, although volume percentage and mass-volume percentage are also known. We will consider only percentage by mass. Accordingly, percentage concentration is

$$\%\ \text{concentration} \equiv \frac{\text{grams solute}}{\text{grams solution}} \times 100$$

$$= \frac{\text{grams solute}}{\text{grams solute + grams solvent}} \times 100 \qquad (14.1)$$

EXAMPLE 14.1: 25.0 grams of sugar are dissolved in 165 grams of water. Calculate the percentage concentration.

GIVEN: 25.0 g sugar (solute), 165 g water (solvent) FIND: %

EQN: $\% = \frac{\text{g solute}}{\text{g solute + g solvent}} \times 100 = \frac{25.0\ \text{g}}{25.0\ \text{g} + 165\ \text{g}} \times 100 = 13.2\%\ \text{sugar}$

It was pointed out in Section 2.6 that percentage gives the ratio of one part of a mixture to 100 parts of the whole mixture. This led to Equation 2.10, which may be applied specifically to solutions. If X is the percentage concentration of a solution, then

$$X\ \% = \frac{X \text{ g solute}}{100 \text{ g solution}} \tag{14.2}$$

This ratio may be used as a conversion factor in the dimensional analysis approach to percentage problems. If percentage concentration is *given*, along with either grams of solute or grams of solvent, the other mass can be calculated with this conversion factor.

EXAMPLE 14.2: Calculate the mass of salt that must be dissolved to prepare 5.00×10^2 grams of a 7.00% solution.

Now the relationship between the starting and wanted quantities is a conversion factor based on Equation 14.2, which you should first rewrite using the proper number(s) from this problem. Also write the unit path by which you will solve the problem, but do not calculate the answer yet.

GIVEN: 5.00×10^2 g solution FIND: g salt (solute)
FACTOR: 7.00 g solute/100 g solution PATH: g solution → g solute

The percentage figure, 7.00, corresponds with X in Equation 14.2.

It is a one-step conversion to grams of solute. Complete the problem.

$$5.00 \times 10^2 \text{ g solution} \times \frac{7.00 \text{ g solute}}{100 \text{ g solution}} = 35.0 \text{ g solute}$$

EXAMPLE 14.3: How many milliliters of a 3.0% saline (salt) solution must be measured out to get 0.25 grams of solute? Assume that the density of the solution is 1.00 g/mL.

Solve the problem for grams of solution and then change to milliliters.

GIVEN: 0.25 g solute FIND: mL solution
FACTOR: 3.0 g solute/100 g solution, 1.00 g solution/mL solution.
PATH: g solute → g solution → mL solution

$$0.25 \text{ g solute} \times \frac{100 \text{ g solution}}{3.0 \text{ g solute}} \times \frac{1 \text{ mL solution}}{1.00 \text{ g solution}} = 8.3 \text{ mL solution}$$

14.2 SOLUTION CONCENTRATION: MOLE FRACTION

Some solution phenomena are related to concentration calculated in **mole fraction**. The mole fraction of a gaseous solution was introduced in Section 12.8. If, in calculating percentage concentration by Equation 14.1, you were to disregard the multiplication by 100, the result would be the "gram fraction" concentration. In Example 14.1, where 25.0 grams of sugar are dissolved in 165 grams of water, the gram fraction of the solution that is sugar is

$$\frac{25.0\text{ g}}{25.0\text{ g} + 165\text{ g}} = 0.132$$

Mole fraction is calculated in exactly the same way, except that the quantities are written in moles rather than grams. The symbol for mole fraction is X. Thus

$$X_{\text{solute}} \equiv \frac{\text{moles solute}}{\text{moles solute} + \text{moles solvent}} \qquad (14.3)$$

Sometimes it is the mole fraction of the solvent that is important. This is

$$X_{\text{solvent}} = \frac{\text{moles solvent}}{\text{moles solute} + \text{moles solvent}}$$

Just as the sum of all the percentages in a sample of something is 100%, the sum of all mole fractions in a sample is 1:

$$X_{\text{solute}} + X_{\text{solvent}} = \frac{\text{mol solute}}{\text{mol solute} + \text{mol solvent}} + \frac{\text{mol solvent}}{\text{mol solute} + \text{mol solvent}}$$

$$= \frac{\text{mol solute} + \text{mol solvent}}{\text{mol solute} + \text{mol solvent}} = 1$$

EXAMPLE 14.4: If the molar mass of the sugar in Example 14.1 is 342.30 grams per mole, what is the mole fraction of the solution (25.0 grams of sugar dissolved in 165 grams of water) that is sugar? Also find the mole fraction that is water.

You have been asked for mole fraction, so you must find the moles of both sugar and water. Begin by making these conversions.

GIVEN: 25.0 g sugar, 165 g H_2O FIND: mol sugar, mol water
FACTOR: MM sugar, MM water PATH: g sugar → mol sugar, g H_2O → mol H_2O

$$25.0\text{ g sugar} \times \frac{1\text{ mol sugar}}{342.30\text{ g sugar}} = 0.0730\text{ mol sugar}$$

$$165\text{ g } H_2O \times \frac{1\text{ mol } H_2O}{18.02\text{ g } H_2O} = 9.16\text{ mol } H_2O$$

The 0.0730 mol sugar and 9.16 mol H_2O may now be substituted directly into Equation 14.3.

$$X_{sugar} = \frac{\text{mol sugar}}{\text{mol sugar + mol } H_2O} = \frac{0.0730 \text{ mol sugar}}{0.0730 \text{ mol sugar} + 9.16 \text{ mol } H_2O} = 0.00791$$

$$X_{H_2O} = \frac{\text{mol water}}{\text{mol sugar + mol } H_2O} = \frac{9.16 \text{ mol water}}{0.0730 \text{ mol sugar} + 9.16 \text{ mol } H_2O} = 0.992$$

Notice that the sum of the mole fractions is 1: 0.00791 + 0.992 = 1.000.

14.3 SOLUTION CONCENTRATION: MOLALITY

Some properties of solutions are related to the ratio of the moles of solute to the mass of solvent. **Molality**, symbolized by the small letter **m**, is the name given to this concentration:

$$m \equiv \frac{\text{moles solute}}{\text{kilograms solvent}} = \frac{\text{mol solute}}{\text{kg solvent}} \tag{14.4}$$

Quantity of solvent is given in grams in most problems, but molality is based on the kilograms of solvent. There are 1000 grams in a kilogram. To change grams to kilograms, you divide by 1000, which is accomplished by moving the decimal three places left. Thus a mass given as 123 grams can also be written 0.123 kilograms. This conversion is required often, and it is easier to make it mentally than to introduce an extra step into a problem setup. This will be done without comment hereafter.

EXAMPLE 14.5: Calculate the molality of the solution described in Example 14.1: 25.0 grams of sugar (342.30 g/mol) dissolved in 165 grams of water.

The quantity to find is molality, the ratio of the moles of solute (sugar) in the numerator to kilograms of solvent (water) in the denominator. One way to solve this problem is to write separate unit paths for both these quantities. Analyze the problem; identify the GIVEN-FIND-FACTOR-PATH from the data, including unit paths for both the numerator and denominator of the molality.

GIVEN: 25.0 g sugar (solute)
165 g water (solvent)

FIND: $m = \frac{\text{mol solute}}{\text{kg solvent}}$

FACTOR: 342.30 g sugar/mol
1000 g/kg

PATH: $\frac{\text{g sugar} \rightarrow \text{mol sugar}}{\text{g water} \rightarrow \text{kg water}}$

Now set up and solve the problem.

$$\frac{25.0 \text{ g sugar} \times \dfrac{1 \text{ mol sugar}}{342.30 \text{ g sugar}}}{0.165 \text{ kg water}} = \frac{0.443 \text{ mol sugar}}{\text{kg water}} = 0.443 \text{ m sugar}$$

Molality shows how moles of solute are related to mass of solvent. An X molal solution means that

$$\text{X m} = \frac{\text{X mol solute}}{\text{kg solvent}} = \frac{\text{X mol solute}}{1000 \text{ g solvent}} \qquad (14.5)$$

Equation 14.5 provides a conversion factor that can be used for the unit path between moles of solute and grams or kilograms of solvent.

EXAMPLE 14.6: How many grams of methanol, CH_3OH, must be dissolved in 2.50×10^2 milliliters of water to produce a 0.325-molal solution?

This problem takes two steps and uses two conversion factors, one new and one old and familiar. You may assume that the density of water is 1.00 g/mL. Write, or at least think through, the GIVEN-FIND-FACTOR-PATH analysis, set up the problem, and calculate the answer.

GIVEN: 2.50×10^2 mL H_2O = 2.50×10^2 g H_2O FIND: g CH_3OH
FACTOR: 0.325 mol CH_3OH/kg H_2O, MM CH_3OH
PATH: mL H_2O → g H_2O → kg H_2O → mol CH_3OH → g CH_3OH

$$0.250 \text{ kg } H_2O \times \frac{0.325 \text{ mol } CH_3OH}{1 \text{ kg } H_2O} \times \frac{32.04 \text{ g } CH_3OH}{1 \text{ mol } CH_3OH} = 2.60 \text{ g } CH_3OH$$

You might have begun your unit path at any of the first three steps, depending on how many conversions you planned to do mentally. The mass of 250 milliliters of water is 250 grams because the density of water is 1.00 g/mL. The 250 grams is converted to 0.250 kilograms by moving the decimal three places left. If you make both of these mental conversions the unit path is shortened to kg H_2O → mol CH_3OH → g CH_3OH, giving the setup shown.

EXAMPLE 14.7: Calculate the volume of water in which 35.0 grams of urea, $CO(NH_2)_2$, should be dissolved to produce a 0.150-molal solution.

The same ideas are used in this example as in the last. Set up the problem and solve for the answer.

GIVEN: 35.0 g $CO(NH_2)_2$ FIND: mL H_2O
FACTOR: MM of $CO(NH_2)_2$, 0.150 mol $CO(NH_2)_2$/1000 g H_2O
PATH: g $CO(NH_2)_2$ → mol $CO(NH_2)_2$ → g H_2O → mL H_2O

$$35.0 \text{ g } CO(NH_2)_2 \times \frac{1 \text{ mol } CO(NH_2)_2}{60.06 \text{ g } CO(NH_2)_2} \times \frac{1000 \text{ g } H_2O}{0.150 \text{ mol } CO(NH_2)_2} \times \frac{1 \text{ mL } H_2O}{1 \text{ g } H_2O} = 3.89 \times 10^3 \text{ mL } H_2O$$

Your first step should match the one above, but there may have been some variation in the second and third. You might have gone to kilograms of water, and you may have solved for grams and then converted mentally to volume. These different approaches are equally correct.

14.4 SOLUTION CONCENTRATION: MOLARITY

The three concentration units discussed so far depend entirely on measurements of mass. Mass measurements fail to take advantage of the simple way to measure liquid quantities: by volume. **Molarity** uses this easily-performed measuring process by defining concentration in **moles of solute per liter of solution**. Using M for molarity,

$$M \equiv \frac{\text{moles solute}}{\text{liter solution}} = \frac{\text{mol solute}}{\text{L solution}} \tag{14.6}$$

Molarity problems and percentage problems have a point in common. In percentage problems, if you are given the masses of solute and solvent, you must use the defining equation (Equation 14.1) to calculate percentage concentration. With molarity, if you are given the amount of solute and volume of solution, you must use the defining equation (Equation 14.6) to calculate molarity. If given the percentage concentration, you may use it as a conversion factor in a dimensional analysis conversion between mass of solute and mass of solution. Similarly, if molarity is given, it may be used as a conversion factor in a dimensional analysis conversion between moles of solute and volume of solution.

According to Equation 14.6, the molarity of a solution is calculated by dividing the number of moles of solute in a sample of the solution by its volume in liters. If 0.60 moles of solute is dissolved in 0.75 liter, the concentration is

$$\frac{0.60 \text{ moles}}{0.75 \text{ liter}} = \frac{0.60 \text{ mol}}{0.75 \text{ L}} = 0.80 \text{ mol/L} = 0.80 \text{ M}$$

Molarity is defined in terms of liters. Volume is often measured in milliliters. Since there are 1000 milliliters in a liter, a milliliter to liter change may be made by moving the decimal point three places to the left, just as grams were changed to kilograms earlier. Thus the molarity of a solution that contains 0.015 moles in 36 milliliters is

$$\frac{0.015 \text{ mol}}{0.036 \text{ L}} = 0.42 \text{ mol/L} = 0.42 \text{ M}$$

When appropriate, the conversion between milliliters and liters will be made without comment.

EXAMPLE 14.8: Calculate the molarity of a solution if 8.00×10^2 milliliters contain 0.200 moles.

This problem may be solved by direct substitution into Equation 14.6. Do not forget that volume must be in liters, so be sure to make the mL → L conversion.

GIVEN: 0.200 mol solute, 0.800 L solution FIND: M

EQN: $M \equiv \frac{\text{mol solute}}{\text{L solution}} = \frac{0.200 \text{ mol solute}}{0.800 \text{ L solution}} = 0.250 \text{ M}$

EXAMPLE 14.9: 30.0 grams of NaCl are dissolved in about 500 milliliters of water and diluted to 7.50×10^2 milliliters. Find the molarity of the solution.

This example describes how most solutions are prepared. The solute is dissolved in less than the needed amount of water, after which the solution is diluted to the final volume. Do not be distracted by the dissolving volume of 500 milliliters. It has nothing to do with the final concentration.

The strategy for solving this problem is: Molarity is moles per liter (Equation 14.6). In the numerator, dimensional analysis can be used to change the grams of solute to moles. In the denominator, the milliliters of solution must be expressed in liters. Analyze the problem, set it up, and solve.

GIVEN: 30.0 g NaCl, 750 mL solution FIND: M
FACTOR: MM of NaCl PATH: g NaCl → mol NaCl

$$\text{EQ: } M = \frac{\text{mol solute}}{\text{L soln}} = \frac{30.0 \text{ g NaCl} \times \dfrac{1 \text{ mol NaCl}}{58.44 \text{ g NaCl}}}{0.750 \text{ L}} = \frac{0.684 \text{ mol NaCl}}{\text{L soln}} = 0.684 \text{ M NaCl}$$

EXAMPLE 14.10: 6.85 grams of Na_2CO_3 are dissolved in water. What is the molarity of the solution if the volume is 174 milliliters?

GIVEN: 6.85 g Na_2CO_3, 174 mL solution FIND: M
FACTOR: MM of Na_2CO_3 PATH: g Na_2CO_3 → mol Na_2CO_3

$$\text{EQN: } M = \frac{\text{mol solute}}{\text{L solution}} = \frac{6.85 \text{ g } Na_2CO_3 \times \dfrac{1 \text{ mol } Na_2CO_3}{105.99 \text{ g } Na_2CO_3}}{0.174 \text{ L}} = 0.371 \text{ M } Na_2CO_3$$

A procedure followed in a college laboratory is the preparation of solutions of known molarity to be used by students. The laboratory technician must decide how much solute to weigh out. Here's how it is done:

EXAMPLE 14.11: How many grams of silver nitrate must a laboratory assistant weigh out to prepare 1.25 liters of 0.100 M $AgNO_3$?

This time the molarity is given. This means it can be used as a conversion factor between moles of solute and volume of solution. A GIVEN-FIND-FACTOR-PATH analysis you will help you see what to do. Write that as the first step.

GIVEN: FIND:

FACTOR: PATH:

GIVEN: 1.25 L solution FIND: g $AgNO_3$
FACTOR: 0.100 mol $AgNO_3$/L, MM of $AgNO_3$ PATH: L solution → mol $AgNO_3$ → g $AgNO_3$

Everything you need to solve the problem is there.

$$1.25 \text{ L} \times \frac{0.100 \text{ mol } AgNO_3}{1 \text{ L}} \times \frac{169.9 \text{ g } AgNO_3}{1 \text{ mol } AgNO_3} = 21.2 \text{ g } AgNO_3$$

A particularly important application of molarity is converting in either direction between volume of solution and moles of solute. "Volume of solution" is a quantity measurement. Does that give you a hint why the procedure is so useful? "Quantity → moles" and "moles → quantity" are the first and third steps of the stoichiometry 3-step.

EXAMPLE 14.12: How many milliliters of 0.645 M H_2SO_4 contain 0.114 mole of solute?

This matches the last step in the stoichiometry 3-step. It is only one step from moles to volume, but watch the units in your answer. Set up and solve.

GIVEN: 0.114 mol H_2SO_4 FIND: mL solution
FACTOR: 0.645 mol H_2SO_4/L PATH: mol H_2SO_4 → L solution → mL solution

$$0.114 \text{ mol } H_2SO_4 \times \frac{1 \text{ L}}{0.645 \text{ mol } H_2SO_4} = 0.177 \text{ L} = 177 \text{ mL}$$

When milliliters are specified, the calculation setup can end with liters, and the final conversion can be added, as shown above. Alternately, the setup can be made to include the L → mL conversion too.

EXAMPLE 14.13: How many moles of solute are in 26.8 mL 0.448 M NaOH?

This is essentially the opposite of Example 14.12, and it corresponds to the first step in the stoichiometry 3-step. Set up and solve.

GIVEN: 26.8 mL solution FIND: mol NaOH
FACTOR: 0.448 mol NaOH/L PATH: mL → L → mol

$$0.0268 \text{ L} \times \frac{0.448 \text{ mol NaOH}}{1 \text{ L}} = 0.0120 \text{ mol NaOH}$$

Example 14.13 shows that the product of the volume of a solution and its molarity is moles of solute:

$$M \times V = \frac{\text{mol solute}}{\text{L solution}} \times \text{L solution} = \text{mole solute} \quad (14.7)$$

We will have occasion to use this relationship in another way in Section 14.6.

14.5 CONVERSION BETWEEN CONCENTRATION UNITS

The greatest guarantee of success in working with solution concentrations is knowing what each concentration term means and using its units effectively. The following summary should help.

NAME (SYMBOL)	DEFINITION	UNITS
Percentage (%)	$\frac{\text{grams solute}}{\text{grams solute + grams solvent}} \times 100$	Unitless
Mole Fraction (X)	$\frac{\text{moles solute}}{\text{moles solute + moles solvent}}$	Unitless
Molality (m)	$\frac{\text{moles solute}}{\text{kilogram solvent}}$	$\frac{\text{mol solute}}{\text{kg solvent}}$
- - - - - -	- - - - - - - - - -	- - - -
Molarity (ℓ)	$\frac{\text{moles solute}}{\text{liter solution}}$	$\frac{\text{mol}}{\text{L}}$

The dashed line that separates molarity from the other concentrations emphasizes that molarity is based on volume of solution while the others are based entirely on masses of solvent and solute.

If you know or can find the concentration of a solution in percentage, mole fraction, or molality, you can also find the other two. In fact, you have already done it in Examples 14.1, 14.4, and 14.5. All three examples were based on a solution of 25.0 grams of sugar (MM = 342.30 g/mol) in 165 grams of water (MM = 18.02 g/mol). The following table summarizes the quantities that were given or calculated (shown in parentheses) in those examples:

	SOLUTE	SOLVENT	SOLUTION
MASS (g)	25.0	165	(190)
MOLES	(0.0730)	(9.16)	(9.23)

The masses of solute and solvent were the starting point for this solution. The numbers of moles of solute and solvent were found by dividing the masses by the molar masses. The "solution" column is for the sum of the masses, which is needed to calculate percentage, or the sum of the moles, which is required to calculate mole fraction.

Here's how you used the information in the table:

In Example 14.1, percentage $= \frac{25.0 \text{ g}}{190 \text{ g}} \times 100 = 13.2\%$

In Example 14.4, mole fraction of solute $= \frac{0.0730 \text{ mol}}{9.23 \text{ mol}} = 0.00791$

In Example 15.5, molality $= \frac{0.0730 \text{ mol}}{0.165 \text{ kg}} = 0.443 \text{ m}$

We will use a similar table in analyzing the examples ahead.

EXAMPLE 14.14: Calculate the mole fraction and percentage concentration of 10.2 m $HC_2H_3O_2$ (MM = 60.05 g/mol).

The solute is acetic acid. The given molality tells you a quantity of acid (solute), expressed in one kind of unit, and a quantity of water (solvent), expressed in another kind of unit. From the definition of molality, write these two starting quantities as entries in the table below.

	SOLUTE	SOLVENT	SOLUTION
MASS (g)			
MOLES			

	SOLUTE	SOLVENT	SOLUTION
MASS (g)		1000	
MOLES	10.2		

Complete the table. Use the molar masses of acetic acid and water to get the grams of acid and moles of water. Combine the masses and moles to get their respective totals.

GIVEN: 10.2 mol $HC_2H_3O_2$, 1000 g H_2O (from definition)
FIND: g $HC_2H_3O_2$, mol H_2O FACTOR: MM of $HC_2H_3O_2$ and H_2O
PATH: mol $HC_2H_3O_2$ → g $HC_2H_3O_2$, g H_2O → mol H_2O

$$10.2 \text{ mol } HC_2H_3O_2 \times \frac{60.05 \text{ g } HC_2H_3O_2}{1 \text{ mol } HC_2H_3O_2} = 613 \text{ g } HC_2H_3O_2$$

$$1000 \text{ g } H_2O \times \frac{1 \text{ mol } H_2O}{18.02 \text{ g } H_2O} = 55.5 \text{ mol } H_2O$$

	SOLUTE	SOLVENT	SOLUTION
MASS (g)	(613)	1000	(1613)
MOLES	10.2	(55.5)	(65.7)

Now you are ready to calculate the percentage $HC_2H_3O_2$ in the solution.

$$\% = \frac{\text{g solute}}{\text{g solution}} \times 100 = \frac{613 \text{ g } HC_2H_3O_2}{1613 \text{ g solution}} \times 100 = 38.0\% \ HC_2H_3O_2$$

Finally calculate the mole fraction of $HC_2H_3O_2$.

$$X_{HC_2H_3O_2} = \frac{\text{moles } HC_2H_3O_2}{\text{total moles}} = \frac{10.2 \text{ mol } HC_2H_3O_2}{65.7 \text{ mol total}} = 0.155$$

EXAMPLE 14.15: Calculate the molality of an aqueous solution whose mole fraction of glucose ($C_6H_{12}O_6$) is 0.013.

This time the problem asks for a conversion to molality. To find molality, you need the moles of solute and mass of solvent in a sample of the solution. You are beginning with mole fraction of solute. What is the mole fraction of solvent in the solution?

GIVEN: $X_{solute} = 0.013$ FIND: $X_{solvent}$ EQN: $X_{solute} + X_{solvent} = 1$

$X_{solvent} = 1 - X_{solute} = 1 - 0.013 = 0.987$

If you take a sample of solution that contains a total of 1 mole of solute + solvent, what fraction of a mole of each is in the sample?

The solution contains 0.013 mol $C_6H_{12}O_6$ and 0.987 mol H_2O.

The essential meaning of mole fraction is the fraction of a mole of solute or solvent in a total of 1 mole of solute + solvent.

What is the mass of water in which the 0.013 mol $C_6H_{12}O_6$ is dissolved?

GIVEN: 0.987 mol H_2O FIND: g H_2O FACTOR: MM of H_2O PATH: mol H_2O → g H_2O

$$0.987 \text{ mol } H_2O \times \frac{18.02 \text{ g } H_2O}{1 \text{ mol } H_2O} = 17.8 \text{ g } H_2O$$

You now have moles of solute and mass of solvent. Calculate molality.

GIVEN: 0.013 mol $C_6H_{12}O_6$, 17.8 g H_2O FIND: molality

EQN: $$m = \frac{\text{mol solute}}{\text{kg solvent}} = \frac{0.013 \text{ mol } C_6H_{12}O_6}{0.0178 \text{ kg } H_2O} = 0.73 \text{ mol } C_6H_{12}O_6/\text{kg } H_2O$$
$$= 0.73 \text{ m } C_6H_{12}O_6$$

We did not use the table in Example 14.15, but if we had, it would have been necessary to use only part of it. Specifically,

	SOLUTE	SOLVENT	SOLUTION
MASS (g)		(17.8)	
MOLES	0.013	(0.987)	1.000

Like other procedures in this book, the table is an aid when you are learning how to solve concentration conversion problems. Use it as long as it is helpful, but abandon it as soon as you become more familiar with the problems and can solve them directly.

It is not possible to calculate molarity from the data in Examples 14.14 and 14.15. They have no information about volume, on which molarity is based. In Example 14.14, the mass of the solution is 1613 grams. To change mass to volume you need one additional bit of information: density. Density is usually given in grams per cubic centimeter, or grams per milliliter when the substance is a liquid. This is numerically equal to density in kilograms per liter, which is convenient in some problems:

$$\frac{\text{grams}}{\text{milliliter}} \times \frac{1000 \text{ milliliters}}{1 \text{ liter}} \times \frac{1 \text{ kilogram}}{1000 \text{ grams}} = \frac{\text{kilograms}}{\text{liter}} \qquad (14.8)$$

EXAMPLE 14.16: The density of the 10.2 m $HC_2H_3O_2$ solution in Example 14.14 is 1.06 g/mL. Calculate the molarity.

For convenience, the tabular summary from Example 14.14 is repeated here:

	SOLUTE	SOLVENT	SOLUTION
MASS (g)	613	1000	1613
MOLES	10.2	55.5	65.7

Molarity is moles per liter of solution. Molality gives you moles of solute per kilogram of solvent. The moles in molality is the same as the moles in molarity. That is what appears in the numerator of the molarity fraction. Now for the denominator, volume: You have already found that the mass of the solution that contains 10.2 moles of solute is 1613 grams. The density of the solution is given: 1.06 g/mL. What is the volume of the solution?

GIVEN: 1613 g solution FIND: L solution FACTOR: d = 1.06 g/mL PATH: g → mL → L

$$1613 \text{ g soln} \times \frac{1 \text{ mL soln}}{1.06 \text{ g soln}} = 1520 \text{ mL soln} = 1.52 \text{ L solution}$$

You have moles of solute and you have liters of solution. Find molarity.

$$M = \frac{\text{mol solute}}{\text{L solution}} = \frac{10.2 \text{ mol } HC_2H_3O_2}{1.52 \text{ L}} = 6.71 \text{ M } HC_2H_3O_2$$

14.6 DILUTION PROBLEMS

Dilution is the process of adding more solvent to a solution. If a solution is diluted, its molarity becomes less. This is because the volume increases (from the added solvent) while there is no change in the moles of solute. The effect is apparent from the defining equation for molarity:

$$M \equiv \frac{\text{mol solute}}{\text{L solution}} \qquad (14.6)$$

If water is added to an aqueous solution, the denominator is increased, but the numerator remains constant.

The relationship between the initial molarity and the final molarity of the diluted solution can be derived from the observation that the moles of solute do not change during dilution:

initial moles of solute = final moles of solute

The moles of solute in each case can be calculated by multiplying the molarity of the solution by its volume (Equation 14.7). It follow that

$$M_1V_1 = M_2V_2 \qquad (14.9)$$

where M_1 and V_1 are the molarity and volume of the initial (more concentrated) solution, and M_2 and V_2 are the molarity and volume of the final (more dilute) solution. Equation 14.9 may be solved for any one of these four values if the other three are given.

One of the most common mistakes made in using Equation 14.9 is to confuse the initial and final volumes. Whenever you do a dilution problem, you should ask yourself if the answer is reasonable: does the initial (more concentrated) solution have a higher molarity than the final (more dilute) solution? If not, check the analysis and setup of the problem to find the mistake.

EXAMPLE 14.17: 2.50×10^2 mL 0.936 M NaOH are diluted to 7.50×10^2 milliliters. Find the molarity of the diluted solution.

Begin by identifying the starting and wanted quantities as M_1, V_1, M_2, and V_2, but do not solve the problem yet.

GIVEN: $M_1 = 0.936$ M, $V_1 = 250$ mL, $V_2 = 750$ mL FIND: M_2 EQN: $M_1V_1 = M_2V_2$

Now solve Equation 14.9 for the wanted quantity, substitute, and calculate the answer. Is your answer reasonable?

$$M_2 = \frac{M_1V_1}{V_2} = \frac{0.936 \text{ M} \times 250 \text{ mL}}{750 \text{ mL}} = 0.312 \text{ M NaOH}$$

The molarity of the diluted solution, 0.312 M, is less than that of the initial solution, 0.936 M, so the answer is reasonable.

EXAMPLE 14.18: 2.7 liters of 6.0 M HCl are added to 5.0 liters of water. What is the molarity of the diluted solution?

For this problem, the initial volume, 2.7 L, has been given, along with the volume of solvent added, 5.0 L. The volume of the final solution may be assumed to be the sum of these two. Set up and solve the problem in one step.

GIVEN: $M_1 = 6.0$ M, $V_1 = 2.7$ L, $V_2 = 2.7$ L + 5.0 L = 7.7 L FIND: M_2

EQN: $$M_2 = \frac{M_1V_1}{V_2} = \frac{6.0 \text{ M} \times 2.7 \text{ L}}{7.7 \text{ L}} = 2.1 \text{ M HCl}$$

Example 14.11 showed how a laboratory assistant calculates the amount of a solid solute to use in preparing a given volume of solution of known concentration. Some commercial chemicals, notably acids and ammonia, are sold in solution or as liquids having a given molarity. In a case like this the laboratory assistant's task is to calculate the volume of the concentrated solution to use in preparing the reagent for classroom use.

EXAMPLE 14.19: A commercial ammonia solution is labeled 7.4 M. How many liters of this solution are needed to prepare 2.5 liters of 3.0 M NH_3 for the laboratory?

Now the unknown is the initial volume of the more concentrated solution. Set up and solve in one step.

GIVEN: $M_1 = 7.4$ M, $M_2 = 3.0$ M, $V_2 = 2.5$ L FIND: V_1

EQN: $$V_1 = \frac{M_2V_2}{M_1} = \frac{3.0 \text{ M} \times 2.5 \text{ L}}{7.4 \text{ M}} = 1.0 \text{ L of } 7.4 \text{ M } NH_3$$

EXAMPLE 14.20: A stock bottle of phosphoric acid is labeled 3.00 M. To how many milliliters of water must 5.00×10^2 milliliters of the stock solution be added to prepare 0.350 M H_3PO_4?

The unknown is now the volume of water to which the initial volume of acid is to be added. What is the relationship among V_1, V_2, and the volume of water, V_w?

$V_2 = V_1 + V_w$

If we assume the volumes are additive—they usually are, but there are exceptions—the total volume is the sum of the two volumes combined.

Now you have all the information needed to solve the whole problem.

GIVEN: $M_1 = 3.00$ M, $M_2 = 0.350$ M, $V_1 = 500$ mL FIND: V_w
EQN: $M_1V_1 = M_2V_2$, $V_2 = V_1 + V_w$

$$V_2 = \frac{M_1V_1}{M_2} = \frac{3.00 \text{ M} \times 500 \text{ mL}}{0.350 \text{ M}} = 4.29 \times 10^3 \text{ mL}$$

$$V_w = V_2 - V_1 = 4290 \text{ mL} - 500 \text{ mL} = 3790 \text{ mL} = 3.79 \text{ L } H_2O$$

You may wonder about not having changed the 500 mL to 0.500 L in the above example. This would have divided the right side of the equation by 1000 and produced an answer in liters. That answer would have to be multiplied by 1000 to get milliliters. The two operations cancel each other. Some chemists avoid the 1000 factor by working in milliunits. Molarity in moles per liter is the same as millimoles per milliliter. Thus the product M × V is

$$\frac{\text{mol}}{\text{L}} \times \text{L} = \text{mol} \quad \text{or} \quad \frac{\text{mmol}}{\text{mL}} \times \text{mL} = \text{mmol}$$

14.7 SOLUTION STOICHIOMETRY

In Section 10.2 you were introduced to the stoichiometry 3-step. The three steps in the solution of a stoichiometry problem are summarized in the unit path

given quantity → moles given → moles to find → quantity to find

"Given" refers to the substance for which the starting quantity appears in the problem, and "find" identifies the substances whose quantity is to be calculated. Up to now you have worked with the mass of any substance and volume of a gas at specified temperature and pressure. Both of these quantity units can be converted to moles, or found from moles. In Examples 14.12 and 14.13 you learned how to convert between moles of solute and volume of solution via molarity. We are now ready to apply this skill to stoichiometry problems.

EXAMPLE 14.21: Calculate the mass of sodium chloride that is required to precipitate all the silver ion in 35.0 milliliters of 0.254 M $AgNO_3$.

All stoichiometry problems begin with a balanced equation:

$NaCl(aq) + AgNO_3(aq) \rightarrow AgCl(s) + NaNO_3(aq)$

To be sure you see clearly where you are starting and where you will end up, as well as the route between, write the starting and wanted quantities and the unit path. Be specific. Use units and formulas.

GIVEN: 35.0 mL $AgNO_3$ FIND: g NaCl
FACTOR: 0.254 mol $AgNO_3$/L, 1 mol NaCl/1 mol $AgNO_3$, MM NaCl
PATH: L $AgNO_3$ → mol $AgNO_3$ → mol NaCl → g NaCl

Write the first step in the solution of the problem just as you did in Example 14.13. Do not calculate the intermediate answer.

$$0.0350 \text{ L } AgNO_3 \times \frac{0.254 \text{ mol } AgNO_3}{1 \text{ L } AgNO_3} \times \underline{\qquad\qquad} \times \underline{\qquad\qquad} =$$

The next two steps are the familiar mole-to-mole conversion that appears in every stoichiometry problem and moles-to-grams via molar mass. Complete the problem.

$$0.0350 \text{ L } AgNO_3 \times \frac{0.254 \text{ mol } AgNO_3}{1 \text{ L } AgNO_3} \times \frac{1 \text{ mol NaCl}}{1 \text{ mol } AgNO_3} \times \frac{58.44 \text{ g NaCl}}{1 \text{ mol NaCl}} = 0.520 \text{ g NaCl}$$

EXAMPLE 14.22: Calculate the milliliters of 0.507 M H_2SO_4 that will neutralize 1.12 grams of NaOH.

The equation is:

$2\ NaOH(aq) + H_2SO_4(aq) \rightarrow Na_2SO_4(aq) + 2\ H_2O(\ell)$

Write the unit path for the entire problem, but set it up only through the first two steps, to moles of H_2SO_4. Do not calculate the intermediate answer.

GIVEN: 1.12 g NaOH FIND: mL H_2SO_4
FACTOR: MM of NaOH, 2 mol NaOH/1 mol H_2SO_4, 0.507 mol H_2SO_4/L
PATH: g NaOH → mol NaOH → mol H_2SO_4 → mL H_2SO_4

$$1.12 \text{ g NaOH} \times \frac{1 \text{ mol NaOH}}{40.00 \text{ g NaOH}} \times \frac{1 \text{ mol } H_2SO_4}{2 \text{ mol NaOH}} \times \frac{\qquad}{\qquad}$$

So far the setup is the same two steps you used in earlier problems. Now you must change moles of solute to volume of solution, as in Example 14.12. Extend the setup with the necessary conversion factor. Don't forget to change the final answer to milliliters, as asked for in the problem.

$$1.12 \text{ g NaOH} \times \frac{1 \text{ mol NaOH}}{40.00 \text{ g NaOH}} \times \frac{1 \text{ mol } H_2SO_4}{2 \text{ mol NaOH}} \times \frac{1 \text{ L } H_2SO_4}{0.507 \text{ mol } H_2SO_4} = 0.0276 \text{ L } H_2SO_4 = 27.6 \text{ mL } H_2SO_4$$

EXAMPLE 14.23: Calculate the volume of hydrogen, measured at STP, that can be released by the reaction of 1.22 liters of 0.403 M HCl with excess zinc.

The equation:

$$Zn(s) + 2\ HCl(aq) \rightarrow H_2(g) + ZnCl_2(aq)$$

You have 1.22 liters of 0.403 M HCl. If you recall that the STP volume of any gas is 22.41 L/mol, you can probably figure out the unit path to the volume of hydrogen, set up the problem completely, and calculate the answer.

GIVEN: 1.22 L HCl, STP FIND: L H_2
FACTOR: 0.403 mol HCl/L, 2 mol HCl/mol H_2, 22.41 L H_2/mol H_2
PATH: L HCl → mol HCl → mol H_2 → L H_2

$$1.22 \text{ L} \times \frac{0.403 \text{ mol HCl}}{1 \text{ L}} \times \frac{1 \text{ mol } H_2}{2 \text{ mol HCl}} \times \frac{22.41 \text{ L } H_2}{1 \text{ mol } H_2} = 5.51 \text{ L } H_2$$

EXAMPLE 14.24: How many milliliters of 0.219 M $AgNO_3$ will react with 25.0 mL 0.184 M $NiCl_2$ in a reaction that precipitates AgCl?

The equation:

$$2\ AgNO_3(aq) + NiCl_2(aq) \rightarrow 2\ AgCl(s) + Ni(NO_3)_2(aq)$$

Begin by writing the unit path. The units of the starting and the wanted quantities may influence your path this time.

GIVEN: 25.0 mL $NiCl_2$ FIND: mL $AgNO_3$
FACTOR: 0.184 mol $NiCl_2$/L, 2 mol $AgNO_3$/1 mol $NiCl_2$, 0.219 mol $AgNO_3$/L
PATH: L $NiCl_2$ → mol $NiCl_2$ → mol $AgNO_3$ → L $AgNO_3$

The procedure is exactly the same as in the previous examples. Write the setup for the entire problem and solve.

$$0.0250\ \text{L}\ NiCl_2 \times \frac{0.184\ \text{mol}\ NiCl_2}{1\ \text{L}\ NiCl_2} \times \frac{2\ \text{mol}\ AgNO_3}{1\ \text{mol}\ NiCl_2} \times \frac{1\ \text{L}\ AgNO_3}{0.219\ \text{mol}\ AgNO_3} = 0.0420\ \text{L}\ AgNO_3 = 42.0\ \text{mL}\ AgNO_3$$

This is another problem that can be treated in milliunits. The unit path becomes mL $NiCl_2$ → mmol $NiCl_2$ → mmol $AgNO_3$ → mL $AgNO_3$. The setup is

$$25.0\ \text{mL}\ NiCl_2 \times \frac{0.184\ \text{mmol}\ NiCl_2}{1\ \text{mL}\ NiCl_2} \times \frac{2\ \text{mmol}\ AgNO_3}{1\ \text{mmol}\ NiCl_2} \times \frac{1\ \text{mL}\ AgNO_3}{0.219\ \text{mmol}\ AgNO_3} = 42.0\ \text{mL}\ AgNO_3$$

14.8 TITRATION

Titration is the controlled addition of one solution into another in a way that permits precise measurement of the volumes of the two solutions that react completely with each other. Neither reactant is in excess; there is no limiting reactant. The quantities are chemically—stoichiometrically—equivalent, with the mole ratios just as they are in the balanced equation. The process is widely used in quantitative analysis.

Analytical chemists often work with a **standardized solution, a solution whose concentration is known.** Two things must be known to standardize a solution: (1) the volume of solution and (2) the number of moles of solute in that volume. Divide moles by liters and you have molarity.

A **primary standard** is often used in determining the molarity of a solution. Suppose you wish to find the molarity of a solution of hydrochloric acid. You dissolve 0.317 grams of sodium carbonate, the primary standard, in water and titrate the solution with the hydrochloric acid. You find that 22.9 milliliters of acid are required to neutralize the sodium carbonate. This is the needed volume, the first of the two items above.

Solution stoichiometry is used to find the second item, moles of HCl in the 22.9 milliliters of acid. $2\ HCl + Na_2CO_3 \rightarrow 2\ NaCl + H_2O + CO_2$ is the equation for the reaction. It takes only two steps of the stoichiometry 3-step to get to moles of HCl: g Na_2CO_3 → mol Na_2CO_3 → mol HCl:

$$0.317 \text{ g } Na_2CO_3 \times \frac{1 \text{ mol } Na_2CO_3}{105.99 \text{ g } Na_2CO_3} \times \frac{2 \text{ mol HCl}}{1 \text{ mol } Na_2CO_3} = 0.00598 \text{ mol HCl}$$

You now divide the calculated number of moles of HCl by the volume that contains it, 0.0229 liter, to get molarity:

$$\frac{0.00598 \text{ mol HCl}}{0.0229 \text{ L}} = 0.261 \text{ mol HCl/L} = 0.261 \text{ M HCl}$$

EXAMPLE 14.25: 32.14 milliliters of a sodium hydroxide solution react with 2.9362 grams of potassium hydrogen phthalate, $KHC_8H_4O_4$ (204.22 g/mol). $NaOH + KHC_8H_4O_4 \rightarrow H_2O + NaKC_8H_4O_4$ is the equation. Calculate the molarity of the NaOH.

Do not be concerned over the "complicated" formula of potassium hydrogen phthalate; you handle it exactly as you would handle HCl. You are looking for the molarity of the NaOH. The volume is given as 32.1 milliliters. You can find the number of moles in that volume by the first two steps in the stoichiometry 3-step. Calculate that quantity.

GIVEN: 2.9362 g $KHC_8H_4O_4$, 32.14 mL NaOH FIND: mol NaOH/L
FACTOR: MM $KHC_8H_4O_4$; 1 mol NaOH/1 mol $KHC_8H_4O_4$
PATH: g $KHC_8H_4O_4 \rightarrow$ mol $KHC_8H_4O_4 \rightarrow$ mol NaOH $\rightarrow$ M NaOH

$$2.9362 \text{ g } KHC_8H_4O_4 \times \frac{1 \text{ mol } KHC_8H_4O_4}{204.22 \text{ g } KHC_8H_4O_4} \times \frac{1 \text{ mol NaOH}}{1 \text{ mol } KHC_8H_4O_4} = 0.014378 \text{ mol NaOH}$$

You now have the moles of sodium hydroxide in 32.14 milliliters of solution. Those are the two things you need to satisfy the definition of molarity, moles per liter. Complete the problem.

$$\frac{0.014378 \text{ mol NaOH}}{0.03214 \text{ L}} = 0.4474 \text{ M NaOH}$$

Another way to find the molarity of a solution is to titrate it with a standardized solution. The idea is the same as the first method, except that you start with volume and molarity of a solution instead of mass and molar mass of a reactant.

Suppose you measure out 10.00 milliliters of acetic acid of unknown concentration and titrate it with the 0.4474 M NaOH of Example 14.25. The equation is $NaOH + HC_2H_3O_2 \rightarrow NaC_2H_3O_2 + H_2O$. You find the titration takes 15.35 milliliters of base. The volume and molarity of the NaOH allow you to find the moles of base. By stoichiometry you can convert to moles of $HC_2H_3O_2$. Divide by the volume of the acid and you have its molarity:

$$0.01535 \text{ L NaOH} \times \frac{0.4474 \text{ mol NaOH}}{1 \text{ L NaOH}} \times \frac{1 \text{ mol } HC_2H_3O_2}{1 \text{ mol NaOH}} = 0.006868 \text{ mol } HC_2H_3O_2$$

$$\frac{0.006868 \text{ mol } HC_2H_3O_2}{0.01000 \text{ L}} = 0.6868 \text{ M } HC_2H_3O_2$$

EXAMPLE 14.26: 25.8 milliliters of 0.396 M KOH are required to titrate a 25.0-milliliter sample of phosphoric acid by the reaction 2 KOH + H_3PO_4 → 2 H_2O + K_2HPO_4. Find the molarity of the phosphoric acid solution.

Solve the problem all the way.

GIVEN: 25.8 mL 0.396 M KOH, 25.0 mL H_3PO_4 FIND: M of H_3PO_4

FACTOR: 0.396 mol KOH/L, 2 mol KOH/1 mol H_3PO_4

PATH: mL KOH → mol KOH → mol H_3PO_4 → M H_3PO_4

$$0.0258 \text{ L KOH} \times \frac{0.396 \text{ mol KOH}}{1 \text{ L KOH}} \times \frac{1 \text{ mol } H_3PO_4}{2 \text{ mol KOH}} = 0.00511 \text{ mol } H_3PO_4$$

$$\frac{0.00511 \text{ mol } H_3PO_4}{0.0250 \text{ L } H_3PO_4} = 0.204 \text{ M } H_3PO_4$$

CHAPTER 14 PROBLEMS

1) A solution of 26.2 grams of ammonium chloride in water has a mass of 355 grams. Calculate the percentage ammonium chloride.

2) What is the percentage concentration of a solution prepared by dissolving 5.48 grams of silver nitrate in 25.0 grams of water?

3) The density of 20.0% formic acid is 1.05 g/mL. Calculate the grams of formic acid in 50.0 mL of the solution.

4) How many grams sodium carbonate are needed to prepare 800 grams of a 18.1% solution?

5) How many grams of potassium nitrate and water must be used to prepare 175 grams of 22.5% solution?

6) Calculate the volume of water in which to dissolve 75.0 grams of calcium chloride if the concentration is to be 15.0%.

7) What is the percentage concentration of an iron(III) chloride solution if 75.0 mL contains 13.8 grams of solute and the solution density is 1.15 g/mL.

8) The density of a solution prepared by dissolving 5.46 grams of copper nitrate in 22.0 mL of water is 1.19 g/mL. Calculate the mass of solute in 15.0 mL of the solution.

9) What is the mole fraction of oxalic acid, $H_2C_2O_4$, in a solution prepared by dissolving 1.89 grams in 15.0 mL of water?

10) Calculate the mole fraction of each component in a mixture of 70.5 grams of C_2H_5OH, 41.0 grams of CH_3OH, and 36.7 grams of water.

11) If 56.0 grams of sodium bromide are dissolved in 1.50 liters of water, what is the molality of the solution?

12) Calculate the molality of a solution made by dissolving 58.1 grams of isopropyl alcohol, C_3H_7OH, in 302 grams of water.

13) How many grams of solute should be used in preparing 0.200 m $C_{12}H_{22}O_{11}$ in 2.50×10^2 grams of water?

14) Find the number of grams of solute having a molar mass of 109 g/mol that should be dissolved in 185 grams of water to make a 0.411 molal solution?

15) Calculate the mass of sodium fluoride to dissolve in 1.00 liter of water in the preparation of 0.230 m NaF.

16) 65.8 grams of potassium carbonate are used to prepare 0.268 m K_2CO_3. How many grams of water are needed?

17) How many milliliters of water should be used to dissolve 61.8 grams of $HC_3H_5O_2$ in preparing a 0.300 molal solution?

18) The density of methanol, CH_3OH, is 0.792 g/mL. In how many milliliters of methanol should 125 grams of water be dissolved to make 0.366 m H_2O? Consider water to be the solute and methanol to be the solvent.

19) 0.750 mole of potassium nitrate is dissolved in water and diluted to a final volume of 500.0 mL. What is the molarity of the solution?

20) How many grams of potassium hydrogen carbonate, $KHCO_3$, must be dissolved and diluted to 2.00 liters to produce a solution that is 0.750 molar?

21) 274 milligrams of lithium chloride are dissolved in water and diluted to 500.0 mL. What is the molarity of the solution?

22) How many grams of potassium permanganate are in 175 mL of 0.281 M $KMnO_4$?

23) Calculate the molarity of a solution prepared by dissolving 9.43 grams of Na_2CO_3 in 150 mL of water and then diluting to a final volume of 250.0 mL.

24) How many milliliters of 1.54 M H_3PO_4 hold 0.249 moles?

25) Calculate the moles of sodium hydroxide in 750.0 mL 1.23 M NaOH.

26) How many milligrams of solute must be dissolved to prepare 100.0 mL of 0.600 molar silver nitrate, $AgNO_3$?

27) How many grams of hydrogen chloride, HCl(g), are dissolved in 150 liters of 12 M HCl?

28) Calculate the volume of ammonia, measured at STP, that must be dissolved in water to prepare 4.50 liters of 3.50 M NH_3.

29) What is the molality of 10.0% NaCl(aq)?

30) What is the mole fraction of 0.500 m $C_{10}H_8$ dissolved in C_6H_6?

31) From the information given about each of four solutions, complete the following table:

SOLUTION	SOLUTE	SOLVENT	PERCENTAGE	MOLE FRACTION	MOLALITY
a	16.8 g $HC_2H_3O_2$	56.3 g H_2O	_____	_____	_____
b	$C_6H_{12}O_6$	H_2O	10.0	_____	_____
c	$NaNO_3$	H_2O	_____	0.128	_____
d	$MgSO_4$	H_2O	_____	_____	1.17

32) What is the molality of 12.8 M aqueous acetic acid, density 1.068 g/mL?

33) What is the molarity of 10.0% NaOH(aq), density 1.11 g/mL?

34) What is the mole fraction of sulfuric acid in a 6.00 molar aqueous solution with a density of 1.339 g/mL?

35) How many milliliters of concentrated sulfuric acid that is 17.8 molar must be diluted to 5.00 liters to prepare a solution that is 3.00 molar?

36) What is the concentration of the solution prepared by diluting 65.0 mL 6.25 M NaOH to 100.0 mL?

37) How many milliliters of concentrated ammonia (7.4 M) must be diluted to 350 mL to obtain a solution that is 0.315 molar?

38) 150 mL 0.772 M H_2SO_4 are diluted to 825 mL. What is the molarity of the diluted solution?

39) How much 15.4 M HNO_3 must be used to prepare 6.5 liters of 6.0 M HNO_3?

40) How much water must be added to 25.0 mL of 3.00 M H_2SO_4 to dilute the solution to 1.00 M?

41) How many grams of phosphorus are required to react with 50.0 milliliters of 4.00 M HNO_3 in $3\ P(s) + 5\ HNO_3(aq) + 2\ H_2O(\ell) \rightarrow 3\ H_3PO_4(aq) + 5\ NO(g)$?

42) Calculate the volume of 1.35 M H_2SO_4 that is needed to react with 7.91 grams of sodium oxide. $Na_2O(s) + H_2SO_4(aq) \rightarrow Na_2SO_4(aq) + H_2O(\ell)$.

43) If 82.8 mL 0.211 M HCl react with iron(III) oxide, how many grams of $FeCl_3$ will be produced? $Fe_2O_3(s) + 6\ HCl(aq) \rightarrow 2\ FeCl_3(aq) + 3\ H_2O(\ell)$.

44) What volume of carbon dioxide, at 27°C and 747 torr, will be released by the reaction between excess ammonium carbonate and 42.7 mL 0.937 M HNO_3? $(NH_4)_2CO_3(aq) + 2\ HNO_3(aq) \rightarrow 2\ NH_4NO_3(aq) + H_2O(\ell) + CO_2(g)$.

45) How many milliliters of 0.0934 M $AgNO_3$ are required to precipitate all of the chloride ion in 25.0 milliliters of 0.112 M NaCl?

46) What volume of 0.224 M NaOH is required to react with 50.0 mL of 0.329 M $MgCl_2$? The equation is $2\ NaOH(aq) + MgCl_2(aq) \rightarrow Mg(OH)_2(s) + 2\ NaCl(aq)$?

47) In the reaction $H_3PO_4(aq) + 2\ NaOH(aq) \rightarrow Na_2HPO_4(aq) + 2\ H_2O(\ell)$, 27.5 milliliters of phosphoric acid were needed to titrate 36.7 mL 0.476 M NaOH. Calculate the molarity of the acid.

48) A student finds that 14.1 mL of 0.437 M NaOH are required to titrate the acetic acid in 22.8 mL of vinegar. What is the molarity of the acid? $HC_2H_3O_2(aq) + NaOH(aq) \rightarrow NaC_2H_3O_2(aq) + H_2O(\ell)$.

49) Calculate the molarity of a calcium chloride solution if 33.8 milliliters of it are needed to react with 41.2 milliliters of 0.877 M $AgNO_3$. The equation is: $CaCl_2(aq) + 2\ AgNO_3(aq) \rightarrow 2\ AgCl(s) + Ca(NO_3)_2(aq)$.

++++++++++++++++++++++++

50) 72.4 grams of copper sulfate are dissolved in 7.50×10^2 grams of water. What is the percent concentration of the solution?

51) How many grams of hydrogen peroxide are in 225 mL of a 14.0% solution if its density is 1.05 g/mL?

52) 245 grams of solution from a salt pond are evaporated, producing 23.6 grams of salt. What is the percent concentration of salt in the pond?

53) How many grams of a 15.0% solution must be weighed out to get 4.50 grams of solute?

54) The density of a 3.4% solution of potassium chlorate is 1.02 g/mL. What volume of that solution must be measured out to obtain 1.47 grams of solute?

55) 8.00 grams of benzene, C_6H_6, are mixed with 9.41 grams of cyclohexane, C_6H_{12}. What is the mole fraction of each component?

56) A solution contains 104 grams of acetone, C_3H_6O, 126 grams of ethanol, C_2H_6O, and 108 grams of water. Find the mole fraction of each component.

57) What is the molal concentration of potassium carbonate if 8.35 grams of the salt are dissolved in 75.0 grams of water?

58) Calculate the molality of ammonium sulfate in a solution containing 13.7 grams of the salt and 157 milliliters of water.

59) How many grams of urea, $CO(NH_2)_2$, must be dissolved in 225 grams of water to produce 0.445 m $CO(NH_2)_2$?

60) Calculate the mass of ammonium chloride to dissolve in 3.50×10^2 mL of water to make a solution that is 0.330 molal.

61) What is the molarity of a solution prepared by dissolving 72.4 grams of Na_2CO_3 in water and diluting to 800.0 mL?

62) The volume of a solution containing 55 grams of NaOH is 850 mL. What is the molarity of the solution?

63) What mass of solute is required to prepare 5.00×10^2 mL 0.400 M $NiCl_2$?

64) 25.0 mL of 0.134 M $Al(NO_3)_3$ contains how many millimoles of solute?

65) What is the molality of 30.0% $H_2O_2(aq)$?

66) The density of 20.0% $MgSO_4(aq)$ is 1.220 g/mL. What is its molarity?

67) How many milliliters of 0.341 M $CuSO_4$ must be diluted to 425 mL to produce a solution that is 0.118 molar?

68) 35.0 mL 0.951 M $AgNO_3$ are diluted to 225 mL. What is the new molarity?

69) A laboratory assistant is to prepare 7.5 liters of 2.5 M HCl from a stock solution that is 10.5 molar. How much concentrated acid should be used?

70) How much 6.00 M HCl should be added to 225 mL of water to produce a solution which is 2.00 M?

71) Calculate the volume of 0.221 M $BaCl_2$ that is needed to precipitate all the sulfate ion from 25.0 mL of 0.184 M Na_2SO_4. How many grams of barium sulfate will be produced?

72) How many mL of 0.270 M HCl are needed to react with 315 mg of Na_2CO_3 by the reaction $2\ HCl(aq) + Na_2CO_3(aq) \rightarrow 2\ NaCl(aq) + H_2O(\ell) + CO_2(g)$?

73) Excess zinc is added to 35.0 mL of 2.46 M HCl. The reaction proceeds until all of the available hydrogen ion is converted to hydrogen gas according to the equation $Zn(s) + 2\ HCl(aq) \rightarrow ZnCl_2(aq) + H_2(g)$. What is the volume of the hydrogen produced at 552 torr and 22°C?

ADDITIONAL PROBLEMS

74) Sea water has a density of 1.025 g/mL and is typically 3.00% NaCl by weight. What are the molarity and molality of NaCl in sea water?

75) A 25.00 mL sample of sulfuric acid of unknown molarity was diluted to 100.00 mL. A 10.00 mL sample of this dilute solution required 21.69 mL of 0.1238 M sodium hydroxide to neutralize. What was the molarity of the original 25.00 mL sample of sulfuric acid?

76) When 3.96 grams of aluminum metal dissolve in 125 mL of 3.00 M HCl, how many liters of H_2 form, measured at 14°C and 482 torr?

77) A 1.072 gram sample required 37.96 mL of 0.01036 M $KMnO_4$ to titrate. What is the %Fe in the ore?

$$5\ Fe^{2+}(aq) + MnO_4^-(aq) + 8\ H^+(aq) \rightarrow 5\ Fe^{3+}(aq) + Mn^{2+}(aq) + 8\ H_2O(\ell)$$

CHAPTER TEST

1) Calculate the percentage concentration of a solution prepared by dissolving 8.19 grams of barium bromide in 123 grams of water.

2) How many milliliters of a 7.9% solution (density = 1.07 g/mL) of sodium chromate must be measured out to obtain 6.74 grams of sodium chromate?

3) Calculate the mole fraction of each component in a solution consisting of 93.1 grams of methanol, CH_3OH, and 50.4 grams of water.

4) How many grams of potassium nitrate should be dissolved in 5.00×10^2 milliliters of water to make 6.70 m KNO_3?

5) How many grams of solute are needed for 7.50×10^2 mL of 0.685 M KI?

6) How many millimoles of potassium permanganate are in 26.2 milliliters of 0.0982 M $KMnO_4$?

7) Find a) the percent concentration and b) the molality of an aqueous solution in which $X_{CO(NH_2)_2} = 0.240$.

8) Find a) the percent concentration and b) the molality of 12.0 M aqueous sulfuric acid, density 1.634 g/mL.

9) Calculate the molarity of the solution formed when 24.0 milliliters of 5.94 M NaOH are diluted to 100.0 mL.

10) What mass of PbI_2 can be formed from 35.0 mL 0.520 M NaI by adding excess $Pb(NO_3)_2$? The equation is $Pb(NO_3)_2(aq) + 2\ NaI(aq) \rightarrow PbI_2(s) + 2\ NaNO_3(aq)$.

11) How many mL of 0.519 M Na_2CO_3 are needed to react completely with 54.1 mL of 0.378 M $Ca(NO_3)_2$ in $Ca(NO_3)_2(aq) + Na_2CO_3(aq) \rightarrow CaCO_3(s) + 2\ NaNO_3(aq)$?

12) The titration of 0.437 grams of anhydrous oxalic acid, $H_2C_2O_4$, uses 19.4 mL of NaOH in the reaction 2 NaOH(aq) + $H_2C_2O_4$(aq) → 2 H_2O(ℓ) + $Na_2C_2O_4$(aq). Calculate the molarity of the NaOH.

13) 17.2 mL of 0.608 M H_2SO_4 are required to titrate 25.0 mL of a $NaHCO_3$ solution: H_2SO_4(aq) + 2 $NaHCO_3$(aq) → 2 H_2O(ℓ) + 2 CO_2(g) + Na_2SO_4(aq). Find the molarity of the sodium hydrogen carbonate solution.

15. NET IONIC EQUATIONS

PREREQUISITES

1. Chapter 7 discussed writing names and formulas of compounds. In this chapter you will often be given names of reactants and/or products and must write their formulas correctly in each equation.

2. Sections 9.6 and 9.7 covered the writing of conventional equations for oxidation-reduction, precipitation-dissolving, and acid-base reactions. These conventional equations provide a starting point for writing net ionic equations for these reaction types.

CHEMICAL SKILLS

You may refer to an approved periodic table and to a solubility table while practicing this skill. Some instructors do not permit reference to a solubility table, but require instead that certain solubility rules be memorized and applied as needed.

1. Given information from which you can write the conventional equation for one of the following types of reactions in aqueous solution, write the net ionic equation:

   (a) Oxidation-reduction reactions that are described by single replacement equations.
   (b) Precipitation or dissolving reactions that are described by double replacement equations.
   (c) Acid-base reactions that are described by double replacement equations.

15.1 WHY NET IONIC EQUATIONS?

Many chemical changes take place in aqueous (water) solution. For example, the reaction that occurs when a piece of solid calcium carbonate is dropped into hydrochloric acid is described by the following conventional (also called molecular or overall) equation:

$$CaCO_3(s) + 2\ HCl(aq) \rightarrow CaCl_2(aq) + CO_2(g) + H_2O(\ell)$$

The conventional equation for a reaction is useful for many purposes, but it does not describe accurately the chemical changes that occur. In the above reaction, for example, there are almost no HCl molecules present, and the resulting solution contains no chemical units having the formula $CaCl_2$. A **net ionic equation** does not have these shortcomings. It identifies precisely the species that react and the species produced. It includes no formula of any substance, real or symbolic, that is not an active participant in the chemical change.

15.2 ELECTROLYTES: STRONG, WEAK, AND NON-

Electrolytes and Nonelectrolytes In Chapter 9 all equations were written using neutral "molecular" formulas only. (Technically, ionic compounds do not exist as distinct molecules and therefore do not have "molecular" formulas. Nevertheless, the word is commonly used this way, and we will do so too.) Some molecular **solutes**—a substance that is dissolved in water or some other medium is called a solute—continue to exist as neutral molecular particles when they dissolve. Other molecular solutes separate, or **dissociate**, into ions. They are no longer present as neutral particles. When an ionic solute dissolves, the ions are released from their crystal structure and are free to move throughout the solution.

It is the purpose of this chapter to show you how to recognize which compounds exist as ions in water and how they should be written when they occur in chemical equations.

First of all, a substance must be **soluble**—it must be able to dissolve—in water before it can release ions. Its formula must be followed by "(aq)" in a conventional equation. If the substance in an equation is written as a gas, liquid, or solid, by using (g), (ℓ), or (s), it does *not* dissolve in water and cannot exist as aqueous ions.

A simple test may be performed to determine if ions are present in a solution. When a solution contains ions, it becomes a better conductor of electricity than pure water. A compound whose aqueous solution has a higher conductivity than water is called an **electrolyte**.

When some molecular compounds dissolve in water, the dissolved particles are present as individual molecules that are electrically neutral. To illustrate, the solution of methanol in water may be shown by the equation

$$CH_3OH(\ell) \xrightarrow{H_2O} CH_3OH(aq) \qquad (15.1)$$

A methanol solution does not conduct electricity. This shows that there are no electrically charged particles—ions—in a methanol solution. Methanol is therefore called a **nonelectrolyte**. Its **solution inventory**—the precise identification of solute particles in the solution—consists simply of methanol molecules.

Soluble Ionic Compounds as Strong Electrolytes An ionic compound usually has the form of a crystal made up of ions arranged in a definite geometric pattern. When the compound dissolves in water, the ions are released from the crystal. They keep their identity as charged particles, but now they are free to move in the solution.

These free-to-move ions make the solution a good conductor of electricity. Therefore, a soluble ionic compound is called a **strong electrolyte**. "Strong," in this case, means **the dissolved substance exists completely or**

almost completely as ions in solution. Very few, if any, electrically neutral particles remain in the solution. The process is illustrated in the following equations for the dissolving of ionic compounds:

$$NaCl(s) \xrightarrow{H_2O} Na^+(aq) + Cl^-(aq) \qquad (15.2)$$

$$Na_2SO_4(s) \xrightarrow{H_2O} 2\ Na^+(aq) + SO_4{}^{2-}(aq) \qquad (15.3)$$

$$Al_2(SO_4)_3(s) \xrightarrow{H_2O} 2\ Al^{3+}(aq) + 3\ SO_4{}^{2-}(aq) \qquad (15.4)$$

The solution inventory of a sodium chloride solution consists entirely of sodium ions and chloride ions; of a sodium sulfate solution, sodium ions and sulfate ions; and of an aluminum sulfate solution, aluminum ions and sulfate ions. CAUTION: Recognize that the formula of a sodium ion is Na^+ regardless of its source. This ion is not $Na_2{}^{2+}$ because each formula unit of Na_2SO_4 happens to yield two sodium ions.

Some salts do not dissolve appreciably in water. Therefore, very few ions are present, and the solution is almost a nonconductor of electricity. When you write equations for chemical reactions, you must recognize salts that are not soluble. Their formulas are followed by (s) to show that they are present as solids, not dissolved in aqueous solution. In writing ionic equations, soluble salts dissolved in water are written as separated aqueous ions. Salts that are not soluble are written as solid neutral compounds.

To determine if a salt is soluble or not you may refer to the solubility table and/or solubility rules in Section 15.4. If your instructor requires you to know the solubility rules, they must be memorized.

EXAMPLE 15.1: Write the dissolving equations for KBr, $(NH_4)_2SO_4$, and Na_3PO_4, showing their solution inventories on the right side.

$KBr(s) \rightarrow K^+(aq) + Br^-(aq)$ $\qquad$ $(NH_4)_2SO_4(s) \rightarrow 2\ NH_4{}^+(aq) + SO_4{}^{2-}(aq)$

$Na_3PO_4(s) \rightarrow 3\ Na^+(aq) + PO_4{}^{3-}(aq)$

It is not really necessary to write an equation to find a solution inventory. If you can recognize the ions in the compound, those ions *are* the solution inventory. In Example 15.1, potassium and bromide ions are present in a 1:1 ratio, so the solution inventory is 1 K^+ ion and 1 Br^- ion. Similarly, in $(NH_4)_2SO_4$ there are 2 $NH_4{}^+$ ions and 1 $SO_4{}^{2-}$ ion; and in Na_3PO_4 there are three sodium ions for each phosphate ion.

Acids as Electrolytes Acids are substances that can release H^+ ions. (Remember that acids have H written first in their formulas.) When dissolved in water, acids donate the H^+ to H_2O molecules. This process results in the formation of ions. The dissociation equations for hydrochloric and sulfuric acids are typical:

$$HCl(g) + H_2O(\ell) \rightarrow H_3O^+(aq) + Cl^-(aq) \quad (15.5)$$

$$H_2SO_4(\ell) + H_2O(\ell) \rightarrow H_3O^+(aq) + HSO_4^-(aq) \quad (15.6)$$

Acids begin as neutral molecules and then *react* with water to produce ions. This is unlike salts which exist as ions in the solid crystal that are merely freed as they dissolve in water.

The ions produced by the reaction of acids with water are H_3O^+, the hydronium ion, and the negative ion formed by removing H^+ from the acid. In the case of HCl this is Cl^-, the chloride ion. In the case of H_2SO_4 this is HSO_4^-, the hydrogen sulfate ion.

For simplicity, many chemists refer to the hydrogen ion, H^+, rather than the hydronium ion, H_3O^+. The equations that correspond to Equations 15.5 and 15.6 are

$$HCl(aq) \rightarrow H^+(aq) + Cl^-(aq) \quad (15.7)$$

$$H_2SO_4(aq) \rightarrow H^+(aq) + HSO_4^-(aq) \quad (15.8)$$

We will follow this simplification in this book, going to the hydronium ion only when it is necessary to explain some detail of the reaction in question. If your instructor prefers to use the hydronium ion, you should follow his or her directions.

Hydrochloric acid and sulfuric acid are strong electrolytes because the reactions described by Equations 15.5 to 15.8 essentially "go to completion." This means that virtually 100% of the acid molecules present form ions. Acids that are strong electrolytes, that dissociate completely, are called **strong acids**.

There are only seven common strong acids. They are:

Three of the best known acids are strong:	Two other hydrohalic acids are strong:	Two oxoacids of chlorine are strong:
nitric, HNO_3	hydrobromic, HBr	chloric, $HClO_3$
sulfuric, H_2SO_4	hydroiodic, HI	perchloric, $HClO_4$
hydrochloric, HCl		

The names and formulas of these seven acids must be memorized.

If an acid is not one of the seven strong acids, it may be assumed to be a **weak acid**. This is precisely how you distinguish between a strong acid and a weak acid: if an acid is not strong, it is weak. Weak acids are **weak electrolytes**. Their water solutions conduct electricity, but very weakly. They dissociate into ions only slightly when dissolved; usually under 10% of the molecules present form ions. The principal solution inventory species is the undissociated molecule.

To illustrate, only about 6% of the HF molecules dissociate in a typical hydrofluoric acid solution:

$$HF(aq) \rightarrow H^+(aq) + F^-(aq) \quad (15.9)$$

This means that the **major species** present in solution is HF molecules—making up about 94% of the HF added. The ions are **the minor species**. This is typical behavior for weak acids: the major species present is undissociated HX molecules, and the minor species present are H^+ and X^- ions.

When we write equations for reactions in aqueous solutions, we would like the formulas to reflect changes in the *major* species present. As a result, *strong acids are written as $H^+(aq)$ and $X^-(aq)$ ions, but weak acids are written as $HX(aq)$ molecules.*

The strength of weak acids varies quite a bit. Some weak acids dissociate more than others. One of the strongest of the weak acids is the hydrogen sulfate ion, HSO_4^-. It is between 10% and 20% dissociated at ordinary concentrations.

$$HSO_4^-(aq) \rightarrow H^+(aq) + SO_4^{2-}(aq) \qquad (15.10)$$

For simplicity, many chemists assume that sulfuric acid dissociates completely to hydrogen ions and sulfate ions. If your instructor so states, you should write its solution inventory as $2\ H^+ + SO_4^{2-}$.

Bases as Electrolytes There are strong and weak bases too. Strong bases are soluble ionic compounds in which the anion is hydroxide ion, OH^-. There are only a few soluble hydroxides. They are the alkali metal hydroxides and the hydroxides of barium, strontium, and calcium in Group IIA (2). The hydroxide ion becomes a part of the solution inventory just like any other anion from a soluble ionic compound.

Ammonia is a weak base that warrants special mention. Its behavior as a base in aqueous solution comes from its ability to accept a hydrogen ion from water and produce the hydroxide ion:

$$NH_3(aq) + H_2O(\ell) \rightarrow NH_4^+(aq) + OH^-(aq) \qquad (15.11)$$

This reaction occurs only slightly—to an extent of about 1% at the common concentrations of the solution. Because the ion concentrations are low, the solution is a weak conductor. Ammonia molecules are the major species in the solution inventory, and the ammonium and hydroxide ions are minor species.

> Aqueous ammonia is sometimes referred to as *ammonium hydroxide* and given the formula NH_4OH. There is no real substance that has this name and formula. The formula does suggest that ammonia is a base, as are other soluble hydroxides, though a weak one whose major solution inventory species is neutral ammonia molecules. If your instructor uses the NH_4OH formula, you should too. We will identify these solutions as $NH_3(aq)$.

EXAMPLE 15.2: Classify each acid and base as weak or strong. Write the solution inventory for each. Include both major and minor species.

	TYPE	MAJOR	MINOR
(a) $HClO_3(aq)$			
(b) $NH_3(aq)$			
(c) $HC_2H_3O_2(aq)$			
(d) $KOH(aq)$			

	TYPE	MAJOR	MINOR
(a) $HClO_3(aq)$	strong acid	$H^+(aq)$, $ClO_3^-(aq)$	none
(b) $NH_3(aq)$	weak base	$NH_3(aq)$	$NH_4^+(aq)$, $OH^-(aq)$
(c) $HC_2H_3O_2(aq)$	weak acid	$HC_2H_3O_2(aq)$	$H^+(aq)$, $C_2H_3O_2^-(aq)$
(d) $KOH(aq)$	strong base	$K^+(aq)$, $OH^-(aq)$	none

Solution Inventory Summary Your skill in writing net ionic equations depends largely on your ability to identify the solution inventory of a dissolved substance. We therefore summarize here the main points in this section:

1. The solution inventory from a soluble ionic compound is always made up of the ions in the compound.
2. The solution inventory from a strong acid consists of the hydrogen ion and the anion derived from removing H^+ from the acid. There are seven common strong acids: nitric, HNO_3; sulfuric, H_2SO_4; hydrochloric, HCl; hydrobromic, HBr; hydroiodic, HI; perchloric, $HClO_4$; and chloric $HClO_3$.
3. The solution inventory of a weak acid or base is the undissociated acid or base molecule. If an acid is not one of the seven common strong acids, it should be treated as a weak acid. Ammonia, $NH_3(aq)$, is the only common neutral molecule which is a weak base.

15.3 OXIDATION-REDUCTION REACTIONS (SINGLE REPLACEMENT EQUATIONS)

In Section 9.6 we considered conventional equations for the kind of oxidation-reduction (redox) reactions that are described by single replacement equations. An example is the reaction between iron metal and a solution of copper(II) sulfate. We will use this reaction to develop the basic principles of writing net ionic equations:

$$Fe(s) + CuSO_4(aq) \rightarrow Cu(s) + FeSO_4(aq) \quad (15.12)$$

The reactants are an element and a compound. The conventional equation makes it appear as if the element iron is replacing the element copper in the compound. Such is not the case at all.

$CuSO_4(aq)$ and $FeSO_4(aq)$ are soluble salts. Their solution inventories tell us that they actually exist in solution as $Cu^{2+}(aq) + SO_4{}^{2-}(aq)$ and $Fe^{2+}(aq) + SO_4{}^{2-}(aq)$, respectively. If we replace the formulas of these compounds with their solution inventories, we get a revised equation, called the **total ionic equation**, which is a more accurate representation of what is in the solution:

$$Fe(s) + Cu^{2+}(aq) + SO_4{}^{2-}(aq) \rightarrow Cu(s) + Fe^{2+}(aq) + SO_4{}^{2-}(aq) \quad (15.13)$$

Careful examination of Equation 15.13 shows that the sulfate ion appears on both sides. Although physically present, it is unchanged in the reaction. Such an ion is called a **spectator**. A total ionic equation is changed into a net ionic equation by eliminating all spectators:

$$Fe(s) + Cu^{2+}(aq) \rightarrow Cu(s) + Fe^{2+}(aq) \quad (15.14)$$

In future discussions of *how* reactions take place, the net ionic equation will be the most useful representation of the reaction. (In Equation 15.14, for example, the reaction may be seen to take place by a transfer of two electrons from Fe to Cu^{2+}.)

The above reaction illustrates the three steps in writing a net ionic equation:

1. Write the balanced conventional equation, including designations of state or solution.
2. Write the total ionic equation by replacing those species that are in aqueous solution, designated (aq), with their solution inventory species.
3. Write the net ionic equation by removing the spectators. If necessary, reduce coefficients to their lowest integral (whole-number) values.

In going from Step 1 to Step 2, it is important that you *do not change the form of any species designated (s), (ℓ), or (g).* Only those species designated (aq) *can* be present as ions. Remember that only soluble ionic compounds and strong acids *will* be present as ions.

Notice that in Equation 15.14 each side carries a net charge of +2. Net ionic equations should always be **balanced in charge** as well as in atoms of each element. This was not a problem with conventional equations because only the formulas of neutral compounds were used.

EXAMPLE 15.3: Write the net ionic equation for the release of hydrogen by the reaction between zinc and sulfuric acid.

The three-step procedure is exactly the same as for the reaction between iron and copper(II) sulfate. Begin by writing the conventional equation.

$$Zn(s) + H_2SO_4(aq) \rightarrow H_2(g) + ZnSO_4(aq)$$

In the space above, separate the soluble strong electrolytes into their inventory species.

$$Zn(s) + H^+(aq) + HSO_4^-(aq) \rightarrow H_2(g) + Zn^{2+}(aq) + SO_4^{2-}(aq)$$

For the strong acid H_2SO_4 the solution inventory species are H^+ and HSO_4^-. This is a result of the dissolving reaction: $H_2SO_4 \rightarrow H^+ + HSO_4^-$.

The third step is to write the net ionic equation by removing spectators.

No spectators are present in this case, so the total ionic equation *is* the net ionic equation.

EXAMPLE 15.4: Chlorine replaces the bromide ion in a solution of sodium bromide. Write the net ionic equation.

This time it is the negative ion that is going to be replaced in the solution. The procedure is exactly the same, however, so take it through all three steps. Assume the reacting chlorine is a gas, but the bromine produced remains in solution. Think about what that means when you are writing your solution inventories. Be sure to keep the equations balanced as you progress.

$$Cl_2(g) + 2\ NaBr(aq) \rightarrow Br_2(aq) + 2\ NaCl(aq)$$
$$Cl_2(g) + 2\ Na^+(aq) + 2\ Br^-(aq) \rightarrow Br_2(aq) + 2\ Na^+(aq) + 2\ Cl^-(aq)$$
$$Cl_2(g) + 2\ Br^-(aq) \rightarrow Br_2(aq) + 2\ Cl^-(aq)$$

In writing the solution inventories for NaBr(aq) and NaCl(aq), the coefficient 2 from the conventional equation must be carried down to keep the total ionic equation balanced. In the case of bromine, its solution inventory is bromine molecules, so bromine appears in the same way in all three equations. Soluble ionic compounds and strong acids are the only dissolved substances that are separated into ions when writing an ionic equation. Weak acids and bases and other molecular solutes are unchanged.

Predicting Single Replacement Redox Reactions Single replacement equations can be written for many redox reactions, but not all of them occur spontaneously. Some elements cannot replace others. Table 15.1 lists some common metals in order of their replacement strength. This list is often called an **activity series**.

TABLE 15.1: SINGLE REPLACEMENT STRENGTHS OF COMMON METALS

Metal	
Li	
K	Very reactive metals
Ba	
Ca	Will replace H_2 from H_2O
Na	
Mg	
Al	Moderately reactive metals
Zn	
Fe	Will replace H_2 from acids
Sn	
Pb	
Cu	
Ag	Weakly reactive metals
Au	

Using Table 15.1, it is possible to predict whether or not three types of metal-metal ion redox reactions will occur. This prediction can be made relative to the conventional equation, and then it can be applied to writing the net ionic equation. Adding a fourth statement for the halogens and halide ions produces a complete summary of the reactions that have been discussed in this section.

1. One metal, M, will replace another, M′, from an aqueous solution of M′X, where X is any anion, if the M is above M′ in the activity series. The conventional equation is

$$M(s) + M'X(aq) \rightarrow M'(s) + MX(aq) \qquad (15.15)$$

2. A metal, M, from the very reactive group will replace hydrogen from liquid water, yielding a solution of MOH. The conventional equation is

$$2\ M(s) + 2\ H_2O(\ell) \rightarrow H_2(g) + 2\ MOH(aq) \qquad (15.16)$$

3. A metal, M, from the very reactive or moderately reactive groups will replace hydrogen from an acid, HX. The conventional equation is

$$M(s) + 2\ HX(aq) \rightarrow H_2(g) + MX_2(aq) \qquad (15.17)$$

4. One halogen, X, will replace another, X′, from an aqueous solution of MX′ if X is above X′ in Group VIIA (17) in the periodic table. The conventional equation is

$$X_2(g,\ l,\ \text{or}\ aq) + 2\ MX'(aq) \rightarrow X'_2(aq) + 2\ MX(aq) \qquad (15.18)$$

EXAMPLE 15.5: Consider the following single replacement redox reactants. Write the net ionic equation for any reaction that will occur.

(a) $Mg(NO_3)_2(aq) + Cu(s) \rightarrow$

(b) $Ba(s) + H_2O(\ell) \rightarrow$

(c) $Zn(s) + HC_2H_3O_2(aq) \rightarrow$

(d) $KF(aq) + Br_2(\ell) \rightarrow$

Begin by completing the conventional equation for those reactions that will occur spontaneously.

(a) Cu is below Mg. No replacement will occur.

(b) Ba is a very reactive metal and will replace H_2 from H_2O:

$$Ba(s) + 2\ H_2O(\ell) \rightarrow H_2(g) + Ba(OH)_2(aq)$$

(c) Zn is a moderately reactive metal and will replace H_2 from acids:

$$Zn(s) + 2\ HC_2H_3O_2(aq) \rightarrow H_2(g) + Zn(C_2H_3O_2)_2(aq)$$

(d) Br_2 is below F_2 in the halogen family. No replacement will occur.

Now write net ionic equations for (b) and (c). Show all three steps.

(a) $Ba(s) + 2\ H_2O(\ell) \rightarrow H_2(g) + Ba(OH)_2(aq)$
$Ba(s) + 2\ H_2O(\ell) \rightarrow H_2(g) + Ba^{2+}(aq) + 2\ OH^-(aq)$
no spectators

(b) $Zn(s) + 2\ HC_2H_3O_2(aq) \rightarrow H_2(g) + Zn(C_2H_3O_2)_2(aq)$
$Zn(s) + 2\ HC_2H_3O_2(aq) \rightarrow H_2(g) + Zn^{2+}(aq) + 2\ C_2H_3O_2^-(aq)$
no spectators

15.4 PRECIPITATION REACTIONS

In Section 9.7 we considered reactions in which solutions of two soluble compounds were mixed and a precipitate formed. Such reactions are described by double replacement equations in which the positive ion and negative ions of the reactants appear to "change partners." At least one of the compounds formed from these new ion combinations must not be soluble. This is the precipitate.

The ability to recognize precipitation reactions depends on your knowledge of the solubilities of the compounds that may form. The solubilities of some compounds are summarized in Table 15.2, which you may use as a reference. Some instructors prefer to have their students learn certain solubility rules which may be applied to possible precipitation reactions. If your instructor states this preference, learn the rules and do not use Table 15.2. The solubility rules are summarized in Table 15.3.

The same procedure is followed in writing a net ionic equation for a precipitation reaction as for a single replacement redox reaction.

EXAMPLE 15.6: Write the net ionic equation for the reaction between solutions of $Pb(NO_3)_2$ and KI.

Begin with the conventional equation. Include state designations of all substances, if you can. If you cannot predict the designations for the products, leave those parentheses empty.

$Pb(NO_3)_2(aq) \;+\; 2\ KI(aq) \;\rightarrow\; PbI_2(\ ) \;+\; 2\ KNO_3(\ )$

Let's think about the state designations for the products. There are three possibilities. Ion combinations can form a precipitate if the compound is not soluble; they can remain in solution as soluble ionic compounds; or they can form molecules, ususally water, a weak acid, or a weak base.

In this example both products are ionic compounds, so we can eliminate molecule formation. If the product is soluble, "aq" should be placed in the

TABLE 15.2: SOLUBILITIES OF IONIC COMPOUNDS

IONS	Br^-	Cl^-	ClO_3^-	CO_3^{2-}	HCO_3^-	F^-	I^-	NO_3^-	NO_2^-	OAc^-	OH^-	PO_4^{3-}	S^{2-}	SO_3^{2-}	SO_4^{2-}
Ag^+	i	i	s	i		s	i	s	i	i	—	i	i	i	i
Al^{3+}	s	s	s	i		i	s	s		—	i	i	s	s	
Ba^{2+}	s	s	s	i		i	s	s	s	s	s	i	s	i	i
Ca^{2+}	s	s	s	i	s	i	s	s	s	s	i	i	s	i	i
Co^{2+}	s	s	s	i		—	s	s	s	s	i	i	i	s	s
Cu^{2+}	s	s	s	i		s	s	s	s	s	i	i	i	s	s
Fe^{2+}	s	s		i		i	s	s		s	i	i	i	i	s
Fe^{3+}	s	s				i		s		—	i	i	s		s
K^+	s	s	s	s	s	s	s	s	s	s	s	s	s	s	s
Li^+	s	s	s	s	s	s	s	s	s	s	s	i	s		s
Mg^{2+}	s	s	s	i		i	s	s	s	s	i	i	s	s	s
Na^+	s	s	s	s	s	s	s	s	s	s	s	s	s	s	s
NH_4^+	s	s	s	s	s	s	s	s	s	s	—	s	s	s	s
Ni^{2+}	s	s	s	i		s	s	s		s	i	i	i	i	s
Pb^{2+}	i	i	s	i		i	i	s	s	s	i	i	i	i	i
Zn^{2+}	s	s	s	i		s	s	s		s	i	i	i	i	s

To determine the solubility of an ionic compound, locate the intersection of the horizontal row for the positive ion and the vertical column for the negative ion. An "s" in that box indicates the compound is soluble in aqueous solution to a molarity of 0.1 M or more at 20°C. An "i" indicates the compound is not soluble to that concentration, but remains in the solid state or precipitates if the ions are combined. A blank space indicates lack of data, and a dash (—) identifies an unstable substance.

Note: $OAc^- = C_2H_3O_2^-$*, the acetate ion.*

parentheses; if it is not soluble, "s" should appear. How do you tell? One way is to look up both compounds in the solubility table, Table 15.2. If your instructor requires that you learn solubility rules, you simply apply the proper rule. When you have decided on the solubility of each compound, insert "aq" or "s" in the parentheses after its formula.

$$Pb(NO_3)_2(aq) \quad + \quad 2\ KI(aq) \quad \rightarrow \quad PbI_2(s) \quad + \quad 2\ KNO_3(aq)$$

By the rules, all halogen salts are soluble *except* Pb^{2+} compounds and a few others, so PbI_2 precipitates. All nitrates and nearly all alkali metal salts are soluble, so KNO_3 is soluble on two counts. The table confirms these conclusions. The intersection of the Pb^{2+} line and the I^- column shows "i," indicating precipitate, and the intersection of the K^+ line and the NO_3^- column shows "s" for a soluble compound.

Now proceed as before, writing the total ionic and net ionic equations.

$$Pb(NO_3)_2(aq) \quad + \quad 2\ KI(aq) \quad \rightarrow \quad PbI_2(s) \quad + \quad 2\ KNO_3(aq)$$

$$Pb^{2+}(aq) + 2\ NO_3^-(aq) + 2\ K^+(aq) + 2\ I^-(aq) \rightarrow PbI_2(s) + 2\ K^+(aq) + 2\ NO_3^-(aq)$$

$$Pb^{2+}(aq) \quad + \quad 2\ I^-(aq) \quad \rightarrow \quad PbI_2(s)$$

The only things left after the spectators are removed are the precipitating ions and the product.

TABLE 15.3: SOLUBILITY RULES*

Soluble in Water	Important Exceptions (Not Soluble)
All NH_4^+ and alkali metal salts	Li_3PO_4
All nitrates and perchlorates	none
All acetates	$(AgC_2H_3O_2)$
All sulfates	$BaSO_4$, $SrSO_4$, $(CaSO_4)$ (Ag_2SO_4), $PbSO_4$, Hg_2SO_4
All chlorides, bromides, iodides	AgX, Hg_2X_2, (X is bromide, chloride, or iodide), PbI_2 $(PbCl_2$, $PbBr_2)$

Not Soluble in Water	Important Exceptions (Soluble)
All carbonates and phosphates	Group IA (1) and NH_4^+ salts except Li_3PO_4
All hydroxides	Group IA (1), and Ba^{2+}
All sulfides	Group IA (1), IIA (2), and NH_4^+ salts

* *A substance is classified as soluble if it dissolves to a concentration of 0.1 molar or more at 20°C. Substances whose formulas are enclosed in parentheses dissolve to concentrations of 0.01 to 0.1 molar. They are classified as slightly soluble. In writing net ionic equations, assume that they precipitate.*

EXAMPLE 15.7: Write the conventional, total ionic, and net ionic equations for the reaction between solutions of $BaBr_2$ and $(NH_4)_2SO_4$.

Check the solubility of the products predicted by the double displacement equation. Complete the exercise without further comment.

$$BaBr_2(aq) + (NH_4)_2SO_4(aq) \rightarrow BaSO_4(s) + 2\ NH_4Br(aq)$$
$$Ba^{2+}(aq) + 2Br^-(aq) + 2NH_4^+(aq) + SO_4^{2-}(aq) \rightarrow BaSO_4(s) + 2NH_4^+(aq) + 2Br^-(aq)$$
$$Ba^{2+}(aq) + SO_4^{2-}(aq) \rightarrow BaSO_4(s)$$

EXAMPLE 15.8: Write the equations that result in the net ionic equation for the reaction between $AgNO_3(aq)$ and $MgCl_2(aq)$.

You know the procedure. Go ahead, but be careful about your final equation. When you get there, you may wish to check the third step in the net ionic equation procedure once again.

$$2\ AgNO_3(aq) + MgCl_2(aq) \rightarrow 2\ AgCl(s) + Mg(NO_3)_2(aq)$$
$$2Ag^+(aq) + 2NO_3^-(aq) + Mg^{2+}(aq) + 2Cl^-(aq) \rightarrow 2AgCl(s) + Mg^{2+}(aq) + 2NO_3^-(aq)$$
$$2\ Ag^+(aq) + 2\ Cl^-(aq) \rightarrow 2\ AgCl(s)$$
$$Ag^+(aq) + Cl^-(aq) \rightarrow AgCl(s)$$

The net ionic equation that results from the elimination of the magnesium and nitrate ion spectators has coefficients of 2 for all species. Step 3 in the procedure says all coefficients should be expressed in their lowest integral terms. Dividing the third equation by 2 yields the final equation.

EXAMPLE 15.9: Write the net ionic equation for the reaction between solutions of $AgNO_3(aq)$ and $NaCl(aq)$.

$$AgNO_3(aq) + NaCl(aq) \rightarrow AgCl(s) + NaNO_3(aq)$$
$$Ag^+(aq) + NO_3^-(aq) + Na^+(aq) + Cl^-(aq) \rightarrow AgCl(s) + Na^+(aq) + NO_3^-(aq)$$
$$Ag^+(aq) + Cl^-(aq) \rightarrow AgCl(s)$$

Examples 15.8 and 15.9 resulted in the same net ionic equation. This shows that every time a solution containing a silver ion is added to a solution containing a chloride ion, silver chloride will precipitate. The same is true of all ion combinations that yield products that are not soluble. The source of the precipitating ions is not important; once they get into the same solution, they precipitate. This makes it possible to predict a net ionic equation without writing either the conventional or total ionic equations. Try it in the next example, but don't be discouraged if you do not fully understand this subtle point at this time.

EXAMPLE 15.10: Write the net ionic equation for the reaction between solutions of Na_2CO_3 and $CaBr_2$.

Try to determine mentally, without writing the conventional equation, the products that will appear on the right side of that equation. Check each for its solubility. Decide which ions will be spectators, and then write the net ionic equation using only those ions that are not spectators. If you don't yet see the single-step approach, write all three equations. Afterward reread the first three sentences of this paragraph while looking at your three equations to see how you might have reached the net ionic equation directly.

$$Ca^{2+}(aq) + CO_3^{2-}(aq) \rightarrow CaCO_3(s)$$

Considering the reactants, sodium carbonate and calcium bromide, the products that form by exchanging ions are sodium bromide and calcium carbonate. The solubility table indicates that sodium bromide is soluble, so the sodium and bromide ions will be spectators that will not appear in the net ionic equation. Calcium carbonate, however, is not soluble. It will precipitate when calcium and carbonate ions enter the same solution from different sources. The net ionic equation is therefore a direct combination of those two ions. The full set of equations is

$$CaBr_2(aq) + Na_2CO_3(aq) \rightarrow CaCO_3(s) + 2\ NaBr(aq)$$
$$Ca^{2+}(aq) + 2Br^-(aq) + 2Na^+(aq) + CO_3^{2-}(aq) \rightarrow CaCO_3(s) + 2Na^+(aq) + 2Br^-(aq)$$
$$Ca^{2+}(aq) + CO_3^{2-}(aq) \rightarrow CaCO_3(s)$$

The next example has a surprise ending.

EXAMPLE 15.11: Write the net ionic equation for what happens when a solution of $NaNO_3$ is poured into a solution of $(NH_4)_2SO_4$.

Proceed, but when you come to something you haven't seen before, think about it.

There is no net ionic equation because there is no reaction! Both products, Na_2SO_4 and NH_4NO_3, are soluble and their solution inventories are their ions. The total ionic equation has nothing but spectators:

$$2\ Na^+(aq) + 2\ NO_3^-(aq) + 2\ NH_4^+(aq) + SO_4^{2-}(aq) \rightarrow 2\ Na^+(aq) + SO_4^{2-}(aq) + 2\ NH_4^+(aq) + 2\ NO_3^-(aq)$$

The solution inventory of the combined solutions consists of all four ions. When you meet a question such as this, simply state, "No reaction."

15.5 ACID-BASE REACTIONS

In Section 9.7 we encountered two types of acid-base reactions: neutralization and the formation of a weak acid from a strong acid. These reactions follow the general pattern:

Neutralization: $$HX(aq) + MOH(aq) \rightarrow MX(aq) + H_2O(\ell) \qquad (15.19)$$

Weak acid formed: $$\underset{\text{strong}}{HX(aq)} + MY(aq) \rightarrow MX(aq) + \underset{\text{weak}}{HY(aq)} \qquad (15.20)$$

Both equations are double replacements. Note the similarities between the two reactions. If water is thought of as HOH, its character as a weak acid becomes apparent. Thus where OH appears in the neutralization reaction, Y appears in the weak acid formation reaction.

The solution inventories of strong acids and bases contain ions as the major species present. The above conventional equations can be changed to net ionic equations to reflect this presence of ions. Weak acids and bases, however, exist primarily as neutral molecules, with ions being present only as minor species. Their formulas will thus remain unchanged in net ionic equations. Keep these facts in mind as you do the following examples.

EXAMPLE 15.12: Write the conventional, total ionic, and net ionic equations for the reaction between hydrochloric acid, HCl, and a solution of sodium hydroxide, NaOH.

The procedure is exactly the same. Be sure to include state designations, and be sure to keep unchanged in the ionic equations any substance that is solid, liquid, or gas in the conventional equation.

$$HCl(aq) + NaOH(aq) \rightarrow NaCl(aq) + H_2(\ell)$$
$$H^+(aq) + Cl^-(aq) + Na^+(aq) + OH^-(aq) \rightarrow Na^+(aq) + Cl^-(aq) + H_2O(\ell)$$
$$H^+(aq) + OH^-(aq) \rightarrow H_2O(\ell)$$

The hydroxide ion is the strongest base that can exist in an aqueous solution. As a result, it is capable of stripping any ionizable hydrogen from almost all other substances. If a hydroxide ion can find a hydrogen ion to combine with to form a water molecule, it will happen. This is why hydroxide ions react with weak acids as well as strong acids. This includes the intermediate "acid ions" in the stepwise dissociation of a polyprotic acid. The next example illustrates this.

EXAMPLE 15.13: Write all three equations that result in the net ionic equation for the reaction between a solution of potassium hydroxide, KOH, and sulfuric acid, H_2SO_4.

Proceed in the usual manner. Be sure to keep all equations balanced.

$$H_2SO_4(aq) + 2\ KOH(aq) \rightarrow K_2SO_4(aq) + 2\ H_2O(\ell)$$
$$H^+(aq) + HSO_4^-(aq) + 2\ K^+(aq) + 2\ OH^-(aq) \rightarrow 2\ K^+(aq) + SO_4^{2-}(aq) + 2\ H_2O(\ell)$$
$$H^+(aq) + HSO_4^-(aq) + 2\ OH^-(aq) \rightarrow SO_4^{2-}(aq) + 2\ H_2O(\ell)$$

When the reactant acid is a weak acid, the major species present are neutral molecules. The net ionic equation for the neutralization of a weak acid is thus different from the equation for a strong acid.

EXAMPLE 15.14: Write the net ionic equation for the reaction between acetic acid, $HC_2H_3O_2$, and a solution of sodium hydroxide, NaOH.

Remember what kind of acid acetic acid is.

$$HC_2H_3O_2(aq) + NaOH(aq) \rightarrow NaC_2H_3O_2(aq) + H_2O(\ell)$$
$$HC_2H_3O_2(aq) + Na^+(aq) + OH^-(aq) \rightarrow Na^+(aq) + C_2H_3O_2^-(aq) + H_2O(\ell)$$
$$HC_2H_3O_2(aq) + OH^-(aq) \rightarrow C_2H_3O_2^-(aq) + H_2O(\ell)$$

Because acetic acid is a weak acid, its solution inventory consists primarily of undissociated acid molecules.

The weak acid in Example 15.14 should alert you to be careful in the next example.

EXAMPLE 15.15: Write the net ionic equation for the reaction between solutions of HCl and NaF.

$$HCl(aq) + NaF(aq) \rightarrow NaCl(aq) + HF(aq)$$
$$H^+(aq) + Cl^-(aq) + Na^+(aq) + F^-(aq) \rightarrow Na^+(aq) + Cl^-(aq) + HF(aq)$$
$$H^+(aq) + F^-(aq) \rightarrow HF(aq)$$

This time you must recognize that hydrofluoric acid, HF, is not one of the seven strong acids. It therefore must be a weak acid, and its solution inventory is undissociated HF molecules.

Be alert to weak acids as molecular products, which must be written with molecular formulas, not ions. *Breaking a molecule of a weak acid into ions is the most common error in writing net ionic equations.*

In describing the properties of a substance, chemical handbooks often list the solubility of that substance in water, in acids, and in one or two common organic solvents. Most hydroxides are insoluble in water, but "soluble" in strong acids. What this really means is that the acid reacts with the hydrox-ide, rather than "dissolving" it. It is that same strong tendency to form water molecules once again. The hydrogen ion from the acid strips the hydrox-ide ion out of a normally insoluble ionic compound in order to produce water.

EXAMPLE 15.16: Write the net ionic equation for the reaction between solid magnesium hydroxide and hydrochloric acid.

Proceed as before, but think through the steps carefully.

$$Mg(OH)_2(s) + 2\ HCl(aq) \rightarrow MgCl_2(aq) + 2\ H_2O(\ell)$$
$$Mg(OH)_2(s) + 2\ H^+(aq) + 2\ Cl^-(aq) \rightarrow Mg^{2+}(aq) + 2\ Cl^-(aq) + 2\ H_2O(\ell)$$
$$Mg(OH)_2(s) + 2\ H^+(aq) \rightarrow Mg^{2+}(aq) + 2\ H_2O(\ell)$$

Notice that the reactant magnesium hydroxide is not soluble in water and is, in fact, identified as a solid. It therefore appears as a solid in the net ionic equation.

Acid-Base Reactions Involving Ammonia In Section 15.2 we introduced ammonia, NH_3, as a weak base. We will now identify two reactions of ammonia: first, its neutralization by an acid, and second, its formation by the reaction of a strong base with an ammonium salt.

The neutralization of ammonia by an acid has the form

$$NH_3(aq) + HX(aq) \rightarrow NH_4X(aq) \quad (15.21)$$

ammonia + acid → ammonium salt

This is not a double replacement equation, but rather a combination. As such, it must be remembered as a special case of neutralization.

EXAMPLE 15.17: Write the net ionic equation for the reaction between solutions of hydrofluoric acid and ammonia.

$$HF(aq) + NH_3(aq) \rightarrow NH_4F(aq)$$
$$HF(aq) + NH_3(aq) \rightarrow NH_4^+(aq) + F^-(aq) \quad \text{(no spectators)}$$

The weak acid hydrofluoric acid has neutralized the weak base ammonia. Neither is written in ionic form. The product, ammonium fluoride, is a soluble salt and so exists as ions in solution.

The formation of ammonia, a weak base, occurs when a strong base is added to an ammonium salt. The conventional equation a has double replacement form:

$$MOH(aq) + NH_4X(aq) \rightarrow MX(aq) + \text{"}NH_4OH\text{"} \quad (15.22)$$

As noted in Section 15.2, the product written "NH_4OH" is really a solution of ammonia, $NH_3(aq)$, plus a water molecule; $NH_3(aq) + H_2O(\ell)$ is the better way to write it in an equation:

$$MOH(aq) + NH_4X(aq) \rightarrow MX(aq) + NH_3(aq) + H_2O(\ell) \quad (15.23)$$

The strong base, MOH, and the salts MX and NH_4X all exist as ions, so a net ionic equation gives a better picture of how the reaction occurs.

EXAMPLE 15.18: Write the balanced net ionic equation for the reaction between solutions of barium hydroxide and ammonium chloride.

$$Ba(OH)_2(aq) + 2\ NH_4Cl(aq) \rightarrow BaCl_2(aq) + 2\ \text{"}NH_4OH\text{"}$$
$$Ba^{2+}(aq) + 2\ OH^-(aq) + 2\ NH_4^+(aq) + 2\ Cl^-(aq) \rightarrow Ba^{2+}(aq) + 2\ Cl^-(aq) + 2\ NH_3(aq) + 2\ H_2O(\ell)$$
$$OH^-(aq) + NH_4^+(aq) \rightarrow NH_3(aq) + H_2O(\ell)$$

From the net ionic equation, it is much clearer how the reaction occurs. An H^+ transfers from NH_4^+ to OH^-. The strong base OH^- forms the weak base NH_3.

One key to working with acid-base reactions of ammonia is the role played by the ammonium ion, NH_4^+. In neutralization, NH_3 forms NH_4^+; in a weak base formation, NH_4^+ forms NH_3. Whenever ammonia appears in an equation, ammonium ion will be on the other side. The above examples illustrate this point.

15.6 UNSTABLE SUBSTANCES

If hydrochloric acid is added to a solution of sodium carbonate, the expected equation is a double replacement:

$$2\ HCl(aq) + Na_2CO_3(aq) \rightarrow 2\ NaCl(aq) + H_2CO_3(aq) \quad (15.24)$$

Carbonic acid, H_2CO_3, is a weak acid, but it is unstable. Its presence in water solutions is limited to quite low concentrations. At concentrations above about 0.03 moles per liter, carbonic acid decomposes into water and carbon dioxide gas, which bubbles off to the atmosphere:

$$H_2CO_3(aq) \rightarrow H_2O(\ell) + CO_2(g) \quad (15.25)$$

The decomposition of carbonic acid may be added to Equation 15.24 to show the end products of the reaction:

$$2\ HCl(aq) + Na_2CO_3(aq) \rightarrow 2\ NaCl(aq) + H_2CO_3(aq) \rightarrow 2\ NaCl(aq) + H_2O(\ell) + CO_2(g)$$

In practice the equation is written to show only the initial reactants and final product:

$$2\ HCl(aq) + Na_2CO_3(aq) \rightarrow 2\ NaCl(aq) + H_2O(\ell) + CO_2(g) \quad (15.26)$$

The two steps to the net ionic equation are

$$2\ H^+(aq) + 2\ Cl^-(aq) + 2\ Na^+(aq) + CO_3^{2-}(aq) \rightarrow 2\ Na^+(aq) + 2\ Cl^-(aq) + H_2O(\ell) + CO_2(g) \quad (15.27)$$

$$2\ H^+(aq) + CO_3^{2-}(aq) \rightarrow H_2O(\ell) + CO_2(g) \quad (15.28)$$

Another unstable ion combination that forms a weak acid occurs when hydrogen ions and sulfite ions get into the same solution. The double replacement product is sulfurous acid, $H_2SO_3(aq)$, which decomposes into $H_2O(\ell)$ and $SO_2(aq)$. If the reactants are hydrochloric acid and sodium sulfite, the conventional equation is strikingly similar to Equation 15.26:

$$2\ HCl(aq) + Na_2SO_3(aq) \rightarrow 2\ NaCl(aq) + H_2O(\ell) + SO_2(aq) \quad (15.29)$$

The only differences between Equations 15.26 and 15.29 are the elements carbon and sulfur, and the fact that SO_2 is shown in aqueous solution because it is more soluble than CO_2. The same differences appear in the net ionic equation:

$$2\ H^+(aq) + SO_3^{2-}(aq) \rightarrow H_2O(\ell) + SO_2(aq) \qquad (15.30)$$

In short, whenever a double replacement equation predicts the formation of H_2CO_3 or H_2SO_3—and we might include "NH_4OH" too—the actual end products are one water molecule and whatever is left after the H_2O is "subtracted" from the predicted formula.

EXAMPLE 15.19: Write the net ionic equation for the reaction between excess stomach acid, HCl, and sodium bicarbonate (sodium hydrogen carbonate) solution, $NaHCO_3$.

Unlike Equation 15.26 above, you are working with the hydrogen carbonate ion rather than the carbonate ion. The only difference is that you start out with the carbonate ion half neutralized. The ion combination still yields unstable carbonic acid. First, write the conventional equation with the final products, just to be sure you begin correctly.

$$HCl(aq) + NaHCO_3(aq) \rightarrow NaCl(aq) + H_2O(\ell) + CO_2(g)$$

The simple double displacement equation is $HCl + NaHCO_3 \rightarrow NaCl + H_2CO_3$. The unstable carbonic acid, H_2CO_3, decomposes to the final products, H_2O and CO_2.

Write the total ionic and net ionic equations in the usual way.

$$H^+(aq) + Cl^-(aq) + Na^+(aq) + HCO_3^-(aq) \rightarrow Na^+(aq) + Cl^-(aq) + H_2O(\ell) + CO_2(g)$$

$$H^+(aq) + HCO_3^-(aq) \rightarrow H_2O(\ell) + CO_2(g)$$

ELECTROLYTES AND NET IONIC EQUATION EXERCISES

Electrolytes. Classify each of the substances below as a strong electrolyte, weak electrolyte, non-electrolyte, or as not soluble. Then give the solution inventory of each, listing both major and minor species.

1) $Fe_2(SO_4)_3$ 2) $C_{12}H_{22}O_{11}$ 3) HNO_2 4) AgCl 5) $Ba(OH)_2$

6) $HClO_2$ 7) $CuCl_2$ 8) NH_3 9) H_2SO_4 10) $Ca(HCO_3)_2$

Net Ionic Equations. The members of each pair of chemicals listed below are possible reactants for one of the kinds of reactions listed in the Chemical Skills for this chapter. For each combination, write the net ionic equation for the reaction, if any, that will actually occur. Use the solubility table or solubility rules, as directed by your instructor.

11) $Pb(NO_3)_2(aq) + MgSO_4(aq)$

12) $Mg(s) + HCl(aq)$

13) $ZnCl_2(aq) + (NH_4)_2CO_3(aq)$

14) $NH_4Cl(aq) + KOH(aq)$

15) $MgCl_2(aq) + Ni(NO_3)_2(aq)$
16) $Mg(C_2H_3O_2)_2(aq) + H_2SO_4(aq)$
17) $Zn(s) + NiSO_4(aq)$
18) $NiCO_3(s) + HNO_3(aq)$
19) $Na(s) + H_2O(l)$
20) $(NH_4)_2SO_3(aq) + HI(aq)$
21) $HCl(aq) + Cu(OH)_2(s)$
22) $CoCl_2(aq) + NaOH(aq)$
23) $AlBr_3(aq) + NH_4F(aq)$
24) $NaC_5H_9O_2(aq) + HCl(aq)$
25) $ZnF_2(aq) + NaClO_3(aq)$
26) $KHCO_3(aq) + HBr(aq)$
27) $Br_2(aq) + NaI(aq)$
28) $AgF(aq) + Pb(C_2H_3O_2)_2(aq)$
29) $HBr(aq) + Na_2SO_3(s)$
30) $NaC_3H_5O_2(aq) + HNO_3(aq)$

31) Magnesium chloride solution is poured into aqueous lithium hydroxide.

32) Solid iron(II) carbonate is dropped into sulfuric acid.

33) Cobalt(II) sulfate and sodium iodide solutions are combined.

34) Hydrochloric acid is poured onto solid zinc hydroxide.

35) Aqueous ammonium carbonate is added to a solution of potassium hydroxide.

36) A solution of magnesium acetate is poured into hydrobromic acid.

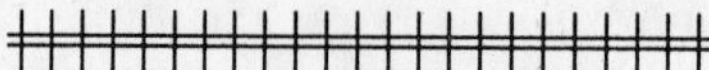

Electrolytes. Classify each of the substances below as a strong electrolyte, weak electrolyte, non-electrolyte, or as not soluble. Then give the solution inventory of each, listing both major and minor species.

37) $HClO_3$ 38) KOH 39) $Hg_2(NO_3)_2$ 40) HF 41) CH_3NH_2 (like ammonia)

42) $PbSO_4$ 43) Na_3PO_4 44) H_2O 45) H_3PO_4 46) C_2H_5OH (alcohol)

Net Ionic Equations. The members of each pair of chemicals listed below are possible reactants for one of the kinds of reactions listed in the Chemical Skills for this chapter. For each combination, write the net ionic equation for the reaction, if any, that will actually occur. Use the solubility table or solubility rules, as directed by your instructor.

47) $KOH(aq) + HNO_3(aq)$
48) $CoSO_4(aq) + Li_2S(aq)$
49) $KNO_2(aq) + H_2SO_4(aq)$
50) $FeCl_3(aq) + Al_2(SO_4)_3(aq)$
51) $Li_2CO_3(aq) + HCl(aq)$
52) $Cu(s) + FeBr_2(aq)$
53) $(NH_4)_3PO_4(aq) + KOH(aq)$
54) $CuSO_4(aq) + Al(s)$
55) $FeCl_2(aq) + K_3PO_4(aq)$
56) $K(s) + H_2O(l)$
57) $H_2SO_4(aq) + Ba(OH)_2(aq)$
58) $H_2SO_4(aq) + NaCHO_2(aq)$

59) Lithium fluoride solution is added to sulfuric acid.

60) Zinc chlorate and sodium phosphate solutions are combined.

61) A strip of magnesium is placed into a silver nitrate solution.

62) Solid potassium sulfite is treated with hydrochloric acid.

ADDITIONAL PROBLEMS

63) A common metallurgical process is to "roast" sulfide ores by heating them in air. Balance these roasting equations.

a) $Cu_2S(s) + O_2(g) \rightarrow Cu(s) + SO_2(g)$

b) $PbS(s) + O_2(g) \rightarrow PbSO_4(s)$

c) $MoS_2(s) + O_2(g) \rightarrow MoO_3(s) + SO_2(g)$

64) The solubility of aluminum sulfide is not listed in Table 15.2. The compound does exist; it can be made by heating solid sulfur and aluminum metal. It cannot, however, be formed as a precipitate from aqueous solution. If solutions of aluminum chloride and sodium sulfide are mixed, the products are a precipitate of aluminum hydroxide and a solution of equal amounts of sodium chloride and sodium hydrogen sulfide. Write the net ionic equation for the reaction. [Hint: $H_2O(\ell)$ is a reactant.]

65) A common laboratory problem is the recycling of student-prepared silver chloride. One effective method first converts the solid silver chloride to solid silver oxide by addition of aqueous sodium hydroxide (also forming water and aqueous sodium chloride) and then dissolves the silver oxide in nitric acid to form silver nitrate and water. The silver nitrate is isolated by evaporation and then recrystallized for purity. Write the net ionic equations for the above two reactions.

66) Aluminum pans corrode easily to form deposits of aluminum oxide. The deposits may be removed by treatment with vinegar (aqueous acetic acid). Excess acetic acid may be neutralized with baking soda (sodium hydrogen carbonate). Write balanced net ionic equations for these three reactions.

CHAPTER TEST

Electrolytes. Classify each of the substances below as a strong electrolyte, weak electrolyte, non-electrolyte, or as not soluble. Then give the solution inventory of each, listing both major and minor species.

1) $Al_2(SO_4)_3$ 2) H_2SO_4 3) HNO_2 4) NH_3

Net Ionic Equations. For each combination of possible reactants below, write the net ionic equation for the reaction that will occur, if any. You may refer to an approved periodic table, and, if your instructor permits, a solubility table.

5) Aluminum nitrate and sodium hydroxide solutions are combined.

6) Ammonium chloride solution is added to a solution of sodium sulfate.

7) Solid calcium hydroxide is dropped into nitric acid.

8) Hydrobromic acid is poured onto solid zinc metal.

9) Sodium sulfite solution is treated with hydrochloric acid.

10) Barium metal is dropped into water.

11) Ammonium bromide solution reacts with aqueous potassium hydroxide.

12) Aluminum metal is added to a solution of copper(II) nitrate.

16. OXIDATION-REDUCTION REACTIONS

PREREQUISITES

1. Chapter 7 covered names and formulas of compounds and ions which will be needed in writing oxidation-reduction equations.

2. Chapter 9 presented the basics of equation writing and introduced oxidation-reduction reactions that are described by single replacement equations.

3. Chapter 15 discussed electrolytes and how their formulas should be written in net ionic equations, including those for oxidation-reduction reactions.

CHEMICAL SKILLS

1. Oxidation number. Given the formula of a chemical substance, state the oxidation number of each element in that formula.

2. Recognizing oxidation and reduction. Distinguish between oxidation and reduction in terms of both electron gain or loss and change in oxidation number.

3. Recognizing oxidizing agents and reducing agents. Given an oxidation-reduction equation, identify the oxidizing agent and reducing agent.

4. Writing half-reaction equations. Given an oxidizing agent or a reducing agent and its reduced or oxidized product, write the half-reaction equation for the change in acidic or basic solution. State whether the change is oxidation or reduction. (You may use either the mass-before-charge or charge-before-mass method.)

5. Writing redox equations: half-reaction method. Given oxidation and reduction half-reaction equations, or information from which they may be written, write the net ionic equation for the redox reaction produced by combining them.

6. Writing redox equations: oxidation number method. Given an oxidizing agent, a reducing agent, and their reduced and oxidized products in a redox reaction, write the net ionic equation for the reaction by the oxidation number method.

When two equations are added algebraically, as they are in this chapter, the process is seen more clearly if each equation is complete on one line and their arrows placed one over the other. This arrangement is often too wide for the page if the state designations (g), (ℓ), (s), and (aq) are used. We will therefore not use them in Chapter 16. You may assume that all substances are in their natural states at room conditions and that all ions are in aqueous solution.

16.1 OXIDATION STATE; OXIDATION NUMBER

Reactions in which electrons are transferred from one species to another are called **oxidation-reduction reactions**. If the transfer is between atoms and monatomic ions, the electron exchange is seen quite clearly. If a polyatomic particle, such as a nitrate ion, NO_3^-, is involved, it is not possible to identify which atom in the particle loses or gains electrons. This problem is solved by assigning to each atom in the particle an **oxidation number** or **oxidation state**. The terms mean the same thing.

The purpose of oxidation numbers is to keep track of the number of electrons being exchanged in a redox reaction. Indeed, their use is often described as an "electron bookkeeping system." The numbers are assigned by an arbitrary system. Within that system there is more than one procedure by which these numbers can be assigned, but they all yield the same result. We recommend the following set of rules:

1. The oxidation state of an uncombined element is zero.
2. The sum of the oxidation numbers in a chemical species is equal to the charge on that species. This means that:
   a. The sum of the oxidation numbers in a neutral compound is zero.
   b. The oxidation state of a monatomic ion is the same as the charge on the ion.
   c. Ionic compounds should be separated into their component ions when assigning oxidation numbers.
3. The oxidation state of combined hydrogen is +1 except in hydrides, in which the oxidation state is -1 by 2b above.
4. The oxidation state of combined oxygen is -2 except in peroxides and superoxides, in which the oxidations states are, respectively, -1 and -½ by 2b and 2c above.

The steps in the above procedure should be applied in order. If they are, the common "exceptions" to the usual oxidation numbers of hydrogen and oxygen (hydrides, peroxides, and superoxides) are provided for. There are a few compounds that require additional rules to complete the oxidation number procedure, but they do not usually appear in an elementary course. The examples that follow will guide you in assigning oxidation numbers.

EXAMPLE 16.1: State the oxidation number of the designated element in each of the following:

(a) Na in Na____; (b) Cl in Cl_2____; (c) Na in NaCl____; (d) Cl in NaCl____

(a) 0; (b) 0; (c) +1; (d) -1

In (a) and (b) the substance is in elemental form, which sets the oxidation number at zero (Rule 1). It makes no difference if the element forms diatomic molecules. In (c) and (d), NaCl contains Na^+ and Cl^- ions. Sodium ion has a charge of +1, so its oxidation number is +1 according to Rule 2. Similarly, chloride ion has a charge of -1, so it is in a -1 oxidation state.

EXAMPLE 16.2: What are the oxidation numbers of nitrogen in

(a) NH_4^+: (b) NO_3^-:

Hydrogen and oxygen have their standard oxidation numbers of +1 and -2 in these ions. This means +1 and -2 *per atom.* The algebraic sum of all the oxidation numbers is equal to the charge on the species. In each case, what oxidation number for nitrogen must be added to the total oxidation number of the other element to produce the charge on the ion?

(a) -3 (b) +5

Neither species is an element or a monatomic ion, so Rules 1 and 2 do not apply. Each hydrogen atom in NH_4^+ contributes +1 (Rule 3) to the total oxidation number of the ion, which is +1 according to Rule 2. If "oxno. of N" is the oxidation number of nitrogen, (oxno. of N) + 4(+1) = +1, and the oxidation number of nitrogen is -3. Similarly, with each O at -2 (Rule 4), (oxno. of N) + 3(-2) = -1 for NO_3^-. This gives oxno. of N = +5.

EXAMPLE 16.3: Calculate the oxidation number of chromium in

(a) Cr_2O_3 and (b) $Cr_2O_7{}^{2-}$.

In Example 16.2, *each* oxygen atom contributed -2 to the total. Similarly, in this example, the oxidation number of chromium will be the contribution of each atom. Work out the answer for (a) first, and save (b) until Cr_2O_3 is checked.

+3

Cr_2O_3 is a neutral compound, so the total of all the oxidation numbers must be zero. Three oxygens contribute 3(-2) = -6 to the total. Chromium must account for +6 to balance the -6. This +6 comes from two chromium atoms, so each chromium atom must account for +3: 2(+3) = +6. For the whole compound, 2(+3) + 3(-2) = 0. Algebraically, solving 2(oxno. of Cr) + 3(-2) = 0 for oxno. of Cr confirms +3 as the answer.

Now use the same approach for the oxidation state of chromium in $Cr_2O_7^{2-}$.

+6

Oxygen accounts for 7(-2) = -14, The total is -2, the charge on the ion. Chromium must therefore contribute +12 to give (+12) + 7(-2) = -2. But that +12 comes from two chromium atoms. Each must provide +6: 2(+6) + 7(-2) = -2. By algebra, 2(oxno. of Cr) + 7(-2) = -2, from which oxno. of Cr = +6.

EXAMPLE 16.4: Find the oxidation number of

(a) hydrogen in LiH: (b) oxygen in Na_2O_2:

These are both ionic compounds. Begin by separating them into ions. Be sure you apply the oxidation number rules in their order of priority and everything will come out all right.

(a) -1 (b) -1

Both species are compounds, so their oxidation number totals are zero.

(a) Lithium ion has a charge of +1. This must be added to the oxidation number (charge) of the ion from hydrogen to produce a total of zero. Hydrogen must therefore have a charge of -1, the charge on the hydride ion, H^-. By algebra, +1 + oxno. of H = 0, and oxno. of H = -1. The compound is lithium hydride.
(b) Sodium ion has its usual +1 oxidation number, and two of them put the total positive charge at +2. Two oxygen atoms in the diatomic oxygen ion are therefore contributing -2 to the total, which means the oxidation number is -1 for each atom. Algebraically, 2(+1) + 2(oxno. of O) = 0 yields oxno. of O = -1. The compound is sodium peroxide, made up of two sodium ions and one diatomic peroxide ion, O_2^{2-}.

16.2 THE LANGUAGE OF OXIDATION-REDUCTION REACTIONS

In Section 15.3 you learned how to write net ionic equations for redox reactions whose conventional equation is a single replacement type. In this section we will examine what happens in such a reaction at the atomic and ionic level.

The net ionic equation for the reaction between metallic iron and a solution of copper sulfate is (see Equation 15.14)

$$Fe + Cu^{2+} \rightarrow Fe^{2+} + Cu \quad (16.1)$$

Consider what happens to the copper "particle" and to the iron "particle" in Equation 16.1. The iron particle on the left is an atom. On the right, it is an ion with a +2 charge. If a neutral atom is to become an ion with a +2 charge, it must lose two electrons. The equation for this change is

$$Fe \rightarrow Fe^{2+} + 2\ e^- \qquad (16.2)$$

This process is known as **oxidation**. In the reaction, iron is said to be *oxidized*. The loss of electrons results in a change in the oxidation number for iron. The oxidation state of Fe increases from 0 to +2. These two observations define oxidation and help us recognize when it is taking place.

Copper in Equation 16.1 begins as an ion with a +2 charge. It ends up as a neutral atom. To do this it gains two electrons:

$$Cu^{2+} + 2\ e^- \rightarrow Cu \qquad (16.3)$$

This process is known as **reduction**. Copper is said to be *reduced*. During reduction, the oxidation number of copper is reduced from +2 to 0.

Oxidation is:	Reduction is:
1. the loss of electrons;	1. the gain of electrons;
2. identified by an increase in oxidation number.	2. identified by a reduction in oxidation number..

Equations 16.2 and 16.3 are called **half-reaction equations** because each equation represents one-half of the electron transfer that makes up a redox reaction. Oxidation and reduction always occur together. At least one element is oxidized and one element is reduced in every oxidation-reduction reaction. In some redox reactions the same element is oxidized and reduced.

The oxidation and reduction **half-reactions** represented by Equations 16.2 and 16.3 can be added algebraically to give the net ionic equation for the redox reaction:

$$\begin{array}{lrr} \text{Oxidation:} & Fe \rightarrow Fe^{2+} + 2\ e^- & (16.2) \\ \text{Reduction:} & Cu^{2+} + 2\ e^- \rightarrow Cu & (16.3) \\ \hline & Fe + Cu^{2+} \rightarrow Fe^{2+} + Cu & (16.1) \end{array}$$

The summed up equations yield Equation 16.1. The two electrons on the right side of Equation 16.2 cancel the two electrons on the left side of Equation 16.3 when the equations are added. Therefore there are no electrons in the summed up equation.

Redox reactions are often called **electron transfer reactions**. Either in fact or by appearance, electrons are transferred from one species to another. In the above reaction, an iron atom transfers two electrons to a copper ion. In doing so, iron atoms are reducing copper ions to copper atoms. Iron is therefore acting as a **reducing agent**. **A reducing agent is the chemical species that loses electrons in a redox change**. Copper ion is taking those electrons, and thereby oxidizing iron. Copper ion is therefore an **oxidizing agent**. **An oxidizing agent is a receiver of electrons in a redox change**.

In summary, the characteristics of oxidizing and reducing agents are:

Oxidizing Agents:

1. cause oxidation;
2. gain electrons;
3. contain the element reduced, which goes down in oxidation number.

Reducing agents:

1. cause reduction;
2. lose electrons;
3. contain the element oxidized, which goes up in oxidation number.

EXAMPLE 16.5: In $MnO_2(s) + 2\ ClO_2(g) \rightarrow Mn^{2+}(aq) + 2\ ClO_3^-(aq)$, identify the following:

Element oxidized ________ Element reduced ________

Oxidizing agent ________ Reducing agent ________

Oxidized, Cl; reduced, Mn; oxidizing agent, MnO_2; reducing agent, ClO_2

Chlorine changes from +4 in ClO_2 to +5 in ClO_3^-, an oxidation. Manganese changes from +4 in MnO_2 to +2 in Mn^{2+}, a reduction. The species that oxidizes chlorine is the other redox reactant, MnO_2, the oxidizing agent. Notice that we do not say that *manganese* is the oxidizing agent. The entire compound or ion containing the element reduced is considered to be the oxidizing agent. The same is true for reducing agents. Thus ClO_2 is the reducing agent. *Elements are oxidized and reduced. Elements, ions, or compounds may be oxidizing and reducing agents.*

16.3 HALF-REACTION EQUATIONS: TWO WAYS TO WRITE THEM

Many redox reactions that occur in water solutions involve polyatomic ions and molecules. Their half-reaction equations are more complex than the simple iron and copper half-reactions shown in Equations 16.2 and 16.3. If the solution is acidic, the hydrogen ion is also involved; if the solution is basic, the hydroxide ion appears. The water molecule is a part of the equation as well. This section addresses the procedures for writing those half-reaction equations.

A half-reaction equation must be balanced in both charge and mass (atoms). These may be balanced in either order. Mass-before-charge is more common, but the procedure for an acidic solution is different from any of the methods used for basic solutions. By contrast, the same charge-before-mass procedure is used for both acidic and basic solutions. Section 16.4 presents the charge-before-mass procedure, and Section 16.5 will discuss the mass-before-charge method. This section closes with three suggestions:

1. If your instructor tells you which procedure to learn, learn that one.
2. If you are free to choose, select the procedure that appeals to you and learn it.
3. Learn only one method; do not try to master both.

16.4 HALF-REACTION EQUATIONS: CHARGE-BEFORE-MASS METHOD

The charge-before-mass procedure for half-reaction equations is

1. Balance the element oxidized or reduced.
2. Determine the change in oxidation number of the element oxidized or reduced. Multiply by the number of atoms experiencing that change. Add that number of electrons to the side on which the redox element has the higher oxidation number.
3. Balance any elements other than hydrogen and oxygen that may be present. (This is rare, but it happens occasionally.)
4. Balance charges by adding H^+ ions if the solution is acidic and OH^- ions if the solution is basic.
5. Balance hydrogen with water molecules.
6. Check the oxygen balance.

To illustrate this procedure in an acidic solution, consider the half-reaction in which sulfate ion is changed to hydrogen sulfide:

	SO_4^{2-}	→	H_2S
Oxidation numbers	+6		-2

Step 1 Sulfur is balanced, so Step 1 is already completed.

Step 2 The oxidation number change for sulfur is from +6 on the left to -2 on the right. To calculate the change, subtract the initial oxidation number from the final: -2 - (+6) = -8. (The sign, + or -, of the oxidation number change is not significant at this time, but it is later.) This indicates that eight electrons are in the half-reaction equation. The electrons are always added to the side on which the redox element has the higher—more positive—oxidation number, thereby balancing the oxidation number change.

$$8\ e^- + SO_4^{2-} \rightarrow H_2S$$

Step 3 We now look for additional elements, other than hydrogen and oxygen, which may be present. There usually are none, as in this example, so we go on to the next step.

Step 4 Charge is balanced by adding H^+ ions to the more negative side. On the left side of the equation there are eight negative charges from the electrons and two more from the sulfate ion, a total of ten. There is no charge on the right side. We must therefore add ten positive charges to the left side to make the total charge zero, just as it is on the right:

$$10\ H^+ + 8\ e^- + SO_4^{2-} \rightarrow H_2S$$

Step 5 If everything has been done correctly, both hydrogen and oxygen can be balanced by adding water molecules to one side of the equation. It is well to work with one element and use the other as a check. Using hydrogen, there are ten hydrogen atoms on the left and two on the right. Eight more are needed on the right. That takes four water molecules:

$$10\ H^+ + 8\ e^- + SO_4^{2-} \rightarrow H_2S + 4\ H_2O \qquad (16.4)$$

Step 6 Now checking with oxygen, we find four on the left and four on the right. The equation is balanced.

EXAMPLE 16.6: Use the charge-before-mass sequence to develop the half-reaction equation for the following change in an acidic solution:

$$ClO_3^- \rightarrow Cl^-$$

Oxidation numbers Change =

Chlorine is balanced, so Step 1 requires no action. Step 2 requires the change in the oxidation number of chlorine. Insert the oxidation numbers above and calculate the change.

$$ClO_3^- \rightarrow Cl^-$$

Oxidation numbers $+5$ -1 Change = $-1 - (+5) = -6$

The number of electrons that corresponds to the oxidation number change is added to the side of the equation on which the redox element has the higher oxidation number. Make that addition.

$$6\ e^- + ClO_3^- \rightarrow Cl^-$$

There are no miscellaneous elements, so Step 3 does not apply. The solution is acidic, so H^+ is used to balance charge (Step 4).

$$6\ H^+ + 6\ e^- + ClO_3^- \rightarrow Cl^-$$

There were a total of seven negative charges on the left and one on the right. Adding six positive charges (H^+ ions) on the left balances the charge at -1 on each side.

Water molecules should now balance both hydrogen and oxygen.

$$6\ H^+ + 6\ e^- + ClO_3^- \rightarrow Cl^- + 3\ H_2O$$

To illustrate the charge-before-mass procedure with basic solutions we will use

$$Cl_2 \rightarrow ClO_3^-$$

Oxidation numbers 0 +5 Change = 5 - 0 = 5

Step 1 This time the element oxidized, chlorine, is not balanced. That must be corrected first:

$$Cl_2 \rightarrow 2\ ClO_3^-$$

Oxidation numbers 2(0) 2(+5) Change = 2(5 - 0) = 10

Step 2 The oxidation number changes from 0 on the left to +5 on the right, a change of 5. Two atoms in the equation undergo that change, so the total number of electrons lost is 2 × 5 = 10. These 10 electrons are added to the side on which the redox element has the higher oxidation number:

$$Cl_2 \rightarrow 2\ ClO_3^- + 10\ e^-$$

Step 3 There are no other elements besides hydrogen and oxygen.

Step 4 Total charge is zero on the left and 2(-1) + (-10) = -12 on the right. In a basic solution charge is balanced by adding negatively charged hydroxide ions to the more positive side of the equation:

$$12\ OH^- + Cl_2 \rightarrow 2\ ClO_3^- + 10\ e^-$$

Step 5 Water molecules should now balance both hydrogen and oxygen. Twelve hydrogens on the left demand six water molecules on the right:

$$12\ OH^- + Cl_2 \rightarrow 2\ ClO_3^- + 10\ e^- + 6\ H_2O$$

Step 6 Checking, there are 12 oxygen atoms on each side of the equation. The equation is balanced.

EXAMPLE 16.7: Write the half-reaction equation for the $ClO_3^- \rightarrow Cl^-$ change in a basic solution.

This is the same reaction as in Example 16.6, except it is in basic rather than acidic solution. Steps 1 to 3 are the same, whether the reaction be acidic or basic. We may therefore start with Step 4 applied to

$$6\ e^- + ClO_3^- \rightarrow Cl^-$$

In a basic solution charge is balanced by adding hydroxide ions.

Step 4: charge -7 → -1 (add 6 OH^- to right)

$$6\ e^- + ClO_3^- \rightarrow Cl^- + 6\ OH^-$$

The starting seven negative charges on the left and one on the right are balanced by adding six more negatives (hydroxide ions) to the right.

Water molecules should complete the balance.

Step 5: hydrogen 0 → 6 (add 3 H_2O to left)

$$3\ H_2O + 6\ e^- + ClO_3^- \rightarrow Cl^- + 6\ OH^-$$

Step 6: A final check of the oxygen shows three atoms on each side.

EXAMPLE 16.8: Complete the half-reaction equation $Sb \rightarrow Sb_2O_5$ by the charge-before-mass method. The solution is acidic. Caution on Step 1.

$$Sb \rightarrow Sb_2O_5$$

$$Sb \rightarrow Sb_2O_5$$

Step 1: antimony $1 \rightarrow 2$ (add 1 Sb to left side)

$$2\ Sb \rightarrow Sb_2O_5$$

Step 2: oxidation number $0 \rightarrow +5$ (add $2 \times 5 = 10\ e^-$ to right side)

$$2\ Sb \rightarrow Sb_2O_5 + 10\ e^-$$

Step 3: no elements other than oxygen

Step 4: charge $0 \rightarrow -10$ (add 10 H^+ to right side)

$$2\ Sb \rightarrow Sb_2O_5 + 10\ e^- + 10\ H^+$$

Step 5: hydrogen $0 \rightarrow 10$ (add 5 H_2O to left side)

$$5\ H_2O + 2\ Sb \rightarrow Sb_2O_5 + 10\ e^- + 10\ H^+$$

Step 6: oxygen $5 \rightarrow 5$

EXAMPLE 16.9: Use the charge-before-mass method to write the half-reaction equation for $TeO_3{}^{2-} \rightarrow Te$ in basic solution.

$$TeO_3{}^{2-} \rightarrow Te$$

$4\ e^- + \overset{+4}{TeO_3{}^{2-}} \rightarrow \overset{0}{Te}$ Change = -4	Balance oxidation number change with electrons.
$4\ e^- + TeO_3{}^{2-} \rightarrow Te + 6\ OH^-$	Balance charge with OH^-.
$3\ H_2O + 4\ e^- + TeO_3{}^{2-} \rightarrow Te + 6\ OH^-$	Balance hydrogen and oxygen with H_2O.

16.5 HALF-REACTION EQUATIONS: MASS-BEFORE-CHARGE METHOD

The steps in the mass-before-charge procedure for writing a half-reaction equation are:

1. Balance the element oxidized or reduced.
2. Balance any elements other than hydrogen and oxygen that may be present. (This is unlikely, but it happens occasionally.)
3. Balance oxygen by adding water molecules as necessary.
4. Balance hydrogen by adding H^+ ions as necessary.
   - 4.1 If (and *only* if) the reaction occurs in a basic solution, "neutralize" the H^+ ions from step 4 by adding a like number of OH^- ions to both sides of the equation.
   - 4.2 Combine H^+ and OH^- ions to form water.
   - 4.3 If water molecules appear on both sides of the equation, subtract the smaller number from each side.
5. Balance charges by adding electrons to the more positive side.
6. Check the electrons by matching them against the change in oxidation number.

To illustrate the procedure in an acidic solution, consider the half-reaction in which sulfate ion is changed to hydrogen sulfide:

$$SO_4{}^{2-} \rightarrow H_2S$$

This is a reduction equation, as sulfur is being reduced from +6 in the sulfate ion to -2 in hydrogen sulfide. We note that the oxidation number change is -2 - (+6) = -8 so we may use this as a check at the end of the procedure (Step 6).

Step 1 Sulfur, the element reduced, is already balanced.

Step 2 There are no other elements except hydrogen and oxygen.

Step 3 In an acidic solution, hydrogen and oxygen are balanced by the proper combination of hydrogen ions and water molecules. Both are present in the solution, and either may be a reactant or a product. Oxygen is balanced first by adding to the right side the number of water molecules needed. There are four oxygen atoms in the sulfate ion, so it takes four water molecules:

$$SO_4{}^{2-} \rightarrow H_2S + 4\ H_2O$$

Step 4 Hydrogen ions can now be added to the left side to bring hydrogen atoms into balance. There are ten hydrogen atoms on the right, so it takes ten hydrogen ions on the left:

$$10\ H^+ + SO_4{}^{2-} \rightarrow H_2S + 4\ H_2O$$

Step 5 At this point the atoms of all three elements are in balance, but the charge is not. The left side has a total charge of +8, and the right side, 0. Charge is balanced by adding the required number of electrons (negative charges) to the positive side of the equation. Eight electrons are needed:

$$8\ e^- + 10\ H^+ + SO_4{}^{2-} \rightarrow H_2S + 4\ H_2O \qquad (16.5)$$

Step 6 There is a quick check to see if the reduction balance is correct. The number of electrons must be the same as the change in oxidation number of the element reduced. We noted in Step 1 that this number is 8, which does, indeed, match the eight electrons in the equation.

EXAMPLE 16.10: Write the half-reaction equation in which ClO_3^- is changed to Cl^- in an acidic solution:

$$ClO_3^- \rightarrow Cl^-$$

By inspection we can see that Steps 1 and 2 of the above procedure are already completed. Chlorine is balanced at one atom on each side, and there are no other elements besides hydrogen and oxygen. There is an oxygen, however, and Step 3 tells us to balance it with water molecules. Add water molecules to the above equation in whatever place and number that will balance oxygen.

$$ClO_3^- \rightarrow Cl^- + 3\ H_2O$$

Three water molecules on the right balance the oxygen in the chlorate ion.

Now take Step 4: Add H^+ ions to the above equation to balance hydrogen.

$$6\ H^+ + ClO_3^- \rightarrow Cl^- + 3\ H_2O$$

Six hydrogen ions balance six hydrogen atoms in three water molecules.

Add some electrons to the equation so the charges balance (Step 5).

$$6\ e^- + 6\ H^+ + ClO_3^- \rightarrow Cl^- + 3\ H_2O$$

Before adding the electrons there was a net charge of +5 on the left and a net of -1 on the right. Adding six negative charges (electrons) to the left sets both sides at -1.

Confirm that the change in oxidation number is six (Step 6).

Chlorine in chlorate ion is +5; in chloride ion, -1. The change is -1 - (+5) = -6.

The procedure is similar for a basic solution, except that hydroxide ions and water molecules are used to balance hydrogen and oxygen. Chlorine can be oxidized to chlorate ion in a basic solution:

$$Cl_2 \rightarrow ClO_3^-$$

Step 1 This time we note that the element oxidized is not balanced. This is corrected by using a coefficient of 2 on the right:

$$Cl_2 \rightarrow 2\ ClO_3^-$$

Step 2 Oxygen is the only other element, so Step 2 does not apply.

Step 3 Oxygen is balanced by adding water molecules:

$$6\ H_2O + Cl_2 \rightarrow 2\ ClO_3^-$$

Step 4 Next hydrogen is balanced by adding H^+ ions on the right:

$$6\ H_2O + Cl_2 \rightarrow 2\ ClO_3^- + 12\ H^+$$

Step 4.1 Now we move to the path for basic solutions, Steps 4.1 to 4.3. The solution is first made basic "on paper" by adding 12 hydroxide ions to both sides of the equation, neutralizing the 12 hydrogen ions on the right:

$$12\ OH^- + 6\ H_2O + Cl_2 \rightarrow 2\ ClO_3^- + 12\ H^+ + 12\ OH^-$$

Step 4.2 The hydrogen and hydroxide ions on the right combine as water:

$$12\ OH^- + 6\ H_2O + Cl_2 \rightarrow 2\ ClO_3^- + 12\ H_2O$$

Step 4.3 We now have water molecules on both sides of the equation. The smaller number, 6, is subtracted from both sides:

$$12\ OH^- + Cl_2 \rightarrow 2\ ClO_3^- + 6\ H_2O$$

Step 5 Steps 5 and 6 are the same as in an acidic solution. The left side of the equation has twelve negative charges, and the right side has two. The right side needs ten more. They come in the form of electrons:

$$12\ OH^- + Cl_2 \rightarrow 2\ ClO_3^- + 6\ H_2O + 10\ e^-$$

Step 6 If everything has been done correctly, the total oxidation number change for the chlorines should also be ten. Chlorine is zero on the left, and +5 on the right. Two chlorines undergo this change: 2(5 - 0) = 10.

EXAMPLE 16.11: The change from ClO_3^- to Cl^- (Example 16.10) can occur in a basic solution too. Write the reduction half-reaction equation.

Steps 1 to 4 are the same for both acidic and basic solutions, giving

$$6\ H^+ + ClO_3^- \rightarrow Cl^- + 3\ H_2O$$

Step 4.1 for the basic solution sequence says you must "neutralize" the hydrogen ions with the same number of hydroxide ions added to both sides.

6 H^+, so add 6 OH^- to each side:

$$6\ OH^- + 6\ H^+ + ClO_3^- \rightarrow Cl^- + 3\ H_2O + 6\ OH^-$$

Now form water with hydrogen and hydroxide ions (Step 4.2).

6 H^+ and 6 OH^- combine to form 6 H_2O:

$$6\ H_2O + ClO_3^- \rightarrow Cl^- + 3\ H_2O + 6\ OH^-$$

Step 4.3 warns of water appearing on both sides of the equation. Subtract the smaller number from each side in the space above.

Subtract 3 H_2O from each side:

$$3\ H_2O + ClO_3^- \rightarrow Cl^- + 6\ OH^-$$

Now add the electrons to balance charge (Step 5).

$$6\ e^- + 3\ H_2O + ClO_3^- \rightarrow Cl^- + 6\ OH^-$$

After adjusting water there was one negative charge on the left and there were seven on the right. The left side needed six more negative charges—six electrons.

The equation should now be balanced—but check it against the change in oxidation number (Step 6).

Chlorine's oxidation state changes from +5 in ClO_3^- to -1 in Cl^-. This is a change of 6, as -1 - (+5) = -6. This matches the 6 electrons in the half-reaction equation.

EXAMPLE 16.12: Develop the half-reaction equation for the following change in an acidic solution:

$$MnO_2(s) \rightarrow MnO_4^-(aq)$$

Manganese is already in balance, and oxygen is the only other element, so no action is needed for Steps 1 and 2. Begin with Step 3 and carry through to Step 6.

$$MnO_2(s) \rightarrow MnO_4^-(aq)$$

$$MnO_2(s) \rightarrow MnO_4^-(aq)$$

Step 3: oxygen 2 → 4 (add 2 H_2O to left side)

$$2\ H_2O(\ell) + MnO_2(s) \rightarrow MnO_4^-(aq)$$

Step 4: hydrogen 4 → 0 (add 4 H^+ to right side)

$$2\ H_2O(\ell) + MnO_2(s) \rightarrow MnO_4^-(aq) + 4\ H^+(aq)$$

Step 5: charge 0 → +3 (add 3 e^- to right side)

$$2\ H_2O(\ell) + MnO_2(s) \rightarrow MnO_4^-(aq) + 4\ H^+(aq) + 3\ e^-$$

Step 6: oxidation number changes from +4 to +7, a difference of 3, matching 3 electrons in equation.

EXAMPLE 16.13: Write the half-reaction equation for the $MnO_2 \rightarrow MnO_4^-$ change in basic solution. The first four steps are the same, so you may begin with Step 4.1 for

$$2\ H_2O(\ell) + MnO_2(s) \rightarrow MnO_4^-(aq) + 4\ H^+(aq)$$

$$2\ H_2O(\ell) + MnO_2(s) \rightarrow MnO_4^-(aq) + 4\ H^+(aq)$$

Step 4.1: neutralize H^+ $\quad 4 \rightarrow 0$ (add 4 OH^- to both sides)

$$4\ OH^-(aq) + 2\ H_2O(\ell) + MnO_2(s) \rightarrow MnO_4^-(aq) + 4\ H^+(aq) + 4\ OH^-(aq)$$

Step 4.2: combine ($4\ OH^- + 4\ H^+ \rightarrow 4\ H_2O$)

$$4\ OH^-(aq) + 2\ H_2O(\ell) + MnO_2(s) \rightarrow MnO_4^-(aq) + 4\ H_2O(\ell)$$

Step 4.3: cancel waters $\quad 2 \rightarrow 4$ (subtract 2 H_2O from both sides)

$$4\ OH^-(aq) + MnO_2(s) \rightarrow MnO_4^-(aq) + 2\ H_2O(\ell)$$

Step 5: charge $\quad -4 \rightarrow -1$ (add 3 e^- to right side)

$$4\ OH^-(aq) + MnO_2(s) \rightarrow MnO_4^-(aq) + 2\ H_2O(\ell) + 3\ e^-$$

Step 6: oxidation number changes from +4 to +7, a difference of 3, matching 3 electrons in equation.

16.6 WRITING REDOX EQUATIONS BY THE HALF-REACTION METHOD

In Section 16.2 we showed that oxidation and reduction half-reactions can be added together to give the net ionic equation for the redox reaction.

Oxidation:	$Fe \rightarrow Fe^{2+} + 2\ e^-$	(16.2)
Reduction:	$Cu^{2+} + 2\ e^- \rightarrow Cu$	(16.3)
Net:	$Fe + Cu^{2+} \rightarrow Fe^{2+} + Cu$	(16.1)

It is essential in the addition of half-reaction equations that the final equation have no electrons on either side. The electrons lost in oxidation must be equal to the electrons gained in reduction. They then cancel in the addition process. In the above example the two electrons to the right of the arrow in the oxidation half-reaction equation cancel the two electrons to the left of the arrow in the reduction half-reaction equation.

Sometimes the electrons in the oxidation and reduction half-reactions are not the same. The reaction between metallic copper and a solution of silver nitrate is an example. The half-reaction equations and their addition give:

Oxidation:	$Cu \rightarrow Cu^{2+} + 2\ e^-$
Reduction:	$Ag^+ + e^- \rightarrow Ag$
	$Cu + Ag^+ \rightarrow Cu^{2+} + Ag + e^-$

The single electron on the left side of the reduction half-reaction does not cancel the two electrons on the right side. To correct this condition whenever it appears, either or both half-reaction equations are multiplied by integers that will make the number of electrons the same. Then the equations will have no electrons when they are added. In the copper-silver ion example, this is accomplished by doubling the reduction equation:

Oxidation: $Cu \rightarrow Cu^{2+} + 2\ e^-$

Reduction: $2\ Ag^+ + 2\ e^- \rightarrow Ag$

$$Cu + 2\ Ag^+ \rightarrow Cu^{2+} + 2\ Ag$$

The electrons now cancel and a balanced net ionic equation is the result.

EXAMPLE 16.14: When dropped into a solution of nickel nitrate, aluminum replaces nickel ion. Write the half-reaction equations and add them to produce the net ionic equations.

The nitrate ion will be a spectator in this reaction. The reactants are therefore elemental aluminum and nickel(II) ion. The products will be aluminum ion and metallic nickel. Write the two half-reaction equations, one above the other.

Reduction: $Ni^{2+} + 2\ e^- \rightarrow Ni$

Oxidation: $Al \rightarrow Al^{3+} + 3\ e^-$

A nickel atom must gain two electrons to become a nickel ion, and an aluminum atom must lose three electrons to become an Al^{3+} ion.

You now have a reduction equation with two electrons and an oxidation equation with three electrons. Can you figure out a multiplication, or perhaps two, that will put the same number of electrons in the two half-reaction equations? If so, complete the equations in the space above and add them.

Reduction: $3\ Ni^{2+} + 6\ e^- \rightarrow 3\ Ni$

Oxidation: $2\ Al \rightarrow 2\ Al^{3+} + 6\ e^-$

$$2\ Al + 3\ Ni^{2+} \rightarrow 2\ Al^{3+} + 3\ Ni$$

A 3-2 combination is adjusted by taking two of the 3's and three of the 2's. Two equations with three electrons gives six electrons in oxidation; three equations with two electrons gives six electrons in reduction. The electrons cancel, and the net ionic equation is the result. The equation is balanced in mass and charge. There are two Al, three Ni, and six + charges on each side.

So far our examples have been limited to elements and monatomic ions. The same procedure is followed when the reaction includes polyatomic species. In such cases, however, the oxidation and/or reduction half-reaction equations must be developed as in Section 16.4 or 16.5.

EXAMPLE 16.15: Write the half-reaction and net ionic equations for the reaction between iodide and sulfate ions in an acidic solution, producing iodine and hydrogen sulfide: $I^- + SO_4{}^{2-} \rightarrow I_2 + H_2S$.

You begin by separating the half reactions:

Oxidation: $I^- \rightarrow I_2$

Reduction: $SO_4{}^{2-} \rightarrow H_2S$

The skeleton equation for each half-reaction must now be expanded and balanced. Do the oxidation reaction first. Balance the redox element, and then use electrons to balance the charge.

Oxidation: $2\ I^- \rightarrow I_2 + 2\ e^-$

Two iodide ions are required by the two atoms in the diatomic iodine molecule. Each iodide ion loses one electron in becoming an iodine atom, giving two electrons on the right. This half-reaction is quite common among halogens.

The reduction half-reaction was used to introduce half-reaction equations with polyatomic species. The result was Equation 16.4 or Equation 16.5. The oxidation half-reaction equation has been rewritten below, and directly under it is the previously developed reduction equation. Can you take it from there? Use whatever multiplier or multipliers that are necessary to equate the electrons, and then add the equations to get the balanced net ionic equation.

Oxidation: $2\ I^- \rightarrow I_2 + 2\ e^-$

Reduction: $8\ e^- + 10\ H^+ + SO_4{}^{2-} \rightarrow H_2S + 4\ H_2O$ (16.4 or 16.5)

Oxidation: $8\ I^- \rightarrow 4\ I_2 + 8\ e^-$

Reduction: $8\ e^- + 10\ H^+ + SO_4{}^{2-} \rightarrow H_2S + 4\ H_2O$

$$8\ I^- + 10\ H^+ + SO_4{}^{2-} \rightarrow 4\ I_2 + H_2S + 4\ H_2O$$

Two electrons in the oxidation half-reaction equation must be multiplied by 4 to match the eight electrons in the reduction equation.

These multiplications and additions provide the opportunity for small errors to creep in. To eliminate these, be sure to recheck mass and charge balance: There are 8 iodine, 10 hydrogen, 1 sulfur, and 4 oxygen atoms on both sides of the equation, and the total charge is zero on the two sides. Mass and charge are balanced.

These are the steps that were taken in the above example:

1. Identify the elements oxidized and reduced. For each, write a skeleton equation, showing the starting form of the element (monatomic ion, polyatomic ion, or elemental) on the left and the oxidized or reduced form on the right.
2. Write balanced half-reaction equations for both oxidation and reduction.
3. If the number of electrons in the oxidation and reduction half-reaction equations are not equal, multiply either or both equations by whatever factors that will make them equal.
4. Add the half-reaction equations to get the net ionic equation.
5. Check to be sure that mass and charge are still balanced.

The mass-before-charge and charge-before-mass methods for writing half-reaction equations appear side-by-side in the next three examples. It may be helpful to look ahead and cross out the method you do *not* use so it does not distract you as you work the examples.

EXAMPLE 16.16: Chromate ion, CrO_4^{2-}, and hydrogen sulfide react to produce chromium(III) ion and sulfur in an acidic solution. Write the half-reaction and net ionic equations.

Begin by writing the skeleton equations (Step 1).

Oxidation: $H_2S \rightarrow S$

Reduction: $CrO_4^{2-} \rightarrow Cr^{3+}$

Both oxidation and reduction include polyatomic species. Using the space around the equations above, expand them to balanced half-reaction equations.

MASS-BEFORE-CHARGE METHOD	CHARGE-BEFORE-MASS METHOD
	<u>Oxidation</u>
	-2 $\quad$ 0 $\qquad$ Change = +2
$H_2S \rightarrow S + 2\ H^+$	$H_2S \rightarrow S + 2\ e^-$
$H_2S \rightarrow S + 2\ H^+ + 2\ e^-$	$H_2S \rightarrow S + 2\ e^- + 2\ H^+$

Reduction

$$CrO_4^{2-} \rightarrow Cr^{3+}$$
$$CrO_4^{2-} \rightarrow Cr^{3+} + 4\ H_2O$$
$$8\ H^+ + CrO_4^{2-} \rightarrow Cr^{3+} + 4\ H_2O$$
$$3\ e^- + 8\ H^+ + CrO_4^{2-} \rightarrow Cr^{3+} + 4\ H_2O$$

+6 +3 Change = -3

$$CrO_4^{2-} \rightarrow Cr^{3+}$$
$$3\ e^- + CrO_4^{2-} \rightarrow Cr^{3+}$$
$$8\ H^+ + 3\ e^- + CrO_4^{2-} \rightarrow Cr^{3+}$$
$$8\ H^+ + 3\ e^- + CrO_4^{2-} \rightarrow Cr^{3+} + 4\ H_2O$$

The oxidation reaction involves two electrons, and the reduction reaction has three. Rewrite the equations so the electrons are equal, but don't add them yet.

Oxidation: $H_2S \rightarrow S + 2\ e^- + 2\ H^+$
Reduction: $8\ H^+ + 3\ e^- + CrO_4^{2-} \rightarrow Cr^{3+} + 4\ H_2O$

Oxidation: $3\ H_2S \rightarrow 3\ S + 6\ e^- + 6\ H^+$
Reduction: $16\ H^+ + 6\ e^- + 2\ CrO_4^{2-} \rightarrow 2\ Cr^{3+} + 8\ H_2O$

Multiplying the oxidation equation by 3 produces six electrons, and multiplying the reduction equation by 2 yields six electrons. The six electrons will now cancel in the addition of the equations.

Complete the addition.

Oxidation: $3\ H_2S \rightarrow 3\ S + 6\ e^- + 6\ H^+$
Reduction: $16\ H^+ + 6\ e^- + 2\ CrO_4^{2-} \rightarrow 2\ Cr^{3+} + 8\ H_2O$

$$3\ H_2S + 16\ H^+ + 2\ CrO_4^{2-} \rightarrow 3\ S + 6\ H^+ + 2\ Cr^{3+} + 8\ H_2O$$

There is something new this time. There are hydrogen ions on both sides of the equation, sixteen on the left and six on the right. Six of those may be subtracted from both sides—they will cancel just as the six electrons canceled. Rewrite the final equation above without the extra hydrogen ions. Check that mass and charge are still balanced.

$$3\ H_2S + 10\ H^+ + 2\ CrO_4^{2-} \rightarrow 3\ S + 2\ Cr^{3+} + 8\ H_2O$$

Mass: H: 16 → 16; S: 3 → 3; Cr: 2 → 2; O: 8 → 8
Charge: $3(0) + 10(+1) + 2(-2) = 3(0) + 2(+3) + 8(0)$
$+6 = +6$

The dual appearance of hydrogen ions, as in the above example, is quite common. If oxygen had been present in the oxidation half-reaction, water would have appeared on both sides too. Watch for these duplications and remove them whenever possible.

EXAMPLE 16.17: If sulfur dioxide is bubbled through an acidic solution of potassium permanganate, the redox reaction shown below occurs. Develop the half-reaction equations and add them to produce the net ionic equation.

$$SO_2 + MnO_4^- \rightarrow SO_4^{2-} + Mn^{2+}$$

MASS-BEFORE-CHARGE METHOD / CHARGE-BEFORE-MASS METHOD

Oxidation

MASS-BEFORE-CHARGE METHOD	CHARGE-BEFORE-MASS METHOD
$SO_2 \rightarrow SO_4^{2-}$	$\overset{+4}{SO_2} \rightarrow \overset{+6}{SO_4^{2-}}$
$2\ H_2O + SO_2 \rightarrow SO_4^{2-}$	$SO_2 \rightarrow SO_4^{2-} + 2\ e^-$
$2\ H_2O + SO_2 \rightarrow SO_4^{2-} + 4\ H^+$	$SO_2 \rightarrow SO_4^{2-} + 2\ e^- + 4\ H^+$
$2\ H_2O + SO_2 \rightarrow SO_4^{2-} + 4\ H^+ + 2\ e^-$	$2\ H_2O + SO_2 \rightarrow SO_4^{2-} + 2\ e^- + 4\ H^+$

Reduction

MASS-BEFORE-CHARGE METHOD	CHARGE-BEFORE-MASS METHOD
$MnO_4^- \rightarrow Mn^{2+}$	$\overset{+7}{MnO_4^-} \rightarrow \overset{+2}{Mn^{2+}}$
$MnO_4^- \rightarrow Mn^{2+} + 4\ H_2O$	$5\ e^- + MnO_4^- \rightarrow Mn^{2+}$
$8\ H^+ + MnO_4^- \rightarrow Mn^{2+} + 4\ H_2O$	$8\ H^+ + 5\ e^- + MnO_4^- \rightarrow Mn^{2+}$
$5\ e^- + 8\ H^+ + MnO_4^- \rightarrow Mn^{2+} + 4\ H_2O$	$8\ H^+ + 5\ e^- + MnO_4^- \rightarrow Mn^{2+} + 4\ H_2O$

Oxidation: $2\ H_2O + SO_2 \rightarrow SO_4^{2-} + 2\ e^- + 4\ H^+$

Reduction: $8\ H^+ + 5\ e^- + MnO_4^- \rightarrow Mn^{2+} + 4\ H_2O$

Oxidation: $10\ H_2O + 5\ SO_2 \rightarrow 5\ SO_4^{2-} + 10\ e^- + 20\ H^+$

Reduction: $16\ H^+ + 10\ e^- + 2\ MnO_4^- \rightarrow 2\ Mn^{2+} + 8\ H_2O$

$$2\ H_2O + 5\ SO_2 + 2\ MnO_4^- \rightarrow 5\ SO_4^{2-} + 4\ H^+ + 2\ Mn^{2+}$$

Mass: H: $4 \rightarrow 4$; S: $5 \rightarrow 5$; O: $20 \rightarrow 20$; Mn: $2 \rightarrow 2$

Charge: $2(0) + 5(0) + 2(-1) = 5(-2) + 4(+1) + 2(+2)$

$-2 = -2$

Ten electrons, sixteen hydrogen ions, and eight water molecules cancel in the final addition of the two half-reaction equations because they appear on one side of one equation and the other side of the other equation.

Sometimes the same element is both oxidized and reduced. In such a case both half-reaction equations begin with the same substance. Chlorine is capable of this kind of reaction in an alkaline solution.

EXAMPLE 16.18: Write the half-reaction and net ionic equations for the reaction $Cl_2 \rightarrow Cl^- + ClO_3^-$ in a basic solution.

To be sure we begin correctly, write the skeleton equations for the oxidation and reduction half-reactions. Include the oxidation numbers on both sides of the equation.

Oxidation: $Cl_2 \rightarrow ClO_3^-$ (0, +5) Reduction: $Cl_2 \rightarrow Cl^-$ (0, -1)

$$\underset{0}{Cl_2} \rightarrow \underset{+5}{ClO_3^-} \qquad \underset{0}{Cl_2} \rightarrow \underset{-1}{Cl^-}$$

Oxidation numbers have been placed below chlorine in both equations to aid in identification of oxidation and reduction.

Now develop the half-reaction equations. The reduction reaction is easy, there being only an element and a monatomic ion. The oxidation reaction is more involved. Watch the redox element. Remember also that the reaction occurs in basic solution.

Reduction: $Cl_2 + 2\ e^- \rightarrow 2\ Cl^-$

Oxidation: MASS-BEFORE-CHARGE

$$Cl_2 \rightarrow 2\ ClO_3^-$$

$$6\ H_2O + Cl_2 \rightarrow 2\ ClO_3^-$$

$$6\ H_2O + Cl_2 \rightarrow 2\ ClO_3^- + 12\ H^+$$

$$12\ OH^- + 6\ H_2O + Cl_2 \rightarrow 2\ ClO_3^- + 12\ H^+ + 12\ OH^-$$

$$12\ OH^- + 6\ H_2O + Cl_2 \rightarrow 2\ ClO_3^- + 12\ H_2O$$

$$12\ OH^- + Cl_2 \rightarrow 2\ ClO_3^- + 6\ H_2O$$

$$12\ OH^- + Cl_2 \rightarrow 2\ ClO_3^- + 6\ H_2O + 10\ e^-$$

Oxidation: CHARGE-BEFORE-MASS

$$\underset{}{\overset{2(0)}{Cl_2}} \rightarrow \overset{2(+5)}{2\ ClO_3^-}$$

$$Cl_2 \rightarrow 2\ ClO_3^- + 10\ e^-$$

$$12\ OH^- + Cl_2 \rightarrow 2\ ClO_3^- + 10\ e^-$$

$$12\ OH^- + Cl_2 \rightarrow 2\ ClO_3^- + 10\ e^- + 6\ H_2O$$

Oxidation: $12\ OH^- + Cl_2 \rightarrow 2\ ClO_3^- + 10\ e^- + 6\ H_2O$
Reduction: $Cl_2 + 2\ e^- \rightarrow 2\ Cl^-$

Oxidation: $12\ OH^- + Cl_2 \rightarrow 2\ ClO_3^- + 10\ e^- + 6\ H_2O$
Reduction: $5\ Cl_2 + 10\ e^- \rightarrow 10\ Cl^-$

$$12\ OH^- + 6\ Cl_2 \rightarrow 2\ ClO_3^- + 6\ H_2O + 10\ Cl^-$$

$$6\ OH^- + 3\ Cl_2 \rightarrow ClO_3^- + 3\ H_2O + 5\ Cl^-$$

In the first step of the reduction balancing it was necessary to multiply ClO_3^- by 2 to balance chlorine atoms. The summed up equations at the end produced coefficients that were all divisible by 2. The division was performed, leaving the final equation with the lowest integral coefficients.

16.7 BALANCING REDOX EQUATIONS BY THE OXIDATION NUMBER METHOD

Redox equations can be balanced without writing half-reaction equations. The method is to find the oxidation number change for both the element oxidized and the element reduced, make them equal numerically, and balance the remainder of the equation by inspection. If you learned the charge-before-mass procedure for developing half-reaction equations, you will find the procedures very similar.

In Example 16.17 you developed the net ionic equation for the following reaction in acidic solution:

$$\overset{+4}{S}O_2 + \underset{+7}{Mn}O_4^- \rightarrow \overset{+6}{S}O_4^{2-} + \underset{+2}{Mn}^{2+}$$

It is convenient to write the oxidation numbers of the redox elements above and below their symbols, as has been done here. The change in oxidation number is the final value minus the initial value: sulfur, +6 - (+4) = +2; manganese, +2 - (+7) = -5. These changes can be shown by writing them between the oxidation numbers above.

$$\overset{+4}{S}O_2 \overset{(+2)}{+} \underset{+7}{Mn}O_4^- \underset{(-5)}{\rightarrow} \overset{+6}{S}O_4^{2-} + \underset{+2}{Mn}^{2+}$$

To make the changes in oxidation number equal, the sulfur oxidation must occur five times while the manganese reduction occurs twice. This is accomplished by multiplying the coefficients of the sulfur-containing species by 5 and the manganese-containing species by 2:

$$5\ \overset{+4}{S}O_2 \overset{(5)(+2)}{+} 2\ \underset{+7}{Mn}O_4^- \underset{(2)(-5)}{\rightarrow} 5\ \overset{+6}{S}O_4^{2-} + 2\ \underset{+2}{Mn}^{2+}$$

The redox elements are now balanced. Their coefficients should not be changed again.

The remainder of the charge and mass balance is done with H^+ ions and water molecules in an acidic reaction. The charge balance comes first. The equation has two negative charges on the left and a net of six negatives on the right. Four positive charges (H^+ ions) added to the right side balances the charges at -2 on each side:

$$5\ SO_2 + 2\ MnO_4^- \rightarrow 5\ SO_4^{2-} + 2\ Mn^{2+} + 4\ H^+$$

$$5\ SO_2 + 2\ MnO_4^- \rightarrow 5\ SO_4^{2-} + 2\ Mn^{2+} + 4\ H^+$$

If all has been done properly, water molecules should complete the balance of both hydrogen and oxygen. The four hydrogens on the right can be balanced by two water molecules on the left:

$$2\ H_2O + 5\ SO_2 + 2\ MnO_4^- \rightarrow 5\ SO_4^{2-} + 2\ Mn^{2+} + 4\ H^+$$

Oxygen is the final check point. It must be balanced. It is: 20 on the left and 20 on the right.

The entire procedure, including a step not required in this example, may be summarized as follows:

1. Balance the elements oxidized and reduced.
2. Determine the oxidation number change for the element oxidized and the element reduced. Note the number of atoms undergoing each change. Use appropriate multiplications to make the total gain in oxidation number numerically equal to the total loss in oxidation number.
3. Balance any element other than hydrogen and oxygen that may be present in the redox species. (This is unlikely, but it happens occasionally.)
4. Balance charge, using H^+ ions if the solution is acidic and OH^- ions if it is basic.
5. Balance hydrogen with water molecules.
6. Check the oxygen balance. (If oxygen is balanced, the entire equation should be balanced.)

EXAMPLE 16.19: Use the oxidation number method to write the net ionic equation for the reaction

$$CrO_4^{2-} + H_2S \rightarrow Cr^{3+} + S$$

No action is required for the first step; both chromium and sulfur are balanced. Begin by determining the oxidation number change for each redox element.

$$\overset{+6}{CrO_4^{2-}} + \overset{(\quad)(-3)}{\underset{-2}{H_2S}} \rightarrow \overset{+3}{\underset{(\quad)(+2)}{Cr^{3+}}} + \underset{0}{S}$$

Multiply the species containing the redox elements by whatever factors will make the oxidation number increase numerically equal to the oxidation number decrease. Parentheses are provided for these factors above.

$$\overset{+6}{2\,CrO_4^{2-}} + \overset{(2)(-3)}{\underset{-2}{3\,H_2S}} \rightarrow \overset{+3}{\underset{(3)(+2)}{2\,Cr^{3+}}} + \underset{0}{3\,S}$$

The change in oxidation number for reduction, -6, is numerically equal to the change in oxidation number for oxidation, +6.

The solution is acidic, so the charges must be balanced by adding H^+ where needed.

$$10\,H^+ + 2\,CrO_4^{2-} + 3\,H_2S \rightarrow 2\,Cr^{3+} + 3\,S$$

Use water molecules to balance hydrogen. Then check oxygen, which must also be balanced if everything has been done correctly.

$$10\,H^+ + 2\,CrO_4^{2-} + 3\,H_2S \rightarrow 2\,Cr^{3+} + 3\,S + 8\,H_2O$$

Sixteen hydrogens on the left call for eight water molecules on the right. Oxygen is balanced at eight atoms on each side. (The half-reaction method for balancing this equation was used in Example 16.16.)

EXAMPLE 16.20: Use the oxidation number method to write the net ionic equation for the reaction of Example 16.18, which occurs in a basic solution:

$$Cl_2 \rightarrow Cl^- + ClO_3^-$$

This is the reaction in which some of the chlorine is oxidized and some is reduced. The two steps must be considered separately, and the oxidation number changes balanced against each other. To see this clearly, it is convenient to write the formula for chlorine twice, and use one for oxidation and the other for reduction:

$$Cl_2 \quad + \quad Cl_2 \quad \rightarrow \quad Cl^- \quad + \quad ClO_3^-$$

Begin by balancing chlorine atoms in both oxidation and reduction (Step 1).

$$Cl_2 \quad + \quad Cl_2 \quad \rightarrow \quad 2\,Cl^- \quad + \quad 2\,ClO_3^-$$

If one Cl_2 molecule is to be used for oxidation, there must be two chlorine atoms in the oxidized form, ClO_3^-. Similarly, if one Cl_2 molecule is to be used for reduction, there must be two reduced chlorine atoms in Cl^-.

Now write the oxidation numbers above and below the elemental symbols, and the oxidation number changes, as before. Do not try to balance the changes at this time.

$$\overset{0}{Cl_2} \quad + \quad \overset{(\quad)(-1)}{\underset{0}{Cl_2}} \quad \rightarrow \quad \overset{-1}{\underset{(\quad)(+5)}{2\,Cl^-}} \quad + \quad \underset{+5}{2\,ClO_3^-}$$

$$\begin{array}{ccccccc} 0 & & (\ \ )(-1) & & -1 & & \\ Cl_2 & + & Cl_2 & \rightarrow & 2\ Cl^- & + & 2\ ClO_3^- \\ & & 0 & & (\ \ )(+5) & & +5 \end{array}$$

The change in oxidation number of an element keeps track of the number of electrons gained or lost by each atom. Each atom loses one electron in the oxidation half-reaction $Cl_2 \rightarrow 2\ Cl^-$. There are two atoms in the equation, so the total number of electrons transferred is (2)(1) = 2. At -1 per atom, the oxidation number change is numerically the same: (2)(-1) = -2. In other words, the oxidation number change must be multiplied by the number of atoms making that change. Thus

$$\begin{array}{ccccccc} 0 & & (2)(-1) & & -1 & & \\ Cl_2 & + & Cl_2 & \rightarrow & 2\ Cl^- & + & 2\ ClO_3^- \\ & & 0 & & (\ \ )(+5) & & +5 \end{array}$$

Insert into the parentheses the proper factor for the oxidation change.

$$\begin{array}{ccccccc} 0 & & (\ \ )(2)(-1) & & -1 & & \\ Cl_2 & + & Cl_2 & \rightarrow & 2\ Cl^- & + & 2\ ClO_3^- \\ & & 0 & & (\ \ )(2)(+5) & & +5 \end{array}$$

The reasoning is identical. The change is from 0 to +5 for each atom, and there are two atoms in the equation so far.

Because the electrons gained and lost are equal, the oxidation number changes in oxidation and reduction must be numerically equal. At this point, the oxidation number change for reduction is (2)(-1) = -2, and for oxidation, (2)(+5) = +10. Decide on and perform the multiplications that make these changes numerically equal. Write the factors both in the parentheses and as coefficients for the species to which they belong.

$$\begin{array}{ccccccc} 0 & & (5)(2)(-1) & & -1 & & \\ 5\ Cl_2 & + & 1\ Cl_2 & \rightarrow & 5{\times}2\ Cl^- & + & 2\ ClO_3^- \\ & & 0 & & (1)(2)(+5) & & +5 \end{array}$$

The chlorine → chloride reduction must be multiplied by 5 to make the oxidation number change for reduction -10: (5)(2)(-1) = -10. (The coefficient for Cl^- is written above as 5×2 to show the multiplication. It will be written 10 Cl^- hereafter.) The oxidation number change for oxidation is numerically equal to the reduction change: (2)(+5) = +10.

$$5\ Cl_2 + Cl_2 \rightarrow 10\ Cl^- + 2\ ClO_3^-$$

The solution is alkaline, so charge must be balanced by OH^- ions. Complete that step. You can also combine the chlorines, now that the redox part of the equation is balanced. Rewrite the equation with these changes.

$$12\ OH^- + 6\ Cl_2 \rightarrow 10\ Cl^- + 2\ ClO_3^-$$

Ten chloride ions and two chlorate ions on the right carry a total -12 charge that must be balanced by 12 OH^- ions on the left.

Now balance hydrogen with water molecules.

$$12\ OH^- + 6\ Cl_2 \rightarrow 10\ Cl^- + 2\ ClO_3^- + 6\ H_2O$$

Twelve hydroxide ions have 12 H atoms on the left. They are balanced by 6 H_2O molecules on the right.

Now check the oxygen. If it is in balance, the equation is balanced. It is. But one final improvement should be made. Write it here.

$$6\ OH^- + 3\ Cl_2 \rightarrow 5\ Cl^- + ClO_3^- + 3\ H_2O$$

All coefficients in the equation two steps above are divisible by 2. You may recall that this appeared in the half-reaction method of balancing too.

CHAPTER 16 PROBLEMS

1) Find the oxidation number of the element whose symbol is underlined in each of the following:

a) $\underline{S}O_3^{2-}$ b) $\underline{Mn}O_2$ c) $Ca\underline{H}_2$ d) $KAl\underline{Si}O_4$ e) $\underline{N}_2O_3$

f) $\underline{Cr}O_4^{2-}$ g) $Mg_3(\underline{P}O_4)_2$ h) $K\underline{O}_2$ i) $\underline{Cl}O_4^-$ j) $Fe(H\underline{S}O_3)_2$

2) Identify which of the following reactions involve oxidation-reduction. For those which are redox, identify the elements oxidized and reduced and the oxidizing and reducing agents.

a) $NH_4HCO_3(s) \rightarrow NH_3(g) + H_2O(g) + CO_2(g)$

b) $2\ CH_3OH(g) + O_2(g) \rightarrow 2\ CH_2O(g) + 2\ H_2O(g)$

c) $CO(g) + H_2O(g) \rightarrow CO_2(g) + H_2(g)$

d) $Hg_2Cl_2(s) + 2\ NH_3(aq) \rightarrow Hg(\ell) + HgNH_2Cl(s) + NH_4Cl(aq)$

3) Write the equations for the following half-reactions, which occur in an acidic solution:

a) $HClO_2 \rightarrow HClO$ b) $H_2MoO_4 \rightarrow Mo$ c) $NO \rightarrow N_2O$ d) $PbO_2 \rightarrow Pb^{2+}$ e) $Sb_2O_3 \rightarrow Sb$

4) Write the equations for the following half-reactions, which occur in a basic solution:

a) $NO_2^- \rightarrow N_2O$ b) $PbO_2 \rightarrow Pb(OH)_2$ c) $P \rightarrow PH_3$ d) $O_2 \rightarrow$

[Hint on 4d: Oxygen must become either OH- or H_2O on the right. Insert one or the other, and proceed as usual. One of them will lead to combining the same species that appears twice on one side of the equation. The other will require subtracting a substance that appears on both sides of the equation. The final answer is the same either way.]

5) For each skeleton equation, write the net ionic equation. The solutions are acidic. Use either the oxidation number or the half-reaction method.

a) $Cr_2O_7^{2-} + Sn^{2+} \rightarrow Cr^{3+} + Sn^{4+}$ b) $Ag + NO_3^- \rightarrow Ag^+ + NO$

c) $MnO_4^- + H_2O_2 \rightarrow O_2 + Mn^{2+}$ d) $Cr_2O_7^{2-} + Br^- \rightarrow Cr^{3+} + Br_2$

e) $Cr_2O_7^{2-} + H_2S \rightarrow Cr^{3+} + S$ f) $Mn^{2+} + BiO_3^- \rightarrow MnO_4^- + Bi^{3+}$

g) $Pb + PbO_2 + SO_4^{2-} \rightarrow PbSO_4$ h) $As + NO_3^- \rightarrow H_3AsO_4 + NO$

i) $Cr_2O_7^- + H_2S \rightarrow Cr^{3+} + HSO_4^-$ j) $Zn + NO_3^- \rightarrow Zn^{2+} + NH_4^+$

6) Write the net ionic equation for the redox reaction described by each of the following skeleton equations. The solutions are basic. Use either the oxidation number method or the half-reaction method.

a) $MnO_4^- + Mn^{2+} \rightarrow MnO_2$ b) $Sn(OH)_4^{2-} + CrO_4^{2-} \rightarrow Sn(OH)_6^{2-} + Cr(OH)_4^-$

c) $Cr(OH)_4^- + HO_2^- \rightarrow CrO_4^{2-}$ d) $MnO_2 + BiO_3^- \rightarrow MnO_4^{2-} + Bi(OH)_3$

e) $I^- + MnO_4^- \rightarrow I_2 + MnO_2$ f) $Fe(OH)_2 + O_2 \rightarrow Fe(OH)_3$

---

7) Find the oxidation number of the element whose symbol is underlined in each of the following:

a) $Al\underline{B}O_3$ b) $\underline{Br}O^-$ c) $H_4\underline{Si}O_4$ d) $H_2\underline{O}_2$

e) $H\underline{P}O_4^{2-}$ f) $\underline{N}_2H_4$ g) $\underline{Mn}SO_4$ h) $\underline{Cu}(H_2\underline{P}O_3)_2$

8) Identify which of the following reactions involve oxidation-reduction. For those which are redox, identify the elements oxidized and reduced and the oxidizing and reducing agents.

a) $2\ KClO_3 \rightarrow 2\ KCl + 3\ O_2$ b) $Ca(HCO_3) + SO_2 \rightarrow CaSO_3 + 2\ CO_2 + H_2O$

c) $2\ Mn(OH)_2 + O_2 + 2\ H_2O \rightarrow 2\ MnO_2$ d) $CaH_2 + 2\ H_2O \rightarrow Ca(OH)_2 + 2\ H_2$

9) Write the equations for the following half-reactions:

Acidic Solutions: a) $FeO_4^{2-} \rightarrow Fe^{3+}$ b) $As_2O_3 \rightarrow H_3AsO_4$

Basic Solutions: c) $ClO_2^- \rightarrow Cl^-$ d) $SO_3^{2-} \rightarrow S_2O_3^{2-}$

10) For each skeleton equation, write the net ionic equation. The solutions are acidic. Use either the oxidation number or half-reaction method.

a) $Cr_2O_7^{2-} + SO_2 \rightarrow SO_4^{2-} + Cr^{3+}$ b) $H_2O_2 \rightarrow H_2O + O_2$

c) $PbO_2 + Cl^- \rightarrow PbCl_2 + Cl_2$ d) $Sn + NO_3^- \rightarrow SnO_2 + NO_2$

e) $MnO_4^- + Cl^- \rightarrow Mn^{2+} + ClO_2$ f) $Ag + NO_3^- \rightarrow Ag^+ + NO_2$

11) For each skeleton equation, write the net ionic equation. The solutions are basic. Use either the oxidation number or half-reaction method.

a) $Zn + OH^- \rightarrow H_2 + Zn(OH)_4^{2-}$ b) $Cu(NH_3)_4^{2+} + S_2O_4^{2-} \rightarrow Cu + SO_4^{2-} + NH_3$

ADDITIONAL PROBLEMS

12) The electrolysis of <u>dilute</u> NaCl(aq) oxidizes water to $O_2(g)$ as the solution becomes acidic at the anode and reduces water to $H_2(g)$ as the solution becomes basic at the cathode. The electrolysis of <u>concentrated</u> NaCl(aq) forms $Cl_2(g)$ at the anode. Write balanced equations for all three of these half-reactions.

13) A common reaction of alcohols is treatment with orange $Cr_2O_7^{2-}(aq)$ to form green Cr^{3+}. For example, $CH_3CH_2OH(aq)$ is oxidized to $CH_3CHO(aq)$ in acidic solution. Heating the solution changes the organic product. Hot acidic $Cr_2O_7^{2-}(aq)$ oxidizes $CH_3CH_2OH(aq)$ to $CH_3CO_2H(aq)$. Write balanced equations for both reactions.

14) Solutions of potassium permanganate may be standardized with pure solid arsenic(III) oxide. In acidic solution $MnO_4^-(aq)$ is reduced to $Mn^{2+}(aq)$ while $As_2O_3(s)$ is oxidized to $H_3AsO_4(aq)$. Write the balanced equation for the reaction.

15) Basic solutions of potassium permanganate are not very stable. The $MnO_4^-(aq)$ oxidizes $H_2O(\ell)$ to $O_2(g)$ while being reduced to $MnO_2(s)$. This product of manganese(IV) oxide is the brown residue you find in glass containers of $KMnO_4$. It is quite hard to clean out. One effective way is to use HCl(aq). The $MnO_2(s)$ is reduced to $Mn^{2+}(aq)$ while $Cl^-(aq)$ is oxidized to $Cl_2(aq)$. Write the balanced equations for both reactions.

CHAPTER TEST

1) What is the oxidation number of:

a) C in H_2CO_3 b) As in AsO_2^- c) Pt in $PtCl_6^{2-}$ d) C in $Na_2C_2O_4$

2) Identify which of the following reactions involve oxidation-reduction. For those which are redox, identify the elements oxidized and reduced and the oxidizing and reducing agents.

a) $2\ H_2C_2O_4(aq) + 2\ H_3AsO_4(aq) \rightarrow 4\ CO_2(g) + 5\ H_2O(\ell) + As_2O_3(s)$

b) $Ca(OCl)_2(aq) + 4\ HCl(aq) \rightarrow 2\ Cl_2(aq) + CaCl_2(aq) + 2\ H_2O(\ell)$

3) Write half-reaction equations for:

a) $SiO_2 \rightarrow Si$ in acidic solution b) $BrO^- \rightarrow Br^-$ in basic solution

4) Write net ionic equations for:

a) $MnO_4^- + CH_3OH \rightarrow MnO_4^{2-} + CO_3^{2-}$ (basic)

b) $H_2C_2O_4 + MoO_4^{2-} \rightarrow CO_2 + Mo^{3+}$ (acidic)

c) $Zn + NO_3^- \rightarrow Zn^{2+} + NH_4^+$ (acidic)

17. COLLIGATIVE PROPERTIES OF SOLUTIONS

PREREQUISITES

1. The distinction between strong electrolytes, weak electrolytes, and nonelectrolytes (Section 15.2) influences the colligative properties of solutions.

2. Osmotic pressure problems use an equation very similar to the ideal gas equation, introduced in Section 12.3. The ideal gas constant, R, is used in exactly the same way.

3. The colligative properties of a solution are related to the solution concentration expressed in mole fraction, molality, and molarity, which were studied in Sections 14.2 to 14.4.

CHEMICAL SKILLS

The following chemical skills are limited to liquid solutions of nonvolatile solutes. Solutes are also nonelectrolytes, unless stated otherwise.

1. Vapor pressure lowering. Given any two of the following, find the third: (a) the vapor pressure of the pure solvent; (b) the mole fraction of the solute (or data from which it can be determined); (c) the vapor pressure of the solution.

2. Boiling point elevation and freezing point depression. Given any three of the following, calculate the fourth: (a) the molal concentration of the solute; (b) the freezing/boiling point of the solvent; (c) the freezing/boiling point of the solution; (d) the molal freezing/boiling point constant for the solvent.

3. Molar mass by freezing/boiling point. Given (a) the mass of solvent and solute in a solution; (b) the normal freezing/boiling point of the solvent; (c) the freezing/boiling point of the solution; and (d) the molal freezing/boiling point constant of the solvent, find the approximate molar mass of the solute.

4. Solution of an electrolyte. Given any four of the following, calculate the fifth: (a) the freezing/boiling point of the solution of an electrolyte; (b) the molality of the solute (or data from which it can be determined); (c) the freezing/boiling point of the solvent; (d) the molal freezing/boiling point constant of the solvent; (e) the van't Hoff i factor.

5. Osmotic pressure. Given any two of the following, calculate the third: (a) the osmotic pressure; (b) the molarity of the solute (or data from which it can be determined); (c) the temperature.

6. Molar mass by osmotic pressure. Given the grams of solute in a solution of known volume, the osmotic pressure of the solution, and the temperature at which it is measured, calculate the molar mass of the solute.

17.1 THE NATURE OF COLLIGATIVE PROPERTIES

A solution is a homogeneous mixture of two or more pure substances. The major component is the solvent; the minor components are the solutes. The solvents in this chapter will be limited to liquids. Solutes will be limited to nonvolatile substances, that is, those which do not have an appreciable vapor pressure. The solutes will be assumed to be nonelectrolytes unless stated otherwise.

The physical properties of a solution are different from the physical properties of its components. In dilute solutions, we tend to compare the properties of the solution with the properties of the solvent that makes up all but a small part of the solution. Two observations may be made about such comparisons:

1. The difference between the value of the property of the solution and its value for the pure solvent is proportional to the concentration of the solute.
2. The concentrations on which the value of the property depends must be expressed in numbers of particles. It makes no difference whether the particles are ionic or molecular.

It is the second of these observations that classifies a property as **colligative**: the property depends entirely on the concentration of solute particles in the solution, and has nothing to do with what those particles are.

The quantitative relationships described in this chapter are accurate for an **ideal solution**. There is no interaction between dissolved particles in an ideal solution. In real solutions, particle interaction is least in dilute solutions of molecular solutes. Unless stated otherwise, we will assume that all solutions discussed in this chapter behave as ideal solutions.

The four major colligative properties of solutions of nonelectrolytes are:

1. Vapor pressure lowering. The vapor pressure of a solution is *lower* than that of the pure solvent.
2. Boiling point elevation. The boiling point of a solution is *higher* than that of the pure solvent.
3. Freezing point depression. The freezing point of a solution is *lower* than that of the pure solvent.
4. Osmotic pressure. This is a property possessed only by solutions. Pure solvents do not have an osmotic pressure.

17.2 THE VAPOR PRESSURE OF A SOLUTION

Liquids vaporize. Some liquids vaporize more readily than others. Substances that vaporize easily are said to be **volatile**. If a volatile liquid is allowed to vaporize in a closed container, its vapor accumulates in the gas space above the liquid. As a gas, this vapor adds its partial pressure to the initial pressure of the gas. As the vapor concentration increases, some of the substance returns to the liquid state, exactly the opposite of vaporization. This vapor-to-liquid change is called **condensation**.

Eventually the rate of evaporation becomes equal to the rate of condensation; a state of **equilibrium** is reached. The vapor concentration then becomes constant, as does its partial pressure. This partial pressure is called the **equilibrium vapor pressure**, or simply the **vapor pressure** of the substance.

It has been found experimentally that the vapor pressure of a solution of a nonvolatile solute is lower than the vapor pressure of the pure solvent. The relationship is known as Raoult's law. Specifically, the vapor pressure of the solution is directly proportional to the mole fraction of the solvent, $X_{solvent}$. The proportionality constant is the vapor pressure of the pure solvent, P°.

Recall that the mole fraction of a component of a solution, $X_{component}$, is the ratio of moles of that component to the total moles of all components. For a two-component solution,

$$X_{solute} = \frac{\text{mol solute}}{\text{mol solute + mol solvent}}; \quad \text{and}$$

$$X_{solvent} = \frac{\text{mol solvent}}{\text{mol solute + mol solvent}} \qquad (17.1)$$

If P is the vapor pressure of the solution, the above proportionality and its resulting equation are

$$P \propto X_{solvent} \quad \text{and} \quad P = P°X_{solvent} \qquad (17.2)$$

Application of Equation 17.2 is a straightforward substitute-and-solve procedure:

EXAMPLE 17.1: The vapor pressure of water is 23.76 torr at 25.0°C. Calculate the vapor pressure of a solution of 7.50 grams of formamide, $HCONH_2$, in 1.50×10^2 grams of water.

The mole fraction of water is needed. Calculate that value first.

GIVEN: 150 g H_2O, 7.50 g $HCONH_2$ FIND: mol H_2O, mol $HCONH_2$
FACTOR: MM of H_2O and $HCONH_2$ PATH: g → mol for H_2O and $HCONH_2$

EQN: $X_{H_2O} = \frac{\text{mol } H_2O}{\text{mol } H_2O + \text{mol } HCONH_2}$

$$1.50 \times 10^2 \text{ g } H_2O \times \frac{1 \text{ mol } H_2O}{18.02 \text{ g } H_2O} = 8.32 \text{ mol } H_2O$$

$$7.50 \text{ g } HCONH_2 \times \frac{1 \text{ mol } HCONH_2}{45.04 \text{ g } HCONH_2} = 0.167 \text{ mol } HCONH_2$$

$$X_{H_2O} = \frac{8.32 \text{ mol } H_2O}{8.32 \text{ mol } H_2O + 0.167 \text{ mol } HCONH_2} = \frac{8.32 \text{ mol}}{8.49 \text{ mol}} = 0.980$$

Actually, the mole fraction of the solvent is an intermediate answer that need not be calculated. Hereafter we will use the fractional setup for this value, beginning as we substitute into Equation 17.2 to find the vapor pressure of the solution. Why don't you do so too?

GIVEN: P° = 23.76 torr, X_{H_2O} = 0.980 FIND: P

EQN: $P = P°X_{H_2O} = 23.76 \text{ torr} \times \frac{8.32 \text{ mol}}{8.49 \text{ mol}} = 23.3 \text{ torr}$

The reduction in vapor pressure from that of the pure solvent to that of the solution is called the **vapor pressure lowering** and symbolized as ΔP. It can be demonstrated experimentally and mathematically that

$$\Delta P \propto X_{solute} \quad \text{and} \quad \Delta P = P°X_{solute} \tag{17.3}$$

Equation 17.3 gives you a way to calculate vapor pressure lowering directly. And by subtracting that value from the vapor pressure of the pure solvent, you have an alternative way to calculate the vapor pressure of the solution:

$$P = P° - \Delta P \tag{17.4}$$

EXAMPLE 17.2: Use Equation 17.3 to calculate the vapor pressure lowering of the solution in Example 17.1: 7.50 g $HCONH_2$ in 1.50×10^2 g H_2O. (P° = 23.76 torr at 25°C.) Then use the result in Equation 17.4 to confirm that the vapor pressure of the solution is 23.3 torr, as found in Example 17.1.

In Example 17.1 you found that this solution contains 8.32 mol H_2O and 0.167 mol $HCONH_2$. Calculate first the difference between the vapor pressure of the solution and the vapor pressure of water (ΔP).

GIVEN: 8.32 mol H_2O, 0.167 mol $HCONH_2$, $P° = 23.76$ torr FIND: ΔP, P

EQN: $\Delta P = P°X_{solute} = (0.167/8.49) \times 23.76 \text{ torr} = 0.467 \text{ torr}$

You now know the vapor pressure of the pure solvent and how much the vapor pressure is reduced by the presence of the solute. Find the vapor pressure of the solution.

$P = P° - \Delta P = 23.76 \text{ torr} - 0.467 \text{ torr} = 23.29 \text{ torr}$

This is from Equation 17.4. The 23.29-torr result obtained by calculating ΔP matches the 23.3 torr found directly from Equation 17.1.

17.3 BOILING POINT ELEVATION; FREEZING POINT DEPRESSION

A liquid boils at the temperature at which its vapor pressure is very slightly higher than the surrounding pressure. If the surrounding pressure is 1.00 atmosphere, the boiling temperature is the **normal boiling point**, T_b°.

We have seen that the vapor pressure of a solution of a nonvolatile solute is lower than the vapor pressure of the solvent. Therefore, it takes a higher temperature to raise the vapor pressure of the solution to the level of the surrounding pressure. This makes the boiling point of the solution, T_b, higher than the boiling point of the pure solvent, T_b°. The difference between these temperatures is called the **boiling point elevation**, ΔT_b:

$$\Delta T_b = T_b(\text{solution}) - T_b^\circ(\text{solvent}) \qquad (17.5)$$

Boiling point elevation is proportional to the molal concentration of the solute. The proportionality constant, K_b, is a property of the solvent called the **molal boiling point constant**. The proportionality and the equation derived from it are

$$\Delta T_b \propto m \qquad \text{and} \qquad \Delta T_b = K_b m \qquad (17.6)$$

The units of K_b, or K_f are found by solving Equation 17.6 for the constant:

$$K_b = \frac{\Delta T_b}{m} = \frac{°C}{m} = \frac{°C}{\text{mol solute/kg solvent}} = \frac{°C \cdot \text{kg solvent}}{\text{mol solute}}$$

The unit, °C•kg solvent/mol solute, is usable with the dimensional analysis method of solving problems, but it is awkward. °C/m is an abbreviated unit which cannot be used in dimensional analysis, but works fine if Equation 17.6 calculations are not combined with other calculations. We will observe this separation.

The freezing point of a solution, T_f, is also different from the freezing point of the pure solvent, T_f°. Solutions freeze at temperatures lower than their pure solvents. The difference between the two freezing points is called the **freezing point depression**, ΔT_f. It, too, is proportional to the molality of the solution. The proportionality constant is another property of the solvent. It is called the **molal freezing point constant**, K_f. The equation that corresponds to Equation 17.6 is

$$\Delta T_f \propto m \qquad \text{and} \qquad \Delta T_f = K_f m \qquad (17.7)$$

Molal boiling and freezing point constants can be determined experimentally from concentrations and boiling and freezing temperatures.

EXAMPLE 17.3: The normal freezing point of benzene, C_6H_6, is 5.50°C. When 7.14 grams of naphthalene, $C_{10}H_8$, are dissolved in 45.0 grams of benzene, the solution freezes at -0.82°C. Calculate the molal freezing point constant of benzene from these data.

Equation 17.7, $\Delta T_f = K_f m$, can be solved for the desired molal freezing point constant:

$$K_f = \frac{\Delta T_f}{m}$$

To find K_f we need both ΔT_f and m. Let's start with ΔT_f.

GIVEN: $T_f^\circ = 5.50°C$, $T_f = -0.82°C$ FIND: ΔT_f

EQN: $\Delta T_f = T_f^\circ - T_f = 5.50°C - (-0.82°C) = 6.32°C$

Molality can be calculated from the grams of solute and grams of solvent that are given in the problem.

GIVEN: 7.14 g $C_{10}H_8$, 45.0 g benzene FIND: m = mol $C_{10}H_8$/kg benzene

FACTOR: 128.16 g/mol $C_{10}H_8$ PATH: g $C_{10}H_8$ → mol $C_{10}H_8$

$$\text{EQ: } m = \frac{\text{mol } C_{10}H_8}{\text{kg benzene}} = \frac{7.14 \text{ g } C_{10}H_8 \times \dfrac{1 \text{ mol } C_{10}H_8}{128.16 \text{ g } C_{10}H_8}}{0.0450 \text{ kg benzene}} = 1.24 \text{ m } C_{10}H_8$$

The units of the freezing point constant are °C/m. You have both numbers. Calculate K_f.

GIVEN: $\Delta T_f = 6.32°C$, m = 1.24 m FIND: K_f

$$\text{EQN: } K_f = \frac{\Delta T_f}{m} = \frac{6.32°C}{1.24 \text{ m}} = 5.10 \text{ °C/m}$$

The molal boiling and freezing point constants for water are 0.512 °C/m and 1.86 °C/m. This means that a 1 molal aqueous solution will boil at 100.512°C, which is 0.512°C above its normal boiling point. The solution will freeze at -1.86°C, which is 1.86°C below the normal freezing point. These constants are used often and should be memorized. The constants for other solvents will be given as they are needed.

EXAMPLE 17.4: What are the boiling and freezing temperatures of the solution of Example 17.1? It contains 7.50 grams of $HCONH_2$ in 1.50×10^2 grams of water.

The molality of the solution appears in both calculations. Find that first.

GIVEN: 7.50 g $HCONH_2$, 150 g H_2O FIND: m
FACTOR: 45.04 g/mol $HCONH_2$ PATH: g $HCONH_2$ → mol $HCONH_2$

$$\text{EQN:} \quad m = \frac{\text{mol } HCONH_2}{\text{kg } H_2O} = \frac{7.50 \text{ g } HCONH_2 \times \dfrac{1 \text{ mol } HCONH_2}{45.04 \text{ g } HCONH_2}}{0.150 \text{ kg } H_2O} = 1.11 \text{ m } HCONH_2$$

Direct substitution into Equations 17.6 and 17.7 yields the changes in the boiling and freezing points. Calculate both of the wanted temperatures.

GIVEN: $K_b = 0.512$ °C/m, $T_b^\circ = 100.000$°C, FIND: T_b, T_f
$K_f = 1.86$ °C/m, $T_f^\circ = 0.000$°C

EQN: $\Delta T_b = K_b m = 0.512 \text{ °C/m} \times 1.11 \text{ m} = 0.568\text{°C}$

$\Delta T_f = K_f m = 1.86 \text{ °C/m} \times 1.11 \text{ m} = 2.06\text{°C}$

$T_b = 100.000\text{°C} + 0.568\text{°C} = 100.568\text{°C}$ $\quad T_f = 0.00\text{°C} - 2.06\text{°C} = -2.06\text{°C}$

Note that ΔT_b is *added* to the boiling point of the solvent to elevate boiling point, while ΔT_f is *subtracted* from the freezing point of the solvent to lower the freezing point.

Freezing or boiling point data may be used to estimate the molar mass of an unknown solute. A solvent is selected whose freezing or boiling point and molal freezing or boiling point constant are known. A measured mass of solute is dissolved in a measured mass of solvent. The freezing or boiling point of the solution is determined. The calculation method is as follows:

1. Use $\Delta T_b = K_b m$ or $\Delta T_f = K_f m$ (Equation 17.6 or 17.7) to calculate the molality of the solution.
2. Calculate the moles of solute from the defining equation for molality, $m \equiv$ mol solute/kg solvent.
3. Divide grams of solute by moles of solute to get molar mass.

EXAMPLE 17.5: Pure benzene freezes at 5.50°C and boils at 80.1°C. 10.9 grams of an unknown solute are dissolved in 75.8 grams of benzene. The solution freezes at 1.44°C and boils at 82.1°C. Estimate the molar mass of the solute by both the boiling point elevation and the freezing point depression. K_b = 2.53 °C/m and K_f = 5.12 °C/m for benzene.

Let's solve the problem first with the freezing point depression. Step 1 is to calculate the molality from Equation 17.7.

GIVEN: T_f = 1.44°C, T_f° = 5.50°C, K_f = 5.12 °C/m FIND: $m = \Delta T_f/K_f$

EQN: $m = \dfrac{\Delta T_f}{K_f} = \dfrac{5.50°\text{C} - 1.44°\text{C}}{5.12\ °\text{C/m}} = 0.793\ \text{m}$

Step 2. You now know that you have 0.793 moles of solute per kilogram of solvent. You also know the mass of solvent in the solution. Calculate the moles of solute present.

GIVEN: 75.8 g = 0.0758 kg benzene (solvent) FIND: mol solute
FACTOR: 0.793 mol solute/kg benzene PATH: kg solvent → mol solute

$$0.0758\ \text{kg solvent} \times \frac{0.793\ \text{mol solute}}{\text{kg solvent}} = 0.0601\ \text{mol solute}$$

Step 3. You now have the mass of a sample of solute and the number of moles in that sample. Calculate molar mass of solute.

GIVEN: 10.9 g solute, 0.0601 mol solute FIND: MM solute

EQN: $MM \equiv \frac{\text{g solute}}{\text{mol solute}} = \frac{10.9 \text{ g solute}}{0.0601 \text{ mol solute}} = 181 \text{ g/mol}$

Now repeat the procedure with the boiling points and the K_b constant. The solvent boils at 80.1°C, the solution at 82.1°C, and K_b = 2.53 °C/m. The solution has 10.9 g solute dissolved in 75.8 g benzene. Carry your work all the way to the molar mass.

GIVEN: 10.9 g solute, 75.8 g benzene, T_b = 82.1°C, T_b° = 80.1°C, K_b = 2.53 °C/m FIND: MM EQN: $m = \Delta T_b/K_b$

FACTOR: mol solute/kg solvent PATH: kg solvent → mol solute

Step 1: $m = \frac{82.1\text{°C} - 80.1\text{°C}}{2.53 \text{ °C/m}} = 0.79 \text{ m}$

Step 2: $0.0758 \text{ kg solvent} \times \frac{0.79 \text{ mol solute}}{\text{kg solvent}} = 0.060 \text{ mole solute}$

Step 3: $MM = \frac{10.9 \text{ g solute}}{0.060 \text{ mol solute}} = 1.8 \times 10^2 \text{ g/mol}$

The boiling temperature data limit the result to two significant figures, whereas the freezing points allow three. Otherwise the molar masses check.

A common application of freezing point lowering is in the winterizing of automobile cooling systems. Enough "antifreeze" must be present to prevent freezing at the lowest anticipated temperature. Ethylene glycol is the substance most widely used for this purpose.

EXAMPLE 17.6: In what proportions should ethylene glycol, $C_2H_6O_2$, be added to water to yield a solution that will freeze at -35°C?

We will find the mass of $C_2H_6O_2$ that must be added to 1 kilogram of water. The molality—moles of $C_2H_6O_2$ per kilogram of water—can be calculated from the data, using Equation 17.7, $\Delta T_f = K_f m$, where K_f = 1.86 °C/m. First, find the molality of the needed solution.

GIVEN: $T_f = -35°C$, $T_f^\circ = 0°C$, $K_f = 1.86\ °C/m$ FIND: m EQN: $\Delta T_f = K_f m$

$$m = \frac{\Delta T_f}{K_f} = \frac{0°C - (-35°C)}{1.86\ °C/m} = 19\ m$$

19 moles of $C_2H_6O_2$ must be dissolved in each kilogram of water. Calculate the mass of the solute.

$$19 \text{ mol } C_2H_6O_2 \times \frac{62.07 \text{ g } C_2H_6O_2}{1 \text{ mol } C_2H_6O_2} = 1.2 \times 10^3 \text{ g } C_2H_6O_2 = 1.2 \text{ kg } C_2H_6O_2$$

1.2 kg of $C_2H_6O_2$ are to be mixed with 1.0 kg H_2O to produce a solution that will freeze at -35°C. If these masses were converted to volumes by dividing by their densities, the volumes would be nearly equal, which is exactly the mixture recommended by one of the major makers of radiator coolant.

17.4 SOLUTES THAT ARE ELECTROLYTES

Nonelectrolytes are the only kinds of solutes we have considered so far in this chapter. Each mole of a nonelectrolyte that dissolves yields a mole of solute particles. The "solute molality" is the same as the "particle molality."

If a strong electrolyte such as NaCl dissolves, the equation for the dissolving reaction, $NaCl(s) \rightarrow Na^+(aq) + Cl^-(aq)$, predicts that 1 mole of solute forms 1 mole of Na^+ ions and 1 mole of Cl^- ions. The solution will contain 2 moles of ions. The particle molality will be twice the solute molality. The freezing point depression and boiling point elevation should therefore be twice as much as would be calculated for a nonelectrolyte of the same solute molality.

Similar reasoning predicts that an electrolyte that yields 3 moles of ions per mole of solute, such as $CaCl_2$, should have three times the boiling and freezing point changes. $AlCl_3$ yields 4 moles of ions per mole of compound, so the boiling and freezing point changes should be four times as great. And so on. . . .

The ratio of moles of particles formed to moles of solute added is known as the **van't Hoff i factor**, i:

$$i \equiv \frac{\text{moles of particles in solution}}{\text{moles of solute dissolved}} \tag{17.8}$$

The particle molality can be obtained by multiplying the solute molality by the van't Hoff i factor. This produces modified forms of the boiling point elevation and freezing point depression relationships (Equations 17.6 and 17.7) as follows:

$$\Delta T_b = imK_b \tag{17.9}$$

$$\Delta T_f = imK_f \tag{17.10}$$

The van't Hoff i factor is approximately equal to the number of ions in the formula of a strong electrolyte. It is equal to 1 for a nonelectrolyte, and is just over 1 for a weak electrolyte.

In actual practice, the van't Hoff i factor, and thus the boiling and freezing point changes, are smaller for most strong electrolytes than the dissolving equations predict. This is because the solutions do not behave as "ideal" solutions. The oppositely charged ions attract each other, forming "ion pairs" that behave as single particles. This reduces the effective particle molality of the solution.

EXAMPLE 17.7: Assuming that NaCl dissociates completely into ions, calculate the expected freezing point of 1.00 m NaCl.

As noted above, 1 mole of NaCl produces 2 moles of ions when it dissolves. If there is *complete* dissociation into ions, the van't Hoff i factor is exactly 2, and the particle concentration of the solution is therefore twice 1.00 molal, or 2.00 m. With this in mind, calculate the expected freezing point depression, and the freezing point of the solution.

GIVEN: m = 1.00 m, i = 2, K_f = 1.86 °C/m

FIND: T_f

EQN: $\Delta T_f = imK_f$, $T_f = T_f^\circ - \Delta T_f$

$\Delta T_f = 2 \times 1.00 \text{ m} \times 1.86\ ^\circ\text{C/m} = 3.72^\circ\text{C}$ $T_f = 0^\circ\text{C} - 3.72^\circ\text{C} = -3.72^\circ\text{C}$

Equations 17.9 and 17.10 may be used to calculate experimental values of the van't Hoff i factor.

EXAMPLE 17.8: Calculate the van't Hoff i factor for 0.0500 m HF if it freezes at -0.103°C.

All that is needed is to rearrange Equation 17.10 to solve for i.

GIVEN: m = 0.0500 m, T_f = -0.103°C, T_f° = 0.000°C, K_f = 1.86 °C/m

FIND: i

EQN: $\Delta T_f = imK_f$

$$i = \frac{\Delta T_f}{mK_f} = \frac{0.000^\circ\text{C} - (-0.103^\circ\text{C})}{0.0500 \text{ m} \times 1.86\ ^\circ\text{C/m}} = 1.11$$

The value is slightly more than 1, as expected for a weak acid.

We might take a moment to notice the significance of the 1.11 van't Hoff i factor in the dissociation of 0.0500 m HF. It indicates that, for each mole of HF dissolved, 1.11 solute particles are present. The particles are H^+ ions, F^- ions, and undissociated HF molecules. The question is, how many of each particle are in the solution?

To answer this question, assume that the solution starts with 1.00 mole of HF molecules. Assume also that x mole dissociates, yielding x mole of H^+ ion and x mole of F^- ion. The fraction of a mole of undissociated HF that remains is 1.00 - x. Using formulas in the dissociation equation as column heads, as we did for limiting reactant problems in Section 10.4, we can show this a bit more clearly:

	HF(aq)	→	H^+(aq)	+	F^-(aq)
Moles at start	1.00		0.00		0.00
Moles lost (-) or formed (+) by dissociation	-x		+x		+x
Moles present in solution	1.00 - x		x		x

The total number of moles present in the final solution is

$$(1.00 - x) \quad + \quad x \quad + \quad x \quad = \quad 1.00 + x$$

$$\text{mol HF} \quad + \text{ mol } H^+ \quad + \text{ mol } F^- \quad = \quad \text{total mol}$$

For a solute that dissociates into two ions, that portion of the i factor after the decimal is the number of moles of solute that dissociate per mole of initial solute. In 0.0500 m HF, for which i = 1.11, x = 0.11. This means that 0.11 mole of HF dissociates per mole of HF solute dissolved. Expressing 0.11 parts per mole as parts per 100 moles gives percentage; the HF is 11% dissociated. This confirms our idea of a weak acid, one that is only slightly dissociated.

17.5 OSMOTIC PRESSURE

A semipermeable membrane is a thin layer of material that small, uncharged particles, such as water molecules, can pass through by a process called **osmosis**. Suppose such a membrane is used to separate pure water from a solution whose solute molecules cannot pass through the membrane. The result will be a net flow of water molecules from the pure water into the solution. This flow can be stopped, and even reversed, by applying a mechanical pressure to the solution. The pressure required to just stop the flow is called **osmotic pressure**.

Osmotic pressure, π, is related to the molarity of the solution, M, and absolute temperature, T, by the equation

$$\pi = MRT \qquad (17.11)$$

R is the universal gas constant introduced in Section 12.3. In fact, the equation is very similar to the ideal gas equation, Equation 12.15. M, molarity, is measured in moles per liter, which is n/V in the gas equation. Osmotic pressure, π, substitutes for gas pressure, P. The equation is used in exactly the same way.

EXAMPLE 17.9: Calculate the osmotic pressure that can be developed by a 0.0100 molar sugar solution at 25°C. Answer in atmospheres.

This is a direct substitution into Equation 17.11. Recall that the ideal gas constant, R, is 0.08205 L•atm/mol•K or 62.36 L•torr/mol•K. Celsius temperature is converted to kelvins by adding 273.

GIVEN: 0.0100 M, 25°C = 298 K FIND: osmotic pressure, π

EQN: $\pi = MRT = \frac{0.0100 \text{ mol}}{1 \text{ L}} \times \frac{0.08205 \text{ L}\cdot\text{atm}}{\text{mol}\cdot\text{K}} \times 298 \text{ K} = 0.245 \text{ atm}$

Osmotic pressure offers a sensitive way to estimate high molar masses from extremely dilute solutions. A measured mass of solute is dissolved in a measured volume of solution. The osmotic pressure of the solution is measured in the laboratory at a certain temperature. The data are then treated in essentially the same way as when finding molar mass by freezing or boiling point changes:

Step 1: Molarity, moles per liter, is found from Equation 17.11, $\pi = MRT$.
Step 2: Moles of solute are calculated from volume × molarity.
Step 3: Molar mass is found by dividing grams by moles.

EXAMPLE 17.10: A solution of 17 grams of a protein in water is diluted to 5.0×10^2 milliliters. At 22°C it has an osmotic pressure of 11 torr. Calculate the molar mass of the solute.

The procedure is given above. Complete all three steps.

Step 1. Concentration of solute.

Step 2. Moles of solute.

Step 3. Molar mass of solute.

Step 1. GIVEN: π = 11 torr, T = 22°C = 295 K FIND: M EQN: π = MRT

$$M = \frac{\pi}{RT} = \frac{11 \text{ torr}}{(62.36 \text{ L·torr/mol·K}) \times 295 \text{ K}} = 6.0 \times 10^{-4} \text{ mol/L}$$

Step 2. GIVEN: 0.50 L FIND: mol protein
FACTOR: 6.0×10^{-4} mol/L PATH: L → mol

$$0.50 \text{ L} \times \frac{6.0 \times 10^{-4} \text{ mol protein}}{\text{L}} = 3.0 \times 10^{-4} \text{ mol protein}$$

Step 3. GIVEN: 17 g protein, 3.0×10^{-5} mol protein FIND: MM

$$\text{EQN: MM} = \frac{17 \text{ g protein}}{3.0 \times 10^{-4} \text{ mol protein}} = 5.7 \times 10^{4} \text{ g protein/mol}$$

CHAPTER 17 PROBLEMS

1) The vapor pressure of water is 26.7 torr at 27°C. Find the vapor pressure at that temperature of a solution of 47.3 g $C_{12}H_{22}O_{11}$ in 5.00×10^2 grams of water.

2) At 20°C the vapor pressure of methanol, CH_3OH, is 94 torr. If 15.4 grams of urea, $CO(NH_2)_2$, are dissolved in 216 grams of methanol, what will be the vapor pressure of that solution?

3) 37.8 grams of oxalic acid, $H_2C_2O_4$, are dissolved in 120.0 grams of ethanol, C_2H_5OH. Find the vapor pressure of the solution at 19°C if the vapor pressure of pure ethanol is 4.0×10^1 torr at the same temperature.

4) 46.2 grams of urea, $CO(NH_2)_2$, are dissolved in 145 grams of water. Calculate the boiling and freezing points of this solution.

5) The molal boiling point constant of ethanol, C_2H_5OH, is 1.22 °C/m, and its normal boiling point is 78.5°C. Find the boiling point of a solution of 15.4 grams of glycerol (MM = 92.1 g/mol) in 64.3 grams of ethanol. Consider glycerol to be a nonvolatile solute.

6) Camphor freezes at 178°C, and 40.0 °C/m is its molal freezing point constant. Calculate the freezing point of a 10.0% solution of a certain solute in camphor if the molar mass of the solute is 135 g/mol.

7) How many grams of glucose, $C_6H_{12}O_6$, must be dissolved in 115 grams of acetic acid to lower its freezing point to 12.0°C? The normal freezing point of acetic acid is 16.6°C, and its molal freezing point constant is 3.90 °C/m.

8) The boiling point of carbon tetrachloride, CCl_4, is 76.5°C, and the molal boiling point constant is 5.03 °C/m. If 5.64 grams of an unknown nonelectrolyte are dissolved in 65.8 grams of carbon tetrachloride, the boiling point increases to 79.1°C. Estimate the molar mass of the solute.

9) A student finds that the freezing point of a solution of 2.78 grams of an unknown nonelectrolyte in 88.1 grams of naphthalene is 76.4°C. Pure naphthalene freezes at 80.1°C, and its molal freezing point constant is 6.9 °C/m. What is the apparent molar mass of the solute?

10) What would be the maximum freezing point depression of a 0.20 molal solution of potassium chloride in water if the solute was 100% ionized?

11) The freezing point depression of 0.545 m aqueous potassium carbonate is 2.40°C. Calculate the van't Hoff i factor. What would the value of i be for 100% dissociation?

12) A 2.00% aqueous lithium chloride solution freezes at -1.72°C. Find the van't Hoff i factor. What would the value of i be for 100% dissociation?

13) The observed freezing point of 0.0200 m $HNO_2(aq)$ is -0.0424°C. Nitrous acid is a weak acid that ionizes but slightly in higher concentrations. Calculate the van't Hoff i factor. What is the apparent percentage ionization?

14) Calculate the osmotic pressure that develops in 0.025 M $C_2H_6O_2$ at 20°C. Answer in torr.

15) Find the osmotic pressure that would exist with 0.30 M KCl at 27°C. Assume 100% dissociation of the solute. (Hint: 100% dissociation means that 1 mole of solute will yield 2 moles of ions, or particles. Recall that colligative properties are related to particle concentration.)

16) A 100.0 mL-solution contains 4.18 grams of an unknown protein. Calculate the molar mass of the protein if the osmotic pressure is 31.2 torr at 18°C.

17) "Sugar" is the name of a whole class of organic compounds. A student was given a bottle labeled "SUGAR" and asked to find its molar mass. She prepared a 2.50×10^2-milliliter solution that contained 4.25 grams of the sugar. She then found that the solution had an osmotic pressure of 2.29 atmospheres at 22°C. Find the molar mass of the sugar.

++++++++++++++++++++++

18) Find the vapor pressure of a solution of 65.1 grams of $C_6H_{12}O_6$ in 115 grams of water at 29°C. The vapor pressure of water at 29°C is 30.0 torr.

19) What is the vapor pressure at -13°C of a solution of 39.2 grams of glutaric acid, $H_2C_3H_2O_4$, in 165 grams of ether, $(C_2H_5)_2O$. The vapor pressure of pure ether is 104 torr at that temperature.

20) Calculate the boiling and freezing points of a solution prepared by dissolving 18.0 grams of analine, $C_6H_5NH_2$, in 109 grams of water.

21) 5.29 grams of formamide, $CHONH_2$, are dissolved in 85.0 grams of acetone, CH_3COCH_3. Assuming formamide to be nonvolatile, calculate the boiling point of the solution. The normal boiling point of acetone is 56.2°C, and its molal boiling point constant is 1.71 °C/m.

22) The molal boiling point constant of carbon disulfide, CS_2, is 2.34 °C/m, and it boils at 46.1°C. The boiling point of a solution of 4.71 grams of an unidentified solute in 71.2 grams of CS_2 is 49.2°C. What is the approximate molar mass of the solute?

23) A 0.123 molal solution of potassium iodide freezes at -0.431°C. What is the van't Hoff i factor? What would the value of i be for 100% dissociation?

24) The freezing point depression of 2.00% iron(III) chloride is 0.753°C. Find the van't Hoff i factor. What would the value of i be for 100% dissociation?

25) Find the osmotic pressure at 24°C of a 250.0 mL-solution that contains 3.81 grams of urea, $CO(NH_2)_2$. Answer in atmospheres.

26) A 5.00×10^2-mL solution of 11.2 grams of an unknown organic compound has an osmotic pressure of 13.1 torr at 21°C. Find the molar mass of the solute.

ADDITIONAL PROBLEMS

27) When 53.2 mg of an unknown nonelectrolyte are dissolved in 5.632 g of water, the vapor pressure of the water is lowered from 21.063 torr to 20.959 torr. a) What is the mole fraction of the water, X_{H_2O}? b) How many moles of unknown, n_{unkn}, are present? Solve the equation below for n_{unkn}.

$$X_{H_2O} = \frac{n_{H_2O}}{n_{H_2O} + n_{unkn}}$$

c) What is the molar mass of the protein?

28) How cold must it be to freeze sea water, approximately 0.55 m NaCl? Answer in both °C and °F.

29) Sulfuric acid, H_2SO_4, is a strong acid and thus a strong electrolyte. However, both H^+'s do not completely dissociate. The first proton dissociates 100%.

$$H_2SO_4(aq) \xrightarrow{100\%} H^+(aq) + HSO_4^-(aq)$$

However, the second proton only partly dissociates.

$$HSO_4^-(aq) \xrightarrow{?\ \%} H^+(aq) + SO_4^{2-}(aq)$$

A 1.13 m H_2SO_4 solution has a boiling point of 101.28°C. a) What is the van't Hoff i factor? b) What is the % dissociation of the second H^+?

30) A certain ethanol/water solution has a freezing point of -25.0°C. What is its weight percent C_2H_5OH?

CHAPTER TEST

1) Acetone, CH_3COCH_3, is a volatile liquid, having a vapor pressure of 112 torr at 9.0°C. To what will the vapor pressure be reduced if 23.6 grams of naphthalene, $C_{10}H_8$, are dissolved in 75.0 grams of acetone?

2) Benzene freezes at 5.50°C. At what temperature does a solution of 21.6 grams of sulfur dichloride, SCl_2, in 350 grams of benzene freeze? The molal freezing point constant of benzene is 5.1 °C/m.

3) Urethane melts at 49.7°C. Its molal freezing point constant is 5.14 °C/m. If 84.2 g of an unknown are dissolved in 307 g of urethane, the solution freezes at 43.8°C. Calculate the approximate molar mass of the unknown.

4) Calculate van't Hoff i factor of 0.253 m $HC_2H_3O_2$ (acetic acid) which freezes at -0.474°C. What is the apparent percentage ionization?

5) A biologist isolates an unknown solid that is slightly soluble in water. He dissolves 8.73 grams of the chemical in 2.00×10^2 milliliters of solution. At 25°C the solution has an osmotic pressure of 45.1 torr. Find the molar mass of the compound.

18. EQUILIBRIUM: BASIC PRINCIPLES

PREREQUISITES

1. Equilibrium problems are usually solved algebraically by the methods described in Sections 1.9 to 1.11.

2. The significant figure rules given in Sections 3.3 to 3.7 are applied throughout equilibrium calculations.

3. The tabular thought process by which limiting reactant problems were solved in Section 10.4 appears again in solving equilibrium problems.

4. The partial pressure (Section 12.8) of a gaseous component in an equilibrium affects the equilibrium system.

5. You will use chemical equations that include energy terms, developed in Section 13.6, to predict the effect of heating or cooling an equilibrium.

CHEMICAL SKILLS

1. The equilibrium constant. Given the equilibrium concentrations of all substances in an equilibrium, calculate the equilibrium constant.

2. Predicting direction of reaction. Given the equilibrium constant of an equilibrium and all the initial concentrations, calculate Q and determine which way, forward or reverse, the reaction will proceed.

3. Le Chatelier's principle: concentration effect. From the equation for a reaction at equilibrium, given the concentration change (increase or decrease) of a reactant or product, or the direction of equilibrium shift (left or right), determine the other and determine the changes (increase or decrease) in all other concentrations.

4. Le Chatelier's principle: temperature effect. From the equation for a reaction at equilibrium, given any two of the following, determine the third: (a) the direction of temperature change (increase or decrease); (b) the sign of ΔH (exothermic or endothermic); (c) the direction of equilibrium shift (left or right) and the changes (increase or decrease) in all concentrations.

5. Le Chatelier's principle: pressure/volume effect. From the equation for a reaction at equilibrium, given the pressure or volume change, or the direction of equilibrium shift (left or right), determine the other.

6. Gas phase equilibrium: one reactant or product. Given the equilibrium constant of a gas phase equilibrium and information from which the initial concentration of a single reactant (or product) can be calculated, find the equilibrium concentrations of the products (or reactants).

7. Gas phase equilibrium: given all initial concentrations. Given the equilibrium constant of a gas phase equilibrium and all initial concentrations, determine which way the reaction will proceed (left or right), and calculate all equilibrium concentrations.

18.1 INTRODUCTION TO EQUILIBRIUM

If a chemical reaction occurs in a closed system, and if its products do not escape into the surroundings, the products may be able to react with each other and reproduce the starting reactants. A reaction that goes in both directions is called a **reversible reaction**. The equation for a reversible reaction is written with a double arrow:

$$A \rightleftharpoons B \tag{18.1}$$

The reaction toward the right, $A \rightarrow B$, is called the **forward reaction**, and the reaction toward the left, $A \leftarrow B$, is the **reverse reaction**.

In the imaginary reversible reaction 18.1, substance A is constantly being used to produce B, while B is constantly being used to produce A. Both substances are always present. When the *rate* of the forward reaction is exactly equal to the *rate* of the reverse reaction, the chemical change appears to stop. The reaction has then reached a state of **equilibrium**.

Notice that both reactions continue at equilibrium, but each substance is used just as fast as it is produced. Consequently, the total amount of each substance and its concentration remain constant in a given system. It should be emphasized that a chemical equilibrium is a dynamic process because the reverse processes continue indefinitely.

The important thing to recognize about equilibrium is that the "things" that are equal are the *rates* of the opposing reactions. The quantities and concentrations of reactants and products may be very different. One concentration may be several orders of magnitude larger than the other, but the system is still at equilibrium if the reaction rates are equal.

An equilibrium is said to be *favored in the forward direction* if the concentrations on the right side of the equation are larger than the concentrations on the left. The equilibrium is *favored in the reverse direction* if the left side concentrations are larger.

The rate of a given chemical reaction depends primarily on reactant concentration, temperature, and, if gases are present, pressure. In the case of gases, concentration may be expressed as partial pressure. In Section 18.3 we will consider how changes in concentration, temperature and pressure affect the position of an equilibrium.

18.2 THE EQUILIBRIUM CONSTANT

The Form of an Equilibrium Constant The following equation may be used to represent any chemical equilibrium:

$$wA + xB \rightleftharpoons yC + zD \qquad (18.2)$$

A, B, C, and D represent formulas of chemicals, and w, x, y, and z are their coefficients in the balanced equation. At any given temperature, it can be demonstrated both theoretically and experimentally that the following ratio is a constant:

$$K = \frac{[C]^y\ [D]^z}{[A]^w\ [B]^x} \qquad (18.3)$$

The symbol K is used to represent an equilibrium constant. A chemical symbol enclosed in square brackets refers to the concentration of the species in moles per liter.

The equilibrium constant expression is always related to a specific equilibrium equation. The numerator is the product of the concentrations of the substances on the *right* side of the equation, each raised to a power equal to its coefficient in the equation. The denominator is a similar product of the concentrations of the substances on the *left* side of the equation, again with each raised to the power equal to its coefficient. If the equation is written in reverse, the left and right sides change places, so the equilibrium constant expression must be inverted.

The equilibrium constant expression includes terms only for those species whose concentrations affect the forward or reverse reaction rates. Generally these are solute and gas concentrations. The amount of a solid or liquid does not affect reaction rates, so solids and liquids never appear in an equilibrium constant. Water, as the solvent in an aqueous solution equilibrium, has no influence on reaction rate, so it is excluded from constants for such solutions. *Only concentrations of solutes and gasses appear in K expressions.*

Concentrations of solutes in an equilibrium constant expression are always in moles per liter, but K itself is written without units. A variation of the equilibrium constant is often used for gases. Terms for gases may be expressed either as concentrations in moles per liter or as partial pressure in atmospheres. This can be done because, according to the ideal gas law, the partial pressure of a gas is proportional to its concentration in moles per liter. We will always use gas concentrations in moles per liter in this book.

Examples of Equilibrium Constants Consider the dissociation of acetic acid in water:

$$HC_2H_3O_2(aq) \rightleftharpoons H^+(aq) + C_2H_3O_2^-(aq) \qquad (18.4)$$

The equilibrium constant expression for this reaction is:

$$K = \frac{[H^+][C_2H_3O_2^-]}{[HC_2H_3O_2]} \qquad (18.5)$$

All substances are in aqueous solution, so all appear in K. All coefficients are 1, so all exponents of [ ] terms are 1.

Now consider the dissolving of solid calcium carbonate in hydrochloric acid. The net ionic equation is

$$CaCO_3(s) \ + \ 2\ H^+(aq) \ \rightleftharpoons \ Ca^{2+}(aq) \ + \ CO_2(g) \ + \ H_2O(\ell) \qquad (18.6)$$

No terms appear in K for $CaCO_3$, a solid, or for H_2O, a liquid.

$$K = \frac{[Ca^{2+}][CO_2]}{[H^+]^2} \qquad (18.7)$$

Because the coefficient of H^+ is 2 in the balanced equation, $[H^+]$ appears squared in the K expression.

EXAMPLE 18.1: Write equilibrium constant expressions for the following reactions:

(a) $CaO(s) + CO_2(g) \rightleftharpoons CaCO_3(s)$ K = ________

(b) $Ba_3(PO_4)_2(s) + 3\ CO_3^{2-}(aq) \rightleftharpoons 3\ BaCO_3(s) + 2\ PO_4^{3-}(aq)$ K = ________

(c) $Pb(s) + 2\ H^+(aq) + 2\ Cl^-(aq) \rightleftharpoons H_2(g) + PbCl_2(s)$ K = ________

(d) $2\ C_6H_6(\ell) + 15\ O_2(g) \rightleftharpoons 12\ CO_2(g) + 6\ H_2O(\ell)$ K = ________

(a) $CaO(s) + CO_2(g) \rightleftharpoons CaCO_3(s)$ $K = \frac{1}{[CO_2]}$

(b) $Ba_3(PO_4)_2(s) + 3\ CO_3^{2-}(aq) \rightleftharpoons 3\ BaCO_3(s) + 2\ PO_4^{3-}(aq)$ $K = \frac{[PO_4^{3-}]^2}{[CO_3^{2-}]^3}$

(c) $Pb(s) + 2\ H^+(aq) + 2\ Cl^-(aq) \rightleftharpoons H_2(g) + PbCl_2(s)$ $K = \frac{[H_2]}{[H^+]^2[Cl^-]^2}$

(d) $2\ C_6H_6(\ell) + 15\ O_2(g) \rightleftharpoons 12\ CO_2(g) + 6\ H_2O(\ell)$ $K = \frac{[CO_2]^{12}}{[O_2]^{15}}$

EXAMPLE 18.2: Consider the reaction $PCl_5(g) \rightleftharpoons PCl_3(g) + Cl_2(g)$. A quantity of PCl_5 was introduced to a closed 12-liter vessel and heated to 250°C until the above equilibrium was reached. At that time the vessel contained 0.21 mol PCl_5, 0.32 mol PCl_3, and 0.32 mol Cl_2. Calculate K for the equilibrium at 250°C.

Begin by writing the equilibrium constant expression from the given equation.

$$K = \frac{[PCl_3][Cl_2]}{[PCl_5]} = \text{\_\_\_\_\_\_\_\_} =$$

The problem gives you the number of moles of each species in the 12-liter vessel at equilibrium. The numbers you must substitute into the above equation are concentrations (not moles) in moles per liter. You may calculate the concentration of each species separately and then substitute into the K equation to find the value of K. Alternately, you can substitute moles/liter expressions for each concentration and calculate K without finding the intermediate answers. Take your choice, and complete the problem.

$$K = \frac{[PCl_3][Cl_2]}{[PCl_5]} = \frac{(0.32/12)(0.32/12)}{0.21/12} = \frac{0.027^2}{0.018} = 0.041$$

The Significance of the Equilibrium Constant An equilibrium constant is a fraction. If the forward reaction (to the right) of an equilibrium is favored, the product (numerator) concentrations are larger than the reactant (denominator) concentrations and $K > 1$. If the reverse reaction (to the left) is favored, $K < 1$.

The extent to which an equilibrium is favored in one direction rather than the other can be overwhelming, almost to the exclusion of the unfavored reaction. Reactions that occur to a very great extent, such as the combustion of hydrocarbons, have equilibrium constants that are very large numbers. Reactions that occur to an extremely small degree, such as the dissolving of a very slightly soluble salt in water (AgCl, for example), have very small equilibrium constants. We will often work with K values that are in the 10^{-14} range, and even smaller. Despite the very small concentrations that yield these constants, those concentrations have a marked effect on many industrial, physiological, and environmental chemical reactions.

The Reaction Quotient It is convenient to identify a product/reactant concentration ratio known as the **reaction quotient, Q**. The reaction quotient has exactly the same form as the equilibrium constant. For the general equation $wA + xB \rightleftharpoons yC + zD$,

$$Q = \frac{[C]^y\,[D]^z}{[A]^w\,[B]^x} \qquad Q = K \text{ only at equilibrium} \qquad (18.8)$$

Concentrations in the Q expression are not necessarily concentrations when the system is at equilibrium. Unlike K, Q does not have a fixed value. When the system is at equilibrium, however, $Q = K$.

Suppose A and B are brought together as reactants at some initial concentrations, and no C or D are present. Both [C] and [D] are zero, so Q is also zero—and smaller than K:

$$Q = \frac{[C]^y\,[B]^z}{[A]^w\,[B]^x} = \frac{0}{\text{large value}} = 0 < K$$

Only the forward reaction can occur. Thus the forward reaction rate is greater than the reverse reaction rate, which is zero. As the reaction begins, C and D are formed, so the reverse reaction begins. [C] and [D] acquire a value greater than zero, and [A] and [B] are reduced somewhat. The numerator of Q increases, the denominator decreases, and the value of the ratio acquires a small positive value:

$$Q = \frac{[C]^y\,[B]^z}{[A]^w\,[B]^x} = \frac{\text{increasing value}}{\text{decreasing value}} > 0 \text{ and increasing, but still} < K$$

The forward reaction rate is still greater than the reverse rate. Therefore the C and D (numerator) concentrations continue to become larger; the A and B (denominator) concentrations continue to become smaller; and Q continues to increase. As long as $Q < K$, the forward rate is greater than the reverse rate and the net reaction proceeds in the forward direction. Eventually the concentrations become such that the forward and reverse rates are equal and equilibrium is reached. At that time $Q = K$.

A similar analysis shows that whenever $Q > K$, exactly the opposite changes occur until equilibrium is reached. At a given temperature, the same Q ratio is reached regardless of the direction from which the equilibrium is approached.

Later in this chapter and in the next, we will refer to the relative values of Q and K to explain why a system moves in the forward or reverse direction. Remember,

> If $Q < K$, the net reaction proceeds in the forward direction—to the right.
>
> If $Q > K$, the net reaction proceeds in the reverse direction—to the left.

EXAMPLE 18.3: PCl_5, PCl_3, and Cl_2 are introduced to a closed flask at the following initial concentrations: $[PCl_5] = 0.060$ mol/L; $[PCl_3] = 0.080$ mol/L; and $[Cl_2] = 0.040$ mol/L. They react until reaching the equilibrium $PCl_5(g) \rightleftharpoons PCl_3(g) + Cl_2(g)$ at a temperature at which $K = 0.041$ (Example 18.2). In which direction, forward or reverse, will the net reaction proceed as equilibrium is approached?

The procedure is to find the value of Q, compare it with the value of K, and then decide which way the net reaction will go. Calculate Q.

GIVEN: $[PCl_5] = 0.060$ mol/L; $[PCl_3] = 0.080$ mol/L; and $[Cl_2] = 0.040$ mol/L

FIND: Q EQN: $Q = \frac{[PCl_3][Cl_2]}{[PCl_5]} = \frac{(0.080)(0.040)}{0.060} = 0.053$

Now you know that the initial value of Q is 0.053, and K is 0.041. In which direction will the reaction proceed? On what do you base your answer?

The reaction will proceed in the reverse direction because $Q > K$.

The product/reactant ratio given by Q is too high. It is reduced by lowering the numerator (product) concentrations and raising the denominator (reactant) concentration. In effect, the product is converted back to the reactant until $Q = K$ (equilibrium).

18.3 LE CHATELIER'S PRINCIPLE

Most of our attention in applying the principles of equilibrium over the next two chapters will be directed to ionic equilibria in water solution. The dissociation of acetic acid is typical:

$$HC_2H_3O_2(aq) \rightleftharpoons H^+(aq) + C_2H_3O_2^-(aq) \qquad (18.4)$$

When molecules of acetic acid are initially placed in water, no products are present. As the acetic acid dissociates, hydrogen and acetate ions form and the reverse reaction begins. Eventually equilibrium is reached, and the concentrations of acetic acid, hydrogen ion, and acetate ion become constant. These values, represented as $[HC_2H_3O_2]$, $[H^+]$, and $[HC_2H_3O_2]$, remain constant unless the equilibrium is disturbed.

Changes that disturb a reaction at equilibrium are: (a) a change in the concentration of a reactant or product in aqueous solution, (b) a change in the temperature of the system, or (c) a change in volume, and hence partial pressure, of an equilibrium system that includes a gas. Predicting these changes is the subject of this section.

Changes in Concentration Suppose hydrochloric acid is added to the equilibrium in Equation 18.4. This strong acid separates into H^+ and Cl^- ions. The possible effects of both ions must be considered. The chloride ion is not present in the equilibrium, nor does it react with anything in the equilibrium. It is a spectator; it does not affect the equilibrium.

The hydrogen ion is different. H^+ *does* appear in the equilibrium. Adding H^+ increases $[H^+]$. $[H^+]$ is now too high. The value of Q is no longer equal to K, so the system is no longer at equilibrium:

$$Q = \frac{[H^+][C_2H_3O_2^-]}{[HC_2H_3O_2]} > K$$

We have seen that when $Q > K$, the reaction proceeds in the reverse direction—to the left. Some of the added H^+ combines with some $C_2H_3O_2^-$ to form $HC_2H_3O_2$. Both $[H^+]$ and $[C_2H_3O_2^-]$ are lowered, and $[HC_2H_3O_2]$ is raised until the value of the ratio is again equal to K. The system is again at equilibrium.

Changes of this kind are summarized in **Le Chatelier's principle**, which states: **If a stress is imposed on a system at equilibrium, the system will respond to that stress by shifting in the direction that will partially relieve the stress, until a new equilibrium is established.**

A "stress" on an equilibrium is anything that changes the forward and/or reverse reaction rates in a way that they are no longer equal. In terms of concentration, the stress changes the product/reactant concentration ratio, Q, away from its equilibrium value, K. To partially relieve the increased $[H^+]$ stress in the above example, the equilibrium shifts to the left, the direction in which H^+ is used up. This reduces $[H^+]$, but it remains larger than it was in the starting equilibrium. Including the changes in $[C_2H_3O_2^-]$ and $[HC_2H_3O_2]$, the whole process may be summarized in a table that includes the equilibrium "equation," written in concentration terms, as column headings:

STRESS	RESPONSE	SHIFT	$[HC_2H_3O_2]$	$\rightleftharpoons$	$[H^+]$	+	$[C_2H_3O_2^-]$
Increase $[H^+]$ by adding HCl	Consume H^+	Left	Increase		Increase (added)		Decrease

If silver nitrate is added to the acetic acid equilibrium, the effects of both the silver and nitrate ions must be considered. Nitrate ion could combine with hydrogen ion to produce nitric acid, but it does not because nitric acid is a strong acid that remains as separated ions. Nitrate ion is a spectator. However, the silver ion can combine with the acetate ion, precipitating silver acetate. If this occurs, some acetate ion will be removed from the equilibrium; $[C_2H_3O_2^-]$ will be reduced. The equilibrium will shift in the direction that raises $[C_2H_3O_2^-]$ somewhat, but not as high as it was originally. That shift is to the right. The net effect will be an increase in $[H^+]$ and a decrease in $[HC_2H_3O_2]$ and $[C_2H_3O_2^-]$. In summary,

STRESS	RESPONSE	SHIFT	$[HC_2H_3O_2]$	$\rightleftharpoons$	$[H^+]$	+	$[C_2H_3O_2^-]$
Decrease $[C_2H_3O_2^-]$; form $AgC_2H_3O_2(s)$	Make more $C_2H_3O_2^-$	Right	Decrease		Increase		Decrease (removed)

The procedure for analyzing an equilibrium in terms of Le Chatelier's principle may be summarized as follows:

1. Write the equilibrium equation. Set up a tabular summary.
2. Identify the stress in the stress column and in the appropriate concentration column. Record the stress in two places in the table.
3. Identify the response, the opposite of the stress.
4. Determine the shift, right or left, caused by the response.
5. Determine the concentration changes of all other species.

EXAMPLE 18.4: Predict the direction of shift in the position of equilibrium, and predict the effect on the equilibrium concentrations of acetic acid, hydrogen ion, and acetate ion in Equation 18.4, $HC_2H_3O_2 \rightleftharpoons H^+ + C_2H_3O_2^-$, caused by adding (a) sodium acetate, (b) sodium chloride, and (c) sodium hydroxide.

Begin with part (a). You have the equation (Step 1). What is the stress? Write it into the two places it goes in the following table (Step 2).

(a) STRESS	RESPONSE	SHIFT	$[HC_2H_3O_2]$	$\rightleftharpoons$	$[H^+]$	+	$[C_2H_3O_2^-]$

(a) STRESS	RESPONSE	SHIFT	$[HC_2H_3O_2]$	$\rightleftharpoons$	$[H^+]$	+	$[C_2H_3O_2^-]$
Increase $[C_2H_3O_2^-]$ by adding $NaC_2H_3O_2$							Increase (added)

Adding solid sodium acetate releases sodium and acetate ions. The Na^+ ion is a spectator, but $C_2H_3O_2^-$ is a product in the equilibrium equation. Adding acetate ion increases $[C_2H_3O_2^-]$.

From Le Chatelier's principle, the response is to resist the stress. What is that response (Step 3) and in which direction does the equilibrium shift to accomplish it (Step 4)? Add your answers to the table.

(a) STRESS	RESPONSE	SHIFT	$[HC_2H_3O_2]$	⇌ $[H^+]$	+ $[C_2H_3O_2^-]$
Increase $[C_2H_3O_2^-]$ by adding $NaC_2H_3O_2$	Use up $C_2H_3O_2^-$	Left			Increase (added)

The stress of increasing $[C_2H_3O_2^-]$ causes a response that decreases $[C_2H_3O_2^-]$. This requires that the equilibrium shift to the left, in which direction the acetate ion, a product as the equation is written, is changed back to acetic acid.

Complete the analysis by summarizing the effects on $[HC_2H_3O_2]$ and $[H^+]$ in the above table (Step 5).

(a) STRESS	RESPONSE	SHIFT	$[HC_2H_3O_2]$	⇌ $[H^+]$	+ $[C_2H_3O_2^-]$
Increase $[C_2H_3O_2^-]$ by adding $NaC_2H_3O_2$	Use up $C_2H_3O_2^-$	Left	Increase	Decrease	Increase (added)

Consuming acetate ion, a product, will also consume H^+, the other product, as well as producing more acetic acid.

Now analyze the effect of adding sodium chloride, part (b).

(b) STRESS	RESPONSE	SHIFT	$[HC_2H_3O_2]$	⇌ $[H^+]$	+ $[C_2H_3O_2^-]$

(b) If NaCl(s) is dissolved in the equilibrium, neither the sodium nor chloride ions will have any effect. Both are spectators. The equilibrium is unchanged.

Finally, what will happen if sodium hydroxide is added to the same equilibrium, part (c)?

(c) STRESS	RESPONSE	SHIFT	$[HC_2H_3O_2]$	⇌ $[H^+]$	+ $[C_2H_3O_2^-]$

(c) If NaOH(s) is dissolved in the equilibrium, the Na^+ ion is again a spectator. The hydroxide ion combines with the hydrogen ion to form water. $[H^+]$ comes down, and the equilibrium shifts right to restore it:

STRESS	RESPONSE	SHIFT	$[HC_2H_3O_2]$	⇌ $[H^+]$	+ $[C_2H_3O_2^-]$
Decrease $[H^+]$: $H^+ + OH^- \rightarrow H_2O$	Make more H^+	Right	Decrease	Decrease (removed)	Increase

Temperature Effect Temperature changes are produced primarily by adding or removing heat. When temperature is raised, the stress on the equilibrium may be identified as adding heat. When temperature is reduced, the stress is removing heat.

Heat may be thought of as a product in exothermic reactions and as a reactant in endothermic processes. Heat may be consumed or produced by shifting the position of equilibrium to the left or right in much the same way that a concentration stress was relieved. Predictions of the direction of shift thus depend on whether the reaction is exothermic or endothermic and on whether temperature is increased or decreased.

As an example, consider the combination of nitrogen and oxygen to make nitrogen monoxide, for which ΔH is +180 kJ:

$$180\ kJ + N_2(g) + O_2(g) \rightleftharpoons 2\ NO(g) \tag{18.9}$$

ΔH is positive, so the reaction is endothermic: 180 kJ of heat energy are required to produce each 2 moles of nitrogen monoxide.

If the temperature is increased by adding heat, the equilibrium will be disturbed. Le Chatelier's principle states that an equilibrium, when stressed, shifts in an attempt to relieve that stress. The response that will help relieve the stress is to absorb or use some of the added heat. Heat is "consumed" in this reaction as it proceeds in the forward direction—a shift to the right. Shifting the equilibrium to the right consumes not only some of the heat added, but also some of the nitrogen and oxygen. Consequently, their concentrations decrease. It also makes more nitrogen monoxide, increasing its concentration. In summary,

STRESS	RESPONSE	SHIFT	HEAT	+	$[N_2]$	+	$[O_2]$	$\rightleftharpoons$	[NO]
Add heat	Absorb heat	Right	Add		Decrease		Decrease		Increase

This effect is one of the main reasons that automobiles produce smog. At ordinary temperatures, the reaction between N_2 and O_2 produces very little NO, a major component of smog. However, at the much higher temperatures of an automobile engine, the N_2 and O_2 from the air react to produce a considerable amount of NO.

EXAMPLE 18.5: When the NO(g) produced in an automobile engine reacts further with $O_2(g)$, $NO_2(g)$ is produced. The ΔH for the reaction is -114 kJ. If the system reaches equilibrium, will more or less red-brown NO_2 be produced from colorless NO at higher temperatures?

Begin by writing the equation for the reaction, showing the heat as a reactant or product.

$$2\ NO(g) + O_2(g) \rightleftharpoons NO_2(g) + 114\ kJ$$

The negative value of ΔH indicates an exothermic reaction, so heat is produced.

We now begin a table like the one above, again showing the stress caused by adding heat. The response is to cancel some of the stress. Which way will the equilibrium shift? Add the response and direction of shift to the table.

STRESS	RESPONSE	SHIFT	[NO]	+	$[O_2]$	⇌	$[NO_2]$	+ HEAT

STRESS	RESPONSE	SHIFT	[NO]	+	$[O_2]$	⇌	$[NO_2]$	+ HEAT
Add heat	Absorb heat	Left						Add

The stress of the added heat will be partially relieved by absorbing some heat. The reaction, as written, produces heat. By going in the reverse direction (to the left), the reaction will consume heat.

What effect will this shift have on the concentrations of the reactants and products? Write your answers in the above table.

STRESS	RESPONSE	SHIFT	[NO]	+	$[O_2]$	⇌	$[NO_2]$	+ HEAT
Add heat	Absorb heat	Left	Increase		Increase		Decrease	Add

Shifting to the left means consuming products and making reactants.

EXAMPLE 18.6: Calcium hydroxide is less soluble in boiling water than in cold water. Is the dissolving of $Ca(OH)_2$ exothermic or endothermic? The dissolving equation is

$$Ca(OH)_2(s) \rightleftharpoons Ca^{2+}(aq) + 2\ OH^-(aq).$$

This time you must find the direction of the Le Chatelier shift when temperature is increased without knowing which side of the equation the energy term is on. This is done by observing the effect of the temperature increase. "Calcium hydroxide is less soluble. . ." in hot water than in cold water. In which direction is the equilibrium shifting when the substance is "less soluble"?

If the substance is "less soluble," some calcium and hydroxide ions must be combining to produce solid $Ca(OH)_2$. This happens when the reaction is shifting in the reverse direction, to the left.

The distrubance to the equilibrium was to increase temperature, or to add heat. The Le Chatelier shift to the left does the opposite; it uses up heat. On which side of the thermochemical equation will does energy appear if a shift to the left consumes heat? Add "heat" to the appropriate side of the equation below. Then classify the reaction, as the equation is written (in the forward direction), as exothermic or endothermic.

$$Ca(OH)_2(s) \rightleftharpoons Ca^{2+}(aq) + 2\ OH^-(aq)$$

The forward reaction is exothermic (__) endothermic (__).

$Ca(OH)_2(s) \rightleftharpoons Ca^{2+}(aq) + 2\ OH^-(aq)$ + heat. As the equation is written, heat is a product, so the reaction is exothermic.

If adding heat causes a shift to the left, the reaction must be endothermic in the reverse direction. That makes it exothermic in the forward direction.

Pressure/Volume Effect There are two ways to change the partial pressure of a gas in a gaseous equilibrium. One is to introduce or remove one gaseous component of the system. This is clearly a change in the concentration—and partial pressure—of that component, and Le Chatelier's principle predictions can be made accordingly. The second way to change the partial pressure of a gaseous equilibrium is to change the volume. But a volume change is different. It changes the partial pressures and concentrations of all the gaseous components in the system. Furthermore, it changes them all in the same direction, up or down.

Consider the following equilibrium:

$$N_2O_4(g) \rightleftharpoons 2\ NO_2(g) \quad (18.10)$$

If the volume of this closed system is increased at constant temperature, the pressure will drop. (This is an example of Boyle's law; P and V are inversely proportional.) The stress on the system is the reduction in total pressure. According to Le Chatelier's principle, the equilibrium will partially relieve this stress by increasing the pressure somewhat. How can the system respond in a way that will increase the total pressure?

The ideal gas equation is $PV = nRT$. Once the new volume is set, V, R, and T are all constant. The two variables, P and n, are directly proportional to each other (Section 12.4). If the system can respond by increasing the number of gas molecules, pressure will rise. This happens if the equilibrium shifts to the right. Each $N_2O_4(g)$ molecule that decomposes is converted to two $NO_2(g)$ molecules. This increases the number of moles of gas present, and the total pressure goes up. Conversely, if the volume of the equilibrium is reduced, raising pressure, the Le Chatelier shift will be to the left, in the direction of fewer gas molecules, which reduces pressure.

In summary, *decreasing volume causes an increase in pressure which is partially relieved by shifting toward the side with the smaller number of gas molecules. In a similar way, increasing volume decreases pressure and favors the side with the larger number of gas molecules.*

EXAMPLE 18.7: The reaction below is at equilibrium. If its volume is reduced, in which direction will the equilibrium shift?

$$4\ NO(g) + 6\ H_2O(\ell) \rightleftharpoons 4\ NH_3(g) + 5\ O_2(g)$$

First, identify the stress on the system.

The decrease in volume causes an increase in pressure.

To relieve the increase in pressure, partially, will the system shift to more gaseous molecules or fewer?

P (pressure) is directly proportional to n (moles), so the system will shift toward fewer molecules to reduce pressure.

What is the total number of gaseous molecules on each side of the equation? In which direction will the equilibrium shift to reduce that total number?

The reaction has 4 moles of gas on the left (water is a liquid) and 9 moles on the right. The left side thus has fewer moles of gas, so the reaction shifts to the left.

18.4 GAS PHASE EQUILIBRIA

This section deals with reactions which occur completely in the gas phase. The values of equilibrium constants for gas phase reactions are often neither very large nor very small, but in a middle range (10^{-3} to 10^{+3}). As such, there are appreciable concentrations of both reactants and products at equilibrium.

When you solve any problem involving an equilibrium constant, you will find it helpful to write two equations: the chemical equation for the equilibrium and the mathematical equation for the equilibrium constant. Whether you write out a full GIVEN-WANTED-EQUATION analysis of the problem or not, be sure that at least these two equations are before you. They keep your numerator and denominator concentrations in the proper positions and aid in making the correct substitutions before calculating.

Let us return to the $PCl_5(g) \rightleftharpoons PCl_3(g) + Cl_2(g)$ equilibrium. In Example 18.2 you found that the value of K = 0.041 at 250°C. Suppose 0.060 mol PCl_5 is placed in the same 12-liter vessel and heated to 250°C. The initial concentration is 0.60/12 = 0.050 mol/L. Right away, some of this PCl_5 will begin decomposing into PCl_3 and Cl_2. The decomposition will continue until equilibrium is reached and $[PCl_3][Cl_2]/[PCl_5] = K = 0.041$. But what are the values of each of these concentrations at equilibrium?

First, let's examine $[PCl_5]$ at equilibrium. Some of the initial 0.050 mol/L will have changed to the product gases at equilibrium. Let's set the moles of PCl_5 per liter that have decomposed equal to x. If we begin with 0.050 mol PCl_5 per liter and x mol per liter react, the concentration of PCl_5 that remains at equilibrium is the initial minus the reacting concentration, 0.050 - x. We can look at the problem set up as follows:

	$PCl_5(g)$	$\rightleftharpoons$	$PCl_3(g)$	+	$Cl_2(g)$
Start	0.050 mol/L		0 mol/L		0 mol/L
Change	-x				
Equilibrium	0.050 - x				

To complete the problem we must also know the values of $[PCl_3]$ and $[Cl_2]$ at equilibrium. We have assumed that x moles/liter of PCl_5 have decomposed. How much PCl_3 and Cl_2 have formed? This is essentially a stoichiometry problem. From the balanced equation for the reaction, 1 mole of PCl_3 forms from 1 mole of PCl_5. If x moles/liter of PCl_5 have reacted, x moles/liter of PCl_3 have formed. The same is true of Cl_2. *The values in the change line are always in the same ratio as the coefficients in the balanced equation.* This can be shown in the table that was started above. From now on we will use S for start, C for change, and E for equilibrium, and we will call this arrangement an S-C-E table.

	$PCl_5(g)$	⇌	$PCl_3(g)$	+	$Cl_2(g)$
S	0.050 mol/L		0 mol/L		0 mol/L
C	-x		+x		+x
E	0.050 - x		x		x

Does this table look familiar? It should. If you will look back to the beginning of Section 10.4 you will find that these same three lines are in the tables used for limiting reactant problems. In Chapter 10 the entries were in moles, whereas here they are in moles per liter.

The desired equilibrium concentrations are now all expressed in terms of x. The relationship among these concentrations is given by the equilibrium constant.

$$K = \frac{[PCl_3][Cl_2]}{[PCl_5]} = 0.041$$

The concentration values from the E line in the S-C-E table may be substituted to give an algebraic equation with x as the only unknown:

$$K = \frac{[PCl_3][Cl_2]}{[PCl_5]} = \frac{(x)(x)}{(0.050 - x)} = 0.041$$

This is a quadratic equation; (x)(x) in the numerator is x^2 The equation may be rewritten in the form $ax^2 + bx + c = 0$ and solved for x (Section 1.11).

$$x^2 + 0.041\ x - 0.00205 = 0 \qquad x = 0.029$$

From the E line in the table and the calculated value of x we can find all the equilibrium concentrations:

$$[PCl_5] = 0.050 - x = 0.050 - 0.029 = 0.021\ \text{mol/L},$$
$$[PCl_3] = [Cl_2] = x = 0.029\ \text{mol/L}$$

The complete S-C-E table is:

	$PCl_5(g)$	⇌	$PCl_3(g)$	+	$Cl_2(g)$
S	0.050 mol/L		0 mol/L		0 mol/L
C	-0.029		+0.029		+0.029
E	0.021		0.029		0.029

This S-C-E method for solving equilibrium problems may be summarized as follows:

1. Write the equilibrium equation and the K equation with its numerical value.
2. Fill in the initial concentrations on the S line.
3. Let x equal a change in concentration as the reaction moves toward equilibrium. Fill in the C line with x values in the same ratio as the coefficients in the balanced equation. Use + values for gain, - for loss.

4. Complete the E line by adding the S and C lines.
5. Substitute algebraic expressions from the E line into the K equation and solve for x.
6. Use the value of x and the expressions from the E line to solve for the equilibrium concentrations.

EXAMPLE 18.8: Calculate the number of moles of each species in a 2.0-liter flask if 0.080 mole of HI is allowed to decompose by the reaction $H_2 + I_2 \rightleftharpoons 2\ HI$ until equilibrium is reached. K = 64 at the equilibrium temperature.

The strategy for solving this problem is first to find the concentrations of all species in moles per liter, and then multiply by liters to find the number of moles. Setting up the S-C-E table is a bit different this time. Remember that K is always related to a specific equation. Watch the change line in your table. Finally, choose x so that the calculations are as simple as possible. Set up the table here.

	$H_2(g)$	+	$I_2(g)$	$\rightleftharpoons$	2 HI(g)
S	0 mol/L		0 mol/L		0.040 mol/L
C	+x		+x		-2x
E	x		x		0.040 - 2x

Again, the moles must be divided by liters to get 0.040 M for the starting [HI]: 0.080 mol HI/2.0 L = 0.040 mol HI/L. The starting quantity is on the right side of the equation this time. That makes no difference, as the reaction is reversible. By setting $[H_2]$ and $[I_2]$ equal to x, their changes become +x in the C line. The change in [HI] is -2x because there are 2 moles of HI for each mole of H_2 and I_2, as shown by the coefficients in the equation. Remember that the numbers in the C line are always in the same ratio as the number of moles in the equation.

You might have set x = [HI] at equilibrium. That would have given you 0.040 - x to be subtracted from 0.040 in the C line for HI. The C-line entries for H_2 and I_2 would be 0.020 - 0.5x. These values give the same ultimate answers to the problem, but the algebra is much more complex. Select variables to keep the algebra as simple as possible.

Now you can write the equilibrium constant expression, substitute the equilibrium concentrations, equate to K, and solve for x. Do this thoughtfully and you will not have to use the quadratic formula. (The S-C-E table is repeated at the top of the next page.)

	$H_2(g)$	+	$I_2(g)$	⇌	2 HI(g)
S	0 mol/L		0 mol/L		0.040 mol/L
C	+x		+x		-2x
E	x		x		0.040 - 2x

$$K = \frac{[HI]^2}{[H_2]\,[I_2]} = \frac{(0.040 - 2x)^2}{x_2} = 64$$

$$\frac{0.040 - 2x}{x} = 8$$

$$x = 0.0040 \text{ mol/L} = [H_2] = [I_2]$$

By taking the square root of both sides of the equation, you can avoid squaring the binomial (0.040 - 2x), and the ultimate use of the quadratic formula.

Now, what is the final concentration of [HI]?

$$[HI] = 0.040 - 2x = 0.040 - 2(0.0040) = 0.032 \text{ mol/L}$$

Now that you have the concentrations of each substance, convert them to moles in the 2.0-liter flask.

$$2.0 \text{ L} \times \frac{0.0040 \text{ mol } H_2}{\text{L}} = 0.0080 \text{ mol } H_2$$ Similarly, 0.0080 mol I_2

$$2.0 \text{ L} \times \frac{0.032 \text{ mol HI}}{\text{L}} = 0.064 \text{ mol HI}$$

EXAMPLE 18.9: K = 64 for $H_2 + I_2 \rightleftharpoons 2\text{ HI}$. If the starting concentrations are 0.050 mol/L for HI and 0.027 mol/L for both H_2 and I_2, what are the equilibrium concentrations of all substances?

Begin by calculating Q and comparing it with K. Which way will the reaction go?

$$Q = \frac{[HI]^2}{[H_2][I_2]} = \frac{(0.050)^2}{(0.027)(0.027)} = 1.4 < 64$$

Q is less than K, so the reaction will go to the right. More product will be made until Q increases to the point that it is equal to K.

Now set up the S-C-E table. Be sure to choose the most convenient expressions for x and that the sign of each term corresponds to the predicted change in the direction of the reaction.

	$H_2(g)$	+	$I_2(g)$	⇌	$2\ HI(g)$
S					
C					
E					

	$H_2(g)$	+	$I_2(g)$	⇌	$2\ HI(g)$
S	0.027 mol/L		0.027 mol/L		0.050 mol/L
C	-x		-x		+2x
E	0.027 - x		0.027 - x		0.050 + 2x

The shift to the right indicates a loss of H_2 and I_2, so the change line shows -x for those terms. The gain in product is indicated by +2x for HI.

Now complete the problem. Write the K expression, substitute values from the E line, and solve for x. Use your result to find the equilibrium concentrations of all three species.

$$K = \frac{[HI]^2}{[H_2][I_2]} = \frac{(0.050 + 2x)^2}{(0.027 - x)^2} = 64, \qquad x = 0.017$$

$[H_2] = 0.010$ mol/L $= [I_2]$ $\qquad$ $[HI] = 0.084$ mol/L

CHAPTER 18 PROBLEMS

1) Write K expressions for the following equations.

a) $Sn(s) + 4\ H^+(aq) + 4\ NO_3^-(aq) \rightleftharpoons SnO_2(s) + 4\ NO_2(g) + 2\ H_2O(\ell)$

b) $4\ NO(g) + 6\ H_2O(\ell) \rightleftharpoons 4\ NH_3(aq) + 5\ O_2(g)$

2) Write K expressions for the following equations.

a) $Pb(s) + PbO_2(s) + 2\ H^+(aq) + 2\ HSO_4^-(aq) \rightleftharpoons 2\ PbSO_4(s) + 2\ H_2O(\ell)$

b) $C_6H_5NO(\ell) + \frac{27}{2}\ O_2(g) \rightleftharpoons 6\ CO_2(g) + \frac{5}{2}\ H_2O(\ell) + \frac{1}{2}\ N_2(g)$

3) For the reaction $2\ SO_2(g) + O_2(g) \rightleftharpoons 2\ SO_3(g)$, the equilibrium concentrations are found to be 0.025 M, 0.15 M, and 0.16 M for SO_2, O_2, and SO_3, respectively. What is K at this temperature?

4) The system $2\ H_2S(g) \rightleftharpoons 2\ H_2(g) + S_2(g)$ reaches equilibrium in a 2.6-liter container. At a certain temperature the container holds 1.82 mol H_2S, 0.62 mol H_2, and 0.36 mol S_2. Calculate the equilibrium constant.

5) 23.7 mol NH_3 are placed in a 3.2-liter reaction vessel. When equilibrium is reached according to the equation $2\ NH_3(g) \rightleftharpoons N_2(g) + 3\ H_2(g)$, 3.6 mol N_2 are formed. Calculate K at the existing temperature.

6) For the reaction $PCl_5(g) \rightleftharpoons PCl_3(g) + Cl_2(g)$, K = 0.041. If a 7.5-liter flask is found to contain 0.16 mol PCl_5 and 0.052 mol PCl_3, how many moles of Cl_2 are present?

7) Rust is flushed from radiators using oxalic acid according to the equation $Fe_2O_3(s) + 6\ H_2C_2O_4(aq) \rightleftharpoons 2\ Fe(C_2O_4)_3{}^{3-}(aq) + 3\ H_2O(\ell) + 6\ H^+(aq)$. What is the effect on the dissolving of the rust if

a) more concentrated oxalic acid is used?
b) the solution is made more acidic?
c) more Fe_2O_3 is added?

8) After flushing a radiator with oxalic acid, excess acid is neutralized with sodium carbonate by the reaction

$$H_2C_2O_4(aq) + Na_2CO_3(s) \rightleftharpoons H_2O(\ell) + CO_2(g) + 2\ Na^+(aq) + C_2O_4{}^{2-}(aq).$$

Will this equilibrium shift left or right if

a) the oxalic acid is made more concentrated.
b) sodium hydroxide is added.
c) more sodium carbonate is added.

9) Gypsum, $CaSO_4$, is slightly soluble in an acid. To increase the solubility of gypsum, should the acid concentration be increased or decreased? Explain using Le Chatelier's principle. The equation is $CaSO_4(s) + H^+(aq) \rightleftharpoons Ca^{2+}(aq) + HSO_4{}^-(aq)$.

10) Potassium hydroxide dissolves exothermically. Is KOH more or less soluble at higher temperatures?

11) ΔH = + 41.2 kJ for the reaction in which sulfur precipitates from volcano fumes: $SO_2(g) + 2\ H_2S(g) \rightleftharpoons 3\ S(s) + 2\ H_2O(g)$. Will this equilibrium be shifted more to the right or to the left at the higher temperatures of a volcano?

12) $NO_2(g)$ is red-brown and $N_2O_4(g)$ is colorless. An equilibrium mix of the two, $2\ NO_2(g) \rightleftharpoons N_2O_4(g)$, is observed to become more red-brown as it is heated. Is AH positive or negative for this reaction?

13) The air around electrical generating equipment often smells of ozone from the endothermic reaction $3\ O_2(g) \rightleftharpoons 2\ O_3(g)$. If the total pressure around the equipment is dropped, will this equilibrium shift left or right?

14) Oxides of nitrogen are often formed in the atmosphere during lightning discharges: $N_2(g) + 2\ O_2(g) \rightleftharpoons 2\ NO_2(g)$. Would this reaction be shifted more to the right or to the left in the upper atmosphere where the total pressure is lower?

15) For each reaction listed, state whether the equilibrium would be shifted to the right by a volume increase or by a volume deacrease.

a) $4\ NO(g) + 6\ H_2O(\ell) \rightleftharpoons 4\ NH_3(g) + 5\ O_2(g)$

b) $2\ CaCO_3(s) + 2\ SO_2(g) + O_2(g) \rightleftharpoons 2\ CaSO_4(s) + 2\ CO_2(g)$

c) $CH_4(g) + 2\ O_2(g) \rightleftharpoons CO_2(g) + 2\ H_2O(g)$

16) For the reaction $2\ NO(g) + Cl_2(g) \rightleftharpoons 2\ NOCl(g)$, K = 10.0 at a certain temperature. If the initial concentrations are 0.15 M NO, 0.050 M Cl_2, and 0.10 M NOCl, will the reaction occur to the right or left?

17) Initially, a 25-liter flask contains 0.050 mol SO_2, 0.10 mol O_2, and 0.075 mol SO_3. Will more or less SO_3 be present at equilibrium? K = 280 for $2\ SO_2(g) + O_2(g) \rightleftharpoons 2\ SO_3(g)$ at this temperature.

18) The nerve gas phosgene decomposes according to $COCl_2(g) \rightleftharpoons CO(g) + Cl_2(g)$, K = 0.32 at 1000 K. If the initial concentration of phosgene is 0.015 M, what are the equilibrium concentrations of all gases?

19) The equilibrium $2\ NO_2(g) \rightleftharpoons N_2O_4(g)$ is reached after 1.50 moles of N_2O_4 are introduced to a 3.6-liter container. K = 0.50 at the temperature of the system. Find the concentrations of both gases.

20) The formation of nitrogen monoxide, $\frac{1}{2}\ N_2(g) + \frac{1}{2}\ O_2(g) \rightleftharpoons NO(g)$, has an equilibrium constant of 0.050. If a flask initially contains 0.100 moles per liter of NO at these conditions, what will the equilibrium concentrations be for all gases?

21) Find the concentrations of all species in a 1.0-liter flask if 0.72 mol H_2 reacts with 0.72 mol CO_2 until the equilibrium $CO_2(g) + H_2(g) \rightleftharpoons CO(g) + H_2O(g)$ is reached. K = 0.34 at the existing temperature.

22) A flask is initially 0.050 M in NO_2 and 0.010 M in N_2O_4. What will the equilibrium concentrations be if K = 0.50 for $2\ NO_2(g) \rightleftharpoons N_2O_4(g)$ at the conditions of the flask?

23) For the reaction $Al(OH)_3(s) + OH^-(aq) \rightleftharpoons Al(OH)_4^-(aq)$, K = 5.4 at room temperature. If the initial concentrations are 0.50 M for OH^- and 0.10 M for $Al(OH)_4^-$, what will the equilibrium concentrations be?

++++++++++++++++++++++

24) Write K expressions for the following equations.

a) $CaCO_3(s) + 2\ H^+(aq) \rightleftharpoons Ca_2(aq) + H_2O(\ell) + CO_2(g)$

b) $4\ Fe(OH)_2(s) + O_2(g) + 2\ H_2O(\ell) \rightleftharpoons 4\ Fe(OH)_3(s)$

25) Write K expressions for the following equations.

a) $3\ NO_2(g) + H_2O(\ell) \rightleftharpoons 2\ HNO_3(aq) + NO(g)$

b) $2\ Al(OH)_3(s) \rightleftharpoons Al_2O_3(s) + 3\ H_2O(\ell)$

26) Calculate K for the equilibrium $CO(g) + H_2O(g) \rightleftharpoons CO_2(g) + H_2(g)$ if a 3.2-liter flask contains 0.21 mol CO, 0.27 mol H_2O, 0.58 mol CO_2, and 0.42 mol H_2.

27) Oxides of nitrogen form from air heated to high temperatures inside auto engines. What is the equilibrium concentration of nitrogen monoxide if the equilibrium concentration of nitrogen is 0.036 M and that of oxygen is 0.010 M? K = 0.0025 for $N_2(g) + O_2(g) \rightleftharpoons 2\ NO(g)$ at engine temperatures.

28) Chlorine for pool disinfection is often formed by mixing solid bleach and hydrochloric acid. $Ca(OCl)_2(s) + 4\ H^+(aq) + 2\ Cl^-(aq)\ Cl_2(aq) + 2\ H_2O(\ell)$. Analyze the changes below for effectiveness in increasing Cl_2 concentration at equilibrium: a) add salt, NaCl; b) add a base; c) add more solid bleach.

29) A major effect on blood acidity is caused by the presence of carbonic acid, an end product of metabolism. Too much acid can be dangerous to the body. However, H_2CO_3 easily decomposes to CO_2 which may be exhaled. Explain the following effect using Le Chatelier's principle: if you breathe into a paper bag, you will feel light-headed and then pass out.

30) The heat of formation of ammonia is -46.1 kJ/mol. Will heating the reaction help or hinder the production of NH_3?

31) The dissociation of acetic acid is more extensive at higher temperatures. Is the reaction exothermic or endothermic?

32) Catalytic converters change harmful carbon monoxide into safer carbon dioxide: $2\ CO(g) + O_2(g) \rightleftharpoons 2\ CO_2(g)$. If this reaction were done in a smaller reaction chamber, would the equilibrium shift right or left?

33) $4\ Fe(s) + 3\ O_2(g) + 6\ H_2O(g) \rightleftharpoons 4\ Fe(OH)_3(s)$ represents the formation of rust. This equilibrium is observed to shift to the left (not as much rust) when in dryer climates, such as deserts. Explain this using Le Chatelier's principle.

34) A flask is initially 0.050 M in both NO_2 and N_2O_4. Which will be present in greater amount at equilibrium? K = 0.50 for $2\ NO_2(g) \rightleftharpoons N_2O_4(g)$.

35) 2.57 moles of NOBr are placed in a 1.50-liter container. They dissociate to reach the equilibrium $2\ NOBr(g) \rightleftharpoons 2\ NO(g) + Br_2(g)$. [NO] = 0.58 M at equilibrium. Find [NOBr], [Br_2], and K.

36) Consider the equilibrium $2\ HI(g) \rightleftharpoons H_2(g) + I_2(g)$. Find the concentrations of all species if 0.542 mole of HI is placed in a 15.0-liter container and equilibrium is reached at the temperature for which K = 0.043.

37) A flask initially contains 0.050 M NO_2 and 0.050 M N_2O_4. (See question 34.) What are their equilibrium concentrations? K = 0.50 for the equilibrium $2\ NO_2(g) \rightleftharpoons N_2O_4(g)$. [Note: $(a + b)^2 = a^2 + 2ab + b^2$.]

ADDITIONAL PROBLEMS

38) When the body rids itself of carbon dioxide, the following equilibrium is established between the blood and the air in the lungs.

$$HCO_3^-(aq) + H^+(aq) \rightleftharpoons CO_2(g) + H_2O(\ell)$$

a) Will the blood become more or less acidic when the pressure of CO_2 is lowered by exhaling? b) Will the body be able to get rid of more CO_2 by making the blood more acidic or more basic?

39) Sea water has a pH of about 8.20 (that is, $[H^+] = 6.3 \times 10^{-9}$ M), $[HCO_3^-]$ = 0.0024 M, and $[Ca^{2+}]$ = 0.010 M. a) What is the concentration of carbonate ion in the equilibrium below?

$$HCO_3^-(aq) \rightleftharpoons H^+(aq) + CO_3^{2-}(aq), \quad K = 4.8 \times 10^{-11}$$

b) Will calcium carbonate precipitate? In other words, is Q less than K for:

$$Ca^{2+}(aq) + CO_3^{2-}(aq) \rightleftharpoons CaCO_3(s), \quad K = 2.1 \times 10^{8}?$$

40) Chloroform, $CHCl_3$, can be converted to carbon tetrachloride, CCl_4, by the reaction below.

$$2\ CHCl_3(g) + Cl_2(g) \rightleftharpoons 2\ CCl_4(g) + H_2(g), \quad K = 0.0018 \text{ at } 100°C$$

a) If you begin with 0.10 M of all substances, will the reaction form more products or more reactants? b) If you begin with 0.10 M $CHCl_3$ and Cl_2 with no product as yet present, what will the equilibrium concentrations be for CCl_4 and H_2? [Hint: Assume that the changes in concentration are much less than 0.10 M.]

41) The pollutant sulfur dioxide can be removed from industrial exhaust by the reaction shown below.

$$2\ CaCO_3(s) + 2\ SO_2(g) + O_2(g) \rightleftharpoons 2\ CaSO_4(s) + 2\ CO_2(g)$$

If $K = 1.1 \times 10^{10}$ at 1200°C, what will be the $[SO_2]$ when $[O_2] = 1.9 \times 10^{-3}$ M and $[CO_2] = 3.8 \times 10^{-5}$ M (corresponding to their typical pressures of 150 torr and 3 torr, respectively)?

CHAPTER TEST

1) What is the K equation for the combustion of ethane, C_2H_6?

2) $2\ HI(g) \rightleftharpoons H_2(g) + I_2(g)$. 4.10 moles of H_2 and 4.10 moles of I_2 are placed in a 12.2-liter vessel. They react until the above equilibrium is reached. At that time the flask contains 6.34 moles of HI. Calculate K at the existing temperature.

3) If HCl is added to the equilibrium

$$BiOCl(s) + 2\ H^+(aq) \rightleftharpoons Bi^{3+}(aq) + H_2O(\ell) + Cl^-(aq)$$

will it shift right or left?

4) "Cold packs" are mixtures of ammonium nitrate and water. Is AH for the dissolving of NH_4NO_3 positive or negative?

5) One step in the production of natural gas (methane) from coal is

$$CO(g) + 3\ H_2(g) \rightleftharpoons CH_4(g) + H_2O(g)$$

Will the production of methane be more or less favorable at higher total pressures?

6) A 5.0-liter flask initially contains 0.20 mol CO_2, 0.20 mol H_2, 0.50 mol CO and 0.50 mol H_2O. Will more or less CO_2 be present at equilibrium? K = 0.34 for $CO_2(g) + H_2(g) \rightleftharpoons CO(g) + H_2O(g)$.

7) K = 17 for $Cl_2(g) + PCl_3(g) \rightleftharpoons PCl_5(g)$ at a certain temperature. A sample of this equilibrium was created from an initial charge of 0.10 mol of PCl_5 into a 2.0-liter flask. Calculate the equilibrium concentrations of PCl_3, Cl_2, and PCl_5.

8) A 5.0-liter flask initially contains 0.20 mol CO_2, 0.20 mol H_2, 0.50 mol CO, and 0.50 mol H_2O. What are the concentrations at equilibrium? K = 0.34 for $CO_2(g) + H_2(g) \rightleftharpoons CO(g) + H_2O(g)$.

19. SOLUBILITY EQUILIBRIA

PREREQUISITES

1. Equilibrium problems are solved algebraically by the methods shown in Sections 1.9 to 1.11.

2. Many equilibrium calculations are performed in exponential notation (Section 2.1).

3. The significant figures rules given in Sections 3.3 to 3.7 are applied throughout equilibrium calculations.

4. Solubility equilibria are described by the net ionic equations for precipitation reactions studied in Section 15.4.

5. The molarity and dilution calculations you learned in Sections 14.4 and 15.6 are used in equilibrium calculations.

6. Equilibrium constants (Section 18.2) are written for solubility equilibria. The reaction quotient, Q, is used to predict whether or not precipitation will occur.

7. Le Chatelier's principle (Section 18.3) is applied to concentration changes in precipitation reactions.

CHEMICAL SKILLS

1. K_{sp} and solubility. Given the solubility product constant or the solubility of a slightly soluble compound, calculate the other.

2. The common ion effect. Given the solubility product constant of a slightly soluble compound and the concentration of a solution that contains a common ion, calculate the solubility of the slightly soluble compound.

3. "Will it precipitate?" problems. Determine if a precipitate will form when given volumes of solutions of known concentration are combined.

19.1 SOLUTE AND ION MOLARITIES IN THE SOLUTION OF AN IONIC SOLUTE

In Section 15.2 it was pointed out that the solution inventory of a solution of a strong electrolyte consists of ions. All soluble ionic compounds are strong electrolytes. If 100% dissociation of the compound is assumed, the concentration of the ions can be calculated. It is these ionic concentrations that are important in solubility equilibria.

Consider 0.10 M Na_2SO_4. The solution "contains" 0.10 mole of Na_2SO_4 per liter. Actually, the solution contains no Na_2SO_4 because it has all dissociated into sodium and sulfate ions:

$$Na_2SO_4(s) \rightleftharpoons 2\ Na^+(aq) + SO_4^{2-}(aq) \tag{19.1}$$

One formula unit of Na_2SO_4 that dissolves produces one SO_4^{2-} ion. The sulfate ion concentration is therefore the same as the molarity of the solution:

$$\frac{0.10\ \text{mol}\ Na_2SO_4}{1\ \text{L}} \times \frac{1\ \text{mol}\ SO_4^{2-}}{1\ \text{mol}\ Na_2SO_4} = 0.10\ \text{mol}\ SO_4^{2-}/\text{L} = [SO_4^{2-}]$$

Twice as many sodium ions are released as sulfate ions when sodium sulfate dissolves. The sodium ion concentration is therefore twice the sulfate ion concentration:

$$\frac{0.10\ \text{mol}\ Na_2SO_4}{1\ \text{L}} \times \frac{2\ \text{mol}\ Na^+}{1\ \text{mol}\ Na_2SO_4} = 0.20\ \text{mol}\ Na^+/\text{L} = [Na^+]$$

In summary, in 0.10 M Na_2SO_4, $[Na^+] = 0.20$ and $[SO_4^{2-}] = 0.10$. The ion concentrations of all solutions of ionic compounds can be found in the same way. In the sections ahead, we will proceed directly from solute molarities to ion molarities such as these with little or no comment.

19.2 THE SOLUBILITY OF A COMPOUND

Some salts are "insoluble" in water. Actually, it is more correct to say that some salts are very, very slightly soluble in water. No salt is completely insoluble.

Lead(II) sulfate is a salt of low solubility. When placed in water, it dissolves until an equilibrium is reached:

$$PbSO_4(s) \rightleftharpoons Pb^{2+}(aq) + SO_4^{2-}(aq) \tag{19.2}$$

The equilibrium constant expression has no denominator, as $PbSO_4$ is a solid:

$$K_{sp} = [Pb^{2+}][SO_4^{2-}] \tag{19.3}$$

The equilibrium constant for every solubility equilibrium has the same form. It is the product of two ion concentrations, with each concentration raised to a power, when appropriate. It is called the **solubility product constant, K_{sp}**.

The lead(II) ion concentration in a saturated solution of lead sulfate is 1.3×10^{-4} moles per liter. Equation 19.2 shows that the sulfate ion is produced in the same molar quantities as the lead ion. If solid lead(II) sulfate is the only source of the two ions, their concentrations must be equal at 1.3×10^{-4} moles per liter. Substituting into Equation 19.3 gives the value of K_{sp} for lead(II) sulfate:

$$K = [Pb^{2+}][SO_4^{2-}] = (1.3 \times 10^{-4})(1.3 \times 10^{-4}) = 1.7 \times 10^{-8}$$

EXAMPLE 19.1: The calcium ion concentration is 6.9×10^{-5} M in a saturated solution of calcium carbonate. Find the solubility product constant for calcium carbonate.

Begin with the analysis of the problem—at least the two equations that are essential for all equilibrium constant problems.

GIVEN: $[Ca^{2+}] = 6.9 \times 10^{-5}$ M FIND: Ksp of $CaCO_3$

EQN: $CaCO_3(s) \rightleftharpoons Ca^{2+}(aq) + CO_3{}^{2-}(aq)$, $K_{sp} = [Ca^{2+}][CO_3{}^{2-}]$

You are given the calcium ion concentration. What is $[CO_3{}^{2-}]$?

$[CO_3{}^{2-}] = 6.9 \times 10^{-5}\ M = [Ca^{2+}]$

The equation shows that, as long as calcium and carbonate ions come from only calcium carbonate, their ion concentrations are equal.

You can now substitute values into the Ksp expression and calculate the constant.

$K_{sp} = [Ca^{2+}][CO_3{}^{2-}] = (6.9 \times 10^{-5})^2 = 4.8 \times 10^{-9}$

EXAMPLE 19.2: If $[Pb^{2+}] = 0.0012$ M in a saturated solution of lead(II) iodide, calculate Ksp.

Again begin with the analysis of the problem.

GIVEN: $[Pb^{2+}] = 0.0012$ M FIND: Ksp

EQN: $PbI_2(s) \rightleftharpoons Pb^{2+}(aq) + 2\ I^-(aq)$, $K_{sp} = [Pb^{2+}][I^-]^2$

What values of $[Pb^{2+}]$ and $[I^-]$ will you substitute into the K expression? Careful on the $[I^-]$!

$[Pb^{2+}] = 0.0012$ M (given); $[I^-] = 2 \times 0.0012\ M = 0.0024\ M$

The dissolving equation for PbI_2 shows 2 moles of I^- for each mole of Pb^{2+}, so the I^- concentration is twice the Pb^{2+} concentration. (See Section 19.1.)

Substitute into the Ksp equation and solve.

$$Ksp = [Pb^{2+}][I^-]^2 = (0.0012)(0.0024)^2 = 6.9 \times 10^{-9}$$

Solubility product constants for many compounds are listed in chemical handbooks. Some are shown in Table 19.1. That makes them available for us to use in finding such things as the solubility of a compound in moles per liter—moles of the compound that dissolve in 1 liter of solution.

The solubility of a compound is numerically equal to an ion concentration whenever 1 mole of solute produces 1 mole of that ion *and the ion is not present from any other source*. Look again at the calcium carbonate and lead(II) iodide dissolving equations:

$$CaCO_3(s) \rightleftharpoons Ca^{2+}(aq) + CO_3{}^{2-}(aq) \quad \text{and} \quad PbI_2(s) \rightleftharpoons Pb^{2+}(aq) + 2\ I^-(aq)$$

In the first equation the concentrations of the calcium and carbonate ions are the same as the number of moles of calcium carbonate that dissolve in a liter—the solubility of the compound. The same is true of $[Pb^{2+}]$ for PbI_2. $[I^-]$, on the other hand, is twice the solubility of PbI_2 because 1 mole of solute produces 2 moles of I^- ions.

TABLE 19.1: SOLUBILITY PRODUCT CONSTANTS

Compound	Ksp	Compound	Ksp	Compound	Ksp
$AgBr$	5.0×10^{-13}	CaF_2	4.0×10^{-11}	$MnCO_3$	1.8×10^{-11}
$AgC_2H_3O_2$	2.3×10^{-3}	$Ca(OH)_2$	5.5×10^{-6}	$Mn(OH)_2$	1.0×10^{-13}
Ag_2CO_3	8.2×10^{-12}	$Ca_3(PO_4)_2$	1.0×10^{-26}	$NiCO_3$	6.6×10^{-9}
$AgCN$	2.0×10^{-12}	$CaSO_4$	1.2×10^{-6}	NiS	3×10^{-21}
$AgCl$	1.7×10^{-10}	$CdCO_3$	5.2×10^{-12}	$PbCl_2$	1.6×10^{-5}
Ag_2CrO_4	2.2×10^{-12}	$Cd(OH)_2$	5.1×10^{-15}	$PbCO_3$	5.6×10^{-14}
AgI	8.3×10^{-17}	CdS	7.8×10^{-27}	$PbCrO_4$	1.8×10^{-14}
$AgIO_3$	2.0×10^{-8}	$CoCO_3$	8.0×10^{-13}	PbF_2	2.7×10^{-8}
Ag_3PO_4	1.3×10^{-20}	$Co(OH)_2$	2.0×10^{-16}	PbI_2	6.9×10^{-9}
$AgSCN$	1.0×10^{-12}	$CuCO_3$	2.4×10^{-10}	$Pb(IO_3)_2$	9.8×10^{-14}
Ag_2S	2.0×10^{-49}	$Cu(IO_3)_2$	7.4×10^{-8}	PbS	2.5×10^{-27}
Ag_2SO_4	1.6×10^{-5}	CuS	8×10^{-36}	$PbSO_4$	1.7×10^{-8}
$Al(OH)_3$	3.0×10^{-32}	$CuSCN$	1.9×10^{-13}	$SrCrO_4$	2.2×10^{-5}
$BaCO_3$	8.1×10^{-9}	FeS	5×10^{-18}	SrC_2O_4	5.6×10^{-8}
$BaCrO_4$	2.4×10^{-10}	HgS	1.6×10^{-52}	SrF_2	2.8×10^{-9}
BaF_2	1.7×10^{-6}	Li_2CO_3	1.7×10^{-3}	$ZnCO_3$	1.4×10^{-11}
$BaSO_4$	1.1×10^{-10}	$MgCO_3$	1.0×10^{-5}	$Zn_3(PO_4)_2$	9.1×10^{-33}
Bi_2S_3	1.0×10^{-96}	MgC_2O_4	8.3×10^{-5}	ZnS	1.6×10^{-24}
$CaCO_3$	4.8×10^{-9}	MgF_2	6.5×10^{-9}		
CaC_2O_4	2.4×10^{-9}	$Mg(OH)_2$	1.2×10^{-11}		

EXAMPLE 19.3: Calculate the solubility of silver chloride in moles per liter and grams per liter.

Write the equilibrium equation and its Ksp expression, including its numerical value from Table 19.1.

EQN: $AgCl(s) \rightleftharpoons Ag^+(aq) + Cl^-(aq)$, $K_{sp} = [Ag^+][Cl^-] = 1.7 \times 10^{-10}$

Assuming AgCl(s) is the only source of either ion, how do the concentrations of the two ions compare?

According to the equilibrium equation, they are equal.

At this point we introduce the algebraic symbol S to represent solubility in moles per liter. The best procedure is to set S equal to an ion concentration that is the same as the moles of solute that dissolve per liter. In this case both ion concentrations are the same as the moles of solute dissolved, so we can say $[Ag^+] = S = [Cl^-]$. Now substitute S for both concentrations in the Ksp equation above and solve for its value.

$$[Ag^+][Cl^-] = (S)(S) = S^2 = 1.7 \times 10^{-10}$$

$$S = 1.3 \times 10^{-5} \text{ mol AgCl/L}$$

If you have a question about the algebra, check Procedure 6, Section 1.10.

You now know the solubility in moles per liter. Convert it to grams per liter. This requires a conventional mole → gram conversion.

GIVEN: 1.3×10^{-5} mol AgCl/L FIND: g AgCl/L PATH: mol → g

$$\frac{1.3 \times 10^{-5} \text{ mol AgCl}}{\text{L}} \times \frac{143.4 \text{ g AgCl}}{1 \text{ mol AgCl}} = 1.9 \times 10^{-3} \text{ g AgCl/L}$$

EXAMPLE 19.4: Lead(II) chloride is among the more soluble of those salts commonly thought of as "insoluble," and its Ksp of 1.6×10^{-5} is among the largest listed in Table 19.1. Calculate the solubility of $PbCl_2$ in moles per liter and grams per liter.

As usual, start by writing the equation and the Ksp expression, including its value.

$$PbCl_2(s) \rightleftharpoons Pb^{2+}(aq) + 2\ Cl^-(aq) \qquad K_{sp} = [Pb^{2+}][Cl^-]^2 = 1.6 \times 10^{-5}$$

As before, let S be the ion concentration that is equal to the solubility of the compound. The equation shows that the number of moles of compound that dissolve per liter is the same as one of the ion concentrations, but not the other. Which ion concentration will be equal to S?

$$S = [Pb^{2+}]$$

One mole of $PbCl_2$ yields one mole of Pb^{2+} ion, so $[Pb^{2+}]$ is equal to the moles of compound that dissolve per liter.

In terms of S, what is $[Cl^-]$?

$$[Cl^-] = 2S$$

The equilibrium equation indicates that 2 moles of Cl^- are formed for each mole of Pb^{2+}. Therefore $[Cl^-]$ is twice as large as $[Pb^{2+}]$. If $[Pb^{2+}]$ is S, then $[Cl^-]$ is 2S.

Now substitute S for $[Pb^{2+}]$ and 2S for $[Cl^-]$ in the Ksp equation. Solve for S. If you are not quite sure how to handle the algebra, refer to Sections 1.9 and 1.10.

$$[Pb^{2+}][Cl^-]^2 = (S)(2S)^2 = 4S^3 = 1.6 \times 10^{-5}$$

$$S^3 = 0.40 \times 10^{-5} = 4.0 \times 10^{-6}$$

$$S = 1.6 \times 10^{-2} = 0.016 \text{ mol } Pb^{2+}/L$$

The square of the product of two numbers is equal to the product of the squares of those numbers (Equation 1.12, Section 1.9):

$$(2S)^2 = (2S)(2S) = 2 \times 2 \times S^2 = 4S^2$$

Students are sometimes concerned about "doubling and squaring the chloride ion concentration." That is not what is done. You first *find* the chloride ion concentration, which happens to be 2S *in this problem because lead chloride is the* **only** *source of the two ions*. Then the concentration is squared. Never is $[Cl^-]$ doubled.

Finally, convert the 0.016 mol $PbCl_2$/L to grams per liter.

$$\frac{0.016 \text{ mol } PbCl_2}{1 \text{ L}} \times \frac{278.1 \text{ g } PbCl_2}{1 \text{ mol } PbCl_2} = 4.4 \text{ g } PbCl_2/\text{L}$$

19.3 SOLUBILITY IN THE PRESENCE OF A COMMON ION

In every solubility equilibrium we have considered so far, both dissolved ions have come from a single compound. This is not always the case.

Equation 19.2 described the equilibrium between lead(II) sulfate and a saturated solution of its ions:

$$PbSO_4(s) \rightleftharpoons Pb^{2+}(aq) + SO_4{}^{2-}(aq) \qquad (19.2)$$

Both $[Pb^{2+}]$ and $[SO_4{}^{2-}]$ were shown to be 1.3×10^{-4} M. This is also the solubility of lead(II) sulfate. What would happen if the sulfate ion concentration was increased from another source, by adding sodium sulfate, for example? How would $[Pb^{2+}]$ change? How would the solubility of lead sulfate change?

Do you recognize this as an example of the concentration effect of Le Chatelier's principle described in Section 18.3? The addition of sodium sulfate increases $[SO_4{}^{2-}]$. This causes the equilibrium to shift to the left, so it precipitates more lead sulfate. The overall effect is to reduce the solubility of lead sulfate. In general, the solubility of a slightly soluble compound is reduced upon the addition of a soluble compound that has an ion in common with the slightly soluble compound. This is called a **common ion**, and the result is known as the **common ion effect**.

We will now find the reduced solubility of lead(II) sulfate in a 0.10 molar solution of sodium sulfate. $[SO_4{}^{2-}]$ starts at 0.10 M. To whatever extent $PbSO_4$ dissolves, it will add to the sulfate ion concentration. We have seen that only 1.3×10^{-4} moles of $PbSO_4$ dissolve in a liter of water. If this full amount dissolved—which it does not, according to the preceding paragraph—$[SO_4{}^{2-}]$ would reach a maximum of 0.10 + 0.00013 moles per liter. According to the rules of significant figures, 0.10 + 0.00013 = 0.10. This is the same sulfate ion concentration that was present before the lead(II) sulfate was added. In other words, the amount of $PbSO_4$ that dissolves is so small that the additional sulfate ion concentration cannot be measured. $[SO_4{}^{2-}]$ remains at 0.10 M.

For the lead(II) sulfate equilibrium, $K_{sp} = [Pb^{2+}][SO_4{}^{2-}] = 1.7 \times 10^{-8}$. The equation can be solved for $[Pb^{2+}]$. This is equal to the solubility of $PbSO_4$ in the solution because $PbSO_4(s)$ is the only source of Pb^{2+}. The sulfate ion concentration, 0.10 mole per liter, is substituted and the result calculated:

$$[Pb^{2+}] = \frac{Ksp}{[SO_4{}^{2-}]} = \frac{1.7 \times 10^{-8}}{[SO_4{}^{2-}]} = \frac{1.7 \times 10^{-8}}{0.10} = 1.7 \times 10^{-7}\ M \qquad (19.4)$$

This is about a factor of 1000 less than the solubility of $PbSO_4$ in water!

EXAMPLE 19.5: Calculate the solubility of silver chloride in 0.20 M $AgNO_3$.

From Table 19.1, Ksp = $[Ag^+][Cl^-]$ = 1.7×10^{-10}. What is the value of $[Ag^+]$ in 0.20 M $AgNO_3$?

$[Ag^+]$ = 0.20 M

From $AgNO_3 \rightarrow Ag^+ + NO_3{}^-$, the moles of silver ion per liter is the same as the moles of silver nitrate dissolved in a liter.

Silver chloride is a slightly soluble salt. To whatever extent it dissolves, it will add to the 0.20 mole per liter $[Ag^+]$. However, that addition will be so small that its sum, when added to 0.20 M, will still be 0.20 M, just as $[SO_4{}^{2-}]$ remained at 0.10 M in the above discussion. Therefore, you can solve the Ksp expression for $[Cl^-]$, substitute Ksp and the silver ion concentration, and calculate $[Cl^-]$. Because AgCl is the only source of chloride ion, the solubility of silver chloride is the same as the chloride ion concentration. Complete the problem.

$$[Cl^-] = \frac{Ksp}{[Ag^+]} = \frac{1.7 \times 10^{-10}}{0.20} = 8.5 \times 10^{-10}\ M$$

Notice that this solubility is lower by 5 orders of magnitude than the solubility of silver chloride in pure water, according to Example 19.3. This is exactly what is predicted by Le Chatelier's principle.

19.4 PREDICTING PRECIPITATION

At the beginning of Section 19.3 we saw that the addition of sulfate ion to a $PbSO_4(s) \rightleftharpoons Pb^{2+}(aq) + SO_4{}^{2-}(aq)$ equilibrium causes lead ion to combine with sulfate ion and precipitate more lead sulfate. This can be related to the **reaction quotient**, Q, which was introduced in Section 18.2. Q has exactly the same form as Ksp; for $PbSO_4$,

$$Q = [Pb^{2+}][SO_4{}^{2-}] \qquad (19.5)$$

The ion concentrations for Q can have any value at a particular instant, and the product may be more than Ksp, less than Ksp, or the same as Ksp. It is only when the system is at equilibrium that Q = Ksp.

Adding sulfate ion to the above system at equilibrium, when Q = Ksp, increases $[SO_4{}^{2-}]$. This makes Q greater than Ksp, or Q > Ksp. When this condition exists, precipitation occurs. Both $[Pb^{2+}]$ and $[SO_4{}^{2-}]$ are reduced until Q = Ksp once again. This is the new equilibrium predicted by Le Chatelier's principle. Notice that at this new equilibrium $[Pb^{2+}] = [SO_4{}^{2-}]$ because the ions come from different sources.

This discussion gives us the criterion on which precipitation predictions are based: *Precipitation occurs whenever Q > Ksp.* If Q is less than or equal to Ksp, precipitation does not occur.

EXAMPLE 19.6: Will a precipitate form if 60.0 mL of 2.0×10^{-3} M NaCl are added to 40.0 mL of 3.0×10^{-5} M $AgNO_3$?

The equilibrium equation and K expression were found in Example 19.5:

$$AgCl(s) \rightleftharpoons Ag^+(aq) + Cl^-(aq) \qquad K_{sp} = [Ag^+][Cl^-] = 1.7 \times 10^{-10}$$

We must find the two concentrations in the *combined* solutions, multiply them to get Q, and then compare that value to Ksp to determine whether or not precipitation will occur. Begin by finding $[Ag^+]$ in the combined solutions. (You may wish to review dilution calculations in Section 14.6 at this point.)

GIVEN: $M_1 = [Ag^+]_1 = 3.0 \times 10^{-5}$ M
$V_1 = 40.0$ mL, $V_2 = 100.0$ mL

FIND: $M_2 = [Ag^+]_2$
EQN: $M_1V_1 = M_2V_2$

$$[Ag^+]_2 = [Ag^+]_1 \times \frac{V_1}{V_2} = 3.0 \times 10^{-5}\ M \times \frac{40.0\ mL}{100.0\ mL} = 1.2 \times 10^{-5}\ M$$

Now calculate the diluted concentration of the chloride ion.

GIVEN: $M_1 = [Cl^-]_1 = 2.0 \times 10^{-3}$ M
$V_1 = 60.0$ mL, $V_2 = 100.0$ mL

FIND: $M_2 = [Cl^-]_2$
EQN: $M_1V_1 = M_2V_2$

$$[Cl^-]_2 = [Cl^-]_1 \times \frac{V_1}{V_2} = 2.0 \times 10^{-3}\ M \times \frac{60.0\ mL}{100.0\ mL} = 1.2 \times 10^{-3}\ M$$

You have both concentrations. Calculate Q.

$Q = [Ag^+][Cl^-] = (1.2 \times 10^{-5})(1.2 \times 10^{-3}) = 1.4 \times 10^{-8}$

Now compare Q with Ksp (1.7×10^{-10}) by writing an inequality statement, using > for greater than or < for less than. Follow with a statement that precipitation will or will not occur.

$1.4 \times 10^{-8} > 1.7 \times 10^{-10}$. Precipitation will occur.

Whenever Q > Ksp, a precipitate will form. Note that the exponents being compared are negative: $10^{-8} > 10^{-10}$.

EXAMPLE 19.7: Will lead chloride (Ksp = 1.6×10^{-5}) precipitate if 15 mL 3.0×10^{-4} M $Pb(NO_3)_2$ are combined with 25 mL 4.0×10^{-2} M $CaCl_2$?

The equation and Ksp expression were found in Example 19.4:

$PbCl_2(s) \rightleftharpoons Pb^{2+}(aq) + 2\ Cl^-(aq)$ Ksp $= [Pb^{2+}][Cl^-]^2 = 1.6 \times 10^{-5}$

Begin by finding the diluted concentration of the lead ion.

GIVEN: $M_1 = [Pb^{2+}]_1 = 3.0 \times 10^{-4}$ M FIND: $M_2 = [Pb^{2+}]_2$
$V_1 = 15$ mL, $V_2 = 40$ mL EQN: $M_1V_1 = M_2V_2$

$$[Pb^{2+}]_2 = [Pb^{2+}]_1 \times \frac{V_1}{V_2} = 3.0 \times 10^{-4}\text{ M} \times \frac{15\text{ mL}}{40\text{ mL}} = 1.1 \times 10^{-4}\text{ M}$$

Be careful in calculating the chloride ion concentration.

GIVEN: $M_1 = [Cl^-]_1 = 2 \times 4.0 \times 10^{-2}$ M FIND: $M_2 = [Cl^-]_2$
$V_1 = 25$ mL, $V_2 = 40$ mL EQN: $M_1V_1 = M_2V_2$

$$[Cl^-]_2 = [Cl^-]_1 \times \frac{V_1}{V_2} = 8.0 \times 10^{-2}\text{ M} \times \frac{25\text{ mL}}{40\text{ mL}} = 5.0 \times 10^{-2}\text{ M}$$

When calcium chloride dissolves, the chloride ion concentration is twice the molarity of the solute. (See Section 19.1.)

Use the values you have found to calculate Q, write it as an inequality with Ksp, and state your conclusion regarding precipitation.

$Q = (1.1 \times 10^{-4})(5.0 \times 10^{-2})^2 = 2.8 \times 10^{-7}$
$2.8 \times 10^{-7} < 1.6 \times 10^{-5}$. Therefore, there will be no precipitate.

CHAPTER 19 PROBLEMS

Use Table 19.1 as the source of solubility product constants for slightly soluble salts whenever they are required to solve problems in this chapter.

1) $[Br^-] = 7.1 \times 10^{-7}$ in a saturated solution of AgBr, and in a saturated solution of BaF_2, $[Ba^{2+}] = 7.5 \times 10^{-3}$. Calculate the solubility product constants for these two solutes.

2) The concentration of a saturated solution of $MgCO_3$ is 0.27 grams per liter, and of $Cd(OH)_2$, 0.0016 grams per liter. Find the solubility product constants of magnesium carbonate and cadmium hydroxide.

3) For each solute listed below, calculate the solubility in moles per liter and grams per liter: a) AgI b) $Ca(OH)_2$ c) $Al(OH)_3$.

4) For each solute below, calculate its solubility in the solution indicated. Express the solubility in both moles per liter and grams per liter.

a) $NiCO_3$ in 0.15 M $NiCl_2$ b) CaF_2 in 0.27 M NaF

5) For each combination of solutions below, (1) calculate the reaction quotient Q for any possible precipitate, (2) write it in an inequality with its solubility product constant, and (3) state whether or not a precipitate will form.

a) 5.0×10^1 mL of 0.024 M $AgNO_3$ + 25 mL of 0.018 M KSCN
b) 1.0×10^2 mL of 0.060 M $CuSO_4$ + 25 mL of 3.2×10^{-3} M $NaIO_3$
c) 1 drop (0.05 mL) of 2 M Na_3PO_4 + 50.00 mL of 5.0×10^{-5} M $AgNO_3$

\#

6) Calculate the solubility product constant of the solute in each of the saturated solutions described: a) $[C_2O_4^{2-}] = 1.7 \times 10^{-4}$ M in saturated $CuC_2O_4(aq)$; b) $[Fe^{2+}] = 1.6 \times 10^{-5}$ M in saturated $Fe(OH)_2(aq)$.

7) The concentration of saturated Ag_2CrO_4 is 0.27 grams per liter at a certain temperature. Find Ksp at that temperature.

8) Calculate the solubility of each of the following solutes, expressed in both moles per liter and grams per liter: a) CdS b) PbF_2 c) Ag_3PO_4.

9) For each solute below, calculate the solubility in the solution indicated. Express the result in both moles per liter and grams per liter.

a) $PbSO_4$ in 0.044 M $CuSO_4$ b) $Mn(OH)_2$ in 0.75 M NaOH

10) Determine if precipitation will occur when the following pairs of solutions are combined. Verify your conclusion by showing the reaction quotient Q in an inequality with the solubility product constant.

a) 35 mL of 0.041 M $LiNO_3$ + 85 mL of 0.021 M Na_2CO_3
b) 72 mL of 0.056 M $Mg(NO_3)_2$ + 16 mL of 0.0013 M $Ba(OH)_2$

ADDITIONAL PROBLEMS

11) Which of the following silver salts have the highest concentration of Ag^+, AgCl, Ag_2CrO_4, or Ag_3PO_4?

12) A common problem with water heaters is the precipitation of calcium sulfate at high temperatures. If the solubility of calcium sulfate at 100°C is 1.62 g/L, what is Ksp at this temperature?

13) Many drain cleaners contain sodium hydroxide. The heat generated by the dissolving of the NaOH is effective in melting/dissolving clogs due to grease and vegetable matter. Drain stoppages due to rust, approximately $Fe(OH)_3$, are not removed, however. By what factor is the solubility of $Fe(OH)_3$ lowered in 3.0 M NaOH compared to its solubility in water? Ksp = 4×10^{-38}.

14) If an industrial pollutant which is 0.030 M in Ag^+ is dumped into an equal volume of ocean water which is 0.028 M in SO_4^{2-}, will silver sulfate precipitate?

CHAPTER TEST

1) If $[SO_4^{2-}]$ = 0.016 M in saturated Ag_2SO_4, calculate Ksp for the solute.

2) Find the Ksp of $BaCO_3$ if its solubility is 0.018 grams per liter.

3) Calculate the solubility in moles per liter and grams per liter of

a) HgS (Ksp = 1.6×10^{-52}) b) Ag_2CO_3 (Ksp = 8.2×10^{-12})

4) What is the solubility of $CaSO_4$ in 0.33 M $Ca(NO_3)_2$? Answer in both moles per liter and grams per liter. Ksp = 1.2×10^{-6} for $CaSO_4$.

5) Will magnesium oxalate precipitate if 35 mL of 0.047 M $Mg(NO_3)_2$ are added to 56 mL of 0.064 M $Na_2C_2O_4$? Calculate the reaction quotient Q of the combined solutions, and write it in an inequality with Ksp for MgC_2O_4 to support your conclusion. (Ksp = 8.3×10^{-5}.)

20. ACID-BASE EQUILIBRIA

PREREQUISITES

1. This is the chapter in which you will use logarithms, described in Chapter 5. Calculator techniques for these operations are given in Section 1.6.

2. The algebraic methods shown in Sections 1.9 to 1.11 appear throughout equilibrium problems.

3. Calculations with very small concentrations and equilibrium constants are performed in exponential notation (Section 2.1).

4. The rules of significant figures given in Sections 3.3 to 3.7, continue to be used in this chapter.

5. The rules of significant figures for logarithms (Section 5.5) appear in the chapter.

6. The principles of weak and strong electrolytes, discussed in Section 15.2, are used in identifying equilibria.

7. Section 15.5 describes net ionic equations for acid-base reactions such as those that appear in this chapter.

8. The tabular organization you used in solving limiting reactant problems in Section 10.4 and for equilibrium problems in Section 18.4 is used again in solving equilibrium problems involving acid-base equilibria.

9. Molarity and dilution calculations (Sections 14.4, 14.5 and 19.4) are used again in this chapter.

10. All of the ideas in Chapter 18 are used in Chapter 20. In particular, Le Chatelier's principle and equilibrium problem solving methods for gas phase equilibria are applied to acid-base equilibria.

CHEMICAL SKILLS

1. The pH scale. Given any one of the following, calculate the others: (a) $[H^+]$, or the concentration of a strong acid; (b) $[OH^-]$, or the concentration of a strong base; (c) pH; (d) pOH.

2. K expressions. Given only the formula of the compound, write the defining equation for K_a of a weak acid or K_b of a weak base.

3. Weak acids in water. Given any two of the following three, calculate the third: (a) K_a for a weak acid HA; (b) the initial value of [HA]; (c) $[H^+]$ or $[A^-]$ at equilibrium, or the pH of the solution, or the percentage dissociation.

4. Weak bases in water. Given any two of the following three, calculate the third: (a) K_b for a weak base B; (b) the initial value of [B]; (c) $[OH^-]$ or $[BH^+]$ at equilibrium, or the pH of the solution.

5. Buffer. Given K_a, [HA], and $[A^-]$, or information from which they can be calculated, determine the pH of the buffer.

6. Making a buffer. Determine amounts of a weak acid and the conjugate base of the same acid that are required to prepare a buffer of a certain pH, given K_a of the acid. The amounts may be the grams of a solid or the volume and molarity of a solution.

20.1 THE WATER EQUILIBRIUM; STRONG ACIDS AND BASES IN WATER

Probably the most critical of all equilibria is the dissociation of water. The equilibrium equation and the equilibrium constant at 25°C are

$$H_2O(\ell) \rightleftharpoons H^+(aq) + OH^-(aq) \qquad K_w = [H^+][OH^-] = 1.0 \times 10^{-14} \qquad (20.1)$$

K_w is used to identify this equilibrium constant. Not only does this equilibrium exist in water, but it is also present in all aqueous solutions.

When you studied electrolytes in Section 15.2, you learned that an electrolyte is a better electrical conductor than water because electrolytes produce aqueous solutions containing ions. Equation 20.1 shows that water, too, forms ions: H^+ and OH^-. However, the very small equilibrium constant, 1.0×10^{-14}, shows that the equilibrium is strongly favored in the reverse direction. The concentration of H^+ and OH^- ions is very low in water, so water is a very poor conductor of electricity. As far as net ionic equations are concerned, the dissociation of water is so slight that it can be ignored.

EXAMPLE 20.1: Calculate the hydrogen and hydroxide ion concentrations in pure water, assuming both ions come only from the dissociation of water as given in Equation 20.1.

What can be said about the relative values of $[H^+]$ and $[OH^-]$ if the only source of the ions is the dissociation of water?

They are equal. (Recall Section 19.1.)

Complete the calculation for the ion concentrations.

$[H^+] = [OH^-] = x$; $\quad x^2 = 1.0 \times 10^{-14}$; $\quad x = 1.0 \times 10^{-7}$ M

EXAMPLE 20.2: Calculate the hydrogen and hydroxide ion concentrations of 0.012 M HCl.

HCl is a strong acid and dissociates almost completely to H^+ and Cl^- ions: $HCl \rightarrow H^+ + Cl^-$. The ion concentrations are determined by this equation. What is the H^+ concentration in 0.012 molar HCl?

$[H^+] = 0.012$ M

From $HCl \rightarrow H^+ + Cl^-$, each mole of HCl yields 1 mole of H^+, so 0.012 mole of HCl yields 0.012 mole H^+.

Now you can use Equation 20.1 for K_w to solve for $[OH^-]$.

GIVEN: $[H^+] = 0.012$ M FIND: $[OH^-]$ EQN: $K_w = [H^+][OH^-]$

$$[OH^-] = \frac{K_w}{[H^+]} = \frac{1.0 \times 10^{-14}}{1.2 \times 10^{-2}} = 8.3 \times 10^{-13} \text{ M}$$

20.2 THE pN NUMBERS

It is not very convenient to speak of or work with molarities such as 8.3×10^{-13} M and 1.0×10^{-7} M. The chemist therefore uses a **pN number** to represent these very small values. **A pN number is defined mathematically as the negative of the base 10 logarithm of a number, N:**

$$\text{For any number N, pN} \equiv -\log N \tag{20.2}$$

The exponential character of a logarithm was described in Section 5.1, which included Equation 5.1:

$$\text{If} \quad N = 10^x, \quad \text{then} \quad \log N = \log 10^x = x \tag{5.1}$$

Combining Equations 5.1 and 20.2,

$$\text{If} \quad N = 10^x, \quad \text{then} \quad pN = -\log N = -\log 10^x = -x \tag{20.3}$$

The summary in Section 5.1 pointed out that the logarithm of a number smaller than 1 is negative. The N values in this chapter are all smaller than 1, so their logarithms are all negative. Thus the pN values in this chapter are the opposites (negatives) of negative numbers, which makes them positive. As an example, if N = 0.001, it can be written $N = 10^{-3}$. The logarithm of N is then -3, and pN is its opposite: pN = 3.

The best known pN number is **pH, the negative logarithm of the hydrogen ion concentration of a solution.** Substitution into Equation 20.3 gives the equation for pH:

$$[H^+] = 10^{-pH} \qquad pH = -\log [H^+] \tag{20.4}$$

Equally useful in working with the water equilibrium is **pOH**, for which

$$[OH^-] = 10^{-pOH} \qquad pOH = -\log [OH^-] \tag{20.5}$$

The use of logarithms in the multiplication of exponentials was described in Equation 5.2:

$$10^y \times 10^z = 10^{y+z} \tag{5.2}$$

Equation 20.1 is repeated for comparison, followed with substitutions for $[H^+]$ and $[OH^-]$ from Equations 20.4 and 20.5:

$$[H^+][OH^-] = 1.0 \times 10^{-14} \tag{20.1}$$

$$10^{-pH} \times 10^{-pOH} = 1.0 \times 10^{-14} \tag{20.6}$$

From Equations 5.2 and 20.6, it follows that

$$pH + pOH = 14.00 \tag{20.7}$$

In summary, the following equations are the ones you should know and be able to use in solving pH problems:

$$[H^+][OH^-] = 1.0 \times 10^{-14} \tag{20.1}$$

$$[H^+] = 10^{-pH} \qquad pH = -\log [H^+] \tag{20.4}$$

$$[OH^-] = 10^{-pOH} \qquad pOH = -\log [OH^-] \tag{20.5}$$

$$pH + pOH = 14.00 \tag{20.7}$$

EXAMPLE 20.3: In Examples 20.1 and 20.2 you found the concentrations shown below. Complete the table by filling in the pH and pOH.

LIQUID	$[H^+]$	$[OH^-]$
Water	1.0×10^{-7}	1.0×10^{-7}
	pH = ______	pOH = ______
0.012 M HCl	1.2×10^{-2}	8.3×10^{-13}
	pH = ______	pOH = ______

LIQUID	$[H^+]$	$[OH^-]$
Water	1.0×10^{-7}	1.0×10^{-7}
	pH = 7.00	pOH = 7.00
0.012 M HCl	1.2×10^{-2}	8.3×10^{-13}
	pH = 1.92	pOH = 12.08

The method of finding the logarithm of a number expressed in exponential notation is explained in Sections 5.3. To change from the logarithm to the pN value, simply change the sign of the logarithm.

Example 20.3 provides the occasion to mention the number of decimals to which pH and pOH should be expressed. Recall that the number of significant figures in a quantity written in exponential notation is the number of significant figures in the coefficient. Also remember from Section 5.5 that the number of significant figures in a logarithm is the same as the number of digits after the decimal. Both pH and pOH are logarithms. Therefore, pH and pOH are generally written with the same number of digits after the decimal as there are significant figures in the coefficient of the concentration.

To find the hydrogen ion concentration from a given pH, change the sign of the pH and then find the antilogarithm. (Check Section 1.6 for the procedure if necessary.) For example, if the pH of a solution is 9.23, $[H^+]$ is found from Equation 20.4:

$$[H^+] = \text{antilog}(-pH) = 10^{-pH} = 10^{-9.23} = 5.9 \times 10^{-10}\ M$$

Because pH is expressed in two significant figures (two digits after the decimal), $[H^+]$ is also given in two significant figures.

EXAMPLE 20.4: The pH of a solution is 5.79. Calculate, in order, $[H^+]$, $[OH^-]$ pOH, and pH. If you do everything correctly, you will return to the starting value, with minor variations because of roundoffs.

There are four steps in this problem, and each one requires the use of one of the boxed-in equations above.

$$[H^+] = 10^{-5.79} = 1.6 \times 10^{-6}\ M \qquad \text{(Equation 20.4)}$$

$$[OH^-] = \frac{1.0 \times 10^{-14}}{1.6 \times 10^{-6}} = 6.3 \times 10^{-9}\ M \qquad \text{(Equation 20.1)}$$

$$pOH = -\log(6.3 \times 10^{-9}) = 8.20 \qquad \text{(Equation 20.5)}$$

$$pH = 14.00 - 8.20 = 5.80 \qquad \text{(Equation 20.7)}$$

Example 20.4 shows how you can get as much practice on pH-type conversions as you wish. You can start at any value among the following and make conversions in either direction until you get back to where you started.

$$pH \leftrightarrow pOH \leftrightarrow [OH^-] \leftrightarrow [H^+] \leftrightarrow pH$$

Your final value should match the starting value, with some allowance for roundoffs along the way. If you do not round off at any point—if you leave your display value in the calculator—you will return to exactly the same number you started with.

20.3 WEAK ACIDS IN WATER

The difference between a strong acid and a weak acid is that the strong acid dissociates completely (or nearly so), but the weak acid dissociates only slightly. This is why the solution inventory of a weak acid is written with neutral molecules as the the major species and ions as the minor species.

For 0.012 M HCl, $[H^+] = 0.012$ M, as you found in Example 20.2. HCl is a strong acid. It is 100% dissociated. Therefore, the concentration of the hydrogen ion is the same as the molarity of the solution. It follows that $[H^+] = 0.15$ in 0.15 M HCl, and the pH is $-\log 0.15 = 0.82$.

In 0.15 molar acetic acid, however, $[H^+]$ is not equal to 0.15 M. This is because only a small percentage of the $HC_2H_3O_2$ molecules dissociates. The equilibrium is described by the equation:

$$HC_2H_3O_2(aq) \rightleftharpoons H^+(aq) + C_2H_3O_2^-(aq) \qquad (20.8)$$

Both $[H^+]$ and the extent of dissociation in 0.15 M $HC_2H_3O_2$ can be determined by measuring the pH of the solution, which is 2.79. Thus

$$[H^+] = 10^{-2.79} = 1.6 \times 10^{-3} \text{ M}.$$

Out of the 0.15 moles of $HC_2H_3O_2$ originally present in 1 liter of solution, only 0.0016 dissociate. To find percentage dissociation, divide the moles that dissociate by the total number of moles at the start, and then multiply by 100:

$$\% \text{ dissociation} = \frac{\text{moles that dissociate}}{\text{total moles}} \times 100 = \frac{0.0016}{0.15} \times 100 = 1.1\% \qquad (20.9)$$

This indicates that 98.9% of the original 0.15 mole present remains as undissociated neutral molecules. Thus the equilibrium concentration of acetic acid is

$$98.9\% \text{ of } 0.15 \text{ M} = 0.989 \times 0.15 \text{ M} = 0.14835 \text{ M} = 0.15 \text{ M}$$

The result is rounded off to the two significant figures given in the original concentration. The equilibrium concentration is the same as the original concentration. In other words, the reduction in acetic acid concentration because of dissociation is so small compared to the acetic acid originally present that it may be disregarded.

The figures given above can be used to calculate the **acid dissociation constant** of acetic acid, K_a. (The subscript "a" identifies the equilibrium constant for the dissociation of a weak acid.) The acid dissociation constant applies specifically to the equation for the dissociation of a weak acid when placed in water:

$$HA(aq) \rightleftharpoons H^+(aq) + A^-(aq), \qquad K_a = \frac{[H^+][A^-]}{[HA]} \qquad (20.10)$$

For acetic acid, the K_a expression is

$$K_a = \frac{[H^+][C_2H_3O_2^-]}{[HC_2H_3O_2]} \qquad (20.11)$$

In Equation 20.8, $HC_2H_3O_2$ is the only source for either H^+ or $C_2H_3O_2^-$ ions, and they are produced in identical molar quantities. (The amount of H^+ from the dissociation of water is negligible compared to the 0.0016 mole per liter from the acid.) It follows that $[H^+]$ and $[C_2H_3O_2^-]$ must be the same, 0.0016 M. Substituting into Equation 20.11,

$$K_a = \frac{(0.0016)(0.0016)}{0.15} = 1.7 \times 10^{-5}$$

The K_a for acetic acid as it is listed in handbooks is 1.8×10^{-5}. The difference lies in calculation roundoffs.

It is helpful to use S-C-E tables (start-change-equilibrium), developed in Section 18.4 for gas phase equilibria, to solve acid-base equilibrium problems. The equilibrium equation is written with enough space between the formulas so the formulas become headings for the table. Separate lines are then added for the starting concentration (S), the change (C), and the equilibrium concentration (E). For the 0.15 M $HC_2H_3O_2$ example just worked above this set up is:

	$HC_2H_3O_2(aq)$	⇌	$H^+(aq)$	+	$C_2H_3O_2^-(aq)$
S	0.15 M		0 M		0 M
C	-0.0016		+0.0016		+0.0016
E	0.15		0.0016		0.0016

The change line in the table (C) is found from the values in the start (S) and equilibrium (E) lines. For H^+ and $C_2H_3O_2^-$, if you start with zero and end with 0.0016, the change must be +0.0016.

EXAMPLE 20.5: A 0.21 molar solution of a certain weak acid, HA, has a pH of 3.14. Find $[H^+]$ for the solution, K_a for the acid, and the percentage dissociation.

$[H^+]$ comes directly from the pH of the solution. Calculate that first.

$[H^+] = 10^{-3.14} = 7.2 \times 10^{-4}$ M

Now write the equilibrium equation for the dissociation of the weak acid HA in water.

	HA(aq)	⇌	$H^+(aq)$	+	$A^-(aq)$
S					

The equation has been written with the formulas separated so they may be used as headings for a table. Place in the S line—the starting line—the numbers that express the concentrations of the acid and the two ions _before_ any dissociation occurs.

	$HA(aq)$	$\rightleftharpoons$	$H^+(aq)$	+	$A^-(aq)$
S	0.21 M		0 M		0 M
C					
E					

The starting concentration of the acid is 0.21 M. $[H^+]$ and $[A^-]$ begin at zero, assuming $[H^+]$ from the dissociation of water is negligible.

You have already calculated $[H^+]$ at equilibrium: 7.2×10^{-4} M. Do you also know $[A^-]$? Enter these values into the E line of the table.

	$HA(aq)$	$\rightleftharpoons$	$H^+(aq)$	+	$A^-(aq)$
S	0.21 M		0 M		0 M
C					
E			0.00072		0.00072

Equal moles of the two ions come from the same source, so the concentrations must be the same.

If the beginning $[H^+]$ and $[A^-]$ are zero, and they end at 0.00072 M, what is the change in concentration? Enter that number, preceded by a plus sign if the change is an increase and a minus sign if it is a decrease.

	$HA(aq)$	$\rightleftharpoons$	$H^+(aq)$	+	$A^-(aq)$
S	0.21 M		0 M		0 M
C			+0.00072		+0.00072
E			0.00072		0.00072

If you begin with 0 and end with 0.00072, the change must be a gain of 0.00072, or +0.00072.

Now figure out what the change is in [HA]. This is a stoichiometry problem. If, according to the equation, 1 mole of HA dissociates for each mole of H^+ and each mole of A^- produced, how many moles dissociate if 0.00072 mole of each ion is produced? Enter that number into the table, including a + sign if the change is a gain or a - sign if it is a loss.

	$HA(aq)$	$\rightleftharpoons$	$H^+(aq)$	+	$A^-(aq)$
S	0.21 M		0 M		0 M
C	-0.00072		+0.00072		+0.00072
E			0.00072		0.00072

The HA that dissociates is a loss so far as [HA] is concerned, so the sign is negative. The three substances are equal in moles in the equation, so the numbers of moles per liter are the same, all 0.00072 M.

Only one blank remains in the table. Fill in the equilibrium concentration of HA. Remember the rules of significant figures.

	$HA(aq)$	⇌	$H^+(aq)$	+	$A^-(aq)$
S	0.21 M		0 M		0 M
C	-0.00072		+0.00072		+0.00072
E	0.21		0.00072		0.00072

According to the rules of significant figures, 0.21 - 0.00072 = 0.20928 must be rounded off to the second decimal place. This takes you right back to 0.21 M. The concentration change in HA is negligible compared to the initial concentration.

You now have the equilibrium concentrations of all species. Write the equilibrium constant expression and calculate the value of K_a.

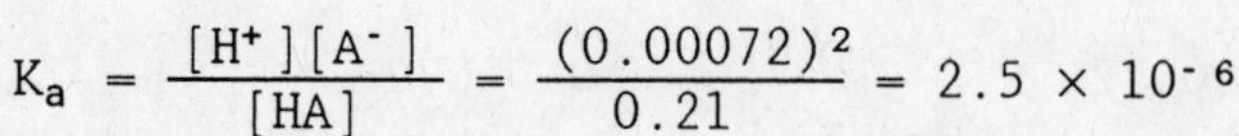

$$K_a = \frac{[H^+][A^-]}{[HA]} = \frac{(0.00072)^2}{0.21} = 2.5 \times 10^{-6}$$

Finally, calculate the percentage dissociation.

GIVEN: [HA] = 0.21 M at start, FIND: % dissociation

[HA] that dissociate = 0.00072 M

$$\text{EQN: } \% = \frac{\text{M that dissociate}}{\text{M at start}} \times 100 = \frac{0.00072}{0.21} \times 100 = 0.34\%$$

Suppose you wish to find the pH and percentage dissociation of 0.27 M $HC_2H_3O_2$. The equilibrium equation is written as the column heads of an S-C-E table. The S line can be completed as before. No equilibrium concentrations are known this time, so algebraic symbols are used. Let y represent $[H^+]$ at equilibrium. Below is the completed table. (AE means "assumed equilibrium." It will be explained shortly.)

	$HC_2H_3O_2(aq)$	⇌	$H^+(aq)$	+	$C_2H_3O_2^-(aq)$
S	0.27 M		0 M		0 M
C	-y (4)		+y (2)		+y (3)
E	0.27 - y (6)		y (1)		y (5)
AE	0.27 (7)		y		y

The circled numbers in the table are keyed to the numbered steps in the procedure below. Study the procedure carefully; be sure you understand it.

Step	Procedure
1	Let y = $[H^+]$ at equilibrium. Place y in E line under H^+.
2	$[H^+]$ was zero at the beginning, and it is y at equilibrium. The change is +y. Write +y in the C line under $[H^+]$.

Step	Procedure
3, 4	All equation coefficients are 1, so all changes in moles per liter are the same. $[C_2H_3O_2^-]$ increased by y, so +y is written under $C_2H_3O_2^-$. $[HC_2H_3O_2]$ decreases, so -y is placed under the acid formula.
5, 6	At equilibrium, $[C_2H_3O_2^-] = 0 + y$ and $[HC_2H_3O_2] = 0.27 - y$. Add these binomials to the E line of the table.
7	See below.

Table 20.1 is a list of acid dissociation constants. The equilibrium concentrations may now be taken from the table, substituted into the equation for K_a, and set equal to the value of the constant. The resulting quadratic equation is then solved for y.

$$K_a = \frac{[H^+][C_2H_3O_2^-]}{[HC_2H_3O_2]} = \frac{y^2}{0.27-y} = 1.8 \times 10^{-5}$$

$$y^2 = 4.86 \times 10^{-6} - 1.8 \times 10^{-5}\ y$$

$$y^2 + 1.8 \times 10^{-5}\ y - 4.86 \times 10^{-6} = 0$$

The quadratic equation has been written in the form $ax^2 + bx + c = 0$, so it can be solved by the quadratic formula (Section 1.11). The result is that the hydrogen ion concentration is 2.2×10^{-3} M.

In the earlier discussion of 0.15 M $HC_2H_3O_2$, we found that the change in $[HC_2H_3O_2]$ caused by the dissociation of the acid was negligible. This appeared also in Example 20.5. Is it negligible for 0.27 M $HC_2H_3O_2$ too? If we assume it is negligible, then $[HC_2H_3O_2]$ is unchanged at 0.27 M. Line AE has been added to the S-C-E table, identifying the "assumed" equilibrium concentration for the acid (Step 7). The other concentrations remain the same. Notice what happens to the calculation:

$$\frac{[H^+][C_2H_3O_2^-]}{[HC_2H_3O_2]} = \frac{y^2}{0.27} = 1.8 \times 10^{-5}$$

$$y^2 = 4.86 \times 10^{-6}$$

$$y = 2.2 \times 10^{-3}$$

By assuming that dissociation has a negligible effect on the concentration of the weak acid, exactly the same answer has been reached *without using the quadratic formula!* Can the quadratic formula be avoided in other problems too? The answer is yes—most of the time.

Because the assumption answer is identical to the quadratic formula answer, we know the assumption is valid. If the answer by the quadratic formula had not been known, it would have been necessary to check the validity of the assumption. The method is to compute the percentage dissociation of the weak acid. If it is less than 5%, the dissociation is negligible and the assumption is valid. The percentage dissociation is:

$$\% = \frac{[C_2H_3O_2^-](\text{equil})}{[HC_2H_3O_2](\text{start})} = \frac{0.0022\ M}{0.27\ M} \times 100 = 0.81\%\ \text{dissociated}$$

The extent of dissociation is less than 5%, so the assumption is valid. The pH may be calculated using $[H^+] = 2.2 \times 10^{-3}$ M.

$$pH = -\log [H^+] = -\log (2.2 \times 10^{-3}) = 2.66$$

TABLE 20.1: DISSOCIATION CONSTANTS OF WEAK ACIDS AND BASES

ACID NAME	EQUATION	K_a
Oxalic	$H_2C_2O_4 \rightleftharpoons H^+ + HC_2O_4^-$	5.9×10^{-2}
Sulfurous	$H_2SO_3 \rightleftharpoons H^+ + HSO_3^-$	1.7×10^{-2}
Hydrogen sulfate ion	$HSO_4^- \rightleftharpoons H^+ + SO_4^{2-}$	1.2×10^{-2}
Phosphoric	$H_3PO_4 \rightleftharpoons H^+ + H_2PO_4^-$	7.5×10^{-3}
Arsenic	$H_3AsO_4 \rightleftharpoons H^+ + H_2AsO_4^-$	5.6×10^{-3}
Phthalic	$H_2C_8H_4O_4 \rightleftharpoons H^+ + HC_8H_4O_4^-$	1.3×10^{-3}
Tartaric	$H_2C_4H_4O_6 \rightleftharpoons H^+ + HC_2H_4O_6^-$	1.0×10^{-3}
Citric	$H_3C_6H_5O_7 \rightleftharpoons H^+ + H_2C_6H_5O_7^-$	8.4×10^{-4}
Hydrofluoric	$HF \rightleftharpoons H^+ + F^-$	6.8×10^{-4}
Nitrous	$HNO_2 \rightleftharpoons H^+ + NO_2^-$	4.6×10^{-4}
Acetoacetic	$HC_3H_5O_3 \rightleftharpoons H^+ + C_3H_5O_3^-$	2.6×10^{-4}
Formic	$HCHO_2 \rightleftharpoons H^+ + CHO_2^-$	1.8×10^{-4}
Lactic	$HC_3H_5O_3 \rightleftharpoons H^+ + C_3H_5O_3^-$	1.5×10^{-4}
Benzoic	$HC_7H_5O_2 \rightleftharpoons H^+ + C_7H_5O_2^-$	6.5×10^{-5}
Hydrogen oxalate ion	$HC_2O_4^- \rightleftharpoons H^+ + C_2O_4^{2-}$	6.4×10^{-5}
Hydrogen tartarate ion	$HC_4H_4O_6^- \rightleftharpoons H^+ + C_4H_4O_6^{2-}$	4.6×10^{-5}
Dihydrogen citrate ion	$H_2C_6H_5O_7^- \rightleftharpoons H^+ + HC_6H_5O_7^{2-}$	1.8×10^{-5}
Acetic	$HC_2H_3O_2 \rightleftharpoons H^+ + C_2H_3O_2^-$	1.8×10^{-5}
Propionic	$HC_3H_5O_2 \rightleftharpoons H^+ + C_3H_5O_2^-$	1.4×10^{-5}
Hydrogen citrate ion	$HC_6H_5O_7^{2-} \rightleftharpoons H^+ + C_6H_5O_7^{3-}$	4.0×10^{-6}
Hydrogen phthalate ion	$HC_8H_4O_4^- \rightleftharpoons H^+ + C_8H_4O_4^{2-}$	3.9×10^{-6}
Carbonic	$H_2CO_3 \rightleftharpoons H^+ + HCO_3^-$	4.3×10^{-7}
Dihydrogen arsenate ion	$H_2AsO_4^- \rightleftharpoons H^+ + HAsO_4^{2-}$	1.7×10^{-7}
Hydrosulfuric	$H_2S \rightleftharpoons H^+ + HS^-$	1.0×10^{-7}
Dihydrogen phosphate ion	$H_2PO_4^- \rightleftharpoons H^+ + HPO_4^{2-}$	6.2×10^{-8}
Hydrogen sulfite ion	$HSO_3^- \rightleftharpoons H^+ + SO_3^{2-}$	5.6×10^{-8}
Hypochlorous	$HClO \rightleftharpoons H^+ + ClO^-$	3.2×10^{-8}
Boric	$H_3BO_3 \rightleftharpoons H^+ + H_2BO_3^-$	5.8×10^{-10}
Ammonium ion	$NH_4^+ \rightleftharpoons H^+ + NH_3$	5.6×10^{-10}
Hydrocyanic	$HCN \rightleftharpoons H^+ + CN^-$	4.9×10^{-10}
Hydrogen carbonate ion	$HCO_3^- \rightleftharpoons H^+ + CO_3^{2-}$	4.8×10^{-11}
Hydrogen arsenate ion	$HAsO_4^{2-} \rightleftharpoons H^+ + AsO_4^{3-}$	4.0×10^{-12}
Hydrogen phosphate ion	$HPO_4^{2-} \rightleftharpoons H^+ + PO_4^{3-}$	4.8×10^{-13}
Water	$HOH \rightleftharpoons H^+ + OH^-$	1.0×10^{-14}
Hydrogen sulfide ion	$HS^- \rightleftharpoons H^+ + S^{2-}$	1.0×10^{-15}

BASE NAME	EQUATION	K_b
Diethylamine	$(C_2H_5)_2NH + H_2O \rightleftharpoons OH^- + (C_2H_5)_2NH_2^+$	1.3×10^{-3}
Ethylamine	$C_2H_5NH_2 + H_2O \rightleftharpoons OH^- + C_2H_5NH_3^+$	5.6×10^{-4}
Dimethylamine	$(CH_3)_2NH + H_2O \rightleftharpoons OH^- + (CH_3)_2NH_2^+$	5.2×10^{-4}
Methylamine	$CH_3NH_2 + H_2O \rightleftharpoons OH^- + CH_3NH_3^+$	5.0×10^{-4}
Trimethylamine	$(CH_3)_3N + H_2O \rightleftharpoons OH^- + (CH_3)_3NH^+$	6.1×10^{-5}
Ammonia	$NH_3 + H_2O \rightleftharpoons OH^- + NH_4^+$	1.8×10^{-5}
Hydrazine	$N_2H_4 + H_2O \rightleftharpoons OH^- + N_2H_5^+$	3.0×10^{-6}

EXAMPLE 20.6: Calculate the pH and percentage dissociation of 0.44 M $HC_2H_3O_2$.

A table is shown below. The S line is completed, and again we have let y represent $[H^+]$ at equilibrium. Procedure reference numbers are shown. See if you can work your way through the table, filling in all the open spaces.

	$HC_2H_3O_2(aq)$	⇌	$H^+(aq)$	+	$C_2H_3O_2^-(aq)$
S	0.44 M		0 M		0 M
C	4		2		3
E	6		1 y		5
AE	7		7		7

	$HC_2H_3O_2(aq)$	⇌	$H^+(aq)$	+	$C_2H_3O_2^-(aq)$
S	0.44 M		0 M		0 M
C	- y		+ y		+ y
E	0.44 - y		y		y
AE	0.44		y		y

This table is exactly the same as the one for 0.27 M $HC_2H_3O_2$, except that the molarity 0.44 has been substituted for 0.27. It is assumed that the dissociation of the acetic acid is so small it will not reduce the molarity below 0.44 M at equilibrium.

Write the equilibrium constant expression for acetic acid, substitute the entries in the AE line of the table, set it equal to K_a, and solve for y, which is $[H^+]$.

$$K_a = \frac{[H^+][C_2H_3O_2^-]}{[HC_2H_3O_2]} = \frac{y^2}{0.44} = 1.8 \times 10^{-5}; \qquad y = 2.8 \times 10^{-3}$$

Now is the time to check the assumption. Calculate the percentage dissociation, and state if the assumption is valid.

$$\% = \frac{[C_2H_3O_2^-](\text{equil})}{[HC_2H_3O_2](\text{start})} \times 100 = \frac{0.0028 \text{ M}}{0.44 \text{ M}} \times 100 = 0.64\% \text{ dissociated}$$

The extent of dissociation is less than 5%, so the assumption is valid.

The pH may be calculated using $[H^+] = 2.8 \times 10^{-3}$ M.

$$pH = -\log [H^+] = -\log (2.8 \times 10^{-3}) = 2.55$$

EXAMPLE 20.7: Find the pH and percentage dissociation of 0.024 M HNO_2. Begin by preparing and filling in the S-C-E table. Include an AE line.

	$HNO_2(aq)$	⇌	$H^+(aq)$	+	$NO_2^-(aq)$
S	0.024 M		0 M		0 M
C	-y		+y		+y
E	0.024 - y		y		y
AE	0.024		y		y

Again the table is the same, except that the acid and molarity are changed.

Calculate the value of y and check the assumption. You will find the K_a for nitrous acid in Table 20.1.

$$K_a = \frac{[H^+][NO_2^-]}{[HNO_2]} = \frac{y^2}{0.024} = 4.6 \times 10^{-4}; \qquad y = 3.3 \times 10^{-3} = 0.0033$$

Assumption check: $(0.0033/0.024) \times 100 = 14\%$. Assumption not valid.

When the assumption is not valid, the answer can be obtained by applying the quadratic formula. Begin by substituting equilibrium concentrations from the E line of the table into the K_a expression. Write and rearrange the equation algebraically so it is in the proper form for the quadratic formula: $ax^2 + bx + c = 0$.

$$\frac{y^2}{0.024 - y} = 4.6 \times 10^{-4}; \qquad y^2 + (4.6 \times 10^{-4})y - (1.1 \times 10^{-5}) = 0$$

Now apply the quadratic formula and solve for y.

$y = 3.1 \times 10^{-3}$; $[H^+] = 3.1 \times 10^{-3}$ M

Finally, calculate the pH and percentage dissociation.

$$pH = -\log [H^+] = -\log (3.1 \times 10^{-3}) = 2.51$$

$$\% = \frac{[NO_2^-](equil)}{[HNO_2](start)} \times 100 = \frac{0.0031 \text{ M}}{0.024 \text{ M}} \times 100 = 13\%$$

20.4 WEAK BASES IN WATER

Some compounds behave as weak bases when placed in water, removing a hydrogen ion from the water molecule and leaving a hydroxide ion. Trimethylamine is such a compound. The equilibrium and K equations are:

$$(CH_3)_3N(aq) + H_2O(\ell) \rightleftharpoons (CH_3)_3NH^+(aq) + OH^-(aq)$$

$$K_b = \frac{[(CH_3)_3NH^+][OH^-]}{[(CH_3)_3N]} = 6.1 \times 10^{-5} \quad (20.12)$$

K_b is the equilibrium constant of a weak base reacting with water. Weak base equilibrium problems are handled in exactly the same way as weak acid problems, except that the unknown, y, is set equal to the hydroxide ion concentration. If the wanted quantity is pH, it must be calculated from $[OH^-]$.

EXAMPLE 20.8: Calculate the pH of 0.29 M $(CH_3)_3N$.

Set up the usual table, including an AE line, with formulas from the equilibrium equation at the top of each column. Remember that no term appears for water in the equilibrium constant expression.

	$(CH_3)_3N(aq)$ +	$H_2O(\ell)$ ⇌	$(CH_3)_3NH^+(aq)$ +	$OH^-(aq)$
S	0.29 M		0 M	0 M
C	-y		+y	+y
E	0.29 - y		y	y
AE	0.29		y	y

The AE line is repeated at the top of the next page. Using entries on that line, set up the equilibrium constant equation and solve for y as before. Check the assumption to be sure it is valid.

	$(CH_3)_3N(aq)$ +	$H_2O(\ell)$ ⇌	$(CH_3)_3NH^+(aq)$ +	$OH^-(aq)$
AE	0.29		y	y

$$K_b = \frac{[(CH_3)_3NH^+][OH^-]}{[(CH_3)_3N]} = \frac{y^2}{0.29} = 6.1 \times 10^{-5}; \qquad y = 4.2 \times 10^{-3}$$

% = (0.0042/0.29) × 100 = 1.4% Assumption valid.

Remember that y represents $[OH^-]$ this time. Calculate the pH.

$$[H^+] = \frac{1.0 \times 10^{-14}}{4.2 \times 10^{-3}} = 2.4 \times 10^{-12}\ M \qquad pOH = -\log(4.2 \times 10^{-3}) = 2.38$$

$$pH = -\log(2.4 \times 10^{-12}) = 11.62 \qquad pH = 14.00 - 2.38 = 11.62$$

There are two ways to convert $[OH^-]$ to pH. Both are shown.

20.5 IONS AS ACIDS AND BASES; HYDROLYSIS

So far in this chapter we have studied compounds that are weak acids and compounds that are weak bases. Ions can be weak acids and bases too. All of the ions in Table 20.1 are either acids or bases, according to the Bronsted-Lowry acid-base theory. Moreover, except for the hydroxide and sulfide ions, they are weak acids or bases.

When a weak acid dissociates to produce a hydrogen ion and another species, the original acid and the other species constitute a conjugate acid-base pair. Thus acetic acid and the acetate ion are a conjugate acid-base pair in Equation 20.8:

$$HC_2H_3O_2(aq) \rightleftharpoons H^+(aq) + C_2H_3O_2^-(aq) \qquad (20.8)$$

The acetate ion is said to be the **conjugate base** of acetic acid; and in the reverse direction, acetate ion is a base and acetic acid is its **conjugate acid**.

Every equation in Table 20.1 has in it a conjugate acid-base pair. In the top section of the table, the acid is on the left and the base is on the right. In the bottom section the bases are on the left and the acids are on the right.

In the top section of Table 20.1, the acids are arranged in order of decreasing strength. This is indicated by the constantly smaller K_a values from the top down. The smaller the K_a, the more the reaction is favored in the reverse direction. The conjugate bases, however, are arranged in order of increasing strength, weaker at the top to stronger at the bottom. The strongest base that can be stable in an aqueous solution is the hydroxide ion.

When a soluble salt, such as sodium hypochlorite, NaClO(s), dissolves in water, $Na^+(aq)$ and $ClO^-(aq)$ are formed:

$$NaClO(s) \rightarrow Na^+(aq) + ClO^-(aq) \qquad (20.13)$$

The sodium ion is a spectator; it is neither an acid nor a base. The hypochlorite ion is a weak base. As such, it can remove an H^+ ion from water:

$$ClO^-(aq) + H_2O(\ell) \rightleftharpoons HClO(aq) + OH^-(aq) \qquad (20.14)$$

Notice the similarity between the above equation and Equation 20.12 for the equilibrium of the weak base $(CH_3)_3N$ in water:

$$(CH_3)_3N(aq) + H_2O(\ell) \rightleftharpoons (CH_3)_3NH^+(aq) + OH^-(aq) \qquad (20.12)$$

Both weak bases react with water to form their conjugate acids and an OH^- ion.

Just as with $(CH_3)_3N$, a K_b expression can be written for ClO^-:

$$K_b = \frac{[HClO][OH^-]}{[ClO^-]} \qquad (20.15)$$

Table 20.1 lists no K_b value for ClO^-, as it does for $(CH_3)_3N$. K_b tables usually list only neutral bases, not ions. However, the value of K_b for ClO^- can be calculated from K_a for HClO. For all weak conjugate acid-base pairs:

$$K_a \times K_b = K_w \qquad (20.16)$$

The validity of this equation can be shown by substituting expressions for K_a, K_b, and K_w, and canceling identical terms. The result is an identity:

$$\frac{[H^+][ClO^-]}{[HClO]} \times \frac{[HClO][OH^-]}{[ClO^-]} = [H^+][OH^-]$$

$$[H^+][OH^-] = [H^+][OH^-]$$

Equation 20.17 can be solved for K_b and its value for ClO^- calculated by substituting K_a for HClO and K_w:

$$K_b = \frac{K_w}{K_a} = \frac{1.0 \times 10^{-14}}{3.2 \times 10^{-8}} = 3.1 \times 10^{-7}$$

The small value of K_b for Equation 20.15 supports the statement that ClO^- is a weak base in water; the equilibrium is favored in the reverse direction.

The reaction of a basic ion with water, such as that shown in Equation 20.15, is called **hydrolysis**. K_b values for such basic ions are identified as **hydrolysis constants, K_h**. Acidic ions, such as $(CH_3)_3NH^+$, can hydrolyze too, and their K_a values are also hydrolysis constants.

EXAMPLE 20.9: Calculate the pH of 0.20 M $NaC_2H_3O_2$.

Sodium acetate contains the basic ion $C_2H_3O_2^-$. The sodium ion is a spectator and does not affect the equilibrium. Write the equilibrium equation for acetate ion reacting with water and the expression for K_b.

$$C_2H_3O_2^-(aq) + H_2O(\ell) \rightleftharpoons HC_2H_3O_2(aq) + OH^-(aq); \qquad K_b = \frac{[HC_2H_3O_2][OH^-]}{[C_2H_3O_2^-]}$$

Now calculate the value of K_b, using Equation 20.17. (You will find the value for K_a for acetic acid in Table 20.1.)

$$K_b = \frac{K_w}{K_a} = \frac{1.0 \times 10^{-14}}{1.8 \times 10^{-5}} = 5.6 \times 10^{-10}$$

You are now ready to set up an S-C-E-AE table and complete the problem as in Example 20.8.

	$C_2H_3O_2^-(aq)$ +	$H_2O(\ell)$ ⇌	$HC_2H_3O_2(aq)$ +	$OH^-(aq)$
S	0.20 M		0 M	0 M
C	-y		+y	+y
E	0.20 - y		y	y
AE	0.20		y	y

$$\frac{y^2}{0.20} = 5.6 \times 10^{-10} \qquad pOH = -\log(1.1 \times 10^{-5}) = 4.96$$

$$y = 1.1 \times 10^{-5} = [OH^-] \qquad pH = 14.00 - 4.96 = 9.04$$

$$\% = (1.1 \times 10^{-5}/0.20) \times 100 = 0.0055\%$$ Assumption valid.

20.6 BUFFERS

Throughout the discussion of weak acid equilibria in Section 20.3, the acid was the only source of both the hydrogen ion and the conjugate base. This is not always the case.

Consider again the acetic acid equilibrium:

$$HC_2H_3O_2(aq) \rightleftharpoons H^+(aq) + C_2H_3O_2^-(aq) \qquad (20.8)$$

What would happen if the acetate ion concentration was increased from another source, by adding sodium acetate, for example? How would $[H^+]$ change? How would the percentage dissociation of acetic acid change?

Do you recognize this as an example of Le Chatelier's principle, and specifically of the common ion effect, described in Section 19.3? The addition of a common ion—an ion already present in the solution ($C_2H_3O_2^-$)—

shifts the equilibrium away from that ion. This reduces the dissociation of the weak acid. It also reduces $[H^+]$ and raises the pH. (If you have already studied Chapter 19, this may all be quite familiar. It helps you to understand concepts such as these if you see them as different applications of the same fundamental idea. It is worth looking for such similarities.)

The addition of small amounts of a strong acid or strong base usually has a sizeable effect on the pH of a solution. However, if the solution has in its inventory a weak acid and a weak base, it can absorb significant quantities of strong acid or base without much change in pH. Such a solution is called a **buffer**; it is said to be **buffered** at whatever its pH is. This simplest kind of a buffer is made up of a weak acid and its weak conjugate base, both present in relatively high concentration.

EXAMPLE 20.10: Calculate the pH of a solution that is 0.35 molar in $HC_2H_3O_2$ and 0.25 molar in $NaC_2H_3O_2$.

This example is the same acetic acid equilibrium you worked with earlier. The problem can be approached just as the others have been. The equilibrium equation is written at the top of the table below. We have furnished only the S line, to be sure we start correctly. Complete that line.

	$HC_2H_3O_2(aq)$	⇌	$H^+(aq)$	+	$C_2H_3O_2^-(aq)$
S					

	$HC_2H_3O_2(aq)$	⇌	$H^+(aq)$	+	$C_2H_3O_2^-(aq)$
S	0.35 M		0 M		0.25 M
C					
E					
AE					

In earlier problems the hydrogen and acetate ions both came from acetic acid. This time the acetate ion is introduced independently of the acid. Sodium acetate is a soluble salt, so its acetate ion concentration is the same as the molar concentration of the salt.

As before, we are looking for the pH of the solution, so you must find $[H^+]$. Complete the rest of the table just as you did earlier.

	$HC_2H_3O_2(aq)$	⇌	$H^+(aq)$	+	$C_2H_3O_2^-(aq)$
S	0.35 M		0 M		0.25 M
C	-y		+y		+y
E	0.35 - y		y		0.25 + y
AE	0.35		y		0.25

If you begin with $[H^+] = 0$ and let $y = [H^+]$ at equilibrium, the change in $[H^+]$ must be +y. This fixes the change line at -y for $[HC_2H_3O_2]$ and +y for $[C_2H_3O_2^-]$. The E line values are found in the usual way: E = S + C. Two assumptions are necessary this time.

	$HC_2H_3O_2(aq)$	⇌	$H^+(aq)$	+	$C_2H_3O_2^-(aq)$
AE	0.35		y		0.25

Recall that $K_a = 1.8 \times 10^{-5}$ for acetic acid. Solve for y and the pH.

$$K_a = \frac{[H^+][C_2H_3O_2^-]}{[HC_2H_3O_2]} = \frac{y(0.25)}{0.35} = 1.8 \times 10^{-5} \qquad pH = -\log(2.5 \times 10^{-5}) = 4.60$$

$$y = 2.5 \times 10^{-5}$$

$(2.5 \times 10^{-5}/0.35) \times 100 = 0.0071\%$ The assumption is valid.

The AE line was really not necessary in Example 20.10. If the dissociation of pure acetic acid is negligible in similar concentrations, as we have seen earlier in this chapter, it will be even smaller—even "more negligible"—in the presence of a common ion.

EXAMPLE 20.11: Find the pH of a solution that is 0.024 molar in HNO_2 and 0.036 molar in NO_2^-. K_a may be found in Table 20.1.

We leave this problem for you to solve without suggestions. The procedure is the same as in Example 20.10.

	$HNO_2(aq)$	⇌	$H^+(aq)$	+	$NO_2^-(aq)$
S	0.024 M		0 M		0.036 M
C	-y		+y		+y
E	0.024 - y		y		0.036 + y
AE	0.024		y		0.036

$$K_a = \frac{[H^+][NO_2^-]}{[HNO_2]} = \frac{y(0.036)}{0.024} = 4.6 \times 10^{-4} \qquad pH = -\log(3.1 \times 10^{-4}) = 3.51$$

$$y = 3.1 \times 10^{-4}$$

$(3.1 \times 10^{-4}/0.024) \times 100 = 1.3\%$ Assumption valid.

In Example 20.7 you found the pH of 0.024 M HNO_2 without the common ion present. The assumption was not valid. In Example 20.11, starting with the same acid concentration, but with the nitrite ion added from another source, the dissociation of the acid was reduced to the point it was negligible. This is almost always true when both acid and conjugate base are present in molarities greater than 0.01 and within a factor of ten of each other, and the equilibrium constant is smaller than 10^{-3}. This leads to a simpler way to solve buffer problems.

The general equation for a weak acid equilibrium and its K_a is

$$HX(aq) \rightleftharpoons H^+(aq) + X^-(aq) \qquad K_a = \frac{[H^+][X^-]}{[HX]} \tag{20.17}$$

Solving for $[H^+]$,

$$[H^+] = K_a \times \frac{[HX]}{[X^-]} \tag{20.18}$$

Using Equation 20.18 in Example 20.11 gives the result directly:

$$[H^+] = 4.6 \times 10^{-4} \times \frac{0.024}{0.036} = 3.1 \times 10^{-4}\ M$$

The message from Equation 20.18 is that *the pH of a buffer made from a given acid and its conjugate base depends entirely on the ratio* $[HX]/[X^-]$.

EXAMPLE 20.12: Find the pH at which a solution is buffered if $[HC_2H_3O_2]$ is 0.12 M and $[C_2H_3O_2^-]$ is 0.16 M.

Use Equation 20.18 to solve this problem.

$$[H^+] = 1.8 \times 10^{-5} \times \frac{0.12}{0.16} = 1.4 \times 10^{-5}\ M; \quad pH = -\log(1.4 \times 10^{-5}) = 4.85$$

EXAMPLE 20.13: 71.0 milliliters of concentrated acetic acid, 17.5 M $HC_2H_3O_2$, and 16.4 grams of sodium acetate, $NaC_2H_3O_2$, are dissolved in water. Calculate the pH at which the solution is buffered.

To find the concentration ratio to substitute into Equation 20.18, the moles of acid and acetate ion must be calculated.

GIVEN: 0.071 L FIND: mol $HC_2H_3O_2$ FACTOR: 17.5 mol $HC_2H_3O_2$/L PATH: L → mol

$$0.071 \text{ L } HC_2H_3O_2 \times \frac{17.5 \text{ mol } HC_2H_3O_2}{1 \text{ L } HC_2H_3O_2} = 1.24 \text{ mol } HC_2H_3O_2$$

GIVEN: 16.4 g $NaC_2H_3O_2$ FIND: mol $C_2H_3O_2^-$ FACTOR: MM $NaC_2H_3O_2$ PATH: g → mol

$$16.4 \text{ g } NaC_2H_3O_2 \times \frac{1 \text{ mol } NaC_2H_3O_2}{82.03 \text{ g } NaC_2H_3O_2} = 0.200 \text{ mol } NaC_2H_3O_2 = 0.200 \text{ mol } C_2H_3O_2^-$$

You now have the moles of each solute which must be put into a ratio of molarities. But the volume of solution is not given. It is a fact that the volume in which 1.24 moles $HC_2H_3O_2$ are dissolved is the same volume in which the 0.200 mole of $NaC_2H_3O_2$ is dissolved. If that volume is V, then

$$[HC_2H_3O_2] = 1.24 \text{ mol } HC_2H_3O_2/V \quad \text{and} \quad [C_2H_3O_2^-] = 0.200 \text{ mol } C_2H_3O_2^-/V$$

If you substitute those expressions into Equation 20.18 you will be able to complete the problem.

$$[H^+] = 1.8 \times 10^{-5} \times \frac{1.24 \text{ mol } HC_2H_3O_2/V}{0.200 \text{ mol } NaC_2H_3O_2/V} = 1.1 \times 10^{-4} \text{ M}$$

$$pH = -\log(1.1 \times 10^{-4}) = 3.96$$

The volume of the solution cancels out in the setup for Example 20.13, no matter what it is. This indicates that the ratio of moles of acid to moles of conjugate base in a single solution is the same as the ratio of concentrations. Equation 20.18 may therefore be extended to include the ratio of moles:

$$[H^+] = K_a \times \frac{[HX]}{[X^-]} = K_a \times \frac{\text{mol HX}}{\text{mol X}^-} \qquad (20.20)$$

The greater usefulness of Equation 20.19 is that it shows what is needed to prepare a buffer of specified pH. Suppose, for example, you wished to prepare a buffer with pH = 5.00, using acetic acid and sodium acetate. What molar ratio of the two substances is needed? Equation 20.19 can be solved for two ratios, one the inverse of the other:

$$\text{(a)} \quad \frac{[HX]}{[X^-]} = \frac{[H^+]}{K_a} = \frac{\text{mol HX}}{\text{mol X}^-} \qquad \text{(b)} \quad \frac{[X^-]}{[HX]} = \frac{K_a}{[H^+]} = \frac{\text{mol X}^-}{\text{mol HX}} \qquad (20.20)$$

If the pH is to be 5.00, $[H^+]$ will be 1.0×10^{-5} M. Therefore,

$$\text{(a)} \quad \frac{[HC_2H_3O_2]}{[C_2H_3O_2^-]} = \frac{\text{mol } HC_2H_3O_2}{\text{mol } C_2H_3O_2^-} = \frac{1.0 \times 10^{-5} \text{ M}}{1.8 \times 10^{-5} \text{ M}} = \frac{1.0}{1.8} = \frac{0.56}{1.0}$$

$$\text{(b)} \quad \frac{[C_2H_3O_2^-]}{[HC_2H_3O_2]} = \frac{\text{mol } C_2H_3O_2^-}{\text{mol } HC_2H_3O_2} = \frac{1.8 \times 10^{-5} \text{ M}}{1.0 \times 10^{-5} \text{ M}} = \frac{1.8}{1.0}$$

The ratios are, of course, reciprocals of each other. Sufficient quantities of acetic acid and sodium acetate to yield the above ratios will buffer a solution at 5.00.

0.56/1 and 1.8/1 are ratios of moles. They still do not tell what amounts of acetic acid and sodium acetate to use. To make the ratios practical, they must be changed to measurable amounts. Another practical question is: How concentrated should the solution be in each solute? It is convenient in preparing a buffer to make the concentrations fall in the range of 0.1 to 1.0 mole per liter. Therefore, using ratio (a) just as it is and (b) at one-tenth of the numbers shown, we need for each liter of buffer either

(a) 0.56 mol $HC_2H_3O_2$ and 1.0 mol $NaC_2H_3O_2$ or
(b) 0.18 mol $NaC_2H_3O_2$ and 0.10 mol $HC_2H_3O_2$

Concentrated acetic acid is 17.5 molar. The molar mass of sodium acetate is 82.03 grams per mole. Consequently, to prepare 1 liter of buffer, you would use for (a) 82.03 g $NaC_2H_3O_2$ and

$$0.56 \text{ mol } HC_2H_3O_2 \times \frac{1 \text{ L } HC_2H_3O_2}{17.5 \text{ mol } HC_2H_3O_2} = 0.032 \text{ L} = 32 \text{ mL } HC_2H_3O_2$$

For (b) the quantities would be

$$0.18 \text{ mol } NaC_2H_3O_2 \times \frac{82.03 \text{ g } NaC_2H_3O_2}{1 \text{ mol } NaC_2H_3O_2} = 14.8 \text{ g } NaC_2H_3O_2$$

and

$$0.10 \text{ mol } HC_2H_3O_2 \; \frac{1 \text{ L } HC_2H_3O_2}{17.5 \text{ mol } HC_2H_3O_2} = 0.0057 \text{ L} = 5.7 \text{ mL } HC_2H_3O_2$$

These quantities are calculated for one liter of solution. The amounts may remain the same for a slightly larger or smaller volume, as it is the *ratio* that determines the pH of the buffer. The quantities should be adjusted proportionally for volumes that are much larger or smaller than 1 liter. As noted, it is desirable to keep both concentrations in the range 0.1 to 1.0 molar.

EXAMPLE 20.14: Specify the grams of sodium acetate and milliliters of 6.0 M $HC_2H_3O_2$ to use in preparing 5.0×10^{-2} L of a solution that is buffered at pH = 4.25.

Begin by determining the hydrogen ion concentration of the solution and the ratio of moles of $HC_2H_3O_2$ and $NaC_2H_3O_2$.

$$[H^+] = 10^{-4.25} = 5.6 \times 10^{-5} \text{ M}$$

$$\frac{[HC_2H_3O_2]}{[C_2H_3O_2^-]} = \frac{[H^+]}{K_a} = \frac{\text{mol } HC_2H_3O_2}{\text{mol } C_2H_3O_2^-} = \frac{5.6 \times 10^{-5}}{1.8 \times 10^{-5}} = \frac{3.1}{1.0}$$

Alternately, the reciprocal, $\frac{[C_2H_3O_2^-]}{[HC_2H_3O_2]} = \frac{\text{mol } C_2H_3O_2^-}{\text{mol } HC_2H_3O_2} = \frac{0.32}{1.0}$

From this point there are several ways to go, all leading to the same result as long as the mole ratio is maintained. Choose what you will, and calculate a mass of sodium acetate and the volume of 6.0 M $HC_2H_3O_2$ that goes with it to produce either of the following ratios:

$$\frac{\text{mol } HC_2H_3O_2}{\text{mol } C_2H_3O_2^-} = \frac{3.1}{1.0} \quad \text{or} \quad \frac{\text{mol } C_2H_3O_2^-}{\text{mol } HC_2H_3O_2} = \frac{0.32}{1.0}$$

This calculation is based on the first ratio, using one-tenth quantities at 0.31 mol $HC_2H_3O_2$/0.10 mol $NaC_2H_3O_2$:

$$0.31 \text{ mol } HC_2H_3O_2 \times \frac{1 \text{ L } HC_2H_3O_2}{6.0 \text{ mol } HC_2H_3O_2} = 0.052 \text{ L} = 52 \text{ mL } HC_2H_3O_2$$

$$0.10 \text{ mol } NaC_2H_3O_2 \times \frac{82.03 \text{ g } NaC_2H_3O_2}{1 \text{ mol } NaC_2H_3O_2} = 8.2 \text{ g } NaC_2H_3O_2$$

Depending on the numbers you chose to work with, your quantities may be different from those above. You may check your result by dividing the milliliters of acid by grams of sodium acetate. You should get

$$\frac{52 \text{ mL } HC_2H_3O_2}{8.2 \text{ g } NaC_2H_3O_2} = 6.3 \text{ mL acid/g } NaC_2H_3O_2$$

If you do, your solution will have a pH of 4.25. If the actual concentrations are to be within the range of 0.1 to 1.0 molar for a 500-mL solution, the mass of sodium acetate should be from 4.1 to 41 grams, and the volume of acid should be from 26 mL to 260 mL.

CHAPTER 20 PROBLEMS

Table 20.1 should be used as the source dissociation constants for weak acids and bases, whenever they are required to solve problems in this chapter.

1) Find the $[H^+]$, $[OH^-]$, pH, and pOH of each solution listed below. Assume complete dissociation of the solute.

a) 7.2×10^{-2} M HNO_3; b) 4.9×10^{-4} M NaOH; c) 5.5×10^{-4} M HCl

2) a) Calculate the pOH, $[OH^-]$, and $[H^+]$ of a solution if pH = 8.26.
b) If pOH = 9.49, find pH, $[OH^-]$, and $[H^+]$.

3) pH = 2.10 for 0.12 M HB, a weak acid. Calculate K_a.

4) Find the acid constant for HY, a weak acid, if pH = 3.42 in 0.32 M HY.

5) 0.16 M HZ is 1.2% dissociated. Calculate the pH of the solution and K_a.

6) Calculate the pH and percent dissociation of each of the following:

a) 0.75 M $HC_2H_3O_2$ (acetic acid)

b) 0.16 M $HCHO_2$ (formic acid)

c) 0.092 M $HC_4H_5O_3$ (acetoacetic acid)

7) Find the pH of 1.2 M methylamine, CH_3NH_2.

8) Determine the pH of a) 0.74 M NaF and b) 0.33 M Na_2CO_3.

9) At what pH will the following solutions be buffered:

a) 0.50 M $HC_2H_3O_2$ and 0.72 M $NaC_2H_3O_2$

b) 0.23 M HF and 0.10 M NaF

c) 0.29 M Na_2CO_3 and 0.12 M $NaHCO_3$

10) 12 mL 26.5 M $HCHO_2$ (formic acid) and 38 g $NaCHO_2$ are dissolved in water and diluted to 750 mL. Calculate the pH of this buffer.

11) 1.75 liters of a buffer are prepared from 35.0 grams of sodium tartarate $Na_2C_4H_4O_6$, and 35.0 grams of sodium hydrogen tartarate, $NaHC_4H_4O_6$. Find the pH of the solution. ($K_a = 4.6 \times 10^{-5}$.)

12) Find the [acid]/[conjugate base] ratio that will yield the pH specified:

a) $\frac{[HC_2H_3O_2]}{[C_2H_3O_2^-]}$ for pH 4.50; b) $\frac{[HNO_2]}{[NO_2^-]}$ for pH 3.41; c) $\frac{[NH_4^+]}{[NH_3]}$ for pH 9.50

13) Determine the quantities of chemicals listed that will yield the following buffers:

a) 500 mL buffered at pH = 4.85, made from 17.4 M $HC_2H_3O_2$ and $NaC_2H_3O_2$

b) 1.20 liters buffered at pH = 6.90, made from $NaHSO_3$ and Na_2SO_3.

++++++++++++++++++++

14) Find the pH, pOH, [H^+], and [OH^-] of a) 0.072 M HNO_3; b) 0.00049 M HCl.

15) a) If [H^+] = 1.8×10^{-6} M, find pH, pOH, and [OH^-].

b) Calculate pH, [H^+], and [OH^-] in a solution in which pOH = 5.17.

16) pH = 3.06 in 0.19 M HA, a weak acid. Calculate K_a.

17) HD is a weak acid. 0.27 M HD is 0.74% dissociated. Calculate the pH of the solution and K_a.

18) Find the pH and percentage dissociation of a) 0.47 M $HC_3H_5O_3$ (lactic acid) and b) 0.31 M $H_2C_4H_4O_6$ (tartaric acid).

19) Calculate the pH of 0.63 M N_2H_4 (hydrazine).

20) Find the pH of a) 0.52 M $NaNO_2$ and b) 0.65 M KCN.

21) What is the pH of the following buffers: a) 0.91 M HNO_2 and 0.77 M $NaNO_2$; b) 0.24 M $NaHSO_3$ and 0.55 M Na_2SO_3.

22) Find the pH of 450 mL of a buffer that is prepared from 29 mL 7.4 M NH_3 and 9.0 grams of NH_4Cl.

23) What [acid]/[conjugate base] ratio will yield the buffer with the pH specified:

a) $\frac{[HC_2O_4^-]}{[C_2O_4^{2-}]}$ for pH 3.90 b) $\frac{[H_2PO_4^-]}{[HPO_4^{2-}]}$ for pH 7.00

24) How many milliliters of 14.6 M H_3PO_4 and grams of NaH_2PO_4 should be used to prepare 750 mL of a buffer at pH = 2.40?

ADDITIONAL PROBLEMS

25) Sodium hydrogen carbonate is used to neutralize both acid and base spills. What is the pH after 25 grams of $NaHCO_3$ are added to a 750 mL spill of 0.20 M HCl? of 0.20 M NaOH? Assume no volume change upon the addition of solid. [Hints: How many moles are present after these reactions: $HCO_3^- + H^+ \rightleftharpoons H_2CO_3$ and $HCO_3^- + OH^- \rightleftharpoons CO_3^{2-} + H_2O$? The remaining solutions will be buffers.

26) Blood is maintained at about pH 7.40 by both phosphate and carbonate buffers. What are the $H_2PO_4^-/HPO_4^{2-}$ and H_2CO_3/HCO_3^- ratios at this pH?

27) What is the $[F^-]$ in a 0.015 M solution of HF? If this solution is also 0.050 M in Ca^{2+}, will CaF_2 precipitate?

28) Urine is basic due to a number of nitrogen bases including urea, NH_2CONH_2. If a 10.00% aqueous solution of urea has a pH of 7.25 and a density of 1.025 g/mL, what is K_b for urea?

CHAPTER TEST

1) Find the $[H^+]$, $[OH^-]$, pH, and pOH of 4.1×10^{-4} M HBr.

2) pOH = 4.78 in a certain solution. Calculate $[H^+]$, $[OH^-]$, and pH.

3) pH = 2.59 for 0.11 M HE, a weak acid. Calculate K_a for that acid.

4) Find the pH and percentage dissociation of 0.028 M $HC_7H_5O_2$ (benzoic acid, $K_a = 6.5 \times 10^{-5}$).

5) Find the pH of 1.1 M dimethylamine, $(CH_3)_2NH$. ($K_b = 5.2 \times 10^{-4}$)

6) Calculate the pH of 0.22 M NaClO if $K_a = 3.2 \times 10^{-8}$ for HClO.

7) What is the pH of a buffer that is 0.012 M $HC_7H_5O_2$ and 0.049 M $NaC_7H_5O_2$? $HC_7H_5O_2$ is benzoic acid, for which $K_a = 6.5 \times 10^{-5}$.

8) What is the pH of a buffer prepared by dissolving 11 grams of $NaC_2H_3O_2$ and 23 mL 17.4 M $HC_2H_3O_2$ in water and diluting to 5.00×10^2 milliliters? The acid dissociation constant of acetic acid is 1.8×10^{-5}.

9) What ratio of $[HC_3H_5O_3]/[C_3H_5O_3^-]$ will yield a buffer with pH = 4.00? $HC_3H_5O_3$ is lactic acid, for which $K_a = 1.5 \times 10^{-4}$.

10) How many milliliters of 7.4 M NH_3 and grams of NH_4Cl will be satisfactory to prepare 2.00 liters of a buffer having a pH of 9.05? $K_a = 5.6 \times 10^{-10}$ for NH_4^+.

ANSWERS TO QUESTIONS AND PROBLEMS

CHAPTER 1

1) a) 18.94 b) -8.81 c) 117.43 d) 131.94046 e) -1.6712741

2) a) 729 b) 8 c) 8.3406054 d) $(560)^{0.25} = 4.8645986$ e) 243 f) 539.57359
g) $(27)^{0.20} = 1.933182$ h) $(0.306)^{0.33333} = 0.6738664$

3) a) $\frac{8(2) + 3(5)}{3 \times 8} = \frac{16 + 15}{24} = \frac{31}{24}$ b) $\frac{14(5) + 4(6) + 13}{28} = \frac{107}{28}$ c) $\frac{3(12) - 5(4)}{15} = \frac{16}{15}$

d) $\frac{5(7) - 10(5) - 4(3) + 2(17)}{40} = \frac{7}{40}$

4) a) $\frac{9}{16} \times \frac{8}{21} \times 24 \times \frac{7}{9} = \frac{1}{16} \times \frac{8}{21} \times 24 \times 7 = \frac{1}{2} \times \frac{1}{21} \times 24 \times 7 = \frac{1}{2} \times \frac{1}{3} \times 24 = 4$

b) $\frac{7}{8} \times \frac{3}{2} \times \frac{1}{5} = \frac{21}{80}$ c) $3 \times 27 \times \frac{1}{3} = 27$ d) $\frac{2}{3} \times \frac{5}{4} \times \frac{27}{28} \times \frac{7}{18} = 2 \times \frac{5}{4} \times \frac{9}{28} \times \frac{7}{18} =$

$\frac{5}{2} \times \frac{9}{28} \times \frac{7}{18} = \frac{5}{2} \times \frac{1}{28} \times \frac{7}{2} = \frac{5}{2} \times \frac{1}{4} \times \frac{1}{2} = \frac{5}{16}$

5) a) x^{-5} b) x^{12} c) 5^{-10}

6) a) $-x = -3$, $x = 3$ b) $7x = 14$, $x = 2$ c) $3x - 6 = 12$, $x = 6$ d) $x = 5/8$ e) $64 = x - 4$, $x = 68$

f) $4x = 5(12)$, $x = 15$ g) $-10x + 5 - 18 = 3x$, $-13 = 13x$, $x = -1$ h) $5 + 6x - 8 = 14x - 21$, $18 = 8x$, $x = 9/4$

i) $x^2 - 2x - 15 = 0$

$$x = \frac{+2 \pm \sqrt{(2)^2 - 4(1)(-15)}}{2}$$

$$x = \frac{+2 \pm 8}{2} = +5 \text{ or } -3$$

j) $5x^2 - 7x + 2 = 0$

$$x = \frac{+7 \pm \sqrt{(7)^2 - 4(5)(-2)}}{2(5)}$$

$$x = \frac{+7 \pm 3}{10} = +1 \text{ or } +2/5$$

7) a) $r = \frac{I}{pt}$ b) $h = \frac{2A}{b + B}$ c) $x = \frac{y - b}{m}$ d) $b = 25-a-c$ e) $b = \sqrt{c^2-a^2}$ f) $M = \frac{gRT}{PV}$

Problems 8 through 11 are "practical" examples. As such, they provide an opportunity to show how meaningless full calculator readouts are as answers to real-life problems. Comments appear with the answers.

8) $b = \frac{2A}{h} = 2 \times \frac{46.7}{5.86} = 15.938567$ cm

The difference between 15.938567 cm and 15.938568 cm, or 0.000001 cm, is about 0.0000004 (4 ten-millionths) of an inch, or about 1/100000 of the thickness of a U. S. dime. This tiny distance has no practical meaning. Correctly rounded off, the answer is 15.9 cm.

9) $A = \pi r^2 = 3.1416 \times 3.79^2 = 45.126257\ cm^2$.

The difference between 45.126257 cm^2 and 45.126258 cm^2, or 0.000001 cm^2, is about 1/30000 (1 thirty-thousandth) of the area of the head of a pin. This microscopic area is meaningless. Properly rounded off, the answer is 45.1 cm^2.

10) $Y = \frac{M}{IQ} \times 100 = \frac{14}{120} \times 100 = 11.666667$ years.

The time interval from 11.1666667 year to 11.1666668 years is 32 seconds. In other words, 32 seconds after the student is 11.1666667 years old, she will be 11.1666668 years old. How old do you think she will be tomorrow? If her age is to be expressed to the same precision, that depends on which of the 2700 thirty-two-second periods of tomorrow we are asking about, because tomorrow she will have 2700 different ages. As a matter of fact, you have probably aged by nearly ten of these periods—0.000001 years—since you began working on this problem. The correctly rounded off answer to the problem is 11 years.

11) $\text{time} = \frac{\text{distance}}{\text{speed}} = \frac{1275}{510} = 2.5$ hours.

CHAPTER 2

1) a) 9.11×10^{4} b) 7.5×10^{-8} c) 6.4×10^{3} d) 1.65×10^{-4}
e) 8.16×10^{-2} f) 9.35×10^{2}

2) a) 0.0000224 b) 930 c) 42,000 d) 0.00295 e) 0.0735 f) 0.0000000000818

3) a) 2.78×10^{5} b) 9.2×10^{-3} c) 6.11×10^{4} d) 4.39 e) 1.65×10^{-4} f) 6.41×10^{4}

4) a) 3.29×10^{11} b) 4.0827×10^{-7} c) 3.63125×10^{-10} d) 8.8065844
e) 1.217687×10^{12} f) 1.820163×10^{-3} g) 58.696042 h) 1.374229×10^{-3}

5) a) 4.826×10^{10} b) 6.861×10^{-2} c) 9.7044×10^{8} d) 2.1926×10^{-4}

The following the answers have been rounded for significant figures. See Chapter 4.

6) $25 \text{ miles} \times \frac{5280 \text{ ft}}{1 \text{ mile}} \times \frac{1 \text{ min}}{1320 \text{ ft}} \times \frac{1 \text{ hr}}{60 \text{ min}} = 1\ 2/3 \text{ hr}$

7) $175 \text{ persons} \times \frac{1.25 \text{ sandwiches}}{1 \text{ person}} \times \frac{\$1.68}{12 \text{ sandwiches}} = \$30.63$

8) $812 \text{ students} \times \frac{9.3 \text{ pencils}}{1 \text{ student}} \times \frac{1 \text{ dozen}}{12 \text{ pencils}} \times \frac{1 \text{ gross}}{12 \text{ dozen}} \times \frac{1 \text{ box}}{1 \text{ gross}} = 52.4 = 53 \text{ boxes}$

9) $403 \text{ miles} \times \frac{1 \text{ gal}}{24.0 \text{ miles}} \times \frac{\$1.44}{1 \text{ gal}} = \$24.18 \times 2 = \$48.36$ (round trip)

10) $\frac{\$2925}{225 \text{ shares}} = \$13.00/\text{share}$

11) $\frac{2088 \text{ lb}}{144 \text{ cartons}} = 14.5 \text{ lb/carton}$

12) $\% = \frac{6.48 \text{ g gold}}{75.0 \text{ g sample}} \times 100 = 8.64\% \text{ gold}$

13) $\% = \frac{13.2 \text{ g ammonia}}{(13.2 + 225) \text{g solution}} \times 100 = 5.54\% \text{ ammonia}$

14) $542 \text{ g solution} \times \frac{37.0 \text{ g acid}}{100.0 \text{ g solution}} = 201 \text{ g acid}$

15) $\% = \frac{44.0 \text{ g salt}}{175 \text{ g solution}} \times 100 = 25.1\% \text{ salt}$

16) $50.0 \text{ g silver} \times \frac{100.0 \text{ g ore}}{23.4 \text{ g silver}} = 214 \text{ g ore}$

17) $520 \text{ kg ore} \times \frac{27.0 \text{ kg iron oxide}}{100.0 \text{ kg ore}} = 140 \text{ kg iron oxide}$

18) $40.4 \text{ g sucrose} \times \frac{100.0 \text{ g solution}}{8.50 \text{ g sucrose}} = 475 \text{ g solution}$

19) $75.5 \text{ g hydrogen} \times \frac{100.0 \text{ g water}}{11.2 \text{ g hydrogen}} = 674 \text{ g water}$

$75.5 \text{ g hydrogen} \times \frac{88.8 \text{ g oxygen}}{11.2 \text{ g hydrogen}} = 599 \text{ g oxygen}$

CHAPTER 3

2) one: 8 three: 7.76 five: 7.7581

two: 7.8 four: 7.758 six: 7.75806

3) 3.1×10^2. Two significant figures set by 0.0042.

4) a)
$$\begin{array}{r} 18.7 \\ -\ 0.56 \\ \hline 18.14 \end{array} = 18.1$$

b)
$$\begin{array}{r} 15.9 \times 10^{-5} \\ -\ 6.42 \times 10^{-5} \\ \hline 9.48 \times 10^{-5} \end{array} = 9.5 \times 10^{-5}$$

5) one: 4 three: 4.11 five: 4.1067

two: 4.1 four: 4.107 six: 4.10674

6) 1.3×10^{-3}. Two significant figures set by 5.6×10^3.

7) a)
$$\begin{array}{r} 294.696 \\ 10.4752 \\ 0.701 \\ 0.0086 \\ 41.61 \\ 17.009 \\ \hline 364.4998 \end{array} = 364.50$$

b)
$$\begin{array}{r} 4{,}390 \\ 126 \\ 58{,}100 \\ \hline 62{,}616 \end{array} = 62{,}600 = 6.26 \times 10^4$$

8) a)

$$6.91 \text{ g} \times \frac{1 \text{ mg}}{10^{-3} \text{ g}} = 6.91 \times 10^3 \text{ mg}$$

$$6.91 \text{ g} \times \frac{1 \text{ kg}}{10^3 \text{ g}} = 6.91 \times 10^{-3} \text{ kg}$$

$$6.91 \text{ g} \times \frac{1 \text{ lb}}{453.6 \text{ g}} \times \frac{16 \text{ oz}}{1 \text{ lb}} = 0.244 \text{ oz}$$

b)

$$8.12 \text{ kg} \times \frac{10^3 \text{ g}}{\text{kg}} = 8.12 \times 10^3 \text{ g}$$

$$8.12 \text{ kg} \times \frac{10^3 \text{ g}}{\text{kg}} \times \frac{1 \text{ mg}}{10^{-3} \text{ g}} = 8.12 \times 10^6 \text{ mg}$$

$$8.12 \text{ kg} \times \frac{10^3 \text{ g}}{\text{kg}} \times \frac{1 \text{ lb}}{453.6 \text{ g}} = 17.9 \text{ lb}$$

c)

$$4.25 \text{ lb} \times \frac{453.6 \text{ g}}{\text{lb}} = 1.93 \times 10^3 \text{ g}$$

$$4.25 \text{ lb} \times \frac{453.6 \text{ g}}{\text{lb}} \times \frac{1 \text{ mg}}{10^{-3} \text{ g}} = 1.93 \times 10^6 \text{ mg}$$

$$4.25 \text{ lb} \times \frac{453.6 \text{ g}}{\text{lb}} \times \frac{1 \text{ kg}}{10^3 \text{ g}} = 1.93 \text{ kg}$$

d)

$$14.8 \text{ oz} \times \frac{1 \text{ lb}}{16 \text{ oz}} \times \frac{453.6 \text{ g}}{\text{lb}} = 4.20 \times 10^2 \text{ g}$$

$$14.8 \text{ oz} \times \frac{1 \text{ lb}}{16 \text{ oz}} \times \frac{453.6 \text{ g}}{\text{lb}} \times \frac{1 \text{ cg}}{10^{-2} \text{ g}} = 4.20 \times 10^4 \text{ cg}$$

$$14.8 \text{ oz} \times \frac{1 \text{ lb}}{16 \text{ oz}} \times \frac{453.6 \text{ g}}{\text{lb}} \times \frac{1 \text{ kg}}{10^3 \text{ g}} = 0.420 \text{ kg}$$

e)

$$301 \text{ mm} \times \frac{10^{-3} \text{ m}}{1 \text{ mm}} \times \frac{1 \text{ cm}}{10^{-2} \text{ m}} = 30.1 \text{ cm}$$

$$301 \text{ mm} \times \frac{10^{-3} \text{ m}}{1 \text{ mm}} = 0.301 \text{ m}$$

$$301 \text{ mm} \times \frac{10^{-3} \text{ m}}{1 \text{ mm}} \times \frac{39.37 \text{ in}}{\text{m}} \times \frac{1 \text{ yd}}{36 \text{ in}} = 0.329 \text{ yd}$$

f)

$$29.4 \text{ in} \times \frac{2.54 \text{ cm}}{\text{in}} = 74.7 \text{ cm}$$

$$29.4 \text{ in} \times \frac{2.54 \text{ cm}}{\text{in}} \times \frac{10^{-2} \text{ m}}{\text{cm}} \times \frac{1 \text{ mm}}{10^{-3} \text{ m}} = 747 \text{ mm}$$

$$29.4 \text{ in} \times \frac{2.54 \text{ cm}}{\text{in}} \times \frac{10^{-2} \text{ m}}{\text{cm}} \times \frac{1 \text{ km}}{10^3 \text{ m}} = 7.47 \times 10^4 \text{ km}$$

g)

$$4.19 \text{ ft} \times \frac{12 \text{ in}}{\text{ft}} \times \frac{2.54 \text{ cm}}{\text{in}} \times \frac{10^{-2} \text{m}}{\text{cm}} = 128 \text{ m}$$

$$4.19 \text{ ft} \times \frac{12 \text{ in}}{\text{ft}} \times \frac{2.54 \text{ cm}}{\text{in}} \times \frac{10^{-2} \text{m}}{\text{cm}} \times \frac{1 \text{ mm}}{10^{-3} \text{m}} = 1280 \text{ mm}$$

$$4.19 \text{ ft} \times \frac{12 \text{ in}}{\text{ft}} \times \frac{2.54 \text{ cm}}{\text{in}} \times \frac{10^{-2} \text{m}}{\text{cm}} \times \frac{1 \text{ cm}}{10^{-2} \text{m}} = 1.28 \text{ cm}$$

h)

$$65.2 \text{ km} \times \frac{10^3 \text{ m}}{\text{km}} = 6.52 \times 10^4 \text{ m}$$

$$65.2 \text{ km} \times \frac{10^3 \text{ m}}{\text{km}} \times \frac{1 \text{ cm}}{10^{-2} \text{ m}} = 6.52 \times 10^6 \text{ cm}$$

$$65.2 \text{ km} \times \frac{1 \text{ mile}}{1.609 \text{ km}} = 40.5 \text{ mi}$$

i)
$2410 \text{ ft} \times \frac{12 \text{ in}}{\text{ft}} \times \frac{2.54 \text{ cm}}{\text{in}} \times \frac{10^{-2}\text{m}}{\text{cm}} = 735 \text{ m}$

$2410 \text{ ft} \times \frac{12 \text{ in}}{\text{ft}} \times \frac{2.54 \text{ cm}}{\text{in}} \times \frac{10^{-2}\text{m}}{\text{cm}} \times \frac{1 \text{ km}}{10^3\text{m}} = 0.735 \text{ km}$

$2410 \text{ ft} \times \frac{1 \text{ mile}}{5280 \text{ ft}} = 0.456 \text{ mi}$

j)
$1.95 \text{ ft}^3 \left(\frac{12 \text{ in}}{\text{ft}}\right)^3 \left(\frac{2.54 \text{ cm}}{\text{in}}\right)^3 \left(\frac{10^{-2}\text{m}}{\text{cm}}\right)^3 = 0.552\text{m}^3$

$1.95 \text{ ft}^3 \times \left(\frac{12 \text{ in}}{\text{ft}}\right)^3 \times \left(\frac{2.54 \text{ cm}}{\text{in}}\right)^3 = 5.52\times10^4 \text{ cm}^3$

$1.95 \text{ ft}^3 \times \left(\frac{12 \text{ in}}{\text{ft}}\right)^3 = 3.37 \times 10^3 \text{ in}^3$

k)
$3070 \text{ cm}^3 \times \left(\frac{10^{-2} \text{ m}}{\text{cm}}\right)^3 = 3.07\times10^{-3} \text{ m}^3$

$3070 \text{ cm}^3 \times \left(\frac{\text{in}}{2.54 \text{ cm}}\right)^3 = 187 \text{ in}^3$

$3070 \text{ cm}^3 \times \left(\frac{10^{-2}\text{m}}{\text{cm}}\right)^3 \times \left(\frac{1 \text{ mm}}{10^{-3}\text{m}}\right)^3 = 3.07\times10^6 \text{ mm}^3$

l)
$7.15 \text{ L} \times \frac{1 \text{ mL}}{10^{-3} \text{ L}} = 7.15 \times 10^3 \text{ mL}$

$7.15 \text{ L} \times \frac{1.057 \text{ qt}}{\text{L}} = 7.56 \text{ qt}$

$7.15 \text{ L} \times \frac{1 \text{ gal}}{3.785 \text{ L}} = 1.89 \text{ gal}$

m)
$119 \text{ mL} \times \frac{1 \text{ cm}^3}{1 \text{ mL}} = 119 \text{ cm}^3$

$119 \text{ mL} \times \frac{10^{-3} \text{ L}}{\text{mL}} = 0.119 \text{ L}$

$119 \text{ mL} \times \frac{10^{-3} \text{ L}}{\text{mL}} \times \frac{1.057 \text{ qt}}{\text{L}} = 0.126 \text{ qt}$

n)
$0.816 \text{ gal} \times \frac{3.785 \text{ L}}{\text{gal}} = 3.09 \text{ L}$

$0.816 \text{ gal} \times \frac{3.785 \text{ L}}{\text{gal}} \times \frac{1 \text{ mL}}{10^{-3}\text{L}} = 3.09 \times 10^3 \text{ mL}$

$0.816 \text{ gal} \times \frac{3.785 \text{ L}}{\text{gal}} \times \frac{1 \text{ mL}}{10^{-3}\text{L}} \times \frac{1 \text{ cm}^3}{1 \text{ mL}} = 3.09\times10^3 \text{ cm}^3$

9) $°\text{C} = (°\text{F} - 32)/1.8 = (110 - 32)/1.8 = 43°\text{C}$, $\quad \text{K} = °\text{C} + 273 = 43 + 273 = 316 \text{ K}$

10) $°\text{F} = 1.8°\text{C} + 32 = 1.8(801) + 32 = 1470 °\text{F}$, $\quad \text{K} = °\text{C} + 273 = 801 + 273 = 1074 \text{ K}$

11) $d = \frac{m}{V} = \frac{39.1 \text{ g}}{5.2 \text{ cm} \times 3.4 \text{ cm} \times 1.7 \text{ cm}} = 1.3 \text{ g/cm}^3$

12) $63.5 \text{ g} \times \frac{1 \text{ mL}}{0.818 \text{ g}} = 77.6 \text{ mL}$

13) $29.5 \text{ cm}^3 \times \frac{7.39 \text{ g}}{1 \text{ cm}^3} = 218 \text{ g}$

14) $d = \frac{m}{V} = \frac{67.6 \text{ g}}{12.8 \text{ mL}} = 5.28 \text{ g/mL}$

15) $326 \text{ mL} \times \frac{1.27 \text{ g}}{1 \text{ mL}} = 414 \text{ g}$

16) $1.35 \text{ kg} \times \frac{10^3 \text{ g}}{1 \text{ kg}} \times \frac{1 \text{ cm}^3}{1.74 \text{ g}} = 776 \text{ cm}^3$

1) a) 3 b) 3 c) 4 d) 3 e) 3 f) 1
g) 4 h) 2 i) 4 j) 2 - be conservative

CHAPTER 4

All answers for this chapter are based on our versions of the graphs drawn. Your graph may differ slightly from ours, and this will produce small differences in our numerical answers. This is to be expected.

1) a) Solubility at 86°C, 18.7 mol/kg; at 37°C, 5.85 mol/kg.
b) Temperature at 7.32 mol/kg, 44.3°C; at 17.7 mol/kg, 83.1°C.

2) a) 11.2 miles. b) 42 torr.
c) Both 30 and 50 miles are beyond the range of the graph, which must therefore be extrapolated to those distances. The present curve is almost flat at this point, only two minor lines (2 torr) above the horizontal axis. To extrapolate into and make estimations in this very limited pressure range is not justified. The only reasonable answer to Question 2c from these data is that the pressures will be less than 2 torr.

CHAPTER 4

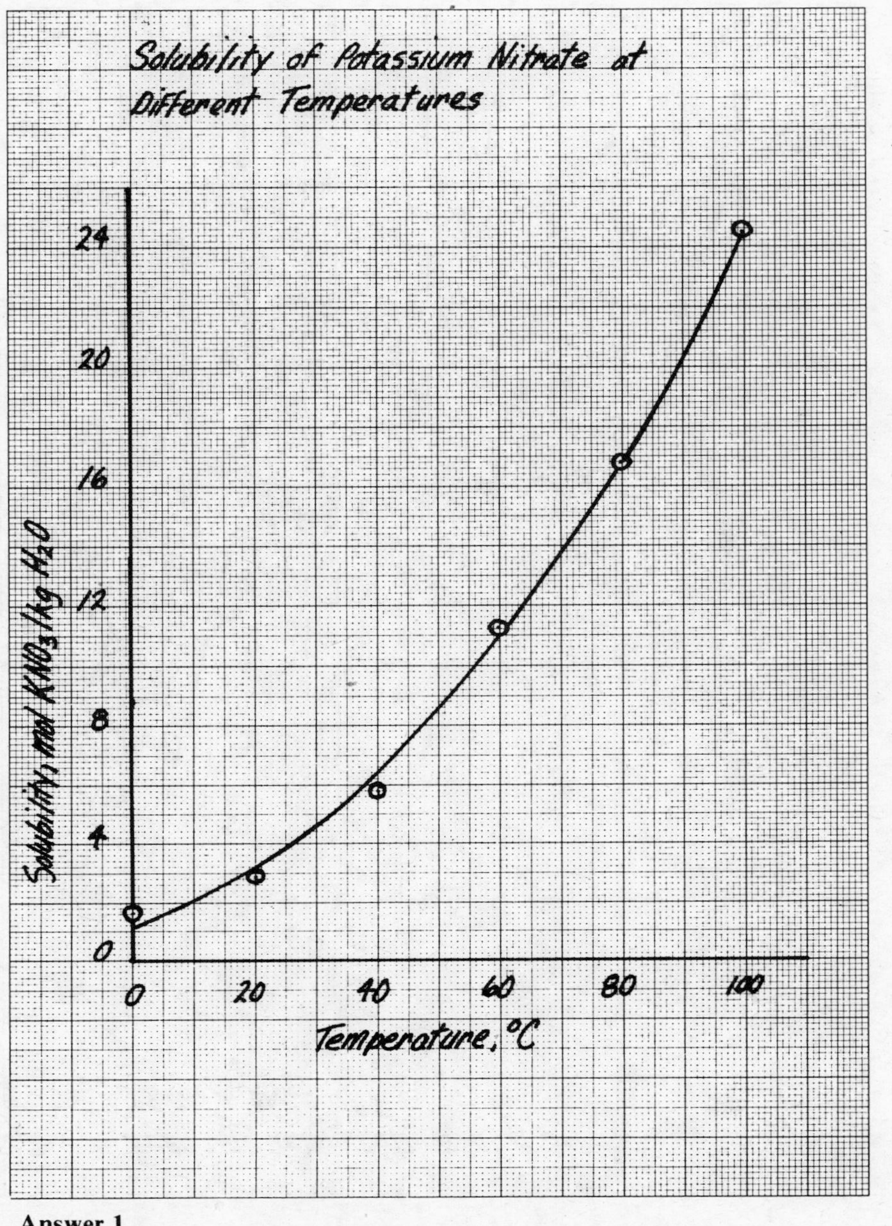

Answer 1

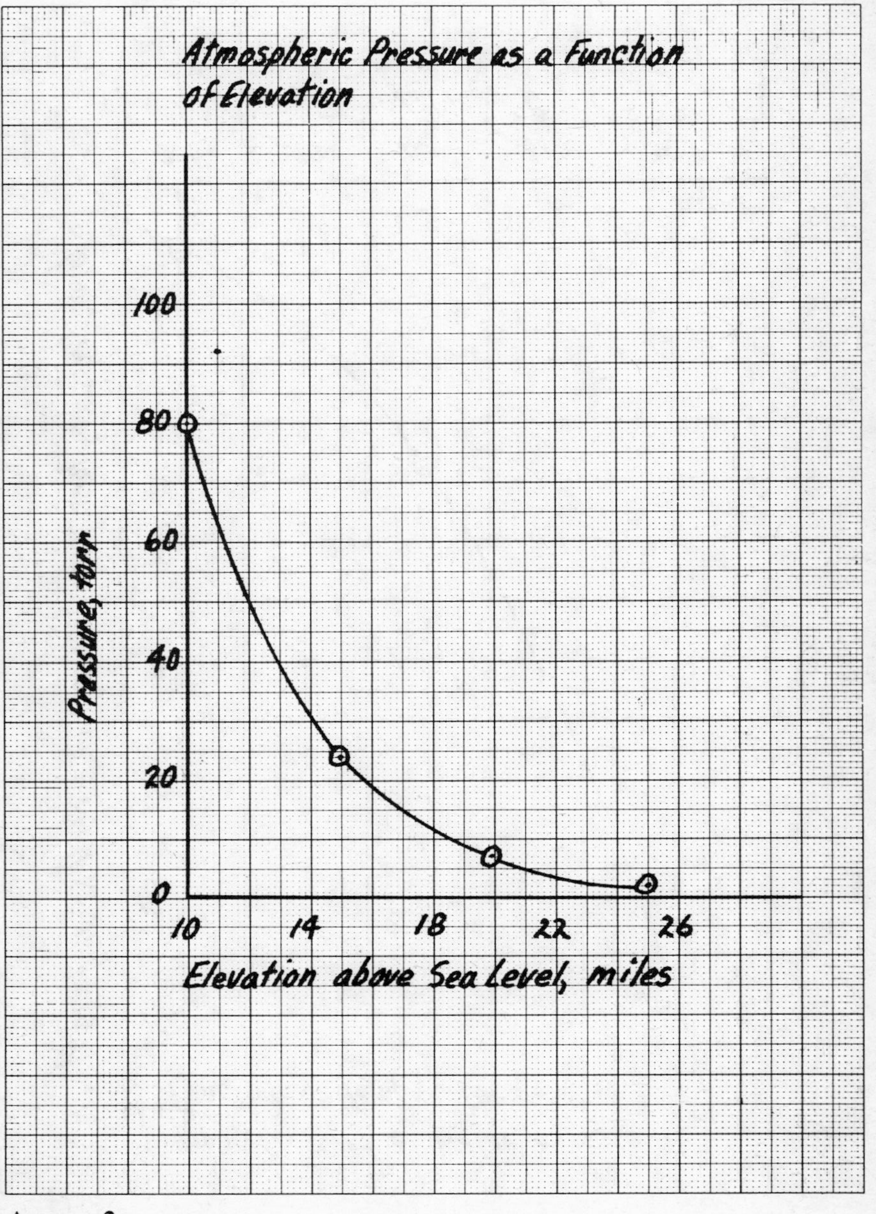

Answer 2

CHAPTER 4

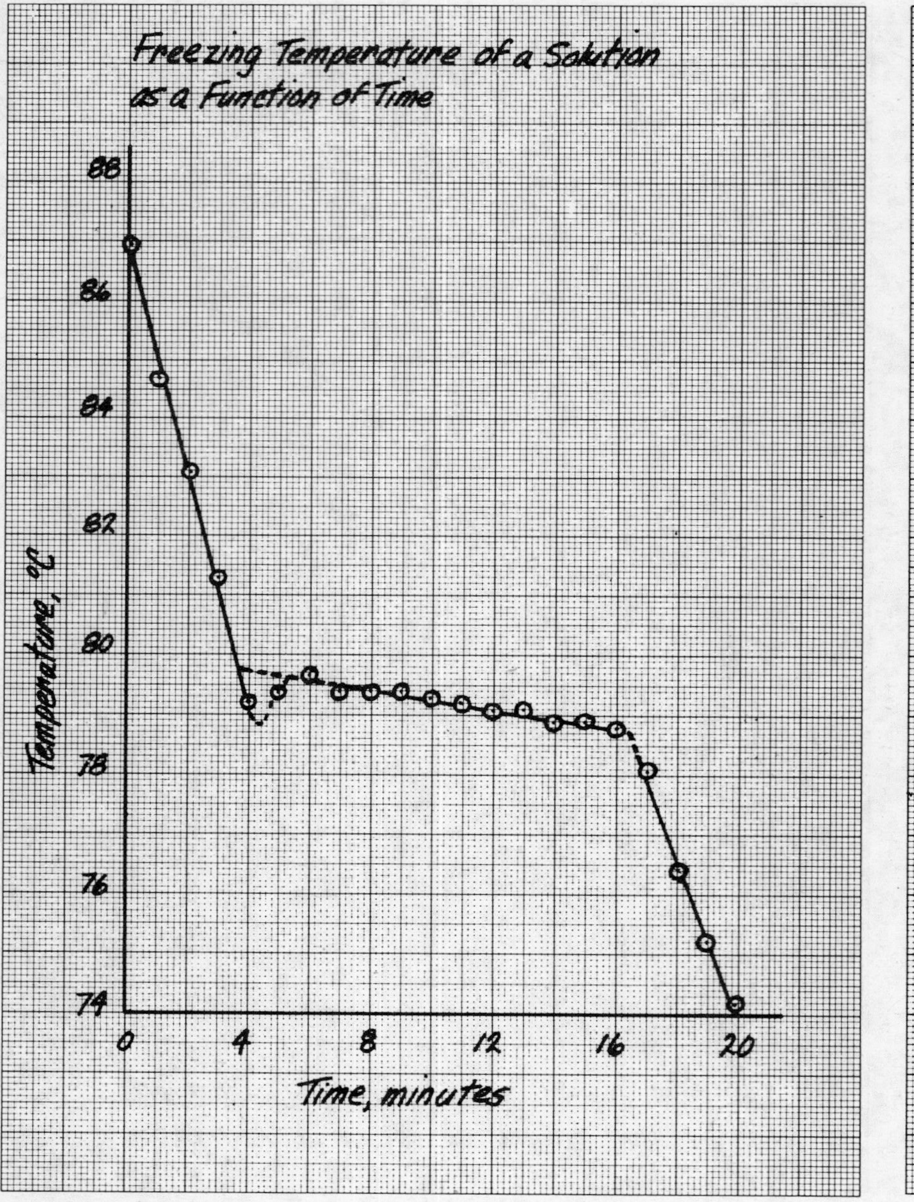

Answer 3

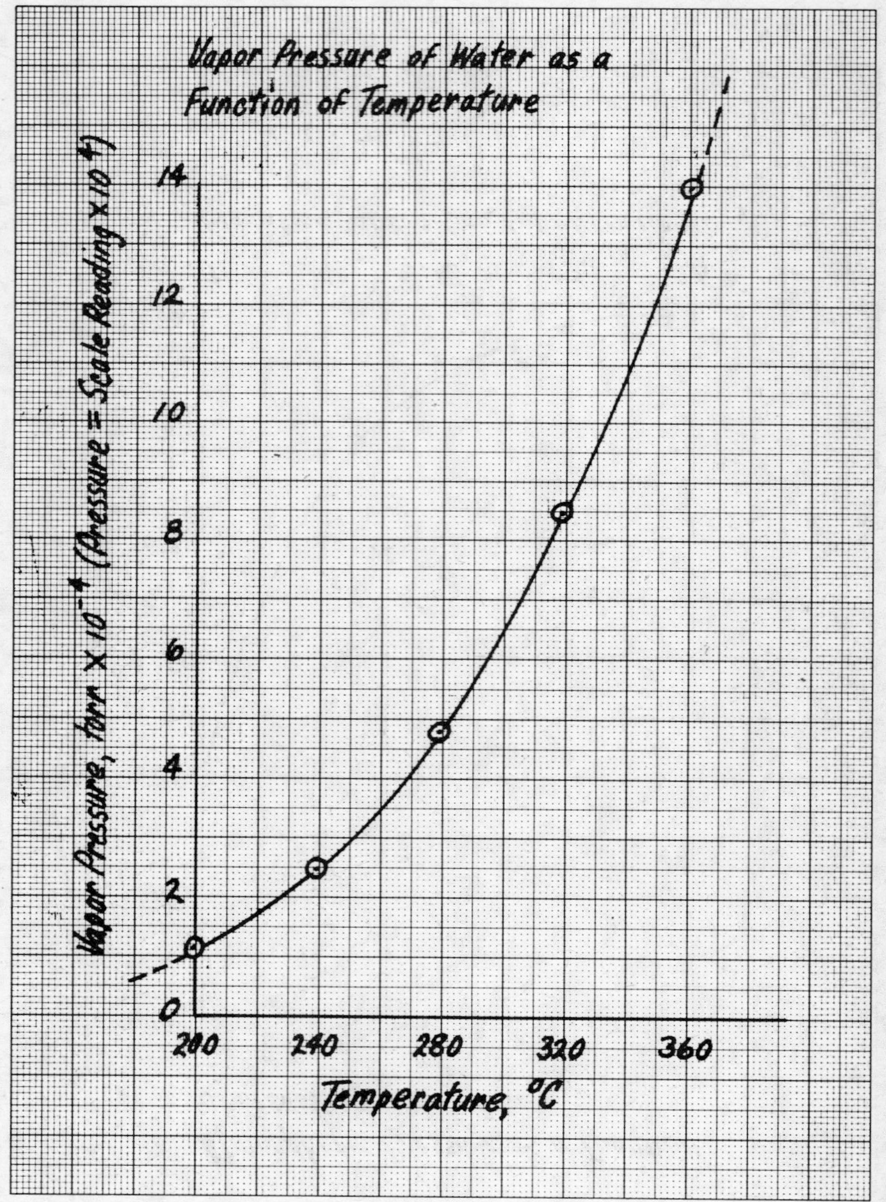

Answer 4

3) There does not appear to be an "abrupt" change between the first and second stages of the graph if the temperatures at 4 and 5 minutes are considered to be valid. If the temperature were to drop to 79.7°C in the first stage, and then change abruptly to the straight-line second stage, you would select 79.7°C as the freezing point. This is the correct answer. The graph is typical of actual measurements, however, and the 4 and 5 minute temperatures are valid. The temperature dip before freezing actually begins is because of supercooling, cooling below the freezing point before a crystal happens to form. When the first crystal appears, the temperature rises immediately to the second stage line. In practice, the second stage line is established from the graph, and extrapolated backward until it intersects the first stage, as it has at 79.7°C in this example. Another note: the temperature irregularities in the second stage are not indications that the temperature holds constant at any time (7 minutes through 9 minutes), and then drops 0.1 degree per minute (9 minutes through 12 minutes). The cooling is gradual. The temperature is measured with a thermometer graduated in 0.1 degree increments, so all readings are ± 0.1°C.

4) a) 7.05×10^4 torr b) 220°C c) 366°C, by extrapolation
d) 6.5×10^3 torr, by extrapolation. If it had been known before the graph was drawn that the 180°C point was to be read, the horizontal scale should have included 180°.

5) a) You have 3.4 minutes.
b) The slope of the tangent to the curve at any point is the instantaneous cooling rate at that point, measured in degrees per minute. The negative slope indicates cooling. The cooling rate is at a maximum at time 0, when the negative slope is at a maximum. Selecting points where the tangents cross the axes as drawn, the cooling rates are:

1.5 minutes: $\frac{69.0 - 20.0}{5.00 - 0.00} = 9.80$°C/min 7 minutes: $\frac{43.5 - 20.0}{9.60 - 0.00} = 2.45$°C/min

6) a) 6.64 L b) 522 K
c) Slope = (8.60-6.20)/(580-428) = 0.0148 L/K. The slope is a measure of the change in volume per kelvin change in temperature.
d) The equation of the curve is not readily apparent without the intercept. To find it requires that the graph includes 0 on the horizontal axis. On redrawing the graph to include zero volume, you would find that the curve passes through the origin, and that the graph is that of a direct proportionality. Its equation is

$$y = mx \quad \text{or} \quad \text{volume (L)} = 0.0148 \times \text{temperature (K)}$$

Without drawing the graph, you might have guessed the direct proportional relationship and therefore the above equation. Without graphical justification, you should check the graph at several points to confirm that the equation does describe the line. For example, at 500 K, volume should be 0.0148 x 500 = 7.4 liters. The point (500,7.4) is on the curve, which it should be if the equation is correct.
e) The volume of a fixed quantity of gas at constant pressure is directly proportional to the temperature measured in kelvins. This is the same absolute temperature developed in the pressure-temperature graph in the chapter. It is, in fact, a second form of experimental evidence supporting the existence of an absolute zero temperature.

7) a) Density is, by definition, mass per unit volume. It is found by dividing the mass of a sample by the volume it occupies. Mass and volume are the variables plotted, and the slope of the line has units of mass over volume, the same as density. But the mass plotted includes the beaker; it is not the mass of a sample of the liquid. However, the <u>difference</u> in mass between any two points and the difference in volume between the same points represent the mass and volume of the sample of liquid between those points. Consequently slope does represent density.

$$\text{slope} = \frac{(240.0 - 200.0)\ \text{g}}{(63.0 - 29.0)\ \text{mL}} = 1.18\ \text{g/mL}$$

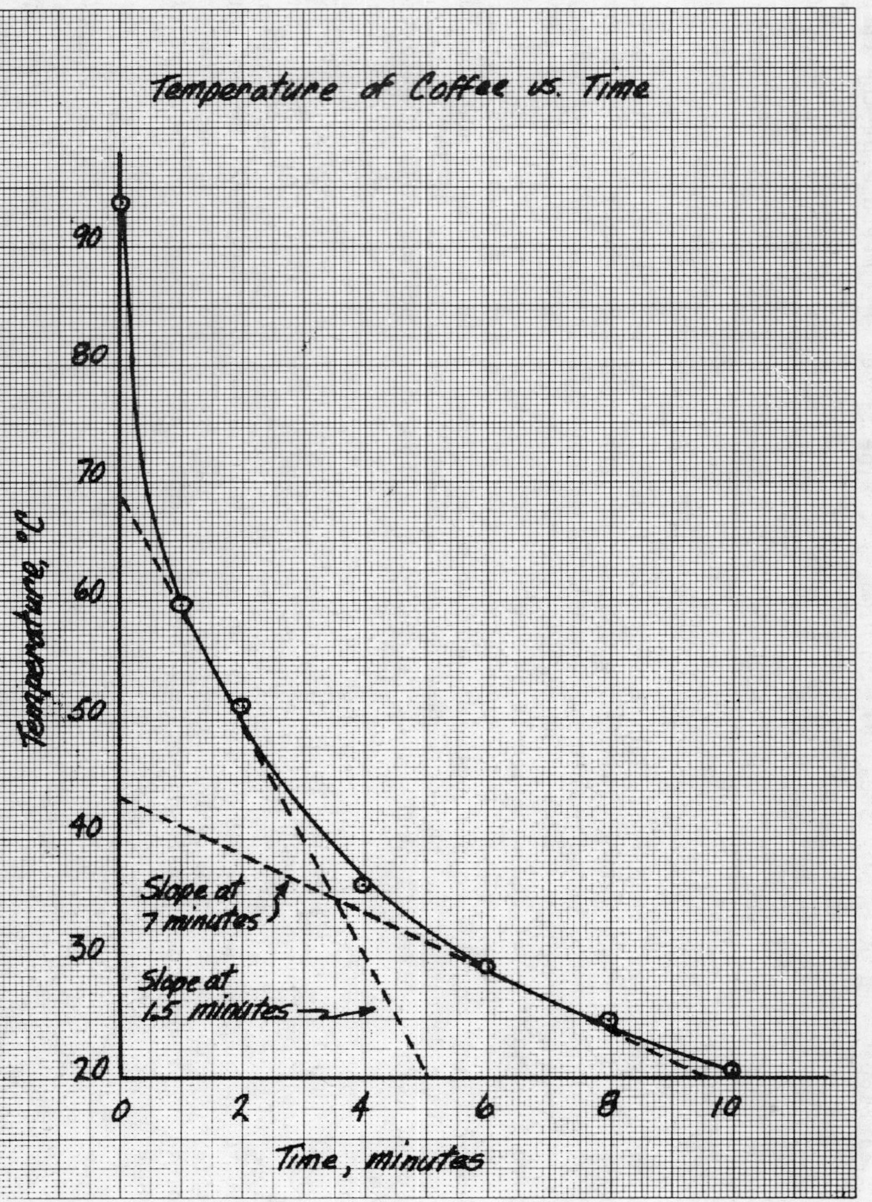

Answer 5

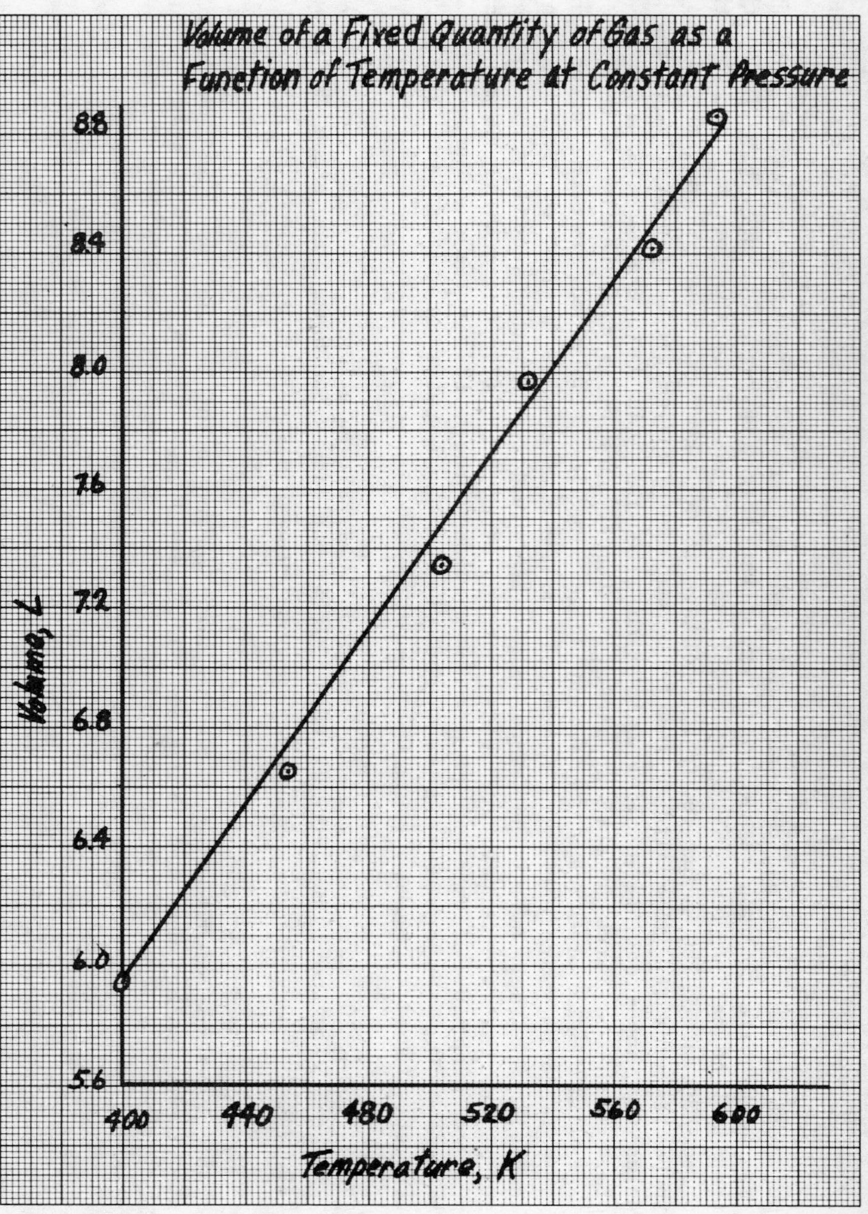

Answer 6

CHAPTER 4

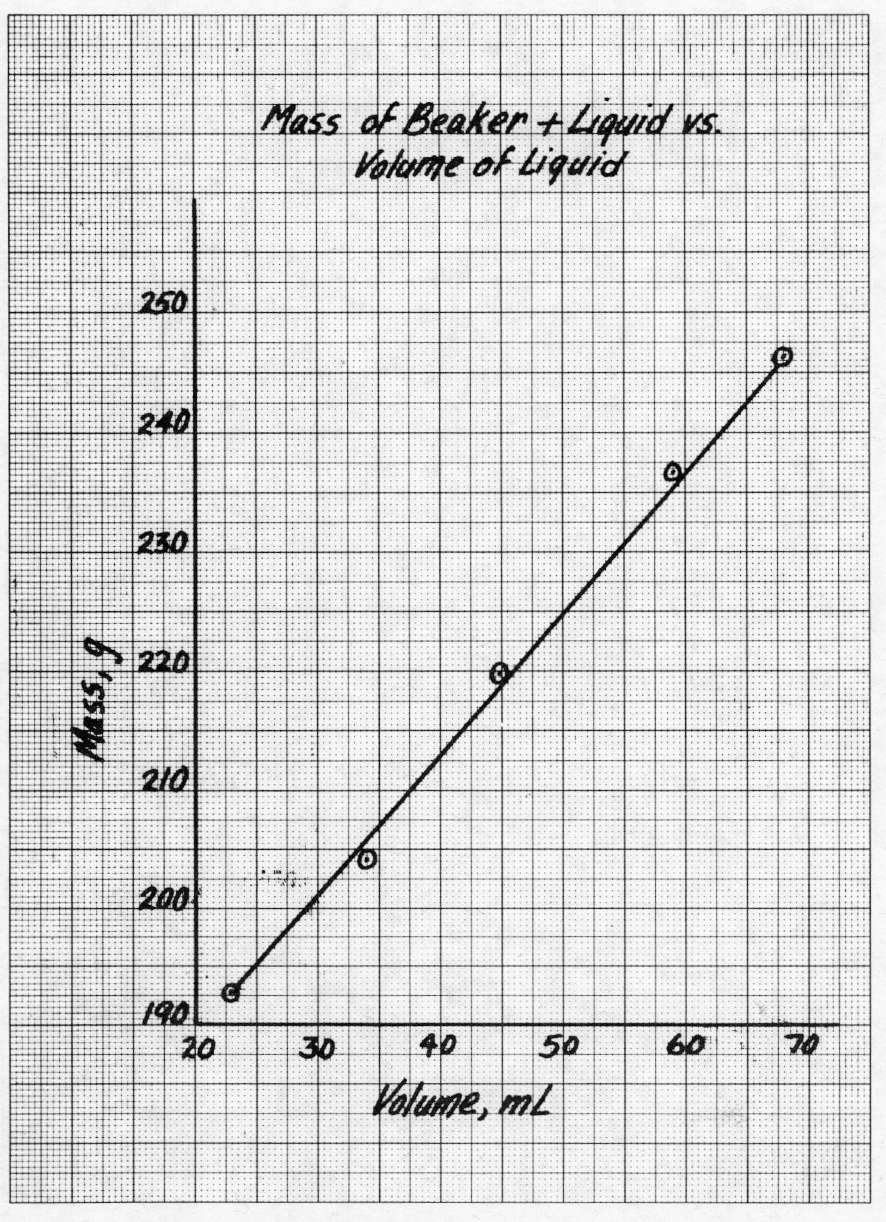

Answer 7.1

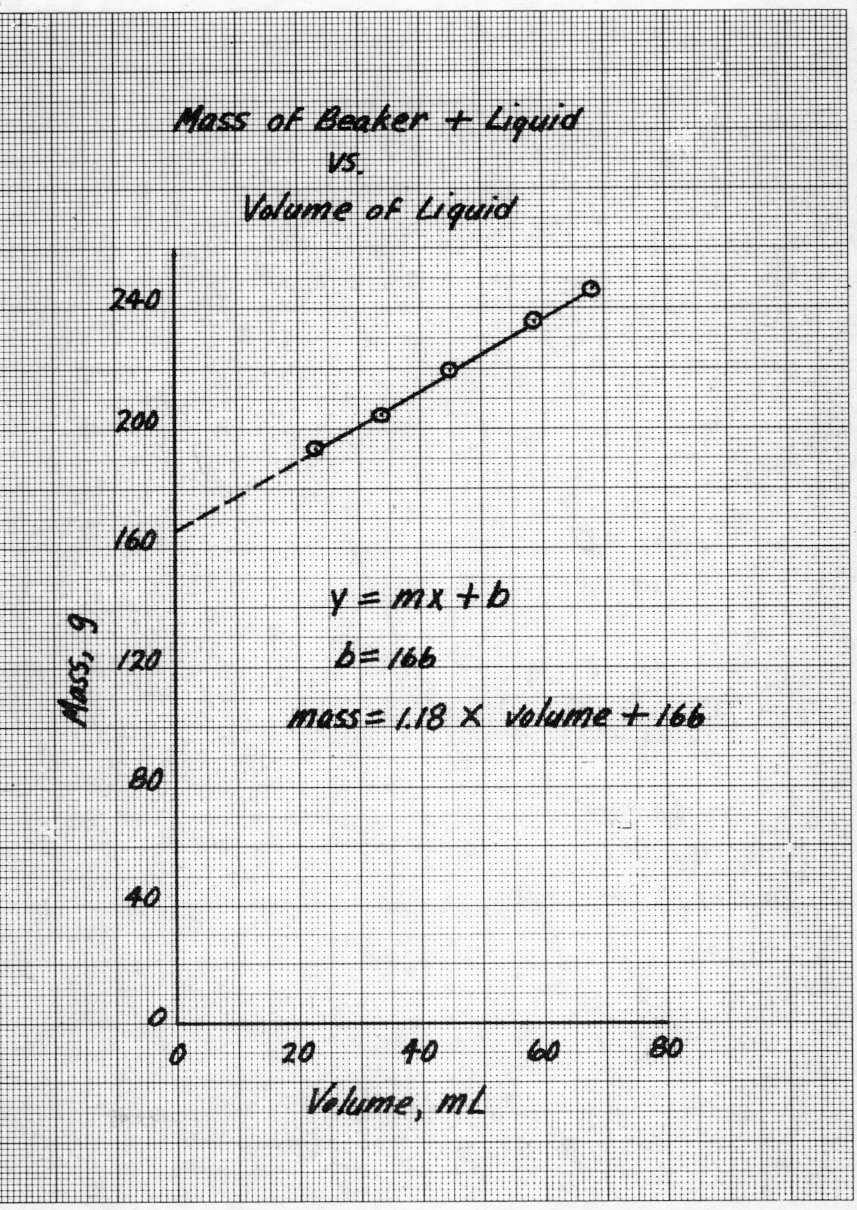

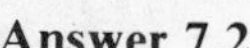
Answer 7.2

CHAPTER 4

CHAPTER 4

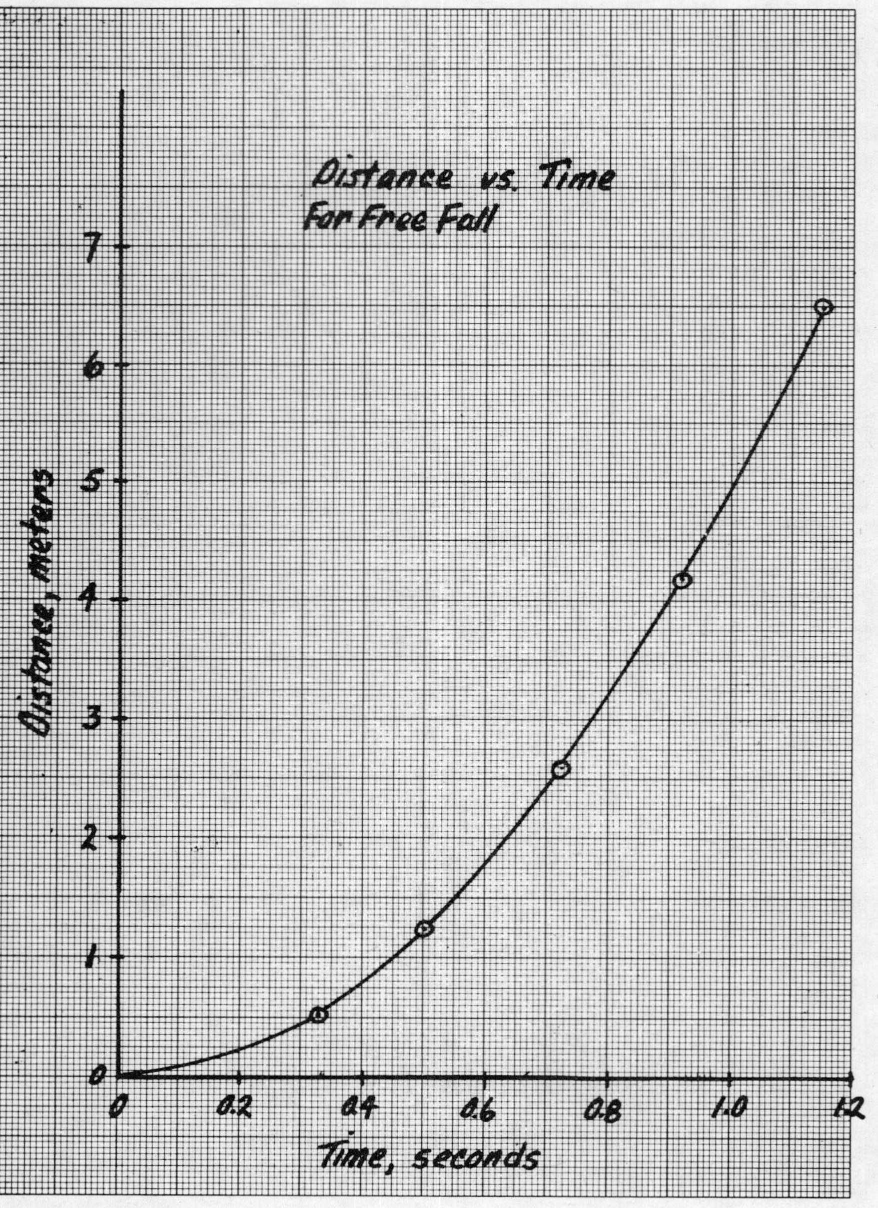

Answer 8.1

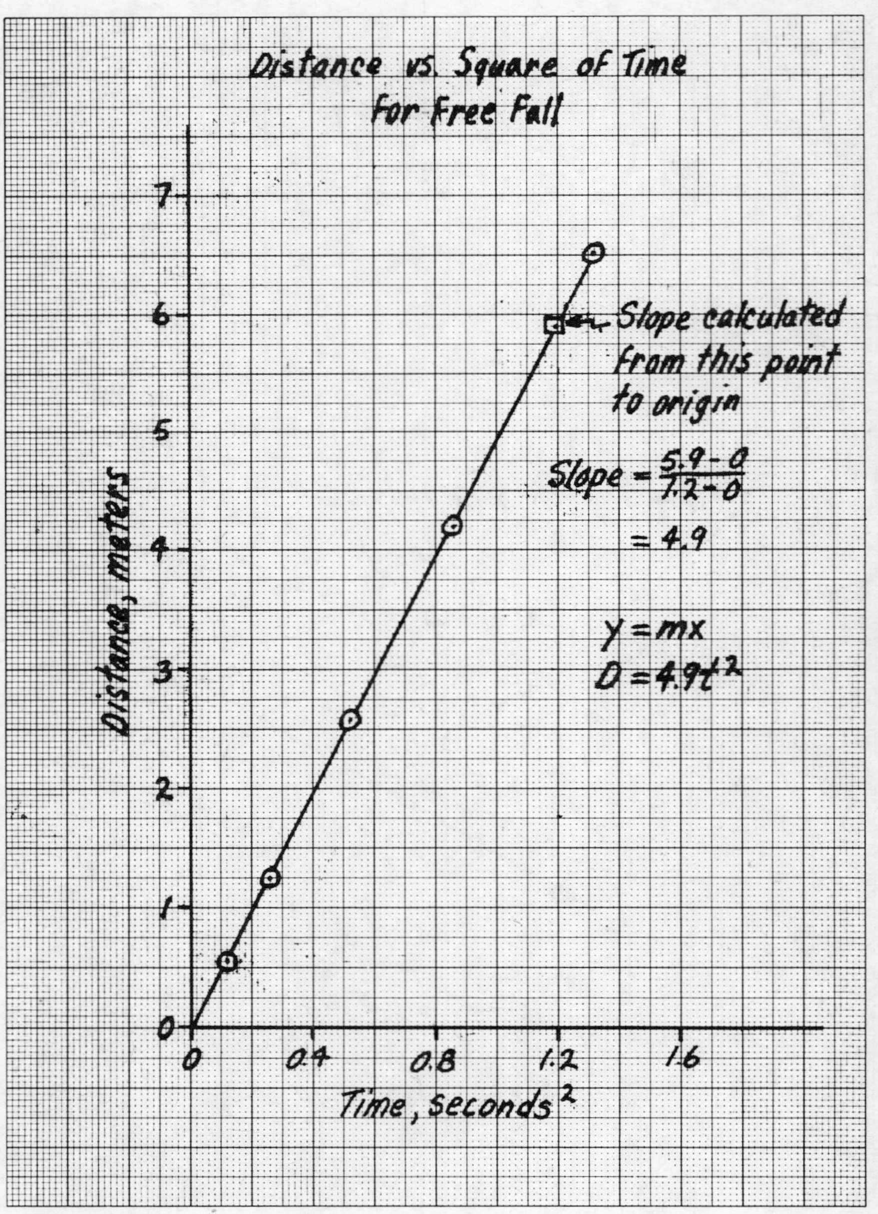

Answer 8.2

b) To write the slope intercept form of the equation of a straight line (Equation 5.2), the y-intercept must be known. This appears on a graph on which zero volume appears. A second graph of the data that includes the origin shows that the y intercept is 166. The equation is developed on the graph.

The y-intercept can also be found by calculation. In $y = mx + b$, m has been found to be 1.18. The coordinates of any point on the line can be substituted for x and y, leaving b as the only unknown. Using the point (65.0, 242.5),

$$242.5 = (1.18)(65.0) + b \qquad b = 165.8, \text{ or } 166$$

The general equation then becomes $y = 1.18x + 166$, or

$$\text{mass} = 1.18 \times \text{volume} + 166$$

The same equation may be reached by substituting the coordinates of any point, such as (65.0,242.5), for x_1 and y_1 in the **point-slope form of the equation of a straight line**, $(y - y_1) = m(x - x_1)$.

c) When volume is zero—when there is no liquid in the beaker—the mass is that of the empty beaker. This is the significance of the y-intercept. The mass of the beaker is 166 grams.

8) The five values for the square of time are 0.11, 0.25, 0.52, 0.85, and 1.32.
a) 3.46 meters b) 1.09 seconds c) $d = 4.9t^2$ d) At 0.83 seconds, $d = 4.9\ (0.83)^2 = 3.4$ m

This matches the 3.46 meters estimated from the graph. When distance is 5.83 meters,

$$t = 5.83/4.9 = 1.1 \text{ seconds}$$

CHAPTER 5

1) a) 4.908 b) 8.0 c) 5.642 (3 sf) d) -4.214 e) -1.62
f) 8.99 g) -11.00 (2 sf) h) -2.441

2) a) 7.26 b) 4.2×10^2 c) 1.295 d) 8.77×10^{-5} e) 95.9 f) 0.28
g) 4.61×10^{13} h) 4.30×10^{-11}

3) a) $y = \log 32.6 + 6.214 = 7.727$ b) $y = \dfrac{\log\ 0.0039}{3.7} = -0.65$
c) $y = \dfrac{733 \log 82}{27} = 7.2$

4) a) $y = \text{antilog}\ \dfrac{8.46}{6.62} = 19$ b) $y = \text{antilog}\ (-304)(-5.06 \times 10^{-5}) = \text{antilog}\ 0.0154 = 1.036$ (4 sf)
c) $y = \text{antilog}\ \dfrac{-5.69 \times 10^{11}}{4.97 \times 10^{10}} = 4 \times 10^{-12}$ (1 sf)

5) $[Cl^-] = 10^{-9.34} = 4.6 \times 10^{-10}$ M

6) $pOH = -\log (6.2 \times 10^{-6}) = 5.21$

7) $K = \text{antilog}\ \dfrac{(1.10)(2.00)}{0.0592} = 1 \times 10^{37}$ (1 sf)

8) $\Delta G° = -[2.303][8.314][325][\log (8.2 \times 10^{12})] = -8.04 \times 10^4$
$\Delta G° = -[8.314]][325][\ln (8.2 \times 10^{12})] = -8.03 \times 10^4$

9) $n = \dfrac{\log (15.234/14.639)}{0.301} = 0.0575$ half-lives

10) $k_2/k_1 = \text{antilog}\ \dfrac{Ea}{2.303R}\ \dfrac{T_2 - T_1}{T_2\ T_1} = \text{antilog}\ \dfrac{(48{,}000)\ (300 - 290)}{(2.303)(8.314)(300)(290)}$
$= 2$ (1 sf from 300 - 290)

CHAPTER 6

1)

Particle	Charge	Mass	Location
electron	-	1/1800	outside
proton	+	1	nucleus
neutron	0	1	nucleus

2) a) C b) Mg c) Ca d) Mn e) Cr f) Fe g) Cl h) I

3) a) potassium b) antimony c) phosphorus d) tin e) beryllium f) selenium g) barium h) tellurium

4) a) family

O: S, Se, Te, Po
Na: H, Li, K, Rb, Cs, Fr
As: N, P, Sb, Bi
Ag: Cu, Au

b) period

Li, Be, B, C, N, F, Ne
Mg, Al, Si, P, S, Cl Ar
K to Kr
Rb to Xe

5) a) main group metal b) non-metal c) transition metal d) rare earth metal e) non-metal f) transition metal

6) Isotopes are atoms of the same element (so they have the same number of protons) but have different mass (because they have a different number of neutrons. Isotopes do not differ in number of electrons (because they have the same number of protons).

7)

Isotope Symbol	Atomic Number	Mass No.	No. protons	No. neutrons
^{67}Zn	30	67	30	37
^{77}As	33	77	33	44
^{90}Sr	38	90	38	52

8) Atomic mass is the average mass of the naturally occurring isotopes of an element, weighted by their percenage abundances, relative to carbon-12 as 12 amu exactly. Atomic mass is reported on a relative scale because the actual masses of atoms are very, very small. Carbon-12 is the reference for the relative atomic mass scale.

9) a)
106.9051 amu × 0.5182 = 55.40 amu
108.9047 amu × 0.4818 = 52.47 amu
107.87 amu

b)
53.9396 amu × 0.0582 = 3.14 amu
55.9349 amu × 0.9166 = 51.27 amu
56.9354 amu × 0.0219 = 1.25 amu
57.9333 amu × 0.0033 = 0.19 amu
55.85 amu

10) a)
34.96885 amu × 0.7553 = 26.41$\underline{2}$ amu
(3) × (1) = (2)
35.453 amu

(1) = 1 - 0.7553 = 0.2447
(2) = 35.453 - 26.41$\underline{2}$ = 9.041
(3) = 9.041 ÷ 0.2447 = 36.95 amu

b)
27.97693 amu × 0.9221 = 25.79$\underline{75}$ amu
(3) × (1) = (2)
29.97376 amu × 0.0309 = 0.926$\underline{2}$ amu
28.0855 amu

(1) = 1 - 0.9221 - 0.0309 = 0.0470
(2) = 28.0855 - 25.79$\underline{75}$ - 0.926$\underline{2}$ = 1.36$\underline{18}$
(3) = 1.36$\underline{18}$ ÷ 0.0470 = 28.9$\underline{7}$ amu

11) Lithium-7 is more abundant because the atomic mass of lithium is very close to 7. Let A equal the fraction of lithium-6 then 1-A equals the fraction of lithium-7.

6.01512 × A + 7.01600 × (1-A) = 6.941
6.01512A + 7.01600 - 7.01600A = 6.941
1.00088A = 0.075 so A = 0.074$\underline{93}$

Thus the two isotope of lithium are: 74.$\underline{93}$% of ^6Li and 25.$\underline{07}$% of ^7Li.

CHAPTER 7

phosphorus, P
zinc ion, Zn^{2+}
sodium bromide, NaBr
oxygen difluoride, OF_2

O_2, oxygen
H_2SO_4, sulfuric acid
OH^-, hydroxide ion
$BaBr_2$, barium bromide

CHAPTERS 6 & 7

potassium ion, K^+
silver chloride, AgCl
iron(III) ion, Fe^{3+}
zinc hydroxide, $Zn(OH)_2$
copper(II) bromide, $CuBr_2$
nitrogen, N_2
nickel sulfide, NiS
calcium hydrogen carbonate, $Ca(HCO_3)_2$
manganese(II) carbonate, $MnCO_3$
iron(III) phosphate, $FePO_4$
phosphate ion, PO_4^{3-}
nitrite ion, NO_2^-
perbromate ion, BrO_4^-
calcium hydrogen sulfate, $Ca(HSO_4)_2$
uranium hexafluoride, UF_6
potassium sulfide, K_2S
hydrogen phosphate ion, HPO_4^{2-}
bromine, Br_2
mercury(I) chloride, Hg_2Cl_2
bromous acid, $HBrO_2$
aluminum oxide, Al_2O_3
hydrogen telluride ion, HTe^-
iron(III) chloride, $FeCl_3$
rubidium sulfate, Rb_2SO_4
sodium hydride, NaH
potassium oxalate, $K_2C_2O_4$
hydrogen carbonate ion, HCO_3^-
silicon dioxide, SiO_2
calcium hydrogen telluride, $Ca(HTe)_2$
sulfide ion, S^{2-}
ammonium phosphate, $(NH_4)_3PO_4$
potassium thiocyanate, KSCN
selenic acid, H_2SeO_4
strontium bromide, $SrBr_2$
ammonium acetate, $NH_4C_2H_3O_2$
magnesium nitride, Mg_3N_2
magnesium hydrogen selenite, $Mg(HSeO_3)_2$
sodium dichromate, $Na_2Cr_2O_7$
hydrosulfuric acid, H_2S

Na_2O, sodium oxide
$CaCl_2$, calcium chloride
P_2O_3, diphosphorus trioxide
Na, sodium
NH_4I, ammonium iodide
$Ba_3(PO_4)_2$, barium phosphate
Li^+, lithium ion
NO_3^-, nitrate ion
Li_3PO_4, lithium phosphate
$Ni(HSO_4)_2$, nickel hydrogen sulfate
C, carbon
KF, potassium fluoride
PbO, lead(II) oxide
HF, hydrofluoric acid
Cs_2CO_3, cesium carbonate
$CuSO_4$, copper(II) sulfate (or copper sulfate)
$HgBr_2$, mercury(II) bromide
IO_2^-, iodite ion
Ca_3P_2, calcium phosphide
Al_2S_3, aluminum sulfide
SO_4^{2-}, sulfate ion
$Sr(OH)_2$, strontium hydroxide
Cl_2, chlorine
$In_2(CO_3)_3$, indium carbonate
$H_2PO_4^-$, dihydrogen phosphate ion
Mn^{2+}, manganese(II)ion
Cl_2O, dichlorine oxide
Na_2O_2, sodium peroxide
$Mg(HSeO_4)_2$, magnesium hydrogen selenate
$Cr(ClO_4)_3$, chromium(III) perchlorate
F^-, fluoride ion
F_2, fluorine
$CaTeO_3$, calcium tellurite
$Ga(HCO_3)_3$, gallium hydrogen carbonate
Na_2CrO_4, sodium chromate
BaH_2, barium hydride
H_2TeO_4, telluric acid
CI_4, carbon tetraiodide
KCN, potassium cyanide

CHAPTER 8

1) 1 C = 1(12.01) = 12.01 g/mol
2 O = 2(16.00) = 32.00 g/mol
44.01 g/mol

3 C = 3(12.01) = 36.03 g/mol
8 H = 8(1.008) = 8.064 g/mol
1 O = 1(16.00) = 16.00 g/mol
60.09 g/mol

2) a) 2 F = 2(19.00) = 38.00 g/mol

b) 1 He = 4.003 g/mol

c) 2 K = 2(39.10) = 78.20 g/mol
1 C = 1(12.01) = 12.01 g/mol
3 O = 3(16.00) = 48.00 g/mol
138.21 g/mol

d) 2 H = 2(1.008) = 2.016 g/mol
1 S = 1(32.06) = 32.06 g/mol
4 O = 4(16.00) = 64.00 g/mol
98.08 g/mol

e) 1 Fe = 55.85 g/mol

f) 1 S = 1(32.06) = 32.06 g/mol
2 O = 2(16.00) = 32.00 g/mol
64.06 g/mol

3) a) 1 Li = 1(6.941) = 6.941 g/mol
1 F = 1(19.00) = 19.00 g/mol
25.94 g/mol

b) 2 K = 2(39.10) = 78.20 g/mol
1 S = 1(32.06) = 32.06 g/mol
110.26 g/mol

c) 1 Na = 1(22.99) = 22.99 g/mol
1 Br = 1(79.90) = 79.90 g/mol
3 O = 3(16.00) = 48.00 g/mol
150.89 g/mol

d) 2 Al = 2(26.98) = 53.96 g/mol
6 Cr = 6(52.00) = 312.00 g/mol
21 O = 21(16.00) = 336.00 g/mol
701.96 g/mol

e) 9 C = 9(12.01) = 108.09 g/mol
11 H = 11(1.008) = 11.088 g/mol
1 N = 1(14.01) = 14.01 g/mol
4 O = 4(16.00) = 64.00 g/mol
197.19 g/mol

f) 3 C = 3(12.01) = 36.03 g/mol
4 H = 4(1.008) = 4.032 g/mol
4 O = 4(16.00) = 64.00 g/mol
104.06 g/mol

4) $\frac{26.98 \text{ g Al}}{121.95 \text{ g AlPO}_4}\times 100 = 22.12\%\text{Al}$ $\frac{30.97 \text{ g P}}{121.95 \text{ g AlPO}_4}\times 100 = 25.40\%\text{P}$ $\frac{64.00 \text{ g O}}{121.95 \text{ g AlPO}_4}\times 100 = 52.48\%\text{O}$

5) $\frac{2(12.01) \text{ g C}}{121.05 \text{ g C}_2\text{H}_3\text{NO}_5} \times 100 = 19.84\% \text{ C}$ $\frac{3(1.008) \text{ g H}}{121.05 \text{ g C}_2\text{H}_3\text{NO}_5} \times 100 = 2.498\% \text{ H}$

$\frac{14.01 \text{ g N}}{121.05 \text{ g C}_2\text{H}_3\text{NO}_5} \times 100 = 11.57\% \text{ N}$ $\frac{80.00 \text{ g O}}{121.05 \text{ g C}_2\text{H}_3\text{NO}_5} \times 100 = 66.09\% \text{ O}$

6) $\frac{2(52.00) \text{ g Cr}}{392.18 \text{ g Cr}_2(\text{SO}_4)_3} \times 100 = 26.52\% \text{ Cr}$ $\frac{3(32.06) \text{ g S}}{392.18 \text{ g Cr}_2(\text{SO}_4)_3} \times 100 = 24.52\% \text{ S}$

$\frac{12(16.00) \text{ g O}}{392.18 \text{ g Cr}_2(\text{SO}_4)_3} \times 100 = 48.96\% \text{ O}$

7) $2.18 \text{ mol H}_2 \times \frac{6.022 \times 10^{23} \text{ H}_2 \text{ molecules}}{1 \text{ mol H}_2} = 1.31 \times 10^{24} \text{ H}_2 \text{ molecules}$

8) $8.71 \times 10^{22} \text{ K atoms} \times \frac{1 \text{ mol K}}{6.022 \times 10^{23} \text{ K atoms}} = 0.145 \text{ mol K}$

9) $0.358 \text{ mol NH}_4\text{Cl} \times \frac{53.49 \text{ g NH}_4\text{Cl}}{1 \text{ mol NH}_4\text{Cl}} = 19.1 \text{ g NH}_4\text{Cl}$

10) $3.79 \times 10^{22} \text{ P}_4\text{O}_{10} \text{ molecules} \times \frac{283.88 \text{ g P}_4\text{O}_{10}}{6.022 \times 10^{23} \text{ P}_4\text{O}_{10} \text{ molecules}} = 17.9 \text{ g P}_4\text{O}_{10}$

11) $16.4 \text{ g Fe}_2\text{O}_3 \times \frac{1 \text{ mol Fe}_2\text{O}_3}{159.70 \text{ g Fe}_2\text{O}_3} = 0.103 \text{ mol Fe}_2\text{O}_3$

12) $17.9 \text{ g CH}_3\text{OH} \times \frac{6.022 \times 10^{23} \text{ CH}_3\text{OH molecules}}{32.04 \text{ g CH}_3\text{OH}} = 3.36 \times 10^{23} \text{ CH}_3\text{OH molecules}$

13) $4.55 \times 10^{25} \text{ NaNO}_3 \text{ units} \times \frac{1 \text{ mol NaNO}_3}{6.022 \times 10^{23} \text{ NaNO}_3 \text{ units}} = 75.6 \text{ mol NaNO}_3$

14) $124 \text{ g BaSO}_4 \times \frac{1 \text{ mol BaSO}_4}{233.4 \text{ g BaSO}_4} = 0.531 \text{ mol BaSO}_4$

15) $8.09\times 10^{21} \text{ Mg(ClO}_3)_2 \times \frac{1 \text{ mol Mg(ClO}_3)_2}{6.022\times 10^{23} \text{ Mg(ClO}_3)_2 \text{ units}} \times \frac{191.21 \text{ g Mg(ClO}_3)_2}{\text{mol Mg(ClO}_3)_2} = 2.57 \text{ g Mg(ClO}_3)_2$

16) $0.512 \text{ mol C}_3\text{H}_8 \times \frac{44.09 \text{ g C}_3\text{H}_8}{1 \text{ mol C}_3\text{H}_8} = 22.6 \text{ g C}_3\text{H}_8$

17) $0.169 \text{ g Na}_2\text{SO}_4 \times \frac{1 \text{ mol Na}_2\text{SO}_4}{142.04 \text{ g Na}_2\text{SO}_4} \times \frac{6.022 \times 10^{23} \text{ Na}_2\text{SO}_4 \text{ units}}{\text{mol Na}_2\text{SO}_4} = 7.17\times 10^{20} \text{ Na}_2\text{SO}_4 \text{ units}$

CHAPTER 8

18) $1.19 \times 10^{23}\ SO_3 \text{ molecules} \times \frac{1 \text{ mol } SO_3}{6.022 \times 10^{23}\ SO_3 \text{ molecules}} = 0.198 \text{ mol } SO_3$

$1.19 \times 10^{23}\ SO_3 \text{ molecules} \times \frac{1 \text{ mol } SO_3}{6.022 \times 10^{23}\ SO_3 \text{ molecules}} \times \frac{80.06 \text{ g } SO_3}{1 \text{ mol } SO_3} = 15.8 \text{ g } SO_3$

19) $80.0 \text{ g } O_2 \times \frac{1 \text{ mol } O_2}{32.00 \text{ g } O_2} \times \frac{6.022 \times 10^{23}\ O_2 \text{ molecules}}{\text{mol } O_2} = 1.51 \times 10^{24}\ O_2 \text{ molecules}$

$80.0 \text{ g } O_2 \times \frac{1 \text{ mol } O_2}{32.00 \text{ g } O_2} \times \frac{6.022 \times 10^{23}\ O_2 \text{ molecules}}{\text{mol } O_2} \times \frac{2 \text{ O atoms}}{1\ O_2 \text{ molecule}} = 3.01 \times 10^{24} \text{ O atoms}$

	ELEMENT	GRAMS	MOLES OF ATOMS	ATOM RATIO	FORMULA RATIO	EMPIRICAL FORMULA
20)	Ni	60.7	$g \times \frac{1 \text{ mol}}{58.70 \text{ g}} = 1.03$ mol	$\frac{1.03}{1.03} = 1.00$	1	
	F	39.3	$g \times \frac{1 \text{ mol}}{19.00 \text{ g}} = 2.07$ mol	$\frac{2.07}{1.03} = 2.01$	2	NiF_2
21)	H	0.292	$g \times \frac{1 \text{ mol}}{1.008 \text{ g}} = 0.290$ mol	$\frac{0.290}{0.146} = 1.99$	4	
	O	3.50	$g \times \frac{1 \text{ mol}}{16.00 \text{ g}} = 0.219$ mol	$\frac{0.219}{0.146} = 2.01$	3	
	N	2.04	$g \times \frac{1 \text{ mol}}{14.01 \text{ g}} = 0.146$ mol	$\frac{0.146}{0.146} = 1.00$	2	$H_4N_2O_3$
22)	C	74.1	$g \times \frac{1 \text{ mol}}{12.01 \text{ g}} = 6.17$ mol	$\frac{6.17}{1.23} = 5.02$	5	
	H	8.64	$g \times \frac{1 \text{ mol}}{1.008 \text{ g}} = 8.57$ mol	$\frac{8.57}{1.23} = 6.97$	7	
	N	17.3	$g \times \frac{1 \text{ mol}}{14.01 \text{ g}} = 1.23$ mol	$\frac{1.23}{1.23} = 1.00$	1	C_5H_7N
23)	C	85.7	$g \times \frac{1 \text{ mol}}{12.01 \text{ g}} = 7.14$ mol	$\frac{7.14}{7.14} = 1.00$	1	CH_2
	H	14.3	$g \times \frac{1 \text{ mol}}{1.008} = 14.3$ mol	$\frac{14.3}{7.14} = 2.00$	2	

$n = \frac{42.0 \text{ g/mol molecular}}{(12.01 + 2.016) \text{ g/mol empirical}} = 3$ Molecular formula: C_3H_6

	ELEMENT	GRAMS	MOLES OF ATOMS	ATOM RATIO	FORMULA RATIO	EMPIRICAL FORMULA
24)	C	10.5	$g \times \frac{1 \text{ mol}}{12.01 \text{ g}} = 0.874$ mol	$\frac{0.874}{0.874} = 1.00$	1	
	F	16.7	$g \times \frac{1 \text{ mol}}{19.00 \text{ g}} = 0.879$ mol	$\frac{0.879}{0.874} = 1.01$	1	CH_3F
	H	2.6	$g \times \frac{1 \text{ mol}}{1.008 \text{ g}} = 2.58$ mol	$\frac{2.58}{0.874} = 2.95$	3	

$n = \frac{34 \text{ g/mol molecular}}{(12.01 + 3.024 + 19.00) \text{ g/mol empirical}} = 1$ Molecular formula: CH_3F

	ELEMENT	GRAMS	MOLES OF ATOMS	ATOM RATIO	FORMULA RATIO	EMPIRICAL FORMULA
25)	C	49.3	$g \times \frac{1 \text{ mol}}{12.01 \text{ g}} = 4.10$ mol	$\frac{4.10}{1.36} = 3.01$	6	
	H	2.1	$g \times \frac{1 \text{ mol}}{1.008 \text{ g}} = 2.08$ mol	$\frac{2.08}{1.36} = 1.53$	3	$C_6H_3Cl_2$
	Cl	48.6	$g \times \frac{1 \text{ mol}}{35.45 \text{ g}} = 1.36$ mol	$\frac{1.36}{1.36} = 1.00$	2	

$n = \frac{292 \text{ g/mol molecular}}{145.98 \text{ g/mol empirical}} = 2$ Molecular formula: $C_{12}H_6Cl_4$

26) $59.7 \text{ g } CaCl_2 \times \frac{1 \text{ mol } CaCl_2}{110.98 \text{ g } CaCl_2} \times \frac{2 \text{ mol Cl}}{1 \text{ mol } CaCl_2} \times \frac{35.45 \text{ g Cl}}{\text{mol Cl}} = 38.1 \text{ g Cl}$

27) a) $162\ g\ NiSO_4 \cdot 6H_2O \times \frac{1\ mol\ NiSO_4 \cdot 6H_2O}{262.86\ g\ NiSO_4 \cdot 6H_2O} \times \frac{1\ mol\ S}{1\ mol\ NiSO_4 \cdot 6H_2O} \times \frac{32.06\ g\ S}{mol\ S} = 19.8\ g\ S$

b) $\%\ H_2O = \frac{(12 \times 1.008 + 6 \times 16.00) g\ H_2O}{262.86\ g\ NiSO_4 \cdot 6H_2O} \times 100 = 41.12\%\ H_2O$

c) $840\ g\ NiSO_4 \cdot 6H_2O \times \frac{1\ mol\ NiSO_4 \cdot 6H_2O}{262.86\ g\ NiSO_4 \cdot 6H_2O} \times \frac{1\ mol\ SO_4{}^{2-}}{1\ mol\ NiSO_4 \cdot 6H_2O} = 3.2\ mol\ SO_4{}^{2-}$

d) $52.8\ g\ Ni \times \frac{1\ mol\ Ni}{58.70\ g\ Ni} \times \frac{1\ mol\ NiSO_4 \cdot 6H_2O}{1\ mol\ Ni} \times \frac{262.86\ g\ NiSO_4 \cdot 6H_2O}{mol\ NiSO_4 \cdot 6H_2O} = 236\ g\ NiSO_4 \cdot 6H_2O$

e) $NiSO_4 \rightarrow mol\ NiSO_4 \cdot 6H_2O \rightarrow mol\ Ni(OH)_2 \rightarrow g\ Ni(OH)_2$

$45.1\ g\ NiSO_4 \cdot 6H_2O \times \frac{1\ mol\ NiSO_4 \cdot 6H_2O}{262.86\ g\ NiSO_4 \cdot 6H_2O} \times \frac{1\ mol\ Ni(OH)_2}{1\ mol\ NiSO_4 \cdot 6H_2O} \times \frac{92.72\ g\ Ni(OH)_2}{mol\ Ni(OH)_2} = 15.9\ g\ Ni(OH)_2$

28) a) $665\ g\ CaCO_3 \cdot MgCO_3 \times \frac{1\ mol\ CaCO_3 \cdot MgCO_3}{184.41\ g\ CaCO_3 \cdot MgCO_3} \times \frac{1\ mol\ Ca}{1\ mol\ CaCO_3 \cdot MgCO_3} \times \frac{40.08\ g\ Ca}{mol\ Ca} = 145\ g\ Ca$

b) $45.0\ g\ Mg \times \frac{1\ mol\ Mg}{24.31\ g\ Mg} \times \frac{1\ mol\ CaCO_3 \cdot MgCO_3}{1\ mol\ Mg} \times \frac{184.41\ g\ CaCO_3 \cdot MgCO_3}{mol\ CaCO_3 \cdot MgCO_3} = 341\ g\ CaCO_3 \cdot MgCO_3$

c) $\frac{g\ Ca}{g\ Mg} = \frac{40.08}{24.31} = 1.649\ g\ Ca/g\ Mg$

29) $\frac{\$505}{lb\ Cu(NO_3)_2 \cdot 6H_2O} \times \frac{1\ lb}{453.6\ g} \times \frac{295.67\ g\ Cu(NO_3)_2 \cdot 6H_2O}{mol\ Cu(NO_3)_2 \cdot 6H_2O} \times \frac{1\ mol\ Cu}{1\ mol\ Cu(NO_3)_2 \cdot 6H_2O}$

$\times \frac{1\ mol\ Cu}{63.55\ g\ Cu} = \$5.18/g\ Cu\ in\ Cu(NO_3)_2 \cdot 6H_2O$

$\frac{\$544}{lb\ CuSO_4 \cdot 5H_2O} \times \frac{1\ lb}{453.6\ g} \times \frac{249.69\ g\ CuSO_4 \cdot 5H_2O}{mol\ CuSO_4 \cdot 5H_2O} \times \frac{1\ mol\ Cu}{1\ mol\ CuSO_4 \cdot 5H_2O}$

$\times \frac{1\ mol\ Cu}{63.55\ g\ Cu} = \$4.71/g\ Cu\ in\ CuSO_4 \cdot 5H_2O$ (better buy)

CHAPTER 9

1)
most complicated: $1\ H_2S(g) + SO_2(g) \rightarrow S(s) + H_2O(g)$
balance H: $1\ H_2S(g) + SO_2(g) \rightarrow S(s) + 1\ H_2O(g)$
balance O: $1\ H_2S(g) + {}^1/_2\ SO_2(g) \rightarrow S(s) + 1\ H_2O(g)$
balance S: $1\ H_2S(g) + {}^1/_2\ SO_2(g) \rightarrow {}^3/_2\ S(s) + 1\ H_2O(g)$
clear fractions: $2\ H_2S(g) + 1\ SO_2(g) \rightarrow 3\ S(s) + 2\ H_2O(g)$

2)
most complicated: $1\ H_2S(g) + O_2(g) \rightarrow S(s) + H_2O(g)$
balance H: $1\ H_2S(g) + O_2(g) \rightarrow S(s) + 1\ H_2O(g)$
balance S: $1\ H_2S(g) + O_2(g) \rightarrow 1\ S(s) + 1\ H_2O(g)$
balance O: $1\ H_2S(g) + {}^1/_2\ O_2(g) \rightarrow 1\ S(s) + 1\ H_2O(g)$
clear fractions: $2\ H_2S(g) + 1\ O_2(g) \rightarrow 2\ S(s) + 2\ H_2O(g)$

3)
most complicated: $H_2S(g) + 1\ HNO_3(aq) \rightarrow NO(g) + H_2O(\ell) + S(s)$
balance N: $H_2S(g) + 1\ HNO_3(aq) \rightarrow 1\ NO(g) + H_2O(\ell) + S(s)$
balance O: $H_2S(g) + 1\ HNO_3(aq) \rightarrow 1\ NO(g) + 2\ H_2O(\ell) + S(s)$
balance H: ${}^3/_2\ H_2S(g) + 1\ HNO_3(aq) \rightarrow 1\ NO(g) + 2\ H_2O(\ell) + S(s)$
balance S: ${}^3/_2\ H_2S(g) + 1\ HNO_3(aq) \rightarrow 1\ NO(g) + 2\ H_2O(\ell) + {}^3/_2\ S(s)$
clear fractions: $3\ H_2S(g) + 2\ HNO_3(aq) \rightarrow 2\ NO(g) + 4\ H_2O(\ell) + 3\ S(s)$

4)
most complicated: $SO_2(g) + O_2(g) + CaO(s) \rightarrow 1\ CaSO_4(s)$
balance Ca: $SO_2(g) + O_2(g) + 1\ CaO(s) \rightarrow 1\ CaSO_4(s)$
balance S: $1\ SO_2(g) + O_2(g) + 1\ CaO(s) \rightarrow 1\ CaSO_4(s)$
balance O: $1\ SO_2(g) + {}^1/_2\ O_2(g) + 1\ CaO(s) \rightarrow 1\ CaSO_4(s)$
clear fractions: $2\ SO_2(g) + 1\ O_2(g) + 2\ CaO(s) \rightarrow 2\ CaSO_4(s)$

5) $4\ P(s) + 3\ O_2(g) \rightarrow P_4O_6(s)$ - Combination
(Alternately $P_4 + 3\ O_2 \rightarrow P_4O_6$ because phosphorus exists as a tetratomic molecule.)
6) $2\ Al(s) + 6\ HCl(aq) \rightarrow 2\ AlCl_3(aq) + 3\ H_2(g)$ - Single Replacement
7) $H_2CO_3(aq) \rightarrow CO_2(g) + H_2O(\ell)$ - Decomposition
8) $Ba(OH)_2(aq) + H_2SO_4(aq) \rightarrow BaSO_4(s) + 2\ H_2O(\ell)$ - Double Replacement (Neutral/Ppt)
9) $BaCl_2(aq) + (NH_4)_2CO_3(aq) \rightarrow BaCO_3(s) + 2\ NH_4Cl(aq)$ - Double Replacement (Pptn)
10) $2\ NH_3(aq) + H_2SO_4(aq) \rightarrow (NH_4)_2SO_4(aq)$ - Combination
11) $C_3H_8(g) + 5\ O_2(g) \rightarrow 3\ CO_2(g) + 4\ H_2O(\ell)$ - Combustion
12) $NaF(aq) + HNO_3(aq) \rightarrow HF(aq) + NaNO_3(aq)$ - Double Replacement (Weak Acid Formation)
13) $2\ KClO_3(s) \rightarrow 2\ KCl(s) + 3\ O_2(g)$ - Decomposition
14) $Zn(s) + Pb(NO_3)_2(aq) \rightarrow Pb(s) + Zn(NO_3)_2(aq)$ - Single Replacement
15) $HBr(aq) + KOH(aq) \rightarrow KBr(aq) + H_2O(\ell)$ - Double Replacement (Neutralization)
16) $Cl_2(g) + 2\ KI(aq) \rightarrow I_2(s) + 2\ KCl(aq)$ - Single Replacement
17) $2\ AgNO_3(aq) + H_2S(g) \rightarrow Ag_2S(s) + 2\ HNO_3(aq)$ - Double Replacement (Precipitation)
18) $2\ P(s) + 3\ I_2(s) \rightarrow 2\ PI_3(s)$ - Combination (Alternately $P_4 + 6\ I_2 \rightarrow 4\ PI_3$)
19) $C_{12}H_{22}O_{11}(s) + 12\ O_2(g) \rightarrow 12\ CO_2(g) + 11\ H_2O(\ell)$ - Combustion
20) $CuO(s) + 2\ HCl(aq) \rightarrow CuCl_2(aq) + H_2O(\ell)$ - Double Replacement (Neutralization)
21) $2\ BaO_2(s) \rightarrow 2\ BaO(s) + O_2(g)$ - Decomposition
22) $2\ C_3H_7CHO(\ell) + 11\ O_2(g) \rightarrow 8\ CO_2(g) + 8\ H_2O(\ell)$ - Combustion
23) $N_2(g) + 3\ H_2(g) \rightarrow 2\ NH_3(g)$ - Combination
24) $2\ HCl(aq) + MgCO_3(s) \rightarrow MgCl_2(aq) + CO_2(g) + H_2O(\ell)$ - Double Replacement (Formation of a Weak Acid) and Decomposition (of H_2CO_3)
25) $2\ NaCHO_2(aq) + H_2SO_4(aq) \rightarrow Na_2SO_4(aq) + 2\ HCHO_2(aq)$ - Double Repl. (Weak Acid Form.)
26) $Zn(s) + 2\ HCl(aq) \rightarrow H_2(g) + ZnCl_2(aq)$ - Single Replacement
27) $2\ C_5H_{11}OH(\ell) + 15\ O_2(g) \rightarrow 10\ CO_2(g) + 12\ H_2O(\ell)$ - Combustion
28) $AgNO_3(aq) + NaIO_3(aq) \rightarrow AgIO_3(s) + NaNO_3(aq)$ - Double Replacement (Precipitation)
29) $H_2SO_4(aq) + Ni(OH)_2(s) \rightarrow NiSO_4(aq) + 2\ H_2O(\ell)$ - Double Replacement (Neutralization)
30) $2\ KNO_2(aq) + H_2SO_4(aq) \rightarrow 2\ HNO_2(aq) + K_2SO_4(aq)$ - Double Replacement (Weak Acid Form)
31) $HgBr_2(aq) + (NH_4)_2S(aq) \rightarrow HgS(s) + 2\ NH_4Br(aq)$ - Double Replacement (Precipitation)

CHAPTER 10

1) $4\ Al(s) + 3\ O_2(g) \rightarrow 2\ Al_2O_3(s)$ $\quad 7.37 \text{ mol } O_2 \times \frac{4 \text{ mol Al}}{3 \text{ mol } O_2} = 9.83 \text{ mol Al};$

$7.37 \text{ mol } O_2 \times \frac{2 \text{ mol } Al_2O_3}{3 \text{ mol } O_2} = 4.91 \text{ mol } Al_2O_3$

2) $2\ KClO_3(s) \rightarrow 2\ KCl(s) + 3\ O_2(g)$

$6.09 \text{ g } KClO_3 \times \frac{1 \text{ mol } KClO_3}{122.55 \text{ g } KClO_3} \times \frac{3 \text{ mol } O_2}{2 \text{ mol } KClO_3} = 0.0745 \text{ mol } O_2$

3) $39.6 \text{ g PbO} \times \frac{1 \text{ mol PbO}}{223.2 \text{ g PbO}} \times \frac{2 \text{ mol PbS}}{2 \text{ mol PbO}} = 0.177 \text{ mol PbS}$

4) $AgNO_3(aq) + NaBr(aq) \rightarrow AgBr(s) + NaNO_3(aq)$

$2.50 \times 10^2 \text{ g AgBr} \times \frac{1 \text{ mol AgBr}}{187.8 \text{ g AgBr}} \times \frac{1 \text{ mol } AgNO_3}{1 \text{ mol AgBr}} \times \frac{169.9 \text{ g } AgNO_3}{1 \text{ mol } AgNO_3} = 226 \text{ g } AgNO_3$

5) $2.7 \text{ kg NaCl} \times \frac{1 \text{ kmol NaCl}}{58.44 \text{ kg NaCl}} \times \frac{1 \text{ kmol } Na_2SO_4}{2 \text{ kmol NaCl}} \times \frac{142.04 \text{ kg } Na_2SO_4}{1 \text{ kmol } Na_2SO_4} = 3.3 \text{ kg } Na_2SO_4$

6) $43.9 \text{ g } LiAlH_4 \times \frac{1 \text{ mol } LiAlH_4}{37.95 \text{ g } LiAlH_4} \times \frac{4 \text{ mol LiH}}{1 \text{ mol } LiAlH_4} \times \frac{7.949 \text{ g LiH}}{1 \text{ mol LiH}} = 36.8 \text{ g LiH}$

7) $11.7 \text{ g } H_2 \times \frac{1 \text{ mol } H_2}{2.016 \text{ g } H_2} \times \frac{2 \text{ mol } NO_2}{7 \text{ mol } H_2} \times \frac{46.01 \text{ g } NO_2}{1 \text{ mol } NO_2} = 76.3 \text{ g } NO_2$

8) $2\ C_4H_{10}(g) + 13\ O_2(g) \rightarrow 8\ CO_2(g) + 10\ H_2O(\ell)$

$816 \text{ g } O_2 \times \frac{1 \text{ mol } C_4H_{10}}{58.12 \text{ g } C_4H_{10}} \times \frac{8 \text{ mol } CO_2}{2 \text{ mol } C_4H_{10}} \times \frac{44.01 \text{ g } CO_2}{1 \text{ mol } CO_2} = 2.47 \times 10^3 \text{ g } CO_2$

9)
$$31.5\text{ g Fe} \times \frac{1\text{ mol Fe}}{55.85\text{ g Fe}} \times \frac{8\text{ mol Al}}{9\text{ mol Fe}} \times \frac{26.98\text{ g Al}}{1\text{ mol Al}} = 13.5\text{ g Al}$$

10)
$$3.00 \times 10^4\text{ kg } CaC_2 \times \frac{1\text{ kmol } CaC_2}{64.10\text{ kg } CaC_2} \times \frac{1\text{ kmol } C_2H_2}{1\text{ kmol } CaC_2} \times \frac{26.04\text{ kg } C_2H_2}{1\text{ kmol } C_2H_2} = 1.22 \times 10^4\text{ kg } C_2H_2\text{ theoretical}$$

$$\%\text{ yield} = \frac{9.13 \times 10^3\text{ kg } C_2H_2\text{ actual}}{1.22 \times 10^4\text{ kg } C_2H_2\text{ theoretical}} \times 100 = 74.8\%\text{ yield}$$

11)
$$75.0\text{ g } C_7H_6O_3 \times \frac{1\text{ mol } C_7H_6O_3}{138.12\text{ g } C_7H_6O_3} \times \frac{1\text{ mol } C_9H_8O_4}{1\text{ mol } C_7H_6O_3} \times \frac{180.15\text{ g } C_9H_8O_4\text{ theo}}{1\text{ mol } C_9H_8O_4} \times \frac{84.6\text{ g } C_9H_8O_4\text{ act}}{100.0\text{ g } C_9H_8O_4\text{ theo}} = 82.8\text{ g } C_9H_8O_4\text{ actual}$$

12) $2\,Al(s) + 2\,H_3PO_4(aq) \rightarrow 3\,H_2(g) + 2\,AlPO_4(s)$

$$73.9\text{ g } AlPO_4\text{ act} \times \frac{100.0\text{ g } AlPO_4\text{ theo}}{62.7\text{ g } AlPO_4\text{ act}} \times \frac{1\text{ mol } AlPO_4}{121.95\text{ g } AlPO_4} \times \frac{2\text{ mol } H_3PO_4}{2\text{ mol } AlPO_4} \times \frac{97.99\text{ g } H_3PO_4}{1\text{ mol } H_3PO_4} = 94.7\text{ g } H_3PO_4$$

13)
$$316\text{ g } NH_3 \times \frac{1\text{ mol } NH_3}{17.03\text{ g } NH_3} \times \frac{6\text{ mol } NH_4Br}{8\text{ mol } NH_3} \times \frac{97.94\text{ g } NH_4Br}{1\text{ mol } NH_4Br} = 1360\text{ g } NH_4Br\text{ theo}$$

$$\%\text{ yield} = \frac{1210\text{ g } NH_4Br\text{ act}}{1360\text{ g } NH_4Br\text{ theo}} \times 100 = 89.0\%\text{ yield}$$

14)
$$46.8\text{ g } HNO_3\text{ act} \times \frac{100.0\text{ g } HNO_3\text{ theo}}{92.3\text{ g } HNO_3\text{ act}} \times \frac{1\text{ mol } HNO_3}{63.02\text{ g } HNO_3} \times \frac{3\text{ mol } NO_2}{2\text{ mol } HNO_3} \times \frac{46.01\text{ g } NO_2}{1\ \mu\rho\omega\ \nu\rho_2} = 55.5\text{ g } NO_2$$

15) $2\,C_2H_6(g) + 7\,O_2(g) \rightarrow 4\,CO_2(g) + 6\,H_2O(\ell)$

$$23.6\text{ g } C_2H_6 \times \frac{1\text{ mol } C_2H_6}{30.07\text{ g } C_2H_6} \times \frac{4\text{ mol } CO_2}{2\text{ mol } C_2H_6} \times \frac{44.01\text{ g } CO_2\text{ theo}}{1\text{ mol } CO_2} \times \frac{79.4\text{ g } CO_2\text{ act}}{100.0\text{ g } CO_2\text{ theo}} = 54.9\text{ g } CO_2\text{ actual}$$

16)

	$6\,KI(aq)$ +	$8\,HNO_3(aq)$ →	$6\,KNO_3(aq)$ +	$2\,NO(g)$ +	$3\,I_2(s)$ +	$4\,H_2O(\ell)$
Grams at start	155	175				
Molar mass	166.0	63.02		30.01		
Moles at start	0.934	2.78				
Moles changing	-0.934	-1.25		+0.311		
Moles at end	0	1.53		0.311		
Grams at end	0	96.4		9.33		
	limiting	excess (unreacted)		produced		

17)

	$2\,H_2(g)$ +	$O_2(g)$ →	$2\,H_2O(\ell)$
Grams at start	50.0	50.0	
Molar mass	2.016	32.00	18.02
Moles at start	24.8	1.56	
Moles changing	-3.12	-1.56	+3.12
Moles at end	21.7	0	3.12
Grams at end	43.7	0	56.2
	unreacted	limiting	produced

18)

	$2\,Al(OH)_3(s)$ +	$3\,H_2SO_4(aq)$ →	$Al_2(SO_4)_3(aq)$ +	$6\,H_2O(\ell)$
Grams at start	54.9	91.0		
Molar mass	78.00	98.08	342.14	
Moles at start	0.704	0.928		
Moles changing	-0.619	-0.928	+0.309	
Moles at end	0.085	0	0.309	
Grams at end	6.6	0	106	
	unreacted	limiting	produced	

CHAPTER 10

19) $98.6 \text{ g LS} \times \frac{88.4 \text{ g } CaCO_3}{100.0 \text{ g LS}} \times \frac{1 \text{ mol } CaCO_3}{100.09 \text{ g } CaCO_3} \times \frac{1 \text{ mol CaO}}{1 \text{ mol } CaCO_3} \times \frac{56.08 \text{ g CaO}}{1 \text{ mol CaO}} = 48.8 \text{ g CaO}$

20) $6.5 \text{ kg HCl} \times \frac{1 \text{ kmol HCl}}{36.46 \text{ kg HCl}} \times \frac{1 \text{ kmol } H_2SO_4}{2 \text{ kmol HCl}} \times \frac{98.08 \text{ kg } H_2SO_4}{1 \text{ kmol } H_2SO_4} \times \frac{100.0 \text{ kg acid}}{79.2 \text{ kg } H_2SO_4}$

$= 11 \text{ kg acid solution}$

21) $62 \text{ L tank} \times \frac{0.85 \text{ mL } (C_2H_5)_4Pb}{1 \text{ L tank}} \times \frac{1.66 \text{ g } (C_2H_5)_4Pb}{1 \text{ mL } (C_2H_5)_4Pb} \times \frac{1 \text{ mol } (C_2H_5)_4Pb}{323.4 \text{ g } (C_2H_5)_4Pb}$

$\times \frac{4 \text{ mol } C_2H_5Cl}{1 \text{ mol } (C_2H_5)_4Pb} \times \frac{64.51 \text{ g } C_2H_5Cl}{1 \text{ mol } C_2H_5Cl} = 7.0 \times 10 \text{ g } C_2H_5Cl$

22) $25 \text{ kg } H_2SO_4 \times \frac{1 \text{ kmol } H_2SO_4}{98.08 \text{ g } H_2SO_4} \times \frac{1 \text{ kmol } SO_3}{1 \text{ kmol } H_2SO_4} \times \frac{2 \text{ kmol } SO_2}{2 \text{ kmol } SO_3} \times \frac{4 \text{ kmol } FeS_2}{8 \text{ kmol } SO_2}$

$\times \frac{119.97 \text{ kg } FeS_2}{1 \text{ kmol } FeS_2} \times \frac{100.0 \text{ kg rock}}{21.6 \text{ kg } FeS_2} = 71 \text{ kg rock}$

23) Let x = g SO_3 in 100 g oleum. (100-x) = g H_2SO_4 in 100 g oleum

$x \text{ g } SO_3 \times \frac{1 \text{ mol } SO_3}{80.06 \text{ g } SO_3} \times \frac{1 \text{ mol } H_2SO_4}{1 \text{ mol } SO_3} \times \frac{98.08 \text{ g } H_2SO_4}{1 \text{ mol } H_2SO_4} = 1.225 \text{ x g } H_2SO_4 \text{ produced}$

H_2SO_4 produced + H_2SO_4 in oleum = 108 g H_2SO_4

1.225 x + (100-x) = 108 x = 35.6 g SO_3

$\frac{35.6 \text{ g } SO_3}{100 \text{ g oleum}} = 35.6\% \ SO_3$ = 35 or 36 rounded off to two significant figures

24) $2 \ KClO_3(s) \rightarrow 2 \ KCl(s) + 3 \ O_2(g)$

$0.895 \text{ g } O_2 \times \frac{1 \text{ mol } O_2}{32.00 \text{ g } O_2} \times \frac{2 \text{ mol } KClO_3}{3 \text{ mol } O_2} \times \frac{122.55 \text{ g } KClO_3}{1 \text{ mol } KClO_3} = 2.29 \text{ g } KClO_3$

$\% \ KClO_3 = \frac{1.71 \text{ g KCl}}{4.09 \text{ g sample}} \times 100 = 56.0\% \ KClO_3$

25) $NaHC_2O_4 + Cl_2 + 5 \ NaOH + 2 \ Ba(NO_3)_2 \rightarrow 2 \ NaCl + 2 \ BaCO_3 + 3 \ H_2O + 4 \ NaNO_3$

	$NaHC_2O_4$	Cl_2	NaOH	$BaCO_3$
Molar mass				197.3
Moles at start	0.10	0.040	0.12	
Moles changing	-0.024	-0.024	-0.12	+0.048
Moles at end	0.08	0.016	0	0.048
Grams at end			limiting	9.5 g produced

CHAPTER 11

1) $\lambda = c/\nu = 2.998 \times 10^8 \text{ m/s} \div 3.75 \times 10^{14} \text{ s}^{-1} = 7.99 \times 10^{-7} \text{ m} = 799 \text{ nm}$

2) $\nu = E/h = 2.25 \times 10^{-18} \text{ J} \div 6.626 \times 10^{-34} \text{ J·s} = 3.39\underline{6} \times 10^{15} \text{ s}^{-1}$

$\lambda = c/\nu = 2.998 \times 10^8 \text{ m/s} \div 3.39\underline{6} \times 10^{15} \text{ s-1} = 8.82\underline{8} \times 10^{-8} \text{ m} = 88.3 \text{ nm}$

3) high frequency: x-rays > green > orange > infrared > microwaves :low frequency

4) $\nu = 3.288 \times 10^{15} \text{ cy/s} \left(\frac{1}{3^2} - \frac{1}{5^2}\right) = 2.338 \times 10^{14} \text{ cy/s}$ (n_i = 3, so Paschen series)

5) See Example 11.3.

6) a) n = 4, ℓ = 1 b) n = 6, ℓ = 0 c) doesn't exist d) n = 5, ℓ = 2

7)

	shorthand	noble gas	valence
Li:	$1s^22s^1$	$[He]2s^1$	$2s^1$
P:	$1s^22s^22p^63s^23p^3$	$[Ne]3s^23p^3$	$3s^23p^3$
Se:	$1s^22s^22p^63s^23p^64s^23d^{10}4p^4$	$[Ar]4s^23d^{10}4p^4$	$4s^24p^4$
Mo:	$1s^22s^22p^63s^23p^64s^23d^{10}4p^65s^14d^5$	$[Kr]5s^14d^5$	$5s^14d^5$
Hg:	$1s^22s^22p^63s^23p^64s^23d^{10}4p^65s^24d^{10}5p^66s^24f^{14}5d^{10}$	$[Xe]6s^24f^{14}5d^{10}$	$6s^25d^{10}$

8) a) $4p^2$

n	ℓ	m	s
4	1	0	+½
4	1	+1	+½

Notes: m values must be different; other possibilities are: {0,-1} and {+1,-1}; s values must be the same: other possibility: {-½,-½}.

b) $6s^2$

n	ℓ	m	s
6	0	0	+½
6	0	0	-½

c) $4d^5$

n	ℓ	m	s
4	2	0	+½
4	2	+1	+½
4	2	-1	+½
4	2	+2	+½
4	2	-2	+½

Note: all s values could be -½.

d) $3p^4$

n	ℓ	m	s
3	1	0	+½
3	1	0	-½
3	1	+1	+½
3	1	-1	+½

Notes: the two paired electrons could both be in an orbital with m = +1 or with m = -1; the two unpaired electrons must be in different orbitals (have different m values) and must have the same spin, although both could have s = -½ instead of +½.

9) See Figure 11.9.

10) a) 6 (half of them)
b) 6 (those in 3p)
c) 9 or 10 (two in each subshell, but 1 <u>or</u> 2 in the 3p subshell depending on whether the paired electrons are in the orbital with m = 0, or the ones with m = +1 or -1).
d) 8 ($2s^2$ and $2p^6$)
e) 10 (those in 3d)

11) a) impossible, Pauli exclusion principle violated, 3s electrons must be spin paired.
b) excited state, m effect, 3p electrons should be in different orbitals, same spin.
c) excited state, s effect, last two 3d electrons should have the same spin.
d) excited state, ℓ effect, the last four electrons should be $4s^23d^2$.

12) a) O: 1s [↑↓] 2s [↑↓] 2p [↑↓|↑|↑]

b) Al: 1s [↑↓] 2s [↑↓] 2p [↑↓|↑↓|↑↓] 3s [↑↓] 3p [↑| |]

c) Ni: 1s [↑↓] 2s [↑↓] 2p [↑↓|↑↓|↑↓] 3s [↑↓] 3p [↑↓|↑↓|↑↓] 4s [↑↓] 3d [↑↓|↑↓|↑↓|↑|↑]

d) Sr: 1s [↑↓] 2s [↑↓] 2p [↑↓|↑↓|↑↓] 3s [↑↓] 3p [↑↓|↑↓|↑↓] 4s [↑↓] 3d [↑↓|↑↓|↑↓|↑↓|↑↓] 4p [↑↓|↑↓|↑↓] 5s [↑↓]

e) Bi: 1s [↑↓] 2s [↑↓] 2p [↑↓|↑↓|↑↓] 3s [↑↓] 3p [↑↓|↑↓|↑↓] 4s [↑↓] 3d [↑↓|↑↓|↑↓|↑↓|↑↓] 4p [↑↓|↑↓|↑↓] 5s [↑↓]
4d [↑↓|↑↓|↑↓|↑↓|↑↓] 5p [↑↓|↑↓|↑↓] 6s [↑↓] 4f [↑↓|↑↓|↑↓|↑↓|↑↓|↑↓|↑↓] 5d [↑↓|↑↓|↑↓|↑↓|↑↓] 6p [↑|↑|↑]

CHAPTER 11

CHAPTER 12

1) a) $830\ \text{torr} \times \frac{1\ \text{atm}}{760\ \text{torr}} = 1.1\ \text{atm}$ $830\ \text{torr} \times \frac{101.3\ \text{kPa}}{760\ \text{torr}} = 110\ \text{kPa}$

b) $2.31\ \text{atm} \times \frac{760\ \text{torr}}{1\ \text{atm}} = 1.76 \times 10^3\ \text{torr}$ $2.31\ \text{atm} \times \frac{101.3\ \text{kPa}}{1\ \text{atm}} = 234\ \text{kPa}$

2) $4.87\ \text{L} \times \frac{(273 + 71)\text{K}}{(273 + 14)\text{K}} = 5.84\ \text{L}$

3) $462\ \text{torr} \times \frac{9.15\ \text{L}}{7.96\ \text{L}} = 531\ \text{torr}$

4) $56.0\ \text{cm}^3 \times \frac{3.61\ \text{atm}}{0.972\ \text{atm}} = 208\ \text{cm}^3$

5) $1.76\ \text{mol} \times \frac{75.0\ \text{L}}{27.5\ \text{L}} = 4.80\ \text{mol}$

6) $P = \frac{nRT}{V} = \frac{1.42\ \text{mol}}{10.0\ \text{L}} \times \frac{0.08205\ \text{L·atm}}{\text{mol·K}} \times 348\ \text{K} = 4.05\ \text{atm}$; or

$= \frac{1.42\ \text{mol}}{10.0\ \text{L}} \times \frac{62.36\ \text{L·torr}}{\text{mol·K}} \times 348\ \text{K} = 3.08 \times 10^3\ \text{torr}$

7) $n = \frac{PV}{RT} = \frac{1.81\ \text{atm}}{298\ \text{K}} \times \frac{\text{mol·K}}{0.08205\ \text{L·atm}} \times 7.07\ \text{L} = 0.523\ \text{mol}$

8) $m = \frac{PV(MM)}{RT} = \frac{790\ \text{torr}}{288\ \text{K}} \times \frac{\text{mol·K}}{62.36\ \text{L·torr}} \times 0.618\ \text{L}\ O_2 \times \frac{32.00\ \text{g}\ O_2}{1\ \text{mol}\ O_2} = 0.87\ \text{g}\ O_2$

9) $V = \frac{nRT}{P} = \frac{1.74\ \text{mol}}{2.90\ \text{atm}} \times \frac{0.08205\ \text{L·atm}}{\text{mol·K}} \times 1016\ \text{K} = 50.0\ \text{L}$

10) $T = \frac{PV(MM)}{mR} = \frac{994\ \text{torr}}{3.12\ \text{g}\ CH_4} \times \frac{\text{mol·K}}{62.36\ \text{L·torr}} \times 10.0\ \text{L} \times \frac{16.04\ \text{g}\ CH_4}{1\ \text{mole}\ CH_4} = 819\ \text{K} = 546\ °\text{C}$

11) $10.6\ \text{L} \times \frac{369\ \text{K}}{356\ \text{K}} \times \frac{638\ \text{torr}}{1076\ \text{torr}} = 6.51\ \text{L}$

12) $392\ \text{cm}^3 \times \frac{258\ \text{K}}{305\ \text{K}} \times \frac{925\ \text{torr}}{775\ \text{torr}} = 396\ \text{cm}^3$

13) $5.79\ \text{L} \times \frac{273\ \text{K}}{298\ \text{K}} \times \frac{518\ \text{torr}}{760\ \text{torr}} = 3.62\ \text{L}$

14) $1.35\ \text{atm} \times \frac{10.0\ \text{L}}{2.32\ \text{L}} \times \frac{327\ \text{K}}{317\ \text{K}} = 6.00\ \text{atm}$

15) $0.193\ \text{L} \times \frac{(273 + 25)\text{K}}{(273 + 80)\text{K}} \times \frac{68.1\ \text{atm}}{750\ \text{torr}} \times \frac{760\ \text{torr}}{1\ \text{atm}} = 11\ \text{L}$

16) $8.75\ \text{L} \times \frac{0.700\ \text{mol}}{0.450\ \text{mol}} \times \frac{785\ \text{torr}}{812\ \text{torr}} = 13.2\ \text{L}$

17) $0.141\ \text{mol} \times \frac{22.41\ \text{L}}{1\ \text{mol}} = 3.16\ \text{L}$

18) $9.22\ \text{g}\ N_2 \times \frac{1\ \text{mol}\ N_2}{28.02\ \text{g}\ N_2} \times \frac{22.41\ \text{L}}{1\ \text{mol}} = 7.37\ \text{L}\ N_2$

19) $0.846\ \text{L} \times \frac{1\ \text{mol}}{22.41\ \text{L}} = 0.0378\ \text{mol}$

20) $4.39\ \text{L} \times \frac{1\ \text{mol He}}{22.41\ \text{L}} \times \frac{4.003\ \text{g He}}{1\ \text{mol He}} = 0.784\ \text{g He}$

21) $\frac{2.28\ \text{g}}{0.905\ \text{L}} \times \frac{22.41\ \text{L}}{1\ \text{mol}} = 56.5\ \text{g/mol}$

22) $3.59\ \text{g}\ H_2 \times \frac{1\ \text{mol}\ H_2}{2.016\ \text{g}\ H_2} \times \frac{22.41\ \text{L}}{1\ \text{mol}} = 39.9\ \text{L}\ H_2$

23) $\frac{44.09\ \text{g}\ C_3H_8}{1\ \text{mol}\ C_3H_8} \times \frac{1\ \text{mol}}{22.41\ \text{L}} = 1.97\ \text{g/L}$

24) $\frac{1.43\ \text{g}}{\text{L}} \times \frac{22.41\ \text{L}}{1\ \text{mol}} = 32.0\ \text{g/mol}$

25) $d = \frac{m}{V} = \frac{P(MM)}{RT} \times \frac{0.725\ \text{atm}}{291\ \text{K}} \times \frac{28.01\ \text{g CO}}{1\ \text{mol CO}} \times \frac{\text{mol·K}}{0.08205\ \text{L·atm}} = 0.851\ \text{g/L}$

26) $MM = \frac{mRT}{PV} = \frac{8.28\ \text{g}}{740\ \text{torr}} \times \frac{300\ \text{K}}{4.41\ \text{L}} \times \frac{62.36\ \text{L·torr}}{\text{mol·K}} = 47\ \text{g/mol}$ or $47.5\ \text{g/mol}$ (The given value 740 torr may have 2 or 3 sf)

27) $42.6 \text{ g } HNO_3 \times \frac{1 \text{ mol } HNO_3}{63.02 \text{ g } HNO_3} \times \frac{1 \text{ mol NO}}{4 \text{ mol } HNO_3} = 0.169 \text{ mol NO}$

$V = \frac{nRT}{P} = \frac{0.169 \text{ mol}}{800 \text{ torr}} \times \frac{62.36 \text{ L}\cdot\text{torr}}{\text{mol}\cdot\text{K}} \times 295 \text{ K} = 3.89 \text{ L NO}$

28) $n = \frac{PV}{RT} = \frac{2.30 \text{ atm}}{348 \text{ K}} \times \frac{0.08205 \text{ L}\cdot\text{atm}}{\text{mol}\cdot\text{K}} \times 9.54 \text{ L} = 0.768 \text{ mol } CO_2$

$0.768 \text{ mol } CO_2 \times \frac{1 \text{ mol } K_2CO_3}{1 \text{ mol } CO_2} \times \frac{138.21 \text{ g } K_2CO_3}{1 \text{ mol } K_2CO_3} = 106 \text{ g } K_2CO_3$

29) $4.51 \text{ g } KMnO_4 \times \frac{1 \text{ mol } KMnO_4}{158.04 \text{ g } KMnO_4} \times \frac{5 \text{ mol } Cl_2}{2 \text{ mol } KMnO_4} = 0.0713 \text{ mol } Cl_2$

$V = \frac{nRT}{P} = \frac{0.0713 \text{ mol}}{751 \text{ torr}} \times \frac{62.36 \text{ L}\cdot\text{torr}}{\text{mol}\cdot\text{K}} \times 304 \text{ K} = 1.80 \text{ L } Cl_2$

30) $n = \frac{PV}{RT} = \frac{7.50 \text{ atm}}{293 \text{ K}} \times \frac{\text{mol}\cdot\text{K}}{0.08205 \text{ L}\cdot\text{atm}} \times 125 \text{ L} = 39 \text{ mol } C_2H_2$

$39 \text{ mol } C_2H_2 \times \frac{1 \text{ mol } CaC_2}{1 \text{ mol } C_2H_2} \times \frac{64.10 \text{ g } CaC_2}{1 \text{ mol } CaC_2} \times \frac{1 \text{ kg}}{10^3 \text{ g}} = 2.5 \text{ kg } CaC_2$

31) $0.0150 \text{ g } C_6H_{12}O_6 \times \frac{1 \text{ mol } C_6H_{12}O_6}{180.16 \text{ g } C_6H_{12}O_6} \times \frac{6 \text{ mol } CO_2}{1 \text{ mol } C_6H_{12}O_6} = 5.00 \times 10^{-4} \text{ mol } CO_2$

$V = \frac{nRT}{P} = \frac{5.00 \times 10^{-4} \text{ mol}}{16 \text{ torr}} \times \frac{62.36 \text{ L}\cdot\text{torr}}{\text{mol}\cdot\text{K}} \times 296 \text{ K} = 0.58 \text{ L } CO_2$

32) $35.8 \text{ L } O_2 \times \frac{16 \text{ L } CO_2}{25 \text{ L } O_2} = 22.9 \text{ L } CO_2$

33) $45.7 \text{ L } CO_2 \times \frac{250 \text{ K}}{318 \text{ K}} \times \frac{1.15 \text{ atm}}{18.5 \text{ atm}} \times \frac{1 \text{ L } C_3H_8}{3 \text{ L } CO_2} = 0.744 \text{ L } C_3H_8$

34) $3.25 \times 10^4 \text{ L } SO_2 \times \frac{298 \text{ K}}{320 \text{ K}} \times \frac{620 \text{ torr}}{740 \text{ torr}} \times \frac{1 \text{ L } O_2}{2 \text{ L } SO_2} = 1.3 \times 10^4 \text{ L } O_2$

35) 351 torr + 581 torr = 932 torr

36) 749 torr - 24.8 torr = 724 torr

37) 4.74 + 0.72 + 0.13 = 5.59 mol total

$p_{He} = 5.49 \text{ atm} \times \frac{4.74}{5.59} = 4.66 \text{ atm}$

$p_{Ne} = 5.49 \text{ atm} \times \frac{0.72}{5.59} = 0.71 \text{ atm}$

$p_{Ar} = 5.49 \text{ atm} \times \frac{0.13}{5.59} = 0.13 \text{ atm}$

38) $98.1 \text{ g } CH_4 \times \frac{1 \text{ mol } CH_4}{16.04 \text{ g } CH_4} = 6.12 \text{ mol } CH_4$

$165 \text{ g } C_2H_6 \times \frac{1 \text{ mol } C_2H_6}{30.07 \text{ g } C_2H_6} = 5.49 \text{ mol } C_2H_6$

6.12 + 5.49 = 11.61 mol total

$p_{CH_4} = 0.954 \text{ atm} \times \frac{6.12}{11.61} = 0.503 \text{ atm}$

$p_{C_2H_6} = 0.954 \text{ atm} \times \frac{5.49}{11.61} = 0.451 \text{ atm}$

39) $P_{CO} = \frac{n_{CO}RT}{V} = \frac{1.73 \text{ mol CO}}{5.50 \text{ L}} \times \frac{0.08205 \text{ L}\cdot\text{atm}}{\text{mol}\cdot\text{K}} \times 290 \text{ K} = 7.48 \text{ atm}$

40) $P_{N_2} = \frac{mRT}{V(MM)} = \frac{26.9 \text{ g } N_2}{6.14 \text{ L}} \times \frac{1 \text{ mol } N_2}{28.02 \text{ g } N_2} \times \frac{0.08205 \text{ L}\cdot\text{atm}}{\text{mol}\cdot\text{K}} \times 314 \text{ K} = 4.03 \text{ atm}$

41) $\frac{r_{Ar}}{r_{Kr}} = \sqrt{\frac{MM_{Kr}}{MM_{Ar}}} = \sqrt{\frac{83.80}{39.95}} = 1.448$ Ar is faster than Kr by a factor of 1.448.

42) $\frac{t_X}{t_{O_2}} = \sqrt{\frac{MM_X}{MM_{O_2}}}$ so $\frac{86 \text{ s}}{73 \text{ s}} = \sqrt{\frac{MM_X}{32.00}}$ thus $MM_X = 44$ g/mol

43) $\frac{t_X}{t_{C_4H_{10}}} = \sqrt{\frac{MM_X}{MM_{C_4H_{10}}}}$ so $\frac{106 \text{ s}}{136 \text{ s}} = \sqrt{\frac{MM_X}{58.12}}$ thus $MM_X = 35.3$ g/mol

CHAPTER 12

CHAPTER 13

1) $195\ g \times \frac{0.13\ J}{g\cdot{}^\circ C} \times (25\text{-}90)^\circ C = -1648\ J = -1.6\ kJ$

2) $\frac{761\ J}{1.50\ g \times (3600\text{-}20)^\circ C} = 0.14\ J/g\cdot{}^\circ C$

3) $\frac{2.51 \times 10^5\ J}{925\ g} \times \frac{g\cdot{}^\circ C}{0.473\ J} = 574^\circ C = \Delta T$ Final T = 18°C + 574°C = 592°C

4) $\frac{2.67 \times 10^3\ J}{(75.0\text{-}20.0)^\circ C} \times \frac{g\cdot{}^\circ C}{0.213\ J} = 228\ g\ Sn$

5) $3.50\ kg \times \frac{0.594\ kJ}{kg\cdot{}^\circ C} \times (82.0\text{-}25.0)^\circ C = 119\ kJ$

6) $\frac{-3.15\ kJ}{(9.8 - 54.8)^\circ C} \times \frac{kg\cdot{}^\circ C}{0.703\ kJ} = 0.0996\ kg = 99.6\ g$

7) a) Cu reaches higher temperatures because less energy is required to raise a given mass of copper by 1°C than is required by the same mass of aluminum.

b) Cu: $\frac{3.5\ kJ}{0.095\ kg} \times \frac{kg\cdot{}^\circ C}{0.38\ kJ} = 97^\circ C$ Al: $\frac{3.5\ kJ}{0.095\ kg} \times \frac{kg\cdot{}^\circ C}{0.88\ kJ} = 42^\circ C$

8) $125\ g \times 4.18\ J/g\cdot{}^\circ C \times (T_2 - 17)^\circ C = -7.5\ g \times 0.88\ J/g\cdot{}^\circ C \times (T_2 - 175)^\circ C$
$522.\underline{5}\ T_2 - 88\underline{82.5} = -6.6\ T_2 + 115\underline{5}$ so $529.\underline{1}\ T_2 = 10{,}0\underline{37.5}$ so $T_2 = 19.0^\circ C$

9) $85\ g \times 4.18\ J/g\cdot{}^\circ C \times (37 - 22)^\circ C = -25\ g \times c \times (37 - 215)^\circ C$ so $c = 1.2\ J/g\cdot{}^\circ C$

10) $85.4\ g \times 519\ J/g = 4.43 \times 10^4\ J = 44.3\ kJ$

11) $453.6\ g \times 2.26\ kJ/g = 1.03\times10^3\ kJ$

12) $\frac{(925\ g)(4.18\ J/g\cdot{}^\circ C)[(28.8 - 23.4)^\circ C]}{773\ g} = 27\ J/g$

13) $Q_1 = 141\ g \times 4.18\ J/g\cdot{}^\circ C \times (100\text{-}62)^\circ C = 2.2 \times 10^4\ J = 22\ kJ$
$Q_2 = 141\ g \times 2.26\ kJ/g = 319\ kJ$
$Q_3 = 141\ g \times 2.0\ J/g\cdot{}^\circ C \times (114\text{-}100)^\circ C = 3948\ J = 3.9\ kJ$
$Q = Q_1 + Q_2 + Q_3 = 22 + 319 + 3.9 = 345\ kJ$

14) $Q_1 = 56\ g \times 0.49\ J/g\cdot{}^\circ C \times (1083\text{-}1420)^\circ C = -9.2 \times 10^3\ J = -9.2\ kJ$
$Q_2 = 56\ g \times -205\ J/g = -11 \times 10^3\ J = -11\ kJ$
$Q_3 = 56\ g \times 0.38\ J/g\cdot{}^\circ C \times (24\text{-}1083)^\circ C = -22 \times 10^3\ J = -22\ kJ$
$Q = Q_1 + Q_2 + Q_3 = -9.2 - 11 - 22 = -42\ kJ$

15) $Q_1 = 24.9\ g \times 0.13\ J/g\cdot{}^\circ C \times (1063\text{-}26)^\circ C = 3.4\times10^3\ J = 3.4\ kJ$
$Q_2 = 24.9\ g \times 64\ J/g = 1.6\times10^3 = 1.6\ kJ$
$Q_3 = 24.9\ g \times 0.15\ J/g\cdot{}^\circ C \times (1140\text{-}1063)^\circ C = 2.9\times10^2\ J = 0.29\ kJ$
$Q = Q_1 + Q_2 + Q_3 = 3.4 + 1.6 + 0.29 = 5.3\ kJ$

16) $325\ g\ C_6H_{12}O_6 \times \frac{1\ mol\ C_6H_{12}O_6}{180\ g\ C_6H_{12}O_6} \times \frac{-2820\ kJ}{1\ mol\ C_6H_{12}O_6} = -5090\ kJ$

17) $649\ kJ \times \frac{3\ mol\ O_2}{89.5\ kJ} \times \frac{22.41\ L\ O_2}{1\ mol\ O_2} = 487\ L\ O_2$

18) $7.38\ L\ HCl \times \frac{1\ mol\ HCl}{22.41\ L\ HCl} \times \frac{-73.3\ kJ}{1\ mol\ HCl} = -24.1\ kJ$

19) $5.08\ kJ \times \frac{1\ mol\ NaCl}{3.89\ kJ} \times \frac{58.5\ g\ NaCl}{1\ mol\ NaCl} = 76.4\ g\ NaCl$

20) $Q = -50.0\ g \times 4.18\ J/g\cdot{}^\circ C \times (15.42 - 23.50)^\circ C = +1690\ J = +1.69\ kJ$
$\Delta H = \frac{Q}{n} = \frac{+1.69\ kJ}{5.00\ g\ KClO_3 \times 1\ mol\ KClO_3/122.55\ g\ KClO_3} = +41.4\ kJ/mol\ KClO_3$

21) $Q = -45.0\ g \times 4.18\ J/g\cdot{}^\circ C \times (43.2 - 17.5)^\circ C = -483\underline{4}\ J = -4.83\underline{4}\ kJ$
$\Delta H = \frac{Q}{n} = \frac{-4.83\underline{4}\ kJ}{2.00\ g\ KOH \times 1\ mol\ KOH/56.11\ g\ KOH} = \frac{-135.\underline{6}\ kJ}{mol\ KOH} \times 2\ mol\ KOH = -271\ kJ$

22) $Q = n \times \Delta H = 1.50\ g\ Ca(OH)_2 \times \frac{1\ mol\ Ca(OH)_2}{74.10\ g\ Ca(OH)_2} \times \frac{-128\ kJ}{1\ mol\ Ca(OH)_2} = -2.59\ kJ = -2{,}590\ J$
$\Delta T = \frac{-Q}{m \times c} = \frac{+2{,}590\ J}{75.0\ g \times 4.18\ J/g\cdot{}^\circ C} = 8.26^\circ C$ $T_2 = T_1 + \Delta T = 21.7^\circ C + 8.26^\circ C = 30.0^\circ C$

23) $\Delta H = 2(-1269.0) - 2(-167.4) = -2203.2\ kJ$

24) $\Delta H = -635.5 - 241.8 - (-986.6) = 109.3\ kJ$

25) $\Delta H^\circ_f + 2(-92.5) - (-399.2) - (-241.8) = -136.0\ kJ$ $\Delta H^\circ_f = -592.0\ kJ/mol$

26) $2\ \Delta H^\circ_f - 2(-296.9) - 2(-285.8) = -457.2$ $\Delta H^\circ_f = -811.3\ kJ/mol$

27) $C_6H_5OH(s) + 7\ O_2(g) \rightarrow 6\ CO_2(g) + 3\ H_2O(\ell)$, $6(-393.5) + 3(-285.8) - \Delta H_f^\circ = -3053.5$ kJ
$\Delta H_f^\circ = -164.9$ kJ/mol

28)
$$\begin{array}{rl} HCN(aq) + OH^-(aq) \rightarrow H_2O(\ell) + CN^-(aq) & \Delta H = -12.1\text{ kJ} \\ H_2O(\ell) \rightarrow H^+(aq) + OH^-(aq) & \Delta H = +55.9\text{ kJ} \\ \hline HCN(aq) \rightarrow H^+(aq) + CN^-(aq) & \Delta H = +43.8\text{ kJ} \end{array}$$

29) $2\ C_4H_{10}(g) + 13\ O_2(g) \rightarrow 8\ CO_2(g) + 10\ H_2O(\ell)$

$\Delta H = 8(-393.5) + 10(-285.8) - 2(-124.7) = -5756.6$ kJ

$$66.4\text{ g }C_4H_{10} \times \frac{1\text{ mol }C_4H_{10}}{58.0\text{ g }C_4H_{10}} \times \frac{5756.6\text{ kJ}}{2\text{ mol }C_4H_{10}} = 3.30 \times 10^3\text{ kJ}$$

CHAPTER 14

1) $\dfrac{26.2\text{ g }NH_4Cl}{355\text{ g solution}} \times 100 = 7.38\%\ NH_4$ 2) $\dfrac{5.48\text{ g }AgNO_3}{(5.48 + 25.0)\text{ g soln}} \times 100 = 18.0\%\ AgNO_3$

3) $50.0\text{ mL} \times \dfrac{1.05\text{ g soln}}{1\text{ mL}} \times \dfrac{20.0\text{ g acid}}{100.0\text{ g soln}} = 10.5\text{ g acid}$

4) $800.0\text{ g soln} \times \dfrac{18.1\text{ g }Na_2CO_3}{100.0\text{ g soln}} = 145\text{ g }Na_2CO_3$

5) $175\text{ g soln} \times \dfrac{22.5\text{ g }KNO_3}{100.0\text{ g soln}} = 39.4\text{ g }KNO_3$ $175\text{ g soln} - 39.4\text{ g }KNO_3 = 136\text{ g }H_2O$

6) $75.0\text{ g }CaCl_2 \times \dfrac{100.0\text{ g soln}}{15.0\text{ g }CaCl_2} = 500\text{ g soln}$ $500\text{ g soln} - 75\text{ g }CaCl_2 = 425\text{ g or mL }H_2O$

7) $75.0\text{ mL} \times \dfrac{1.15\text{ g soln}}{1\text{ mL}} = 86.3\text{ g soln}$ $\dfrac{13.8\text{ g }FeCl_3}{86.3\text{ g soln}} \times 100 = 16.0\%\ FeCl_3$

8) $15.0\text{ mL soln} \times \dfrac{1.19\text{ g soln}}{1\text{ mL soln}} \times \dfrac{19.9\text{ g }Cu(NO_3)_2}{(5.46 + 22.0)\text{ g soln}} = 3.55\text{ g }Cu(NO_3)_2$

9) $1.89\text{ g }H_2C_2O_4 \times \dfrac{1\text{ mol }H_2C_2O_4}{90.04\text{ g }H_2C_2O_4} = 0.0210\text{ mol }H_2C_2O_4$

$15.0\text{ g }H_2O \times \dfrac{1\text{ mol }H_2O}{18.02\text{ g }H_2O} = 0.832\text{ mol }H_2O$ $X_{H_2C_2O_4} = \dfrac{0.0210\text{ mol }H_2C_2O_4}{(0.833+0.0210)\text{ mol total}} = 0.0246$

10) $70.5\text{ g }C_2H_5OH \times \dfrac{1\text{ mol }C_2H_5OH}{46.07\text{ g }C_2H_5OH} = 1.53\text{ mol }C_2H_5OH$ $X_{C_2H_5OH} = \dfrac{1.53}{4.85} = 0.315$

$41.0\text{ g }CH_3OH \times \dfrac{1\text{ mol }CH_3OH}{32.04\text{ g }CH_3OH} = 1.28\text{ mol }CH_3OH$ $X_{CH_3OH} = \dfrac{1.28}{4.85} = 0.264$

$36.7\text{ g }H_2O \times \dfrac{1\text{ mol }H_2O}{18.02\text{ g }H_2O} = 2.04\text{ mol }H_2O$ $X_{H_2O} = \dfrac{2.04}{4.85} = 0.421$

Total moles = 1.53 + 1.28 + 2.04 = 4.85

11) $\dfrac{56.0\text{ g NaBr} \times \dfrac{1\text{ mol NaBr}}{102.89\text{ g NaBr}}}{1.50\text{ kg }H_2O} = 0.363\text{ m NaBr}$ 12) $\dfrac{58.1\text{ g }C_3H_7OH \times \dfrac{1\text{ mol }C_3H_7OH}{60.09\text{ g }C_3H_7OH}}{0.302\text{ kg }H_2O} = 3.20\text{ m }C_3H_7OH$

13) $0.250\text{ kg }H_2O \times \dfrac{0.200\text{ mol }C_{12}H_{22}O_{11}}{1\text{ kg }H_2O} \times \dfrac{342.30\text{ g }C_{12}H_{22}O_{11}}{1\text{ mol }C_{12}H_{22}O_{11}} = 17.1\text{ g }C_{12}H_{22}O_{11}$

14) $0.185\text{ kg }H_2O \times \dfrac{0.411\text{ mol solute}}{1\text{ kg }H_2O} \times \dfrac{109\text{ g solute}}{1\text{ mol solute}} = 8.29\text{ g solute}$

15) $1.00\text{ kg }H_2O \times \dfrac{0.230\text{ mol NaF}}{1\text{ kg }H_2O} \times \dfrac{41.99\text{ g NaF}}{1\text{ mol NaF}} = 9.66\text{ g NaF}$

16) $65.8\text{ g }K_2CO_3 \times \dfrac{1\text{ mol }K_2CO_3}{138.21\text{ g }K_2CO_3} \times \dfrac{1000\text{ g }H_2O}{0.268\text{ mol }K_2CO_3} = 1.78 \times 10^3\text{ g }H_2O$

17) $61.8 \text{ g } HC_3H_5O_2 \times \frac{1 \text{ mol } HC_3H_5O_2}{74.08 \text{ g } HC_3H_5O_2} \times \frac{1000 \text{ mL } H_2O}{0.300 \text{ mol } HC_3H_5O_2} = 2.78 \times 10^3 \text{ mL}$

18) $125 \text{ g } H_2O \times \frac{1 \text{ mol } H_2O}{18.02 \text{ g } H_2O} \times \frac{1000 \text{ g } CH_3OH}{0.366 \text{ mol } H_2O} \times \frac{1 \text{ mL } CH_3OH}{0.792 \text{ g } CH_3OH} = 2.40 \times 10^3 \text{ mL } CH_4OH$

19) $0.750 \text{ mol } KNO_3 \div 0.5000 \text{ L solution} = 1.50 \text{ M } KNO_3$

20) $2.00 \text{ L soln} \times \frac{0.750 \text{ mol } KHCO_3}{1 \text{ L soln}} \times \frac{100.12 \text{ g } KHCO_3}{1 \text{ mol } KHCO_3} = 1.50 \times 10^2 \text{ g } KHCO_3$

21) $\frac{0.274 \text{ g LiCl} \times \frac{1 \text{ mol LiCl}}{42.39 \text{ g LiCl}}}{0.500 \text{ L soln}} = 0.0129 \text{ M LiCl}$

22) $175 \text{ mL soln} \times \frac{0.281 \text{ mol } KMnO_4}{1000 \text{ mL soln}} \times \frac{158.04 \text{ g } KMnO_4}{1 \text{ mol } KMnO_4} = 7.77 \text{ g } KMnO_4$

23) $\frac{9.43 \text{ g } Na_2CO_3 \times \frac{1 \text{ mol } Na_2CO_3}{105.99 \text{ g } Na_2CO_3}}{0.2500 \text{ L soln}} = 0.356 \text{ M } Na_2CO_3$

24) $0.249 \text{ mol } H_3PO_4 \times \frac{1000 \text{ mL soln}}{1.54 \text{ mol } H_3PO_4} = 162 \text{ mL soln}$

25) $750.0 \text{ mL soln} \times \frac{1.23 \text{ mol NaOH}}{1000 \text{ mL soln}} = 0.923 \text{ mol NaOH}$

26) $0.1000 \text{ L soln} \times \frac{0.600 \text{ mol } AgNO_3}{1 \text{ L soln}} \times \frac{169.9 \text{ g } AgNO_3}{1 \text{ mol } AgNO_3} = 10.2 \text{ g } AgNO_3$

27) $150 \text{ L soln} \times \frac{12.0 \text{ mol HCl}}{1 \text{ L soln}} \times \frac{36.46 \text{ g HCl}}{1 \text{ mol HCl}} = 6.6 \times 10^4 \text{ g HCl}$ (or 6.56×10^4 g HCl if 150 L is 3 sig figs)

28) $4.50 \text{ L soln} \times \frac{3.50 \text{ mol } NH_3}{\text{L soln}} \times \frac{22.41 \text{ L } NH_3}{\text{mol } NH_3} = 353 \text{ L } NH_3$

29) 100.0 g of 10.0% NaCl(aq) contains 10.0 g NaCl and 90.0 g H_2O.

$\frac{10.0 \text{ g NaCl} \times \frac{1 \text{ mol NaCl}}{58.44 \text{ g NaCl}}}{0.0900 \text{ kg } H_2O} = 1.90 \text{ m NaCl}$

30) $1 \text{ kg } C_6H_6 \times \frac{10^3 \text{ g}}{1 \text{ kg}} \times \frac{1 \text{ mol } C_6H_6}{78.11 \text{ g } C_6H_6} = 12.8 \text{ mol } C_6H_6 \quad X_{C_{10}H_8} = \frac{0.500 \text{ mol } C_{10}H_8}{(0.500+12.8) \text{ mol tot}} = 0.0376$

31) a) $\frac{16.8 \text{ g } HC_2H_3O_2}{(16.8 + 56.3) \text{ g total}} \times 100 = 23.0\% \; HC_2H_3O_2$

$16.8 \text{ g } HC_2H_3O_2 \times \frac{1 \text{ mol } HC_2H_3O_2}{60.05 \text{ g } HC_2H_3O_2} = 0.280 \text{ mol } HC_2H_3O_2$

$56.3 \text{ g } H_2O \times \frac{1 \text{ mol } H_2O}{18.02 \text{ g } H_2O} = 3.12 \text{ mol } H_2O \quad X_{HC_2H_3O_2} = \frac{0.280}{0.280 + 3.12} = 0.0824$

$m = \frac{0.280 \text{ mol } HC_2H_3O_2}{0.0563 \text{ kg } H_2O} = 4.97 \text{ m } HC_2H_3O_2$

b) $10.0 \text{ g } C_6H_{12}O_6 \times \frac{1 \text{ mol } C_6H_{12}O_6}{180.16 \text{ g } C_6H_{12}O_6} = 0.0555 \text{ mol } C_6H_{12}O_6$

$90.0 \text{ g } H_2O \times \frac{1 \text{ mol } H_2O}{18.02 \text{ g } H_2O} = 4.99 \text{ mol } H_2O \quad X_{C_6H_{12}O_6} = \frac{0.0555}{4.99 + 0.0555} = 0.0110$

$m = \frac{0.0555 \text{ mol } C_6H_{12}O_6}{0.0900 \text{ kg } H_2O} = 0.617 \text{ m } C_6H_{12}O_6$

c) $0.128 \text{ mol } NaNO_3 \times \frac{85.00 \text{ g } NaNO_3}{1 \text{ mol } NaNO_3} = 10.9 \text{ g } NaNO_3$

$(1 - 0.128) \text{ mol } H_2O \times \frac{18.02 \text{ g } H_2O}{1 \text{ mol } H_2O} = 15.7 \text{ g } H_2O$

$\frac{10.9 \text{ g } NaNO_3}{(10.9 + 15.7) \text{ g solution}} \times 100 = 41.0\% \; NaNO_3 \quad m = \frac{0.128 \text{ mol } NaNO_3}{0.0157 \text{ kg } H_2O} = 8.15 \text{ m } NaNO_3$

d) $1.17 \text{ mol } MgSO_4 \times \frac{120.37 \text{ g } MgSO_4}{1 \text{ mol } MgSO_4} = 141 \text{ g } MgSO_4$

$1000 \text{ g } H_2O \times \frac{1 \text{ mol } H_2O}{18.02 \text{ g } H_2O} = 55.5 \text{ mol } H_2O$

$\frac{141 \text{ g } MgSO_4}{(141+1000) \text{ g soln}} \times 100 = 12.4\% \ MgSO_4$ $\quad X_{MgSO_4} = \frac{1.17 \text{ mol } MgSO_4}{(1.17+55.5) \text{ mol total}} = 0.0206$

32) $1000.0 \text{ mL soln} \times \frac{1.068 \text{ g soln}}{\text{mL soln}} = 1068 \text{ g soln}$ $\quad 12.8 \text{ mol HOAc} \times \frac{60.05 \text{ g HOAc}}{1 \text{ mol HOAc}} = 769 \text{ g HOAc}$

$1068 \text{ g soln} - 769 \text{ g HOAc} = 299 \text{ g } H_2O$ $\quad m = \frac{12.8 \text{ mol HOAc}}{0.299 \text{ kg } H_2O} = 42.8 \text{ m HOAc}$

33) $10.0 \text{ g NaOH} \times \frac{1 \text{ mol NaOH}}{40.00 \text{ g NaOH}} = 0.250 \text{ mol NaOH}$ $\quad 100.0 \text{ g soln} \times \frac{1 \text{ mL soln}}{1.11 \text{ g soln}} = 90.1 \text{ mL soln}$

$m = \frac{0.250 \text{ mol NaOH}}{0.0901 \text{ L soln}} = 2.77 \text{ M NaOH}$

34) $1000 \text{ mL soln} \times \frac{1.339 \text{ g soln}}{1 \text{ mL soln}} = 1339 \text{ g soln}$ $\quad 6.00 \text{ mol } H_2SO_4 \times \frac{98.08 \text{ g } H_2SO_4}{1 \text{ mol } H_2SO_4} = 588 \text{ g } H_2SO_4$

$1339 \text{ g soln} - 588 \text{ g } H_2SO_4 = 751 \text{ g } H_2O$

$751 \text{ g } H_2O \times \frac{1 \text{ mol } H_2O}{18.02 \text{ g } H_2O} = 41.7 \text{ mol } H_2O$ $\quad X_{H_2SO_4} = \frac{6.00 \text{ mol } H_2SO_4}{(41.7 + 6.00) \text{ mol total}} = 0.126$

35) $V_1 = \frac{M_2V_2}{M_1} = \frac{3.00 \text{ M} \times 5.00 \text{ L}}{17.8 \text{ M}} = 0.843 \text{ L}$ $\quad$ 36) $M_2 = \frac{M_1V_1}{V_2} = \frac{6.25 \text{ M} \times 65.0 \text{ mL}}{100.0 \text{ mL}} = 4.06 \text{ M}$

37) $V_1 = \frac{M_2V_2}{M_1} = \frac{0.315 \text{ M} \times 350 \text{ mL}}{7.4 \text{ M}} = 15 \text{ mL}$ $\quad$ 38) $M_2 = \frac{M_1V_1}{V_2} = \frac{0.772 \text{ M} \times 150 \text{ mL}}{825 \text{ mL}} = 0.14 \text{ M}$

39) $V_1 = \frac{M_2V_2}{M_1} = \frac{6.0 \text{ M} \times 6.5 \text{ L}}{15.4 \text{ M}} = 2.5 \text{ L}$ $\quad$ 40) $V_2 = \frac{M_1V_1}{M_2} = \frac{3.00 \text{ M} \times 25.0 \text{ mL}}{1.00 \text{ M}} = 75.0 \text{ mL}$

$V_W = V_2 - V_1 = 75.0 - 25.0 = 50.0 \text{ mL } H_2O$

41) $50.0 \text{ mL } HNO_3 \times \frac{4.00 \text{ mol } HNO_3}{1000 \text{ mL } HNO_3} \times \frac{3 \text{ mol P}}{5 \text{ mol } HNO_3} \times \frac{30.99 \text{ g P}}{1 \text{ mol P}} = 3.72 \text{ g P}$

42) $7.91 \text{ g } Na_2O \times \frac{1 \text{ mol } Na_2O}{61.98 \text{ g } Na_2O} \times \frac{1 \text{ mol } H_2SO_4}{1 \text{ mol } Na_2O} \times \frac{1000 \text{ mL } H_2SO_4}{1.35 \text{ mol } H_2SO_4} = 94.5 \text{ mL } H_2SO_4$

43) $0.0828 \text{ L HCl} \times \frac{0.211 \text{ mol HCl}}{1 \text{ L HCl}} \times \frac{2 \text{ mol } FeCl_3}{6 \text{ mol HCl}} \times \frac{162.20 \text{ g } FeCl_3}{1 \text{ mol } FeCl_3} = 0.945 \text{ g } FeCl_3$

44) $42.7 \text{ mL } HNO_3 \times \frac{0.937 \text{ mol } HNO_3}{1000 \text{ mL } HNO_3} \times \frac{1 \text{ mol } CO_2}{2 \text{ mol } HNO_3} = 0.0200 \text{ mol } CO_2$

$V = \frac{nRT}{P} = \frac{0.0200 \text{ mol}}{747 \text{ torr}} \times \frac{62.36 \text{ L·torr}}{\text{mol·K}} \times 300 \text{ K} = 0.501 \text{ L } CO_2$

45) $AgNO_3(aq) + NaCl(aq) \rightarrow AgCl(s) + NaNO_3(aq)$

$25.0 \text{ mL NaCl} \times \frac{0.112 \text{ mol NaCl}}{1000 \text{ mL NaCl}} \times \frac{1 \text{ mol } AgNO_3}{1 \text{ mol NaCl}} \times \frac{1000 \text{ mL } AgNO_3}{0.0934 \text{ mol } AgNO_3} = 30.0 \text{ mL } AgNO_3$

46) $50.0 \text{ mL } MgCl_2 \times \frac{0.329 \text{ mol } MgCl_2}{1000 \text{ mL } MgCl_2} \times \frac{2 \text{ mol NaOH}}{1 \text{ mol } MgCl_2} \times \frac{1000 \text{ mL NaOH}}{0.224 \text{ mol NaOH}} = 147 \text{ mL NaOH}$

47) $36.7 \text{ mL NaOH} \times \frac{\frac{0.476 \text{ mol NaOH}}{1000 \text{ mL NaOH}} \times \frac{1 \text{ mol } H_3PO_4}{2 \text{ mol NaOH}}}{0.0275 \text{ L } H_3PO_4} = 0.318 \text{ M } H_3PO_4$

48) $14.1 \text{ mL NaOH} \times \frac{\frac{0.437 \text{ mol NaOH}}{1000 \text{ mL NaOH}} \times \frac{1 \text{ mol } HC_2H_3O_2}{1 \text{ mol NaOH}}}{0.0228 \text{ L } HC_2H_3O_2} = 0.270 \text{ M } HC_2H_3O_2$

49) $41.2 \text{ mL } AgNO_3 \times \frac{\frac{0.877 \text{ mol } AgNO_3}{1000 \text{ mL } AgNO_3} \times \frac{1 \text{ mol } CaCl_2}{2 \text{ mol } AgNO_3}}{0.0338 \text{ L } CaCl_2} = 0.535 \text{ M } CaCl_2$

CHAPTER 14

CHAPTER 15

1) strong, major: $2\ Fe^{3+}(aq) + 3\ SO_4^{2-}(aq)$, minor: none
2) non, major: $C_{12}H_{22}O_{11}(aq)$, minor: none
3) weak, major: $HNO_2(aq)$, minor: $H^+(aq) + NO_2^-(aq)$
4) not soluble
5) strong, major: $Ba^{2+}(aq) + 2\ OH^-(aq)$, minor: none
6) weak, major: $HClO_2(aq)$, minor: $H^+(aq) + ClO_2^-(aq)$
7) strong, major: $Cu^{2+}(aq) + 2\ Cl^-(aq)$, minor: none
8) weak, major: $NH_3(aq)$, minor: $NH_4^+(aq) + OH^-(aq)$
9) strong, major: $H^+(aq) + HSO_4^-(aq)$, minor: SO_4^{2-}
10) strong, major: $Ca^{2+}(aq) + 2\ HCO_3^-(aq)$, minor: $H^+(aq) + CO_3^{2-}(aq)$ (dissoc of HCO_3^-)

For each reaction below, the equations listed are: conventional, total ionic, net ionic.

11) $Pb(NO_3)_2(aq) + MgSO_4(aq) \rightarrow PbSO_4(s) + Mg(NO_3)_2(aq)$ - Precipitation
$Pb^{2+}(aq) + 2\ NO_3^-(aq) + Mg^{2+}(aq) + SO_4^{2-}(aq) \rightarrow PbSO_4(s) + Mg^{2+}(aq) + 2\ NO_3^-(aq)$
$Pb^{2+}(aq) + SO_4^{2-}(aq) \rightarrow PbSO_4(s)$

12) $Mg(s) + 2\ HCl(aq) \rightarrow MgCl_2(aq) + H_2(g)$ - Single Replacement
$Mg(s) + 2\ H^+(aq) + 2\ Cl^-(aq) \rightarrow Mg^{2+}(aq) + 2\ Cl^-(aq) + H_2(g)$
$Mg(s) + 2\ H^+(aq) \rightarrow Mg^{2+}(aq) + H_2(g)$

13) $ZnCl_2(aq) + (NH_4)_2CO_3(aq) \rightarrow ZnCO_3(s) + 2\ NH_4Cl(aq)$ - Precipitation
$Zn^{2+}(aq) + 2\ Cl^-(aq) + 2\ NH_4^+(aq) + CO_3^{2-}(aq) \rightarrow ZnCO_3(s) + 2\ NH_4^+(aq) + 2\ Cl^-(aq)$
$Zn^{2+}(aq) + CO_3^{2-}(aq) \rightarrow ZnCO_3(s)$

14) $NH_4Cl(aq) + KOH(aq) \rightarrow NH_3(aq) + H_2O(\ell) + KCl(aq)$ - Weak Base Formation
$NH_4^+(aq) + Cl^-(aq) + K^+(aq) + OH^-(aq) \rightarrow NH_3(aq) + H_2O(\ell) + K^+(aq) + Cl^-(aq)$
$NH_4^+(aq) + OH^-(aq) \rightarrow NH_3(aq) + H_2O(\ell)$

15) $MgCl_2(aq) + Ni(NO_3)_2(aq) \rightarrow Mg(NO_3)_2(aq) + NiCl_2(aq)$
No reaction

16) $Mg(C_2H_3O_2)_2(aq) + H_2SO_4(aq) \rightarrow MgSO_4(aq) + 2\ HC_2H_3O_2(aq)$ - Weak Acid Formation
$Mg^{2+}(aq) + 2\ C_2H_3O_2^-(aq) + H^+(aq) + HSO_4^-(aq) \rightarrow Mg^{2+}(aq) + SO_4^{2-}(aq) + 2\ HC_2H_3O_2(aq)$
$2\ C_2H_3O_2^-(aq) + H^+(aq) + HSO_4^-(aq) \rightarrow SO_4^{2-}(aq) + 2\ HC_2H_3O_2(aq)$

17) $Zn(s) + NiSO_4(aq) \rightarrow ZnSO_4(aq) + Ni(s)$ - Single Replacement
$Zn(s) + Ni^{2+}(aq) + SO_4^{2-}(aq) \rightarrow Zn^{2+}(aq) + Ni(s) + SO_4^{2-}(aq)$
$Zn(s) + Ni^{2+}(aq) \rightarrow Zn^{2+}(aq) + Ni(s)$

18) $NiCO_3(s) + 2\ HNO_3(aq) \rightarrow Ni(NO_3)_2(aq) + H_2O(\ell) + CO_2(g)$ - Weak Acid H_2CO_3 Decomposes
$NiCO_3(s) + 2\ H^+(aq) + 2\ NO_3^-(aq) \rightarrow Ni^{2+}(aq) + 2\ NO_3^-(aq) + H_2O(\ell) + CO_2(g)$
$NiCO_3(s) + 2\ H^+(aq) \rightarrow Ni^{2+}(aq) + H_2O(\ell) + CO_2(g)$

19) $2\ Na(s) + 2\ H_2O(\ell) \rightarrow 2\ NaOH(aq) + H_2(g)$ - Single Replacement
$2\ Na(s) + 2\ H_2O(\ell) \rightarrow 2\ Na^+(aq) + 2\ OH^-(aq) + H_2(g)$

20) $(NH_4)_2SO_3(aq) + 2\ HI(aq) \rightarrow 2\ NH_4I(aq) + H_2O(\ell) + SO_2(aq)$ - Weak Acid H_2SO_3 Decomposes
$2\ NH_4^+(aq) + SO_3^{2-}(aq) + 2\ H^+(aq) + 2\ I^-(aq) \rightarrow$
$2\ NH_4^+(aq) + 2\ I^-(aq) + H_2O(\ell) + SO_2(aq)$
$SO_3^{2-}(aq) + 2\ H^+(aq) \rightarrow H_2O(\ell) + SO_2(aq)$

21) $2\ HCl(aq) + Cu(OH)_2s) \rightarrow CuCl_2(aq) + 2\ H_2O(\ell)$ - Neutralization
$2\ H^+(aq) + 2\ Cl^-(aq) + Cu(OH)_2(s) \rightarrow Cu^{2+}(aq) + 2\ Cl^-(aq) + 2\ H_2O(\ell)$
$2\ H^+(aq) + Cu(OH)_2(s) \rightarrow Cu^{2+}(aq) + 2\ H_2O(\ell)$

22) $CoCl_2(aq) + 2\ NaOH(aq) \rightarrow Co(OH)(s) + 2\ NaCl(aq)$ - Precipitation
$Co^{2+}(aq) + 2\ Cl^-(aq) + 2\ Na^+(aq) + 2\ OH^-(aq) \rightarrow Co(OH)(s) + 2\ Na^+(aq) + 2\ Cl^-(aq)$
$Co^{2+}(aq) + 2\ OH^-(aq) \rightarrow Co(OH)_2(s)$

23) $AlBr_3(aq) + 3\ NH_4F(aq) \rightarrow AlF_3(s) + 3\ NH_4Br(aq)$ - Precipitation
$Al^{3+}(aq) + 3\ Br^-(aq) + 3\ NH_4^+(aq) + 3\ F^-(aq) \rightarrow AlF_3(s) + 3\ NH_4^+(aq) + 3\ Br^-(aq)$
$Al^{3+}(aq) + 3\ F^-(aq) \rightarrow AlF_3(s)$

24) $NaC_5H_9O_2(aq) + HCl(aq) \rightarrow HC_5H_9O_2(aq) + NaCl(aq)$ - Weak Acid Formation
$Na^+(aq) + C_5H_9O_2^-(aq) + H^+(aq) + Cl^-(aq) \rightarrow HC_5H_9O_2(aq) + Na^+(aq) + Cl^-(aq)$
$C_5H_9O_2^-(aq) + H^+(aq) \rightarrow HC_5H_9O_2(aq)$

25) $ZnF_2(aq) + 2\ NaClO_3(aq) \rightarrow Zn(ClO_3)_2(aq) + 2\ NaF(aq)$
No reaction - All Soluble

26) $KHCO_3(aq) + HBr(aq) \rightarrow H_2O(\ell) + CO_2(g) + KBr(aq)$ - Weak Acid H_2CO_3 Forms/Decomposes
$K^+(aq) + HCO_3^-(aq) + H^+(aq) + Br^-(aq) \rightarrow H_2O(\ell) + CO_2(g) + K^+(aq) + Br^-(aq)$
$HCO_3^-(aq) + H^+(aq) \rightarrow H_2O(\ell) + CO_2(g)$

27) $Br_2(aq) + 2\ NaI(aq) \rightarrow 2\ NaBr(aq) + I_2(s)$ - Single Replacement
$Br_2(aq) + 2\ Na^+(aq) + 2\ I^-(aq) \rightarrow 2\ Na^+(aq) + 2\ Br^-(aq) + I_2(s)$
$Br_2(aq) + 2\ I^-(aq) \rightarrow 2\ Br^-(aq) + I_2(s)$

28) $2\ AgF(aq) + Pb(C_2H_3O_2)_2(aq) \rightarrow 2\ AgC_2H_3O_2(s) + PbF_2(s)$ - Precipitation
$2\ Ag^+(aq) + 2\ F^-(aq) + Pb^{2+}(aq) + 2\ C_2H_3O_2^-(aq) \rightarrow 2\ AgC_2H_3O_2(s) + PbF_2(s)$

29) $2\ HBr(aq) + Na_2SO_3(aq) \rightarrow H_2O(\ell) + SO_2(aq) + 2\ NaBr(aq)$ - Weak Acid H_2SO_3 Form/Decomp
$2\ H^+(aq) + 2\ Br^-(aq) + 2\ Na^+(aq) + SO_3^{2-}(aq) \rightarrow H_2O(\ell) + SO_2(aq) + 2\ Na^+(aq) + 2\ Br^-(aq)$
$2\ H^+(aq)(aq) + SO_3^{2-}(aq) \rightarrow H_2O(\ell) + SO_2(aq)$

30) $NaC_3H_5O_2(aq) + HNO_3(aq) \rightarrow HC_3H_5O_2(aq) + NaNO_3(aq)$ - Weak Acid Formation
$Na^+(aq) + C_3H_5O_2^-(aq) + H^+(aq) + NO_3^-(aq) \rightarrow HC_3H_5O_2(aq) + Na^+(aq) + NO_3^-(aq)$
$C_3H_5O_2^-(aq) + H^+(aq) \rightarrow HC_3H_5O_2(aq)$

31) $MgCl_2(aq) + 2\ LiOH(aq) \rightarrow Mg(OH)_2(s) + 2\ LiCl(aq)$ - Precipitation
$Mg^{2+}(aq) + 2\ Cl^-(aq) + 2\ Li^+(aq) + 2\ OH^-(aq) \rightarrow Mg(OH)_2(s) + 2\ Li^+(aq) + 2\ Cl^-(aq)$
$Mg^{2+}(aq) + 2\ OH^-(aq) \rightarrow Mg(OH)_2(s)$

32) $FeCO_3(s) + H_2SO_4(aq) \rightarrow FeSO_4(aq) + H_2O(\ell) + CO_2(g)$ - Weak Acid H_2CO_3 Form/Decomp
$FeCO_3(s) + H^+(aq) + HSO_4^-(aq) \rightarrow Fe^{2+}(aq) + SO_4^{2-}(aq) + H_2O(\ell) + CO_2(g)$

33) $CoSO_4(aq) + 2\ NaI(aq) \rightarrow CoI_2(aq) + Na_2SO_4(aq)$
No reaction - All Soluble

34) $2\ HCl(aq) + Zn(OH)_2(s) \rightarrow ZnCl_2(aq) + 2\ H_2O(\ell)$ - Neutralization
$2\ H^+(aq) + 2\ Cl^-(aq) + Zn(OH)_2(s) \rightarrow Zn^{2+}(aq) + 2\ Cl^-(aq) + 2\ H_2O(\ell)$
$2\ H^+(aq) + Zn(OH)_2(s) \rightarrow Zn^{2+}(aq) + 2\ H_2O(\ell)$

35) $(NH_4)_2CO_3(aq) + 2\ KOH(aq) \rightarrow 2\ NH_3(aq) + 2\ H_2O(\ell) + K_2CO_3(aq)$ - Weak Base Formation
$2\ NH_4^+(aq) + CO_3^{2-}(aq) + 2\ K^+(aq) + 2\ OH^-(aq) \rightarrow$
$2\ NH_3(aq) + 2\ H_2O(\ell) + 2\ K^+(aq) + CO_3^{2-}(aq)$
$NH_4^+(aq) + OH^-(aq) \rightarrow NH_3(aq) + H_2O(\ell)$

36) $Mg(C_2H_3O_2)_2(aq) + 2\ HBr(aq) \rightarrow MgBr_2(aq) + 2\ HC_2H_3O_2(aq)$ - Weak Acid Formation
$Mg^{2+}(aq) + 2\ C_2H_3O_2^-(aq) + 2\ H^+(aq) + 2\ Br^-(aq) \rightarrow Mg^{2+}(aq) + 2\ Br^-(aq) + 2\ HC_2H_3O_2(aq)$
$C_2H_3O_2^-(aq) + H^+(aq) \rightarrow HC_2H_3O_2(aq)$

CHAPTER 16

1) a) SO_3^{2-}
$x+3(-2)=-2$
$x - 6 = -2$
$x = +4$

b) MnO_2
$x+2(-2)=0$
$x - 4 = 0$
$x = +4$

c) CaH_2
$(+2)+2x=0$
$2x = -2$
$x = -1$

d) $K^+ + Al^{3+} + SiO_4^{4-}$
$x+3(-2)=-4$
$x - 8 = -4$
$x = +4$

e) N_2O_3
$2x+3(-2)=0$
$2x - 6 = 0$
$x = +3$

f) CrO_4^{2-}
$x+4(-2)=-2$
$x - 8 = -2$
$x = +6$

g) $3\ Mg^{2+} + 2\ PO_4^{3-}$
$x+4(-2)=-3$
$x - 8 = -3$
$x = +5$

h) KO_2
$(+1)+2x=0$
$2x = -1$
$x = -1/2$

i) ClO_4^-
$x+4(-2)=-1$
$x - 8 = -1$
$x = +7$

j) $Fe^{2+} + 2\ HSO_3^-$
$(+1)+x+3(-2)=-1$
$x - 5 = -1$
$x = +4$

2) a) $NH_4HCO_3(s) \rightarrow NH_3(g) + H_2O(g) + CO_2(g)$
ox. nos. -3 +1 +1 +4 -2   -3 +1   +1 -2   +4 -2
Not redox - no change in oxidation numbers

b) $2\ CH_3OH(g) + O_2(g) \rightarrow 2\ CH_2O(g) + 2\ H_2O(g)$

ox. nos. 2- 0 0 2-

C is oxidized from -2 to 0, CH_3OH is the reducing agent.

O is reduced from 0 to -2, O_2 is the oxidizing agent.

c) $CO(g) + H_2(g) \rightarrow CO_2(g) + H_2(g)$

ox. nos. +2 +1 +4 0

C is oxidized from +2 to +4, CO is the reducing agent.

H is reduced from +1 to 0, H_2O is the oxidizing agent.

d) $Hg_2Cl_2(s) + 2\ NH_3(aq) \rightarrow Hg(\ell) + HgNH_2Cl(s) + NH_4Cl(aq)$

ox. nos. +1 0 +2

Hg is reduced from +1 to 0 and oxidized from +1 to +2.

Hg_2Cl_2 is both the oxidizing agent and the reducing agent.

3) Charge-Before-Mass / Mass-Before-Charge

a) Charge-Before-Mass:

+3 +1

$HClO_2 \rightarrow HClO$

e⁻: $2e^- + HClO_2 \rightarrow HClO$

ch: $2\ e^- + 2\ H^+ + HClO_2 \rightarrow HClO$

H: $2\ e^- + 2\ H^+ + HClO_2 \rightarrow HClO + H_2O$

Mass-Before-Charge:

$HClO_2 \rightarrow HClO$

O: $HClO_2 \rightarrow HClO + H_2O$

H: $2\ H^+ + HClO_2 \rightarrow HClO + H_2O$

ch: $2\ e^- + 2\ H^+ + HClO_2 \rightarrow HClO + H_2O$

b) Charge-Before-Mass:

+6 0

$H_2MoO_4 \rightarrow Mo$

e⁻: $6\ e^- + H_2MoO_4 \rightarrow Mo$

ch: $6\ e^- + 6\ H^+ + H_2MoO_4 \rightarrow Mo$

H: $6\ e^- + 6\ H^+ + H_2MoO_4 \rightarrow Mo + 4\ H_2O$

Mass-Before-Charge:

$H_2MoO_4 \rightarrow Mo$

O: $H_2MoO_4 \rightarrow Mo + 4\ H_2O$

H: $6\ H^+ + H_2MoO_4 \rightarrow Mo + 4\ H_2O$

ch: $6\ e^- + 6\ H^+ + H_2MoO_4 \rightarrow Mo + 4\ H_2O$

c) Charge-Before-Mass:

2(+2) 2(+1)

N: $2\ NO \rightarrow N_2O$

e⁻: $2\ e^- + 2\ NO \rightarrow N_2O$

ch: $2\ e^- + 2\ H^+ + 2\ NO \rightarrow N_2O$

H: $2\ e^- + 2\ H^+ + 2\ NO \rightarrow N_2O + H_2O$

Mass-Before-Charge:

N: $2\ NO \rightarrow N_2O$

O: $2\ NO \rightarrow N_2O + H_2O$

H: $2\ H^+ + 2\ NO \rightarrow N_2O + H_2O$

ch: $2\ e^- + 2\ H^+ + 2\ NO \rightarrow N_2O + H_2O$

d) Charge-Before-Mass:

+4 +2

$PbO_2 \rightarrow Pb^{2+}$

e⁻: $2\ e^- + PbO_2 \rightarrow Pb^{2+}$

ch: $2\ e^- + 4\ H^+ + PbO_2 \rightarrow Pb^{2+}$

H: $2\ e^- + 4\ H^+ + PbO_2 \rightarrow Pb^{2+} + 2\ H_2O$

Mass-Before-Charge:

$PbO_2 \rightarrow Pb^{2+}$

O: $PbO_2 \rightarrow Pb^{2+} + 2\ H_2O$

H: $4\ H^+ + PbO_2 \rightarrow Pb^{2+} + 2\ H_2O$

ch: $2\ e^- + 4\ H^+ + PbO_2 \rightarrow Pb^{2+} + 2\ H_2O$

e) Charge-Before-Mass:

2(+3) 2(0)

Sb: $Sb_2O_3 \rightarrow 2\ Sb$

e⁻: $6\ e^- + Sb_2O_3 \rightarrow 2\ Sb$

ch: $6\ e^- + 6\ H^+ + Sb_2O_3 \rightarrow 2\ Sb$

H: $6\ e^- + 6\ H^+ + Sb_2O_3 \rightarrow 2\ Sb + 3\ H_2O$

Mass-Before-Charge:

Sb: $Sb_2O_3 \rightarrow 2\ Sb$

O: $Sb_2O_3 \rightarrow 2\ Sb + 3\ H_2O$

H: $6\ H^+ + Sb_2O_3 \rightarrow 2\ Sb + 3\ H_2O$

ch: $6\ e^- + 6\ H^+ + Sb_2O_3 \rightarrow 2\ Sb + 3\ H_2O$

4) a) Charge-Before-Mass:

2(+3) 2(+1)

N: $2\ NO_2^- \rightarrow N_2O$

e⁻: $4\ e^- + 2\ NO_2^- \rightarrow N_2O$

ch: $4\ e^- + 2\ NO_2^- \rightarrow N_2O + 6\ OH^-$

H: $3\ H_2O + 4\ e^- + 2\ NO_2^- \rightarrow N_2O + 6\ OH^-$

Mass-Before-Charge:

N: $2\ NO_2^- \rightarrow N_2O$

O: $2\ NO_2^- \rightarrow N_2O + 3\ H_2O$

H: $6\ H^+ + 2\ NO_2^- \rightarrow N_2O + 3\ H_2O$

OH⁻: $3\ H_2O + 2\ NO_2^- \rightarrow N_2O + 6\ OH^-$

ch: $3\ H_2O + 4\ e^- + 2\ NO_2^- \rightarrow N_2O + 6\ OH^-$

b) Charge-Before-Mass:

+4 +2

$PbO_2 \rightarrow Pb(OH)_2$

e⁻: $2\ e^- + PbO_2 \rightarrow Pb(OH)_2$

ch: $2\ e^- + PbO_2 \rightarrow Pb(OH)_2 + 2\ OH^-$

H: $2\ H_2O + 2\ e^- + PbO_2 \rightarrow Pb(OH)_2 + 2\ OH^-$

Mass-Before-Charge:

$PbO_2 \rightarrow Pb(OH)_2$

O: $PbO_2 \rightarrow Pb(OH)_2$

H: $2\ H^+ + PbO_2 \rightarrow Pb(OH)_2$

OH⁻: $2\ H_2O + PbO_2 \rightarrow Pb(OH)_2 + 2\ OH^-$

ch: $2\ H_2O + 2\ e^- + PbO_2 \rightarrow Pb(OH)_2 + 2\ OH^-$

CHAPTER 16

c)

$\;\;0 \quad\; -3$
$P \rightarrow PH_3$

e⁻: $3\,e^- + P \rightarrow PH_3$

ch: $3\,e^- + P \rightarrow PH_3 + 3\,OH^-$

H: $3\,H_2O + 3\,e^- + P \rightarrow PH_3 + 3\,OH^-$

$P \rightarrow PH_3$

O: $P \rightarrow PH_3$

H: $3\,H^+ + P \rightarrow PH_3$

OH⁻: $3\,H_2O + P \rightarrow PH_3 + 3\,OH^-$

ch: $3\,H_2O + 3\,e^- + P \rightarrow PH_3 + 3\,OH^-$

d)

$2(0) \quad 2(-2)$

O: $O_2 \rightarrow 2\,OH^-$

e⁻: $4\,e^- + O_2 \rightarrow 2\,OH^-$

ch: $4\,e^- + O_2 \rightarrow 4\,OH^-$

H: $2\,H_2O + 4\,e^- + O_2 \rightarrow 4\,OH^-$

O: $O_2 \rightarrow 2\,OH^-$

H: $2\,H^+ + O_2 \rightarrow 2\,OH^-$

OH⁻: $2\,H_2O + O_2 \rightarrow 4\,OH^-$

ch: $2\,H_2O + 4\,e^- + O_2 \rightarrow 4\,OH^-$

For problems 5 and 6, the first solution shown is by the oxidation number method. The second solution shown is by the half-reaction method.

5) a)

Cr: $Cr_2O_7{}^{2-} + Sn^{2+} \rightarrow 2\,Cr^{3+} + 3\,Sn^{4+}$

$+2 \xrightarrow{3(+2)} +4$

e⁻: $Cr_2O_7{}^{2-} + 3\,Sn^{2+} \rightarrow 2\,Cr^{3+} + 3\,Sn^{4+}$

$2(+6) \xrightarrow[(-6)]{} 2(+3)$

ch: $14\,H^+ + Cr_2O_7{}^{2-} + 3\,Sn^{2+} \rightarrow 2\,Cr^{3+} + 3\,Sn^{4+}$

H: $14\,H^+ + Cr_2O_7{}^{2-} + 3\,Sn^{2+} \rightarrow 2\,Cr^{3+} + 7\,H_2O + 3\,Sn^{4+}$

$$6\,e^- + 14\,H^+ + Cr_2O_7{}^{2-} \rightarrow 2\,Cr^{3+} + 7\,H_2O$$
$$3Sn^{2+} \rightarrow 3\,Sn^{4+} + 6\,e^-$$
$$\overline{14\,H^+ + Cr_2O_7{}^{2-} + 3\,Sn^{2+} \rightarrow 2\,Cr^{3+} + 7\,H_2O + 3\,Sn^{4+}}$$

b)

$0 \xrightarrow{3(+1)} +1$

e⁻: $3\,Ag + NO_3{}^- \rightarrow 3\,Ag^+ + NO$

$+5 \xrightarrow[(-3)]{} +2$

ch: $3\,Ag + 4\,H^+ + NO_3{}^- \rightarrow 3\,Ag^+ + NO$

H: $3\,Ag + 4\,H^+ + NO_3{}^- \rightarrow 3\,Ag^+ + NO + 2\,H_2O$

$$3\,Ag \rightarrow 3\,Ag^+ + 3\,e^-$$
$$3\,e^- + 4\,H^+ + NO_3{}^- \rightarrow NO + 2\,H_2O$$
$$\overline{3\,Ag + 4\,H^+ + NO_3{}^- \rightarrow 3\,Ag^+ + NO + 2\,H_2O}$$

c)

$2(-1) \xrightarrow{5(+2)} 2(0)$

e⁻: $2\,MnO_4{}^- + 5\,H_2O_2 \rightarrow 2\,Mn^{2+} + 5\,O_2$

$+7 \xrightarrow[2(-5)]{} +2$

ch: $6\,H^+ + 2\,MnO_4{}^- + 5\,H_2O_2 \rightarrow 2\,Mn^{2+} + 5\,O_2$

H: $6\,H^+ + 2\,MnO_4{}^- + 5\,H_2O_2 \rightarrow 2\,Mn^{2+} + 8\,H_2O + 5\,O_2$

$$10\,e^- + 16\,H^+ + 2\,MnO_4{}^- \rightarrow 2\,Mn^{2+} + 8\,H_2O$$
$$5\,H_2O_2 \rightarrow 5\,O_2 + 10\,H^+ + 10\,e^-$$
$$\overline{6\,H^+ + 2\,MnO_4{}^- + 5\,H_2O_2 \rightarrow 2\,Mn^{2+} + 8\,H_2O + 5\,O_2}$$

d)

$2(-1) \xrightarrow{3(+2)} 2(0)$

Cr + Br: $Cr_2O_7{}^{2-} + 2\,Br^- \rightarrow 2\,Cr^{3+} + Br_2$

$2(+6) \xrightarrow[(-6)]{} 2(+3)$

e⁻: $Cr_2O_7{}^{2-} + 6\,Br^- \rightarrow 2\,Cr^{3+} + 3\,Br_2$

ch: $14\,H^+ + Cr_2O_7{}^{2-} + 6\,Br^- \rightarrow 2\,Cr^{3+} + 3\,Br_2$

H: $14\,H^+ + Cr_2O_7{}^{2-} + 6\,Br^- \rightarrow 2\,Cr^{3+} + 7\,H_2O + 3\,Br_2$

$$6\,e^- + 14\,H^+ + Cr_2O_7{}^{2-} \rightarrow 2\,Cr^{3+} + 7\,H_2O$$
$$6\,Br^- \rightarrow 3\,Br_2 + 6\,e^-$$
$$\overline{14\,H^+ + Cr_2O_7{}^{2-} + 6\,Br^- \rightarrow 2\,Cr^{3+} + 7\,H_2O + 3\,Br_2}$$

CHAPTER 16

$-2 \xrightarrow{3(+2)} 0$

e) Cr: $Cr_2O_7^{2-} + H_2S \rightarrow 2\ Cr^{3+} + 3\ S$

$2(+6) \xrightarrow[(-6)]{} 2(+3)$

e⁻: $Cr_2O_7^{2-} + 3\ H_2S \rightarrow 2\ Cr^{3+} + 3\ S$

ch: $8\ H^+ + Cr_2O_7^{2-} + 3\ H_2S \rightarrow 2\ Cr^{3+} + 3\ S$

H: $8\ H^+ + Cr_2O_7^{2-} + 3\ H_2S \rightarrow 2\ Cr^{3+} + 7\ H_2O + 3\ S$

$$\begin{array}{r} 6\ e^- + 14\ H^+ + Cr_2O_7^{2-} \rightarrow 2\ Cr^{3+} + 7\ H_2O \\ 3\ H_2S \rightarrow 3\ S + 6\ H^+ + 6\ e^- \\ \hline 8\ H^+ + Cr_2O_7^{2-} + 3\ H_2S \rightarrow 2\ Cr^{3+} + 7\ H_2O + 3\ S \end{array}$$

$+5 \xrightarrow{5(-2)} +3$

f) e⁻: $5\ BiO_3^- + 2\ Mn^{2+} \rightarrow 5\ Bi^{3+} + 2\ MnO_4^-$

$+2 \xrightarrow[2(+5)]{} +7$

ch: $14\ H^+ + 5\ BiO_3^- + 2\ Mn^{2+} \rightarrow 5\ Bi^{3+} + 2\ MnO_4^-$

H: $14\ H^+ + 5\ BiO_3^- + 2\ Mn^{2+} \rightarrow 5\ Bi^{3+} + 7\ H_2O + 2\ MnO_4^-$

$$\begin{array}{r} 2\ Mn^{2+} + 8\ H_2O \rightarrow 10\ e^- + 16\ H^+ + 2\ MnO_4^- \\ 10\ e^- + 30\ H^+ + 5\ BiO_3^- \rightarrow 5\ Bi^{3+} + 15\ H_2O \\ \hline 14\ H^+ + 5\ BiO_3^- + 2\ Mn^{2+} \rightarrow 5\ Bi^{3+} + 7\ H_2O + 2\ MnO_4^- \end{array}$$

g) Pb: $SO_4^{2-} + Pb + PbO_2 \rightarrow 2\ PbSO_4$

$+4 \xrightarrow{-2} +2$

e⁻: $SO_4^{2-} + Pb + PbO_2 \rightarrow 2\ PbSO_4$

$0 \xrightarrow[+2]{} +2$

SO_4^{2-}: $2\ SO_4^{2-} + Pb + PbO_2 \rightarrow 2\ PbSO_4$

ch: $2\ SO_4^{2-} + Pb + 4\ H^+ + PbO_2 \rightarrow 2\ PbSO_4$

H: $2\ SO_4^{2-} + Pb + 4\ H^+ + PbO_2 \rightarrow 2\ PbSO_4 + 2\ H_2O$

$$\begin{array}{r} SO_4^{2-} + Pb \rightarrow PbSO_4 + 2\ e^- \\ 2\ e^- + 4\ H^+ + SO_4^{2-} + PbO_2 \rightarrow PbSO_4 + 2\ H_2O \\ \hline 2\ SO_4^{2-} + Pb + 4\ H^+ + PbO_2 \rightarrow 2\ PbSO_4 + 2\ H_2O \end{array}$$

$0 \xrightarrow{3(+5)} +5$

h) e⁻: $3\ As + 5\ NO_3^- \rightarrow 3\ H_3AsO_4 + 5\ NO$

$+5 \xrightarrow[5(-3)]{} +2$

ch: $3\ As + 5\ H^+ + 5\ NO_3^- \rightarrow 3\ H_3AsO_4 + 5\ NO$

H: $2\ H_2O + 3\ As + 5\ H^+ + 5\ NO_3^- \rightarrow 3\ H_3AsO_4 + 5\ NO$

$$\begin{array}{r} 12\ H_2O + 3\ As \rightarrow 3\ H_3AsO_4 + 15\ H^+ + 15\ e^- \\ 15\ e^- + 20\ H^+ + 5\ NO_3^- \rightarrow 5\ NO + 10\ H_2O \\ \hline 2\ H_2O + 3\ As + 5\ H^+ + 5\ NO_3^- \rightarrow 3\ H_3AsO_4 + 5\ NO \end{array}$$

$-2 \xrightarrow{3(+8)} +6$

i) Cr: $Cr_2O_7^{2-} + H_2S \rightarrow 2\ Cr^{3+} + HSO_4^-$

$2(+6) \xrightarrow[4(-6)]{} 2(+3)$

e⁻: $4\ Cr_2O_7^{2-} + 3\ H_2S \rightarrow 8\ Cr^{3+} + 3\ HSO_4^-$

ch: $29\ H^+ + 4\ Cr_2O_7^{2-} + 3\ H_2S \rightarrow 8\ Cr^{3+} + 3\ HSO_4^-$

H: $29\ H^+ + 4\ Cr_2O_7^{2-} + 3\ H_2S \rightarrow 8\ Cr^{3+} + 16\ H_2O + 3\ HSO_4^-$

$$\begin{array}{r} 24\ e^- + 56\ H^+ + 4\ Cr_2O_7^{2-} \rightarrow 8\ Cr^{3+} + 28\ H_2O \\ 12\ H_2O + 3\ H_2S \rightarrow 3\ HSO_4^- + 27\ H^+ + 24\ e^- \\ \hline 29\ H^+ + 4\ Cr_2O_7^{2-} + 3\ H_2S \rightarrow 8\ Cr^{3+} + 16\ H_2O + 3\ HSO_4^- \end{array}$$

CHAPTER 16

$0 \xrightarrow{4(+2)} +2$

j) e^-: $4\ Zn + NO_3^- \rightarrow 4\ Zn^{2+} + NH_4^+$

$+5 \xrightarrow[(-8)]{} -3$

ch: $4\ Zn + 10\ H^+ + NO_3^- \rightarrow 4\ Zn^{2+} + NH_4^+$

H: $4\ Zn + 10\ H^+ + NO_3^- \rightarrow 4\ Zn^{2+} + NH_4^+ + 3\ H_2O$

$4\ Zn \rightarrow 4\ Zn^{2+} + 8\ e^-$

$8\ e^- + 10\ H^+ + NO_3^- \rightarrow NH_4^+ + 3\ H_2O$

$4\ Zn + 10\ H^+ + NO_3^- \rightarrow 4\ Zn^{2+} + NH_4^+ + 3\ H_2O$

$+2 \xrightarrow{3(+2)} +4$

6) a) Mn: $MnO_4^- + Mn^{2+} \rightarrow MnO_2 + MnO_2$

$+7 \xrightarrow[2(-3)]{} +4$

e^-: $2\ MnO_4^- + 3\ Mn^{2+} \rightarrow 2\ MnO_2 + 3\ MnO_2$

ch: $2\ MnO_4^- + 4\ OH^- + 3\ Mn^{2+} \rightarrow 5\ MnO_2$

H: $2\ MnO_4^- + 4\ OH^- + 3\ Mn^{2+} \rightarrow 5\ MnO_2 + 2\ H_2O$

$4\ H_2O + 6\ e^- + 2\ MnO_4^- \rightarrow 2\ MnO_2 + 8\ OH^-$

$12\ OH^- + 3\ Mn^{2+} \rightarrow 3\ MnO_2 + 6\ e^- + 6\ H_2O$

$2\ MnO_4^- + 4\ OH^- + 3\ Mn^{2+} \rightarrow 5\ MnO_2 + 2\ H_2O$

$+2 \xrightarrow{3(+2)} +4$

b) e^-: $3\ Sn(OH)_4^{2-} + 2\ CrO_4^{2-} \rightarrow 3\ Sn(OH)_6^{2-} + 2\ Cr(OH)_4^-$

$+6 \xrightarrow[2(-3)]{} +3$

ch: $3\ Sn(OH)_4^{2-} + 2\ CrO_4^{2-} \rightarrow 3\ Sn(OH)_6^{2-} + 2\ Cr(OH)_4^- + 2\ OH^-$

H: $3\ Sn(OH)_4^{2-} + 8\ H_2O + 2\ CrO_4^{2-} \rightarrow 3\ Sn(OH)_6^{2-} + 2\ Cr(OH)_4^- + 2\ OH^-$

$6\ OH^- + 3\ Sn(OH)_4^{2-} \rightarrow 3\ Sn(OH)_6^{2-} + 6\ e^-$

$8\ H_2O + 6\ e^- + 2\ CrO_4^{2-} \rightarrow 2\ Cr(OH)_4^- + 8\ OH^-$

$3\ Sn(OH)_4^{2-} + 8\ H_2O + 2\ CrO_4^{2-} \rightarrow 3\ Sn(OH)_6^{2-} + 2\ Cr(OH)_4^- + 2\ OH^-$

$2(-1) \xrightarrow{3(-2)} 2(-2)$

c) O: $Cr(OH)_4^- + HO_2^- \rightarrow CrO_4^{2-} + 2\ H_2O$

$+3 \xrightarrow[2(+3)]{} +6$

e^-: $2\ Cr(OH)_4^- + 3\ HO_2^- \rightarrow 2\ CrO_4^{2-} + 6\ H_2O$

ch: $2\ Cr(OH)_4^- + 3\ HO_2^- \rightarrow 2\ CrO_4^{2-} + 6\ H_2O + OH^-$

H: $2\ Cr(OH)_4^- + 3\ HO_2^- \rightarrow 2\ CrO_4^{2-} + 5\ H_2O + OH^-$

$8\ OH^- + 2\ Cr(OH)_4^- \rightarrow 2\ CrO_4^{2-} + 6\ e^- + 8\ H_2O$

$3\ H_2O + 6\ e^- + 3\ HO_2^- \rightarrow 9\ OH^-$

$2\ Cr(OH)_4^- + 3\ HO_2^- \rightarrow 2\ CrO_4^{2-} + 5\ H_2O + OH^-$

$+5 \xrightarrow{(-2)} +3$

d) e^-: $MnO_2 + BiO_3^- \rightarrow MnO_4^{2-} + Bi(OH)_3$

$+4 \xrightarrow[(+2)]{} +6$

ch: $MnO_2 + BiO_3^- + OH^- \rightarrow MnO_4^{2-} + Bi(OH)_3$

H: $MnO_2 + BiO_3^- + H_2O + OH^- \rightarrow MnO_4^{2-} + Bi(OH)_3$

$4\ OH^- + MnO_2 \rightarrow MnO_4^{2-} + 2\ e^- + 2\ H_2O$

$2\ e^- + BiO_3^- + 3\ H_2O \rightarrow Bi(OH)_3 + 3\ OH^-$

$MnO_2 + BiO_3^- + H_2O + OH^- \rightarrow MnO_4^{2-} + Bi(OH)_3$

CHAPTER 16

e) I:

$$\overset{2(-1)\xrightarrow{3(+2)}2(0)}{}\quad 2\,I^- + MnO_4^- \rightarrow I_2 + MnO_2 \quad \underset{+7\xrightarrow[2(-3)]{}+4}{}$$

e⁻: $6\,I^- + 2\,MnO_4^- \rightarrow 3\,I_2 + 2\,MnO_2$

ch: $6\,I^- + 2\,MnO_4^- \rightarrow 3\,I_2 + 2\,MnO_2 + 8\,OH^-$

H: $6\,I^- + 4\,H_2O + 2\,MnO_4^- \rightarrow 3\,I_2 + 2\,MnO_2 + 8\,OH^-$

$$\begin{array}{r} 6\,I^- \rightarrow 3\,I_2 + 6\,e^- \\ 4\,H_2O + 6\,e^- + 2\,MnO_4^- \rightarrow 2\,MnO_2 + 8\,OH^- \\ \hline 6\,I^- + 4\,H_2O + 2\,MnO_4^- \rightarrow 3\,I_2 + 2\,MnO_2 + 8\,OH^- \end{array}$$

f) O:

$$\overset{2(0)\xrightarrow{(-4)}2(-2)}{}\quad Fe(OH)_2 + O_2 \rightarrow Fe(OH)_3 + 2\,H_2O \quad \underset{+2\xrightarrow[4(+1)]{}+3}{}$$

e⁻: $4\,Fe(OH)_2 + O_2 \rightarrow 4\,Fe(OH)_3 + 2\,H_2O$

H: $4\,Fe(OH)_2 + O_2 + 2\,H_2O \rightarrow 4\,Fe(OH)_3$

$$\begin{array}{r} 4\,Fe(OH)_2 + 4\,OH^- \rightarrow 4\,Fe(OH)_3 + 4\,e^- \\ O_2 + 4\,e^- + 2\,H_2O \rightarrow 4\,OH^- \\ \hline 4\,Fe(OH)_2 + O_2 + 2\,H_2O \rightarrow 4\,Fe(OH)_3 \end{array}$$

CHAPTER 17

1)

$$5.00 \times 10^2\ g\ H_2O \times \frac{1\ mol\ H_2O}{18.02\ g\ H_2O} = 27.7\ mol\ H_2O$$

$$47.3\ g\ C_{12}H_{22}O_{11} \times \frac{1\ mol\ C_{12}H_{22}O_{11}}{342.30\ g\ C_{12}H_{22}O_{11}} = 0.138\ mol\ C_{12}H_{22}O_{11}$$

$$P = P°X = 26.7\ torr \times \frac{27.7}{27.7 + 0.138} = 26.6\ torr$$

$$\text{or } P = P° - \Delta P = P° - P°X = 26.7\ torr - 26.7\ torr \times \frac{0.138}{27.7 + 0.138} = 26.6\ torr$$

2)

$$15.4\ g\ CO(NH_2)_2 \times \frac{1\ mol\ CO(NH_2)_2}{60.04\ g\ CO(NH_2)_2} = 0.256\ mol\ CO(NH_2)_2$$

$$216\ g\ CH_3OH \times \frac{1\ mol\ CH_3OH}{32.04\ g\ CH_3OH} = 6.74\ mol\ CH_3OH$$

$$P = P°X = 94\ torr \times \frac{6.74}{6.74 + 0.256} = 91\ torr$$

$$\text{or } P = P° - \Delta P = P° - P°X = 94\ torr - 94\ torr \times \frac{0.256}{6.74 + 0.256} = 91\ torr$$

3)

$$37.8\ g\ H_2C_2O_4 \times \frac{1\ mol\ H_2C_2O_4}{90.04\ g\ H_2C_2O_4} = 0.420\ mol\ H_2C_2O_4$$

$$120.0\ g\ C_2H_5OH \times \frac{1\ mol\ C_2H_5OH}{46.07\ g\ C_2H_5OH} = 2.605\ mol\ C_2H_5OH$$

$$P = P°X = 40\ torr \times \frac{2.605}{2.605 + 0.420} = 34\ torr$$

$$\text{or } P = P° - \Delta P = P° - P°X = 40\ torr - 40\ torr \times \frac{2.605}{2.605 + 0.420} = 34\ torr$$

4)

$$\frac{46.2\ g\ CO(NH_2)_2 \times \dfrac{1\ mol\ CO(NH_2)_2}{60.04\ g\ CO(NH_2)_2}}{0.145\ kg\ H_2O} = 5.31\ m\ CO(NH_2)_2$$

$\Delta T_b = K_b m = 0.512\ °C/m \times 5.31\ m = 2.72°C$ $\Delta T_f = K_f m = 1.86\ °C/m \times 5.31\ m = 9.88°C$

$T_b(soln) = 100.00°C + 2.72°C = 102.72°C$ $T_f(soln) = 0.00°C - 9.88°C = -9.88°C$

5)

$$\frac{15.4\ g\ glycerol \times \dfrac{1\ mol\ glycerol}{92.1\ g\ glycerol}}{0.0643\ kg\ C_2H_5OH} = 2.60\ m\ glycerol$$

$\Delta T_b = K_b m = 1.22\ °C/m \times 2.60\ m = 3.17°C$ $T_b(soln) = 78.5°C + 3.17°C = 81.7°C$

6) Each 100.0 g of solution contains 90.0 g camphor + 10.0 g solute.

$$\frac{10.0 \text{ g solute} \times \frac{1 \text{ mol solute}}{135 \text{ g solute}}}{0.0900 \text{ kg camphor}} = 0.823 \text{ m solute}$$

$\Delta T_f = K_f m = 40.0\ °C/m \times 0.823\ m = 32.9°C$

$T_f(\text{soln}) = 178°C - 32.9°C = 145°C$

7) $m = \frac{\Delta T_f}{K_f} = \frac{16.6°C - 12.0°C}{3.90\ °C/m} = 1.2\ m\ C_6H_{12}O_6$

$0.115 \text{ kg acetic acid} \times \frac{1.2 \text{ mol } C_6H_{12}O_6}{\text{kg acetic acid}} \times \frac{180.16 \text{ g } C_6H_{12}O_6}{1 \text{ mol } C_6H_{12}O_6} = 25 \text{ g } C_6H_{12}O_6$

8) $m = \frac{\Delta T_b}{K_b} = \frac{79.1°C - 76.5°C}{5.03\ °C/m} = 0.52 \text{ m unk}$ $\quad 0.0658 \text{ kg } CCl_4 \times \frac{0.52 \text{ mol unk}}{\text{kg } CCl_4} = 0.034 \text{ mol unk}$

MM of unknown = 5.64 g unknown ÷ 0.034 mol unknown = 170 g/mol

9) $m = \frac{\Delta T_f}{K_f} = \frac{80.1°C - 76.4°C}{6.9\ °C/m} = 0.54 \text{ m unk}$ $\quad 0.0881 \text{ kg naph} \times \frac{0.54 \text{ mol unk}}{\text{kg naph}} = 0.048 \text{ mol unk}$

MM of unknown = 2.78 g unknown ÷ 0.048 mol unknown = 58 g/mol

10) $\Delta T_f = imK_f = 2 \times 0.20\ m \times 1.86\ °C/m = 0.74°C$ $\quad T_f(\text{soln}) = 0.00°C - 0.74°C = -0.74°C$

11) $i = \frac{\Delta T_f}{mK_f} = \frac{2.40°C}{0.545\ m \times 1.86\ °C/m} = 2.37$ (i = 3 if K_2CO_3 is 100% dissociated.)

12) Each 100.00 g solution contains 2.00 g LiCl and 98.00 g H_2O.

$$\frac{2.00 \text{ g LiCl} \times \frac{1 \text{ mol LiCl}}{42.39 \text{ g LiCl}}}{0.09800 \text{ kg } H_2O} = 0.481 \text{ m LiCl}$$

$i = \frac{\Delta T_f}{mK_f} = \frac{1.72°C}{0.481\ m \times 1.86\ °C/m} = 1.92$ (i = 2 if LiCl is 100% dissociated.)

13) $i = \frac{\Delta T_f}{mK_f} = \frac{0.0424°C}{0.0200\ m \times 1.86\ °C/m} = 1.14$ (14% ionized)

14) $\pi = MRT = \frac{0.025 \text{ mol}}{1 \text{ L}} \times \frac{62.36 \text{ L·torr}}{\text{mol·K}} \times (273 + 20)K = 460 \text{ torr}$

15) $\pi = iMRT = 2 \times \frac{0.30 \text{ mol}}{1 \text{ L}} \times \frac{0.08205 \text{ L·atm}}{\text{mol·K}} \times 300 \text{ K} = 15 \text{ atm}$

16) $M = \frac{\pi}{RT} = \frac{31.2 \text{ torr}}{62.36 \text{ L·torr/mol·K} \times 291 \text{ K}} = 1.72 \times 10^{-3} \text{ M}$

$0.1000 \text{ L soln} \times \frac{1.72 \times 10^{-3} \text{ mol protein}}{1 \text{ L soln}} = 1.72 \times 10^{-2} \text{ mol protein}$

MM of protein = 4.18 g ÷ 1.72×10^{-4} mol = 2.43×10^{4} g/mol

17) $M = \frac{\pi}{RT} = \frac{2.29 \text{ atm}}{0.08205 \text{ L·atm/mol·K} \times 295 \text{ K}} = 0.0946 \text{ M}$

$0.250 \text{ L soln} \times \frac{0.0946 \text{ mol sugar}}{1 \text{ L soln}} = 0.0237 \text{ mol sugar}$

MM of sugar = 4.25 g ÷ 0.0237 mol = 179 g/mol

CHAPTER 18

1) a) $K = \frac{[NO_2]^4}{[H^+]^4[NO_3^-]^4}$ $\quad$ b) $K = \frac{[NH_3]^4[O_2]^5}{[NO]^4}$

2) a) $K = \frac{1}{[H^+]^2[HSO_4^-]^2}$ $\quad$ b) $K = \frac{[CO_2]^6[N_2]^{1/2}}{[O_2]^{27/2}}$

3) $K = \frac{[SO_3]^2}{[SO_2]^2[O_2]} = \frac{(0.16)^2}{(0.025)^2(0.15)} = 2.7 \times 10^2$

4) $K = \frac{[H_2]^2[S_2]}{[H_2S]^2} = \frac{(0.62/2.6)^2(0.36/2.6)}{(1.82/2.6)} = 0.016$

5)

	$2\ NH_3(g)$	$=\ N_2(g)$	$+\ 3\ H_2(g)$
S	7.4 M	0 M	0 M
C	-2.2	1.1	3.3
E	5.2	1.1	3.3

$$K = \frac{[N_2][H_2]^2}{[NH_3]^3} = \frac{(1.1)(3.3)^3}{5.2^2} = 1.5$$

6) $$K = \frac{[PCl_3][Cl_2]}{[PCl_5]} = \frac{(0.052/7.5)[Cl_2]}{(0.16/7.5)} = 0.041$$

$[Cl_2] = 0.13$ M

7.5 L × 0.13 mol Cl_2/L = 0.98 mol Cl_2

7)

	STRESS	RESPONSE	SHIFT	$[H_2C_2O_4]$ =	$[Fe(C_2O_4)_3{}^{2-}]$ +	$[H^+]$	
a)	increase $[H_2C_2O_4]$	use up $H_2C_2O_4$	RIGHT	increase (added)	increase	increase	**MORE RUST DISSOLVES**
b)	increase $[H^+]$	use up H^+	LEFT	increase	decrease	increase (added)	**LESS RUST**

c) **NO EFFECT** - Fe_2O_3 is a solid.

8)

	STRESS	RESPONSE	SHIFT	$[H_2C_2O_4]$ =	$[CO_2]$ +	$[Na^+]$ +	$[C_2O_4]$
a)	increase $[H_2C_2O_4]$	use up $H_2C_2O_4$	**RIGHT**	increase (added)	increase	increase	increase
b)	decrease $[H_2C_2O_4]$ by $H_2C_2O_4$ + OH^-	make $H_2C_2O_4$	**LEFT**	decrease (removed)	decrease	decrease	decrease

c) **NO EFFECT** - Na_2CO_3 is a solid.

9)

STRESS	RESPONSE	SHIFT	$CaSO_4(s)$ +	$[H^+]$ =	$[Ca^{2+}]$ +	$[HSO_4{}^-]$
increase $[H^+]$	use up H^+	RIGHT		**INCREASE(ADD)**	increase	increase

10)

STRESS	RESPONSE	SHIFT	$KOH(s)$ =	$[K^+]$ +	$[OH^-]$ +	HEAT
raise T by adding heat	use up heat	LEFT **LESS SOLUBLE**		decrease	decrease	increase (added)

11)

STRESS	RESPONSE	SHIFT	$[SO_2]$ +	$2[H_2S]$ +	HEAT =	$2\ S(s)$ +	$2[H_2O]$
raise T by adding heat	use up heat	**RIGHT**	decrease	decrease	increase (added)		increase

12)

STRESS	RESPONSE	SHIFT	$2[NO_2]$ =	$[N_2O_4]$ +	HEAT	
raise T by adding heat	use up heat	LEFT	increase	decrease	increase (added)	**ΔH IS NEGATIVE**

13)

STRESS	RESPONSE	SHIFT	$3\ O_2(g) = 2\ O_3(g)$
decrease P by raising V	increase P by raising n_{gas}	**LEFT**	Lowering P favors the side with more moles of gas.

14)

STRESS	RESPONSE	SHIFT	$N_2(g) + 2\ O_2(g) = 2\ NO_2(g)$
decrease P by raising V	increase P by raising n_{gas}	**LEFT**	Lowering P favors the side with more moles of gas.

15) Goal: Change V to shift to the right.

	STRESS	RESPONSE	SHIFT	moles reactant gases → moles product gases
a)	decrease P by **RAISING V**	increase P by raising n_{gas}	RIGHT	4 moles of gas → 9 moles of gas
b)	increase P by **LOWERING V**	decrease P by lowering n_{gas}	RIGHT	3 moles of gas → 2 moles of gas

c) **NO CHANGE** in pressure or volume will cause a shift: 3 moles of gas → 3 moles of gas.

16) $K = 10.0; \quad Q = \frac{[NOCl]^2}{[NO]^2[Cl_2]} = \frac{(0.10)^2}{(0.15)^2(0.050)} = 8.9$

Q < K. Initially, the product/reactant ratio is too low. More product must be formed to reach equilibrium. Reaction will occur to the RIGHT.

17) $K = 280; \quad Q = \frac{[SO_3]^2}{[SO_2]^2[O_2]} = \frac{(0.075/25)^2}{(0.050/25)^2(0.10/25)} = 560$

Q > K. Too much product is present initially. The reaction must shift left to convert product back to reactant. LESS SO_3 will be present.

18) $COCl_2(g) \rightleftharpoons CO(g) + Cl_2(g)$

	$COCl_2(g)$	$CO(g)$	$Cl_2(g)$
S	0.015 M	0 M	0 M
C	-y	+y	+y
E	0.015-y	y	y
	0.001	0.014	0.014

$K = \frac{y^2}{0.015-y} = 0.32$

y = 0.014 (by quadratic)

19) $2\,NO_2(g) \rightleftharpoons N_2O_4(g)$

	$2\,NO_2(g)$	$N_2O_4(g)$
S	0 M	1.50/3.6 M
C	+2y	-y
E	2y	1.50/3.6 - y
	0.54	0.15

$K = \frac{(1.50/3.6)-y}{(2y)^2} = 0.50$

y = 0.27 (by quadratic)

20) $\frac{1}{2}\,N_2(g) + \frac{1}{2}\,O_2(g) \rightleftharpoons NO(g)$

	$\frac{1}{2}\,N_2(g)$	$\frac{1}{2}\,O_2(g)$	$NO(g)$
S	0 M	0 M	0.100 M
C	+y	+y	-2y
E	y	y	0.100-2y
	0.049	0.049	0.002

$K = \frac{0.100-2y}{y} = 0.050$

y = 0.049

21) $CO_2(g) + H_2(g) \rightleftharpoons CO(g) + H_2O(g)$

	$CO_2(g)$	$H_2(g)$	$CO(g)$	$H_2O(g)$
S	0.72 M	0.72 M	0 M	0 M
C	-y	-y	+y	+y
E	0.72-y	0.72-y	y	y
	0.46	0.46	0.26	0.26

$K = \frac{y^2}{(0.72-y)^2} = 0.34, \quad \frac{y}{0.72-y} = 0.58$

y = 0.26

22) $K = 0.50; \quad Q = \frac{[N_2O_4]}{[NO_2]^2} = \frac{(0.010)}{(0.050)^2} = 4.0;$ Q < K. The reaction goes to the left.

$2\,NO_2(g) \rightleftharpoons N_2O_4(g)$

	$2\,NO_2(g)$	$N_2O_4(g)$
S	0.050 M	0.010 M
C	+2y	-y
E	0.050+2y	0.010-y
	0.066	0.002

$K = \frac{0.010-y}{(0.050+2y)^2} = 0.50$

y = 0.0078 (by quadratic)

23) $Al(OH)_3(s) + OH^-(aq) \rightleftharpoons Al(OH)_4^-(aq)$

	$Al(OH)_3(s)$	$OH^-(aq)$	$Al(OH)_4^-(aq)$
S		0.50 M	0.10 M
C		-y	+y
E		0.50-y	0.10+y
		0.09	0.51

$K = \frac{0.10+y}{0.50-y} = 5.4$

y = 0.41

CHAPTER 19

1) AgBr: $K_{sp} = [Ag^+][Br^-] = (7.1 \times 10^{-7})^2 = 5.0 \times 10^{-13}$
BaF_2: $K_{sp}[Ba^{2+}][F^-]^2 = (7.5 \times 10^{-3})(1.5 \times 10^{-2})^2 = 1.7 \times 10^{-6}$

2) $S = 0.27\text{ g } MgCO_3 \times \frac{1\text{ mol } MgCO_3}{84.32\text{ g } MgCO_3} = 3.2\times10^{-3}\text{ M}, \quad K_{sp} = S^2 = (3.2\times10^{-3})^2 = 1.0 \times 10^{-5}$

$S = 0.0016\text{ g } Cd(OH)_2 \times \frac{1\text{ mol } Cd(OH)_2}{146.42\text{ g } Cd(OH)_2} = 1.1 \times 10^{-5}\text{ M}$

$K_{sp} = (S)(2S)^2 = 4(1.1 \times 10^{-5})^3 = 5.3 \times 10^{-15}$

3) a) $Ksp = S^2 = 8.3 \times 10^{-17}$, $S = 9.1 \times 10^{-9}$ M

$$\frac{9.1 \times 10^{-9} \text{ mol AgI}}{\text{L}} \times \frac{234.8 \text{ g AgI}}{\text{mol}} = 2.1 \times 10^{-6} \text{ g AgI/L}$$

b)
$Ksp = (S)(2S)^2 = 4S^3 = 5.5\times10^{-6}$, $S = 0.011$ M $\quad \frac{0.011 \text{ mol Ca(OH)}_2}{\text{L}} \times \frac{74.10 \text{ g}}{\text{mol}} = 0.82$ g/L

$$\frac{0.011 \text{ mol Ca(OH)}_2}{\text{L}} \times \frac{74.10 \text{ g Ca(OH)}_2}{1 \text{ mol Ca(OH)}_2} = 0.82 \text{ g Ca(OH)}_2\text{/L}$$

c) $Ksp = (S)(3S)^3 = 27S^4 = 3.0 \times 10^{-32}$, $S = 5.8 \times 10^{-9}$ M

$$\frac{5.8 \times 10^{-9} \text{ mol Al(OH)}_3}{\text{L}} \times \frac{78.00 \text{ g Al(OH)}_3}{1 \text{ mol Al(OH)}_3} = 4.5 \times 10^{-7} \text{ g Al(OH)}_3\text{/L}$$

4) a) $NiCO_3(s) \rightleftharpoons Ni^{2+}(aq) + CO_3{}^{2-}(aq)$
$Ksp = [Ni^{2+}][CO_3{}^{2-}] = (0.15)(S) = 6.6 \times 10^{-9}$, $S = 4.4 \times 10^{-8}$ M

$$\frac{4.4 \times 10^{-8} \text{ mol NiCO}_3}{\text{L}} \times \frac{118.71 \text{ g NiCO}_3}{1 \text{ mol NiCO}_3} = 5.2 \times 10^{-6} \text{ g NiCO}_3\text{/L}$$

b) $CaF_2(s) \rightleftharpoons Ca^{2+}(aq) + 2\ F^-(aq)$
$Ksp = [Ca^{2+}][F^-]^2 = (S)(0.27)^2 = 4.0 \times 10^{-11}$, $S = 5.5 \times 10^{-10}$ M

$$\frac{5.5 \times 10^{-10} \text{ mol CaF}_2}{\text{L}} \times \frac{78.08 \text{ g CaF}_2}{1 \text{ mol CaF}_2} = 4.3 \times 10^{-8} \text{ g CaF}_2\text{/L}$$

5) a) $AgSCN(s) \rightleftharpoons Ag^+(aq) + SCN^-(aq)$, $Ksp = 1.0 \times 10^{-12}$

$$[Ag^+]_2 = \frac{[Ag^+]_1 \times V_1}{V_2} = \frac{0.024 \text{ M} \times 50 \text{ mL}}{75 \text{ mL}} = 0.016 \text{ M Ag}^+$$

$$[SCN^-]_2 = \frac{[SCN^-]_1 \times V_1}{V_2} = \frac{0.018 \text{ M} \times 25 \text{ mL}}{75 \text{ mL}} = 0.0060 \text{ M SCN}^-$$

$Q = [Ag^+][SCN^-] = (0.016)(0.0060) = 9.6 \times 10^{-5} > 1.0 \times 10^{-12}$
Q > Ksp, so precipitation will occur.

b) $Cu(IO_3)_2(s) \rightleftharpoons Cu^{2+}(aq) + 2\ IO_3{}^-(aq)$, $Ksp = 7.4 \times 10^{-8}$

$$[Cu^{2+}]_2 = \frac{[Cu^{2+}]_1 \times V_1}{V_2} = \frac{0.060 \text{ M} \times 100 \text{ mL}}{125 \text{ mL}} = 0.048 \text{ M Cu}^{2+}$$

$$[IO_3]_2 = \frac{[IO_3{}^-]_1 \times V_1}{V_2} = \frac{0.0032 \text{ M} \times 25 \text{ mL}}{125 \text{ mL}} = 0.00064 \text{ M IO}_3{}^-$$

$Q = [Cu^{2+}][IO_3{}^-]_2 = (0.048)(0.00064)^2 = 2.0 \times 10^{-8} < 7.4 \times 10^{-8}$
Q < Ksp, so there will be no precipitate.

c) $Ag_3PO_4(s) \rightleftharpoons 3\ Ag^+(aq) + PO_4{}^{3-}(aq)$, $Ksp = 1.3 \times 10^{-20}$

$$[Ag^+]_2 = \frac{[Ag^+]_1 \times V_1}{V_2} = \frac{1.2 \times 10^{-5} \text{ M} \times 50.00 \text{ mL}}{50.05 \text{ mL}} = 1.2 \times 10^{-5} \text{ M Ag}^+$$

$$[PO_4{}^{3-}]_2 = \frac{[PO_4{}^{3-}]_1 \times V_1}{V_2} = \frac{2 \text{ M} \times 0.05 \text{ mL}}{50.05 \text{ mL}} = 0.002 \text{ M PO}_4{}^{3-}$$

$Q = [Ag^+]^3[PO_4{}^{3-}] = (1.2 \times 10^{-5})^3(0.002) = 3 \times 10^{-18} > 1.3 \times 10^{-20}$
Q > Ksp, so precipitation will occur.

CHAPTER 20

		$[H^+]$, M	$[OH^-]$, M	pH	pOH
1)	a)	7.2×10^{-2}	1.4×10^{-13}	1.14	12.86
	b)	2.0×10^{-11}	4.9×10^{-4}	10.69	3.31
	c)	5.5×10^{-4}	1.8×10^{-11}	3.26	10.74
2)	a)	5.5×10^{-9}	1.8×10^{-6}	8.26	5.74
	b)	3.1×10^{-5}	3.2×10^{-10}	4.51	9.49

3) $HB(aq) \rightleftharpoons H^+(aq) + B^-(aq)$ $\quad [H^+] = [B^-] = 10^{-2.10} = 7.9 \times 10^{-3}$

	HB	H^+	B^-
S	0.12 M	0 M	0 M
C	-0.0079	+0.0079	+0.0079
E	0.11	0.0079	0.0079

$$Ka = \frac{(7.9 \times 10^{-3})^2}{0.11} = 5.7 \times 10^{-4}$$

4) $[H^+] = [Y^-] = 10^{-3.42} = 3.8 \times 10^{-4}$ $\quad Ka = \dfrac{(3.8 \times 10^{-4})^2}{0.32} = 4.5 \times 10^{-7}$

5) $[Z^-] = 0.16\ M\ HZ \times \dfrac{1.2\ M\ Z^-}{100\ M\ HZ} = 0.0019\ M\ Z^- = 0.0019\ M\ H^+$

$pH = -\log 0.0019 = 2.72$ $\quad Ka = \dfrac{(0.0019)^2}{0.16} = 2.3 \times 10^{-5}$

6) a)

	$HC_2H_3O_2(aq)$	⇌ $H^+(aq)$	+ $C_2H_3O_2^-(aq)$	
S	0.75 M	0 M	0 M	$\dfrac{y^2}{0.75} = 1.8 \times 10^{-5}$, y = 0.0037
C	-y	+y	+y	
E	0.75-y	y	y	$\% = \dfrac{0.0037}{0.75} \times 100 = 0.49\%$, pH = 2.43
A	0.75	y	y	

b)

	$HCHO_2(aq)$	⇌ $H^+(aq)$	+ $CHO_2^-(aq)$	
S	0.16 M	0 M	0 M	$\dfrac{y^2}{0.16} = 1.8 \times 10^{-4}$, y = 0.0054
C	-y	+y	+y	
E	0.16-y	y	y	$\% = \dfrac{0.0054}{0.16} \times 100 = 3.4\%$, pH = 2.27
A	0.16	y	y	

c)

	$HC_4H_5O_3(aq)$	⇌ $H^+(aq)$	+ $C_4H_5O_3^-(aq)$	
S	0.092 M	0 M	0 M	$\dfrac{y^2}{0.092} = 2.6 \times 10^{-4}$, y = 0.0049
C	-y	+y	+y	
E	0.092-y	y	y	$\% = \dfrac{0.0049}{0.092} \times 100 = 5.3\%$, use quadratic
A	0.092	y	y	

$y^2 + (2.6 \times 10^{-4})y - (2.5 \times 10^{-5}) = 0$, y = 0.0048, $\% = \dfrac{0.0048}{0.092} \times 100 = 5.2\%$, pH = 2.32

7)

	$CH_3NH_2(aq) + H_2O(\ell)$	⇌ $CH_3NH_3^+(aq)$	+ $OH^-(aq)$	
S	1.2 M	0 M	0 M	$\dfrac{y^2}{1.2} = 5.0 \times 10^{-4}$, y = 0.024
C	-y	+y	+y	
E	1.2-y	y	y	$\% = \dfrac{0.024}{1.2} \times 100 = 2.0\%$, pOH = 1.62
A	1.2	y	y	pH = 12.38

8) a)

	$F^-(aq) + H_2O(\ell)$	⇌ $HF(aq)$	+ $OH^-(aq)$,	Kb = Kw/Ka
S	0.74 M	0 M	0 M	$\dfrac{y^2}{0.74} = \dfrac{1.0 \times 10^{-14}}{6.8 \times 10^{-4}}$ y = 3.3×10^{-6}
C	-y	+y	+y	
E	0.74-y	y	y	pOH = 5.48, pH = 8.52
A	0.74	y	y	

b)

	$CO_3^{2-}(aq) + H_2O(\ell)$	⇌ $HCO_3^-(aq)$	+ $OH^-(aq)$,	Kb = Kw/Ka
S	0.33 M	0 M	0 M	$\dfrac{y^2}{0.33} = \dfrac{1.0 \times 10^{-14}}{4.8 \times 10^{-11}}$ y = 8.3×10^{-3}
C	-y	+y	+y	
E	0.33-y	y	y	pOH = 2.08, pH = 11.92
A	0.33	y	y	

9) a) $[H^+] = Ka \times \dfrac{[HC_2H_3O_2]}{[C_2H_3O_2^-]} = 1.8 \times 10^{-5} \times \dfrac{0.50}{0.72} = 1.3 \times 10^{-5}\ M$, pH = 4.89

b) $[H^+] = Ka = \dfrac{[HF]}{[F^-]} = 6.8 \times 10^{-4} \times \dfrac{0.23}{0.10} = 1.6 \times 10^{-3}\ M$, pH = 2.80

c) $[H^+] = Ka = \dfrac{[HCO_3^-]}{[CO_3^{2-}]} = 4.8 \times 10^{-11} \times \dfrac{0.12}{0.29} = 2.0 \times 10^{-11}\ M$, pH = 10.70

10) $[HCHO_2]_2 = \dfrac{[HCHO_2]_1 \times V_1}{V_2} = \dfrac{26.5\ M \times 12\ mL}{750\ mL} = 0.42\ M\ HCHO_2$

$$[CHO_2^-] = \dfrac{38\ g\ NaCHO_2 \times \dfrac{1\ mol\ NaCHO_2}{68.01\ g\ NaCHO_2}}{0.75\ L} = 0.74\ M\ CHO_2^-$$

$[H^+] = Ka \times \dfrac{[HCHO_2]}{[CHO_2^-]} = 1.8 \times 10^{-4} \times \dfrac{0.42}{0.74} = 1.0 \times 10^{-4}\ M$, pH = 4.00

CHAPTER 20

11)
$$[C_4H_4O_6{}^{2-}] = \frac{35.0\text{ g } Na_2C_4H_4O_6 \times \frac{1\text{ mol } Na_2C_4H_4O_6}{194.05\text{ g } Na_2C_4H_4O_6}}{1.75\text{ L}} = 0.103\text{ M } C_4H_4O_6{}^{2-}$$

$$[HC_4H_4O_6{}^{-}] = \frac{35.0\text{ g } NaHC_4H_4O_6 \times \frac{1\text{ mol } NaHC_4H_4O_6}{172.07\text{ g } NaHC_4H_4O_6}}{1.75\text{ L}} = 0.116\text{ M } HC_4H_4O_6{}^{-}$$

$$[H^+] = Ka \times \frac{[HC_4H_4O_6{}^{-}]}{[C_4H_4O_6{}^{2-}]} = 4.6 \times 10^{-5} \times \frac{0.116}{0.103} = 5.2 \times 10^{-5}\text{ M},\quad pH = 4.28$$

12) a) $[H^+] = 10^{-4.50} = 3.2 \times 10^{-5}$ $\quad \frac{[HC_2H_3O_2]}{[C_2H_3O_2{}^-]} = \frac{3.2 \times 10^{-5}}{1.8 \times 10^{-5}} = 1.8/1$

b) $[H^+] = 10^{-3.41} = 3.9 \times 10^{-4}$ $\quad \frac{[HNO_2]}{[NO_2{}^-]} = \frac{3.9 \times 10^{-4}}{4.6 \times 10^{-4}} = 0.85/1$

c) $[H^+] = 10^{-9.50} = 3.2 \times 10^{-10}$ $\quad \frac{[NH_4{}^+]}{[NH_3]} = \frac{3.2 \times 10^{-10}}{5.6 \times 10^{-10}} = 0.56/1$

13) a) $[H^+] = 10^{-4.85} = 1.4 \times 10^{-5}$ $\quad \frac{1.4 \times 10^{-5}}{1.8 \times 10^{-5}} \times \frac{[HC_2H_3O_2]}{[C_2H_3O_2{}^-]} = 0.78/1$

Let's use 1.0 M $NaC_2H_3O_2$ and 0.78 M $HC_2H_3O_2$.

$$0.50\text{ L} \times \frac{1.0\text{ mol } NaC_2H_3O_2}{\text{L}} \times \frac{82.03\text{ g } NaC_2H_3O_2}{1\text{ mol } NaC_2H_3O_2} = 41\text{ g } NaC_2H_3O_2$$

$$0.50\text{ L} \times \frac{0.78\text{ mol } HC_2H_3O_2}{\text{L}} \times \frac{1000\text{ mL}}{17.4\text{ mol } HC_2H_3O_2} = 23\text{ mL } 17.4\text{ M } HC_2H_3O_2$$

Use 41 g $NaC_2H_3O_2$ and 23 mL 17.4 M $HC_2H_3O_2$, or quantities in the same ratio.

b) $[H^+] = 10^{-6.90} = 1.3 \times 10^{-7}$ $\quad \frac{[HSO_3{}^-]}{[SO_3{}^-]} = \frac{1.3 \times 10^{-7}}{5.6 \times 10^{-8}} = 2.2$

Let's use 1.0 M Na_2SO_3 and 2.2 M $NaHSO_3$.

$$1.20\text{ L} \times \frac{2.2\text{ mol } NaHSO_3}{\text{L}} \times \frac{104.06\text{ g } NaHSO_3}{1\text{ mol } NaHSO_3} = 2.7 \times 10^2\text{ g } NaHSO_3$$

$$1.20\text{ L} \times \frac{1.0\text{ mol } Na_2SO_3}{\text{L}} \times \frac{126.04\text{ g } Na_2SO_3}{1\text{ mol } Na_2SO_3} = 1.5 \times 10^2\text{ g } Na_2SO_3$$

Use 270 g $NaHSO_3$ and 150 g Na_2SO_3, or quantities in the same ratio.

ANSWERS TO ADDITIONAL PROBLEMS

CHAPTER 1

22) $(125 + 37.9)(25.0 - 4.20) = 10.8T$, so $\boxed{T = 314}$

23) $0.1875 = 1/n^2 - 1/16$, so $1/n^2 = 0.25$, and $\boxed{n = 2}$

24) $$-1.035 = \frac{44{,}000}{8.314}\left[\frac{1}{318} - \frac{1}{T}\right],\ -1.956\times10^{-4} = \frac{1}{318} - \frac{1}{T},\ -3.34\times10^{-3} = -\frac{1}{T},\ \boxed{T = 299}$$

25) $$0.113 = -\frac{0.05916}{2}\log\frac{1.5}{P^2},\ \log\frac{1.5}{P^2} = -3.82,\ \frac{1.5}{P^2} = 1.5\underline{13}\times10^{-4},\ \boxed{P = 99.\underline{6} = 100}$$

CHAPTER 2

35) a)
$$\begin{array}{r} 1.18 \\ 1.28 \\ 1.26 \\ 1.22 \\ \underline{1.16} \\ 6.10 \div 5 = \boxed{1.22\text{ g}} \end{array}$$

b)
$$\begin{array}{r} |1.18 - 1.22| = 0.04 \\ |1.28 - 1.22| = 0.06 \\ |1.26 - 1.22| = 0.04 \\ |1.22 - 1.22| = 0.00 \\ |1.16 - 1.22| = \underline{0.06} \\ d = 0.20 \div 5 = \boxed{0.04\text{ g}} \end{array}$$

c) $$\text{RAD} = \frac{0.04\text{ g}}{1.22\text{ g}} \times 100 = \boxed{3\%}$$

36) $$1\text{ yr} \times \frac{365.25\text{ d}}{1\text{ yr}} \times \frac{24\text{ hr}}{1\text{ d}} \times \frac{60\text{ min}}{1\text{ hr}} \times \frac{60\text{ sec}}{1\text{ min}} = \boxed{3.15576 \times 10^7\text{ sec}}$$

37) a) direct b) inverse c) inverse d) direct

38) %error = (2.5 - 2.0)/2.0 × 100 = 25%

CHAPTER 3

27) a)
$$\begin{array}{r} 55.4\text{ mL} \\ \underline{-47.3\text{ mL}} \\ 8.1\text{ mL} \end{array}$$
b)
$$\begin{array}{r} 123.79\text{ g} \\ \underline{-\ 89.62\text{ g}} \\ 34.17\text{ g} \end{array}$$
c) $$\frac{34.17\text{ g}}{8.1\text{ mL}} = 4.2\underline{185}\text{ g/mL} = 4.2\text{ g/mL}$$

28) $$0.48\text{ g salt} \times \frac{50.00\text{ mL}}{5.00\text{ mL}} = 4.8\text{ g salt in original}$$
$$\begin{array}{r|l} 5.6 & 2\text{ g} \\ -4.8 & \ \ \text{g} \\ \hline 0.8 & 2\text{ g} = \boxed{0.8\text{ g sand}} \end{array}$$

29) $$\frac{67\text{ lb} \times \dfrac{453.6\text{ g}}{1\text{ lb}} \times \dfrac{1\text{ mg}}{10^{-3}\text{ g}}}{1\text{ ft}^3 \times \dfrac{12^3\text{ in}^3}{1\text{ ft}^3} \times \dfrac{2.54^3\text{ cm}^3}{1\text{ in}^3} \times \dfrac{10^3\text{ mm}^3}{1\text{ cm}^3}} = \frac{30.\underline{3912}\text{ mg}}{2.8\underline{317}\times10^7\text{ mm}^3} = \boxed{1.1\ \frac{\text{mg}}{\text{mm}^3}}$$

30) $°\text{S} = 1.2°\text{C} - 35 = 1.2(112) - 35 = 134.\underline{4} - 35 = 99.\underline{4} = \boxed{99°\text{S}}$

CHAPTER 4

9) and 10) See next two pages.

CHAPTER 5

19) $k = e^{-\Delta G°/RT} = e^{-(4450)/(8.314)(298)} = e^{-1.80} = \boxed{0.17}$

9)

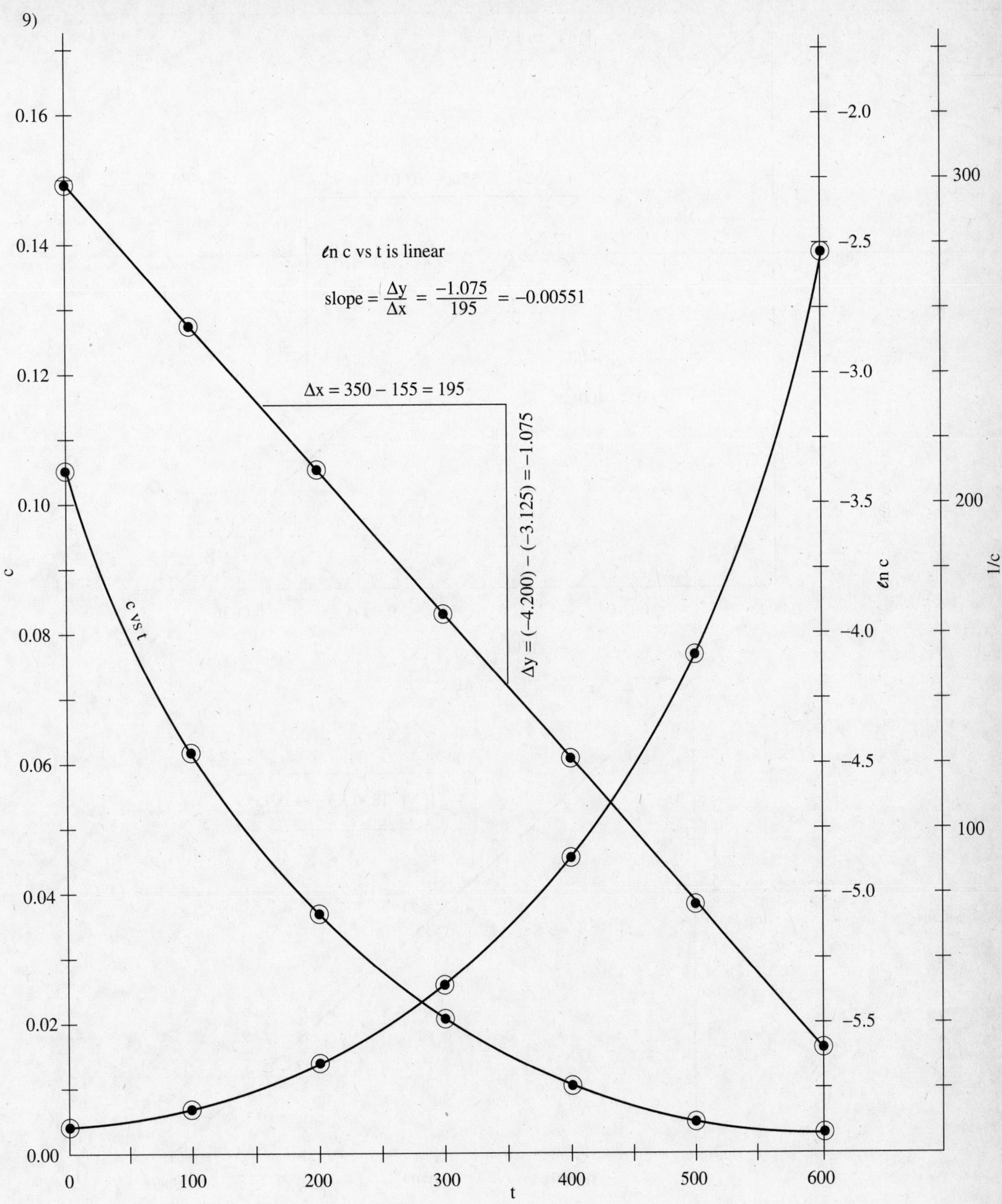

ℓn c vs t is linear
slope = Δy/Δx = −1.075/195 = −0.00551
Δx = 350 − 155 = 195
Δy = (−4.200) − (−3.125) = −1.075
c vs t
c
t
ℓn c
1/c
0.16
0.14
0.12
0.10
0.08
0.06
0.04
0.02
0.00
0
100
200
300
400
500
600
−2.0
−2.5
−3.0
−3.5
−4.0
−4.5
−5.0
−5.5
300
200
100

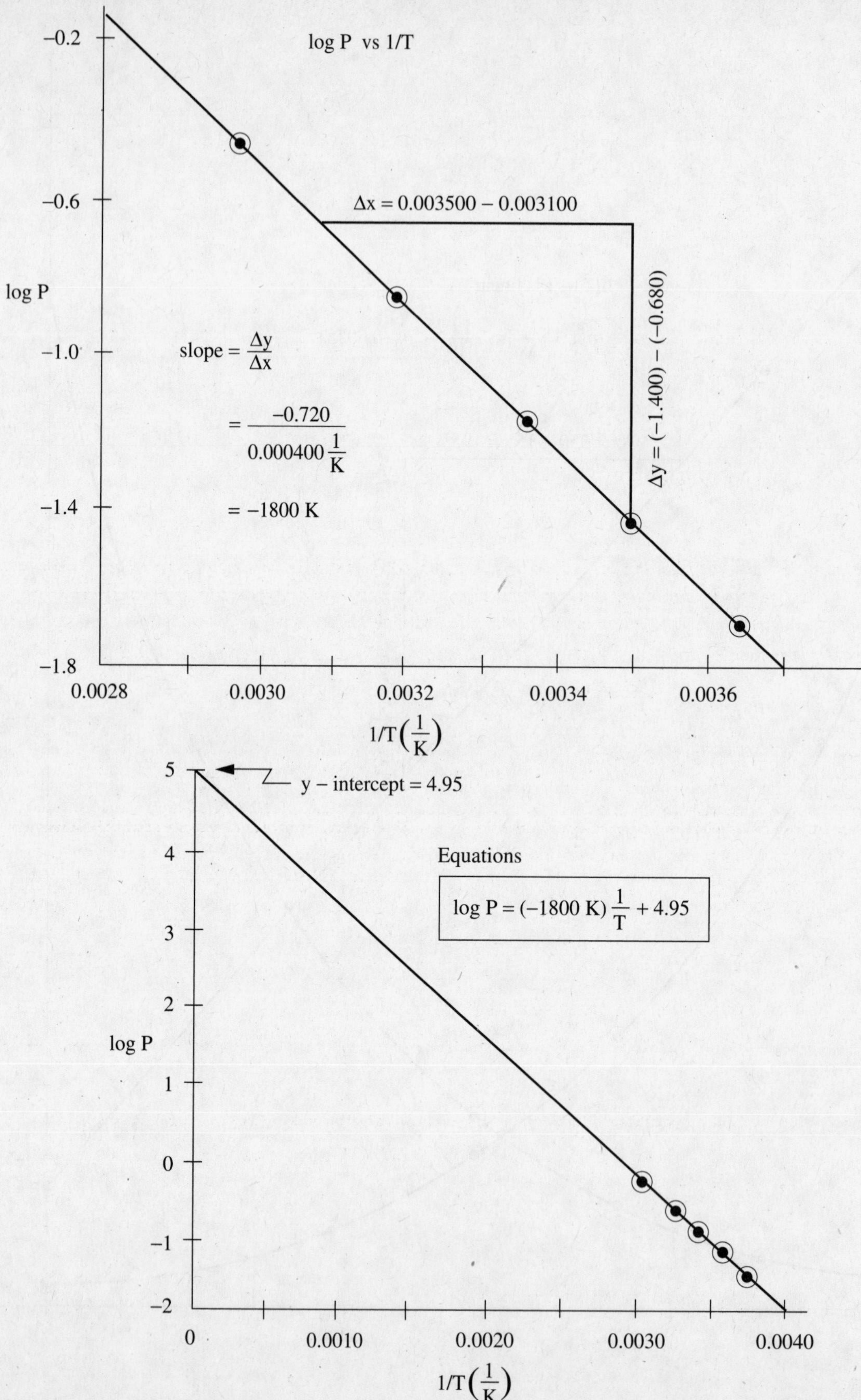

log P vs 1/T
−0.2
−0.6
−1.0
−1.4
−1.8
log P
Δx = 0.003500 − 0.003100
Δy = (−1.400) − (−0.680)
slope = Δy/Δx
= −0.720 / 0.000400 1/K
= −1800 K
0.0028
0.0030
0.0032
0.0034
0.0036
1/T (1/K)
5
4
3
2
1
0
−1
−2
y − intercept = 4.95
Equations
log P = (−1800 K) 1/T + 4.95
0
0.0010
0.0020
0.0030
0.0040
1/T (1/K)

20) $\frac{n(E° - E)}{0.05916} = \log\frac{P}{C^2}$ so $\frac{P}{C^2} = 10^{n(E°-E)/0.05916}$ and $P = C^2 10^{n(E°-E)/0.05916}$

so $P = (5.00)^2 10^{2.00(0.760-0.462)/0.05916} = 25.0 \cdot 10^{10.07} = \boxed{3 \times 10^{11}}$

21) log of both sides: $\log(R_2/R_1) = \log(A_2/A_1)^a$ so $\log(R_2/R_1) = a\log(A_2/A_1)$

$a = \frac{\log(R_2/R_1)}{\log(A_2/A_1)} = \frac{\log(0.562/0.397)}{\log(0.0260/0.0130)} = \boxed{0.501}$

22) $\frac{C_2}{C_1} = e^{-kt}$ so $C_2 = C_1 e^{-kt}$ and $C_2 = 4.0\times10^{15} e^{-(0.795)(14.70)} = \boxed{3 \times 10^{10}}$

CHAPTER 6

23) M has x + 2 protons and T has x + 10 protons

24) a) $I_2(s)$ b) $Li(s)$ c) $B(s)$ d) $Au(s)$ e) $Br_2(\ell)$

25)
0.9221 × 27.9769 = 25.7975
0.0470 × 28.9765 = 1.3619
(1) × (3) = (2)
28.0855

(1) 1-0.9221-0.0470 = 0.0309 = $\boxed{3.09\%}$

(2) 28.0855 - 25.7975 - 1.3619 = 0.9261

(3) 0.9261 amu ÷ 0.0309 = $\boxed{29.97 \text{ amu}}$

26) A × 62.9298 = 62.9298A
(1-A) × 64.9278 = 64.9278 - 64.9278A
63.546

or 62.9298A + 64.9278 - 64.9278A = 63.546

so -1.998A = -1.3818

thus A = 0.6916

$\boxed{69.16\%\ ^{63}Cu \text{ and } 30.84\%\ ^{65}Cu}$

CHAPTER 7

chromium(II) bromide $CrBr_2$	chromium(II) hydrogen phosphite $CrHPO_3$	chromium(II) selenide $CrSe$	chromium(II) hypochlorite $Cr(ClO)_2$	chromium(II) oxalate CrC_2O_4	chromium(II) chromate $CrCrO_4$
ammonium bromide NH_4Br	ammonium hydrogen phosphite $(NH_4)_2HPO_3$	ammonium selenide $(NH_4)_2Se$	ammonium hypochlorite NH_4ClO	ammonium oxalate $(NH_4)_2C_2O_4$	ammonium chromate $(NH_4)_2CrO_4$
hydrobromic acid HBr	phosphorous acid H_3PO_3	hydroselenic acid H_2Se	hypochlorous acid $HClO$	oxalic acid $H_2C_2O_4$	chromic acid H_2CrO_4
strontium bromide $SrBr_2$	strontium hydrogen phosphite $SrHPO_3$	strontium selenide $SrSe$	strontium hypochlorite $Sr(ClO)_2$	strontium oxalate SrC_2O_4	strontium chromate $SrCrO_4$
iron(III) bromide $FeBr_3$	iron(III) hydrogen phosphite $Fe_2(HPO_3)_3$	iron(III) selenide Fe_2Se_3	iron(III) hypochlorite $Fe(ClO)_3$	iron(III) oxalate $Fe_2(C_2O_4)_3$	iron(III) chromate $Fe_2(CrO_4)_3$
zinc bromide $ZnBr_2$	zinc hydrogen phosphite $ZnHPO_3$	zinc selenide $ZnSe$	zinc hypochlorite $Zn(ClO)_2$	zinc oxalate ZnC_2O_4	zinc chromate $ZnCrO_4$

CHAPTER 8

39) $Cr(NO_3)_3 \cdot xH_2O$, x = ? MM $Cr(NO_3)_3$ = 52.00 + 3(14.01) + 9(16.00) = 238.03

400 - 238 = 162 g H_2O ÷ 18 g/mol H_2O = 9 mol H_2O, so $\boxed{Cr(NO_3)_3 \cdot 9H_2O}$

40)

$$\frac{25.0 \text{ pg } X_4 \times \frac{10^{-12} \text{ g } X_4}{1 \text{ pg } X_4}}{50.0\times10^9\ X_4 \text{ molecules} \times \frac{1 \text{ mol } X_4}{6.022\times10^{23}\ X_4 \text{ molecules}}} = \boxed{301 \text{ g/mol } X_4 \div 4 = 75.3 \text{ g/mol X, arsenic}}$$

41) a)

$$2723 \text{ mg } CO_2 \times \frac{1 \text{ mmol } CO_2}{44.01 \text{ mg } CO_2} \times \frac{1 \text{ mmol C}}{1 \text{ mmol } CO_2} \times \frac{12.01 \text{ mg C}}{1 \text{ mmol C}} = \boxed{743.1 \text{ mg C}}$$

$$892 \text{ mg } H_2O \times \frac{1 \text{ mmol } H_2O}{18.02 \text{ mg } H_2O} \times \frac{2 \text{ mmol H}}{1 \text{ mmol } H_2O} \times \frac{1.008 \text{ mg H}}{1 \text{ mmol H}} = \boxed{99.8 \text{ mg H}}$$

b) $1239 \text{ mg } C_xH_yO_z - 743.1 \text{ mg C} - 99.8 \text{ mg H} = \boxed{396 \text{ mg O}}$

c)

	MGRAMS	MMOLES ATOMS	ATOM RATIO	WHOLE # RATIO	EMPIRICAL FORMULA
C	$743.1 \text{ mg} \times \frac{1 \text{ mmol}}{12.01 \text{ mg}}$	= 61.87 mmol	$\frac{61.87}{24.75} = 2.500$	$= 2\frac{1}{2} \times 2 = 5$	
H	$99.8 \text{ mg} \times \frac{1 \text{ mmol}}{1.008 \text{ mg}}$	= 99.01 mmol	$\frac{99.01}{24.75} = 4.000$	$= 4 \times 2 = 8$	$\boxed{C_5H_8O_2}$
O	$396 \text{ mg} \times \frac{1 \text{ mmol}}{16.00 \text{ mg}}$	= 24.75 mmol	$\frac{24.75}{24.75} = 1.000$	$= 1 \times 2 = 2$	

42)

$Ca_3(C_6H_5O_7)_2$		$CaCO_3$		$CaHPO_4$	
3 Ca =	3(40.08)	1 Ca =	40.08	1 Ca =	40.08
12 C =	12(12.01)	1 C =	12.01	1 H =	1.008
10 H =	10(1.008)	3 O =	3(16.00)	1 P =	30.97
14 O =	14(16.00)			4 O =	4(16.00)
	498.44		100.09		136.05$\underline{8}$ = 136.06

$$1000 \text{ mg Ca} \times \frac{1 \text{ mmol Ca}}{40.08 \text{ mg Ca}} \times \frac{1 \text{ mmol } Ca_3(C_6H_5O_7)_2}{3 \text{ mmol Ca}} \times \frac{498.44 \text{ mg } Ca_3(C_6H_5O_7)_2}{1 \text{ mmol } Ca_3(C_6H_5O_7)_2} = \boxed{4145 \text{ mg } Ca_3(C_6H_5O_7)_2}$$

$1000 \times 1/40.08 \times 1/1 \times 100.09/1 = \boxed{2497 \text{ mg } CaCO_3}$

$1000 \times 1/40.08 \times 1/1 \times 136.06/1 = \boxed{3395 \text{ mg } CaHPO_4}$

CHAPTER 9

53) Calcium. $Ca(s) + 2\ H_2O(\ell) \rightarrow H_2(g) + Ca(OH)_2(s)$

54) $(NH_4)_2C_2O_4(s) \rightarrow 2\ NH_3(g) + H_2O(g) + CO(g) + CO_2(g)$
$2\ (NH_4)_2C_2O_4(s) \rightarrow 4\ NH_3(g) + 2\ H_2O(g) + C(s) + 3\ CO_2(g)$

55) $Pb(s) + PbO_2(s) + 2\ H_2SO_4(aq) \rightarrow 2\ PbSO_4(s) + 2\ H_2O(\ell)$
$2\ MnO_2(s) + Zn(s) + 4\ H_2O(\ell) \rightarrow 2\ Mn(OH)_3(s) + Zn(OH)_2(s)$

56) $Cu_2S(s) + O_2(g) \rightarrow 2\ Cu(s) + SO_2(g)$
$PbS(s) + 2\ O_2(g) \rightarrow PbSO_4(s)$
$2\ MoS_2(s) + 7\ O_2(g) \rightarrow 2\ MoO_3(s) + 4\ SO_2(g)$

CHAPTER 10

36) $NaHCO_3(s) + HCl(aq) \rightarrow NaCl(aq) + H_2O(\ell) + CO_2(g)$

$$1.0 \text{ pt} \times \frac{1 \text{ gal}}{8 \text{ pt}} \times \frac{3.785 \text{ L}}{1 \text{ gal}} \times \frac{0.010 \text{ mol HCl}}{\text{L}} \times \frac{1 \text{ mol } NaHCO_3}{1 \text{ mol HCl}} \times \frac{84.01 \text{ g } NaHCO_3}{1 \text{ mol } NaHCO_3} = \boxed{0.40 \text{ g } NaHCO_3}$$

37) $MCl_3(aq) + 3\ AgNO_3(aq) \rightarrow 3\ AgCl(s) + M(NO_3)_3(aq)$

$$4.61 \text{ g AgCl} \times \frac{1 \text{ mol AgCl}}{143.4 \text{ g AgCl}} \times \frac{1 \text{ mol } MCl_3}{3 \text{ mol AgCl}} = 0.0107 \text{ mol } MCl_3$$

$2.37 \text{ g } MCl_3 \div 0.0107 \text{ mol } MCl_3 = 221 \text{ g/mol } MCl_3$
$-3(35.45 \text{ g/mol})$ for 3 mol Cl^-
$\boxed{115 \text{ g/mol M (metal is indium)}}$

38)

$$\left[\frac{x}{142.04} + \frac{1.96-x}{174.26}\right]\times 233.4 = 2.82 \text{ so } \frac{174.26x}{142.04\times 174.26} + \frac{142.04(1.96-x)}{142.04\times 174.26} = 0.0120\underline{8}$$

thus $174.26x + 278.\underline{40} - 142.04x = 299$

and $32.22x = 20.\underline{6}$ therefore $x = 0.63\underline{94}$ g Na_2SO_4

$$\%Na_2SO_4 = \frac{0.63\underline{94} \text{ g } Na_2SO_4}{1.96 \text{ g mixture}} \times 100 = \boxed{32.\underline{62}\% \ Na_2SO_4}$$

39) $Ca_3(PO_4)_2(s) + 6\ HF(aq) \rightarrow 3\ CaF_2(s) + 2\ H_3PO_4(aq)$

$$35\ kg\ Ca_3(PO_4)_2 \times \frac{1\ kmol\ Ca_3(PO_4)_2}{310.18\ kg\ Ca_3(PO_4)_2} \times \frac{2\ kmol\ H_3PO_4}{1\ kmol\ Ca_3(PO_4)_2} \times \frac{97.99\ kg\ H_3PO_4}{1\ kmol\ H_3PO_4} \times \frac{74\ kg\ H_3PO_4(act)}{100\ kg\ H_3PO_4(theo)} = \boxed{16\ kg\ H_3PO_4}$$

$$17\ kg\ HF \times \frac{1\ kmol\ HF}{20.01\ kg\ HF} \times \frac{2\ kmol\ H_3PO_4}{3\ kmol\ HF} \times \frac{97.99\ kg\ H_3PO_4}{1\ kmol\ H_3PO_4} \times \frac{74\ kg\ H_3PO_4(act)}{100\ kg\ H_3PO_4(theo)} = 21\ kg\ H_3PO_4$$

CHAPTER 11

25) $\nu = c/\lambda = 2.998 \times 10^8\ m/s \div 434.2 \times 10^{-9}\ m = 6.905 \times 10^{14}$ cycles/s

$\nu = R_o\left[\frac{1}{n_i^2} - \frac{1}{n_o^2}\right]$ n_o and n_i are whole numbers and $n_i < n_o$

Try $n_i = 1$ and solve for n_o:

$6.905 \times 10^{14}\ cy/s = 3.288 \times 10^{15}\ cy/s\left[\frac{1}{1^2} - \frac{1}{n_o^2}\right]$

gives $0.21 = 1/1^2 - 1/n_o^2$, $1/n_o^2 = 0.79$, and $n_o = 1.12$

Not an integer (and not greater than n_i), so try $n_i = 2$:

gives $0.21 = 1/2^2 - 1/n_o^2$, $1/n_o^2 = 0.04$ and $\boxed{n_o = 5 \text{ and } n_i = 2}$

26) The Bohr model does not consider the interaction between electrons.

27) Hund's rule: Electrons in the same subshell are predicted to occupy different orbitals but to have the same spin.

m effect violation for C: $1s^2$ [↑↓] $2s^2$ [↑↓] $2p^2$ [↑↓][ ][ ]

Electrons in the same orbital area repel one another.

s effect violation for C: $1s^2$ [↑↓] $2s^2$ [↑↓] $2p^2$ [↑][↓][ ]

Electrons with magnetic spins aligned have lower energy.

28) $4s^1$ [↑] $3d^5$ [↑][↑][↑][↑][↑] lower E than $4s^2$ [↑↓] $3d^4$ [↑][↑][↑][↑][ ]

$4s^13d^5$ has less e^-/e^- repulsion (m effect) and more spins aligned (s effect) than does $4s^23d^4$.

CHAPTER 12

74) $1.25\ atm \times \frac{1248\ K}{295\ K} = \boxed{5.29\ atm}$

75) $V = m_{tot}/d_{air} = 2.25 \times 10^5\ g \div 1.10\ g/L = 2.04\underline{5} \times 10^5\ L$

$$n = \frac{PV}{RT} = \frac{635\ torr \times 2.04\underline{5} \times 10^5\ L}{62.36\ L{\cdot}torr/mol{\cdot}K \times 266\ K} = 7.83 \times 10^3\ mol\ He$$

$$7.83 \times 10^3\ mol\ He \times \frac{4.003\ g\ He}{1\ mol\ He} = \boxed{3.13 \times 10^4\ g\ He\ \ (69\ pounds)}$$

76) a)

$$n_{CO_2} = \frac{PV}{RT} = \frac{743\ torr \times 1.225\ L}{62.36\ L{\cdot}torr/mol{\cdot}K \times 292\ K} = 0.0500\ mol\ CO_2$$

$$0.0500\ mol\ CO_2 \times \frac{1\ mol\ C}{1\ mol\ CO_2} \times \frac{12.01\ g\ C}{1\ mol\ C} \times \frac{1\ mg}{10^{-3}\ g} = \boxed{600\ mg\ C}$$

$$n_{H_2O} = \frac{PV}{RT} = \frac{752\ torr \times 1.002\ L}{62.36\ L{\cdot}torr/mol{\cdot}K \times 387\ K} = 0.0312\ mol\ H_2O$$

$$0.0312\ mol\ H_2O \times \frac{2\ mol\ H}{1\ mol\ H_2O} \times \frac{1.008\ g\ H}{1\ mol\ H} \times \frac{1\ mg}{10^{-3}\ g} = \boxed{62.9\ mg\ H}$$

b) $963\ mg\ C_xH_yO_z - 600\ mg\ C - 62.9\ mg\ H = \boxed{300\ mg\ O}$

76) c) ELEMENT GRAMS MOLES OF ATOMS WHOLE NUMBER RATIO EMPIRICAL FORMULA

C $600\ mg \times \frac{1\ mmol}{12.01\ mg} = 49.96\ mmol \quad \frac{49.96}{18.75} = 2.665 = 2\frac{2}{3} \times 3 = 8$

H $62.9\ mg \times \frac{1\ mmol}{1.008\ mg} = 62.40\ mmol \quad \frac{62.40}{18.75} = 3.328 = 3\frac{1}{3} \times 3 = 10$ $\boxed{C_8H_{10}O_3}$

O $300\ mg \times \frac{1\ mmol}{16.00\ mg} = 18.75\ mmol \quad \frac{18.75}{18.75} = 1.000 = 1 \times 3 = 3$

77) $\sqrt{\frac{M_X}{M_{NH_3}}} = \sqrt{\frac{d_{NH_3}}{d_X}}$ so $\sqrt{\frac{}{17.04}} = \frac{}{37.2\ cm}$ so $\boxed{M_X = 86.7\ g/mol}$

CHAPTER 13

46) raise ice from 15°C to 0°C: $Q_1 = mc_s\Delta T = 2.50\ g \times 2.1\ J/g{\cdot}°C \times 15°C = 79\ J$
$4500\ J - 79\ J = 4421\ J$ left

melt ice: $Q_2 = m\Delta H_{fus} = 2.50\ g \times 335\ J/g = 838\ J$
$4421\ J - 838\ J = 3583\ J$ left

heat water from 0°C to 100°C: $Q_3 = mc_l\Delta T = 2.50\ g \times 4.18\ J/g{\cdot}°C \times 100°C = 104\underline{5}\ J$
$3583\ J - 104\underline{5}\ J = 253\underline{8}\ J$ left

find water mass to be vaporized: $m = Q \div \Delta H_{vap} = 253\underline{8}\ J \div 2260\ J/g = 1.12\ g$

$\boxed{\text{Present: } 1.12\ g\ H_2O(g) \text{ and } 1.38\ g\ H_2O(\ell) \text{, both at } 100°C}$

47) $Q_{cold} + Q_{styro} = -Q_{hot}$ so $Q_{styro} = -Q_{cold} - Q_{hot}$

$Q = mc\Delta T$ so $Q_{styro} = (-mc\Delta T_{cold} - mc\Delta T_{hot})$ or $Q_{styro} = -mc(\Delta T_{cold} + \Delta T_{hot})$

$\Delta T_{cold} = 17.0°C - 43.7°C = +26.7°C \quad \Delta T_{hot} = -43.7°C - 73.0°C = -29.3°C$

so $Q_{styro} = -50.0\ g \times 4.18\ J/g{\cdot}°C(+26.7°C - 29.3°C) = \boxed{+54\underline{3}\ J \text{ absorbed}}$

48) $Q = -mc\Delta T = -75.0\ g \times 4.18\ J/g{\cdot}°C \times (32.9\text{-}21.5)°C = -357\underline{4}\ J = -3.57\underline{4}\ kJ$

$\Delta H° = \frac{Q}{n} = \frac{-3.57\underline{4}\ kJ}{0.112\ g\ Al \times 1\ mol\ Al/26.98\ g\ Al} = \boxed{-861\ kJ/mol\ Al}$

$2\ Al(s) + 6\ HCl(aq) \rightarrow 3\ H_2(g) + 2\ AlCl_3(aq)$ so 2 moles of Al react

$2\ mol\ Al \times \frac{-861\ kJ}{1\ mol\ Al} = \boxed{-1722\ kJ}$

49)

Reaction	kJ
$CH_2{=}CH_2(g) + 3\ O_2(g) \rightarrow 2\ CO_2(g) + 2\ H_2O(\ell)$	-1411 kJ
$CO(g) \rightarrow C(s) + \frac{1}{2}\ O_2(g)$	+ 110,5 kJ
$H_2(g) + \frac{1}{2}\ O_2 \rightarrow H_2O(\ell)$	- 285.8 kJ
$3\ CO_2(g) + 3\ H_2O(\ell) \rightarrow CH_3\text{-}CH_2\text{-}CHO(g) + 4\ O_2(g)$	+1806 kJ
$C(s) + O_2(g) \rightarrow CO_2(g)$	- 393.5 kJ
$CH_2{=}CH_2(g) + CO(g) + H_2(g) \rightarrow CH_3\text{-}CH_2\text{-}CHO(g)$	$\boxed{-174\ kJ}$

CHAPTER 14

74)

$$\frac{3.00\ g\ NaCl \times \frac{1\ mol\ NaCl}{58.44\ g\ NaCl}}{100.0\ g\ soln \times \frac{1\ mL\ soln}{1.025\ g\ soln} \times \frac{10^{-3}\ L}{1\ mL}} = \boxed{0.526\ M\ NaCl}$$

$$\frac{3.00\ g\ NaCl \times \frac{1\ mol\ NaCl}{58.44\ g\ NaCl}}{97.0\ g\ H_2O \times \frac{1\ kg\ H_2O}{10^3\ g\ H_2O}} = \boxed{0.529\ m\ NaCl}$$

75) $H_2SO_4(aq) + 2\ NaOH(aq) \rightarrow Na_2SO_4(aq) + 2\ H_2O(\ell)$

$$\frac{0.02169\ L \times \frac{0.1238\ mol\ NaOH}{L} \times \frac{1\ mol\ H_2SO_4}{2\ mol\ NaOH}}{0.01000\ L} = 0.1343\ M\ H_2SO_4 = M_2$$

$$M_1 = \frac{M_2 \times V_2}{V_1} = \frac{0.1343\ M \times 100.00\ mL}{25.00\ mL} = \boxed{0.5370\ M\ H_2SO_4}$$

76) $2\ Al(s) + 6\ HCl(aq) \rightarrow 3\ H_2(g) + 2\ AlCl_3(aq)$

$$3.96\ g\ Al \times \frac{1\ mol\ Al}{26.98\ g\ Al} \times \frac{3\ mol\ H_2}{2\ mol\ Al} = 0.220\underline{2}\ mol\ H_2$$

$$0.125\ L \times \frac{3.00\ mol\ HCl}{1\ L} \times \frac{3\ mol\ H_2}{6\ mol\ HCl} = \boxed{0.187\underline{5}\ mol\ H_2}$$

$$V = \frac{nRT}{P} = \frac{0.187\underline{5}\ mol \times 62.36\ L{\cdot}torr/mol{\cdot}K \times 287\ K}{482\ torr} = \boxed{6.96\ L\ H_2}$$

77)

$$0.03796\ L \times \frac{0.01036\ mol\ MnO_4^-}{L} \times \frac{5\ mol\ Fe^{2+}}{1\ mol\ MnO_4^-} \times \frac{55.85\ g\ Fe}{1\ mol\ Fe} = 0.1098\ g\ Fe$$

$$\%Fe = \frac{g\ Fe}{g\ sample} \times 100 = \frac{0.1098\ g\ Fe}{1.072\ g\ sample} \times 100 = \boxed{10.24\%\ Fe}$$

CHAPTER 15

63) a) $Cu_2S(s) + O_2(g) \rightarrow 2\ Cu(s) + SO_2(g)$

b) $PbS(s) + 2\ O_2(g) \rightarrow PbSO_4(s)$

c) $2\ MoS_2(s) + 7\ O_2(g) \rightarrow 2\ MoO_3(s) + 4\ SO_2(g)$

64) $AlCl_3(aq) + 3\ Na_2S(aq) + 3\ H_2O(\ell) \rightarrow Al(OH)_3(s) + 3\ NaHS(aq) + 3\ NaCl(aq)$

$Al^{3+}(aq) + 3\ Cl^-(aq) + 6\ Na^+(aq) + 3\ S^{2-}(aq) + 3\ H_2O(\ell) \rightarrow$
$Al(OH)_3(s) + 3\ Na^+(aq) + 3\ HS^-(aq) + 3\ Na^+(aq) + 3\ Cl^-(aq)$

$\boxed{Al^{3+}(aq) + 3\ S^{2-}(aq) + 3\ H_2O(\ell) \rightarrow Al(OH)_3(s) + 3\ HS^-(aq)}$

65) $2\ AgCl(s) + 2\ NaOH(aq) \rightarrow Ag_2O(s) + 2\ NaCl(aq) + H_2O(\ell)$

$2\ AgCl(s) + 2\ Na^+(aq) + 2\ OH^-(aq) \rightarrow Ag_2O(s) + 2\ Na^+(aq) + Cl^-(aq) + H_2O(\ell)$

$\boxed{2\ AgCl(s) + 2\ OH^-(aq) \rightarrow Ag_2O(s) + 2\ Cl^-(aq) + H_2O(\ell)}$

$Ag_2O(s) + 2\ HNO_3(aq) \rightarrow 2\ AgNO_3(aq) + H_2O(\ell)$

$Ag_2O(s) + 2\ H^+(aq) + 2\ NO_3^-(aq) \rightarrow 2\ Ag^+(aq) + 2\ NO_3^-(aq) + H_2O(\ell)$

$\boxed{Ag_2O(s) + 2\ H^+(aq) \rightarrow 2\ Ag^+(aq) + H_2O(\ell)}$

66) $\boxed{4\ Al(s) + 3\ O_2(g) \rightarrow 2\ Al_2O_3(s)}$

$Al_2O_3(s) + 6\ HC_2H_3O_2(aq) \rightarrow 3\ H_2O(\ell) + 2\ Al(C_2H_3O_2)_3(aq)$

$\boxed{Al_2O_3(s) + 6\ HC_2H_3O_2(aq) \rightarrow 3\ H_2O(\ell) + 2\ Al^{3+}(aq) + 6\ C_2H_3O_2^-(aq)}$

$HC_2H_3O_2(aq) + NaHCO_3(aq) \rightarrow NaC_2H_3O_2(aq) + H_2O(\ell) + CO_2(g)$

$HC_2H_3O_2(aq) + Na^+(aq) + HCO_3^-(aq) \rightarrow Na^+(aq) + C_2H_3O_2^-(aq) + H_2O(\ell) + CO_2(g)$

$\boxed{HC_2H_3O_2(aq) + HCO_3^-(aq) \rightarrow C_2H_3O_2^-(aq) + H_2O(\ell) + CO_2(g)}$

CHAPTER 16

12) $2\ H_2O(\ell) \rightarrow O_2(g) + 4\ H^+(aq) + 4\ e^-$

$2\ H_2O(\ell) + 2\ e^- \rightarrow H_2(g) + 2\ OH^-(aq)$

$2\ Cl^-(aq) \rightarrow Cl_2(g) + 2\ e^-$

13) $Cr_2O_7^{2-}(aq) + 14\ H^+(aq) + 6\ e^- \rightarrow 2\ Cr^{3+}(aq) + 7\ H_2O(\ell)$

$3\ [\ CH_3CH_2OH(aq) \rightarrow CH_3CHO(aq) + 2\ H^+(aq) + 2\ e^-\ ]$

$Cr_2O_7^{2-}(aq) + 8\ H^+(aq) + 3\ CH_3CH_2OH(aq) \rightarrow 2\ Cr^{3+}(aq) + 7\ H_2O(\ell) + 3\ CH_3CHO(aq)$

$2\ [\ Cr_2O_7^{2-}(aq) + 14\ H^+(aq) + 6\ e^- \rightarrow 2\ Cr^{3+}(aq) + 7\ H_2O(\ell)\ ]$

$3\ [\ CH_3CH_2OH(aq) + H_2O(\ell) \rightarrow CH_3CO_2H(aq) + 4\ H^+(aq) + 4\ e^-\ ]$

$2\ Cr_2O_7^{2-}(aq) + 16\ H^+(aq) + 3\ CH_3CH_2OH(aq) \rightarrow 4\ Cr^{3+}(aq) + 11\ H_2O(\ell) + 3\ CH_3CO_2H(aq)$

14) $4\ [\ MnO_4^-(aq) + 8\ H^+(aq) + 5\ e^- \rightarrow Mn^{2+}(aq) + 4\ H_2O(\ell)\ ]$

$5\ [\ As_2O_3(s) + 5\ H_2O(\ell) \rightarrow 2\ H_3AsO_4(aq) + 4\ e^- + 4\ H^+(aq)\ ]$

$5\ As_2O_3(s) + 4\ MnO_4^-(aq) + 12\ H^+(aq) + 9\ H_2O(\ell) \rightarrow 10\ H_3AsO_4(aq) + 4\ Mn^{2+}(aq)$

15) $4\ [\ MnO_4^-(aq) + 3\ e^- + 2\ H_2O(\ell) \rightarrow MnO_2(s) + 4\ OH^-(aq)\ ]$

$3\ [\ 4\ OH^-(aq) \rightarrow O_2(g) + 2\ H_2O(\ell) + 4\ e^-\ ]$

$4\ MnO_4^-(aq) + 2\ H_2O(\ell) \rightarrow 4\ MnO_2(s) + 3\ O_2(g) + 4\ OH^-(aq)$

$MnO_2(s) + 2\ e^- + 4\ H^+(aq) \rightarrow Mn^{2+}(aq) + 2\ H_2O(\ell)$

$2\ Cl^-(aq) \rightarrow Cl_2(aq) + 2\ e^-$

$MnO_2(s) + 4\ H^+(aq) + 2\ Cl^-(aq) \rightarrow Mn^{2+}(aq) + Cl_2(aq) + 2\ H_2O(\ell)$

CHAPTER 17

27) a) $X_{H_2O} = P_{H_2O}/P°_{H_2O} = 20.959\ \text{torr} \div 21.063\ \text{torr} = \boxed{0.99506}$

b) $n_{H_2O} = 5.632\ g\ H_2O \times \dfrac{1\ mol\ H_2O}{18.02\ g\ H_2O} = 0.3125\ mol\ H_2O$

$\dfrac{0.3125}{0.3125 + n_{unkn}} = 0.99506$ so $0.3125 = 0.3110 + 0.99506 n_{unkn}$

thus $0.99506 n_{unkn} = 0.0015$ and $n_{unkn} = \boxed{0.0015\ mol}$

c) $MM_{unkn} = 0.0532\ g \div 0.0015\ mol = \boxed{3500\ g/mol}$

28) $\Delta T_f = imK_f = 2 \times 0.55\ m \times 1.86\ °C/m = \boxed{-2.0°C}$

$°F = 1.8°C + 32 = 1.8(-2.0) + 32 = -3.6 + 32 = \boxed{28.4°F}$

29) a) $i = \dfrac{\Delta T_b}{mK_b} = \dfrac{101.28 - 100.00°C}{1.13\ m \times 0.512\ °C/m} = \boxed{2.21}$

b) $H_2SO_4 \xrightarrow{100\%} H^+ + HSO_4^-$ would give i = 2. The additional 0.21 above 2 is due to $HSO_4^- \rightarrow H^+ + HSO_4^-$ which must be occurring at 21%.

30) $m = \Delta T_f \div K_f = 25.0°C \div 1.86\ °C/m = 13.4\underline{4}\ m\ C_2H_5OH$

$13.4\underline{4}\ mol\ C_2H_5OH \times \dfrac{46.07\ g\ C_2H_5OH}{1\ mol\ C_2H_5OH} = 619\ g\ C_2H_5OH$

So $13.4\underline{4}$ m C_2H_5OH is a mixture of 619 g C_2H_5OH and 1000 g H_2O.

$\%C_2H_5OH = \dfrac{g\ C_2H_5OH}{g\ C_2H_5OH + g\ H_2O} \times 100 = \dfrac{619\ g}{1619\ g} \times 100 = \boxed{38.2\%\ C_2H_5OH}$

CHAPTER 18

38) STRESS RESPONSE SHIFT $HCO_3^-(aq) + H^+(aq) \rightleftarrows CO_2(g) + H_2O(\ell)$

a) Decrease P_{CO_2} Increase P_{CO_2} RIGHT Decrease Decrease Removed

The decrease in the $[H^+]$ indicates that the blood will be <u>less acidic</u>.

STRESS RESPONSE SHIFT $HCO_3^-(aq) + H^+(aq) \rightleftarrows CO_2(g) + H_2O(\ell)$

b) Increase $[H^+]$ Decrease $[H^+]$ RIGHT Decrease Added Increase

The body will get rid of more CO_2 by making the blood <u>more acidic</u>.

39) a) $K = \frac{[H^+][CO_3^{2-}]}{[HCO_3^-]}$, $4.8 \times 10^{-11} = \frac{6.3\times10^{-9}[CO_3^{2-}]}{0.0024}$, $[CO_3^{2-}] = 1.8 \times 10^{-5}$ M

b) $Q < K?$, $\frac{1}{[Ca^{2+}][CO_3^{2-}]} < K?$, $\frac{1}{(0.010)(1.8 \times 10^{-5})} < 2.1 \times 10^8?$

$5.6 \times 10^6 < 2.1 \times 10^8?$ Yes, $CaCO_3$ <u>will precipitate</u>

40) $2\ CHCl_3(g) + Cl_2(g) \rightleftharpoons 2\ CCl_4(g) + H_2(g)$, $K = 0.0018$ at 100°C

$Q = \frac{[CCl_4]^2[H_2]}{[CHCl_3]^2[Cl_2]} = \frac{(0.10)^2(0.10)}{(0.10)^2(0.10)} = 1.0 > K$ so <u>more reactant</u>

	$2\ CHCl_3(g)$	$+\ Cl_2(g)$	$\rightleftharpoons 2\ CCl_4(g)$	$+\ H_2(g)$
S	0.10 M	0.10 M	--	--
C	-2x	-x	+2x	+x
E	0.10-2x	0.10-x	2x	x

$K = \frac{[CCl_4]^2[H_2]}{[CHCl_3]^2[Cl_2]} = \frac{(2x)^2(x)}{(0.10-2x)^2(0.10-x)} = 0.0018$

Assume 0.10 - 2x = 0.10 and 0.10 - x = 0.10, then:

$\frac{4x^3}{(0.10)^3} = 0.0018$ so $x = 0.0077$

thus $[CCl_4] = 2x = 0.0154$ M
and $[H_2] = x = 0.0077$ M

41) $K = \frac{[CO_2]^2}{[SO_2]^2[O_2]}$ so $1.1\times10^{10} = \frac{(3.8\times 10^{-5})^2}{[SO_2]^2(1.9\times10^{-3})}$

so $[SO_2] = 8.3 \times 10^{-9}$ M
a P of just 0.00066 torr

CHAPTER 19

11) $AgCl(s) \rightleftarrows Ag^+(aq) + Cl^-(aq)$, $K_{sp} = [Ag^+][Cl^-] = 1.7 \times 10^{-10}$

$S^2 = 1.7 \times 10^{-10}$, $S = 1.3 \times 10^{-5}$ M, $[Ag^+] = S = 1.3 \times 10^{-5}$ M

$Ag_2CrO_4(s) \rightleftarrows 2\ Ag^+(aq) + CrO_4^{2-}(aq)$, $K_{sp} = [Ag^+]^2[CrO_4^{2-}] = 2.2 \times 10^{-12}$

$4S^3 = 2.2 \times 10^{-12}$, $S = 8.2 \times 10^{-5}$ M, $[Ag^+] = 2S =$ 1.6×10^{-4} M

$Ag_3PO_4(s) \rightleftarrows 3\ Ag^+(aq) + PO_4^{3-}(aq)$, $K_{sp} = [Ag^+]^3[PO_4^{3-}] = 1.3 \times 10^{-20}$

$27S^4 = 1.3 \times 10^{-20}$, $S = 4.7 \times 10^{-6}$ M, $[Ag^+] = 3S = 1.4 \times 10^{-5}$ M

<u>Silver chromate gives the highest</u> $[Ag^+]$.

12) $S = \frac{1.62\ g}{L} \times \frac{1\ mol}{136.14\ g} = 0.0119$ M

$K_{sp} = [Ca^{2+}][SO_4^{2-}] = S^2 = (0.0119)^2 =$ 1.42×10^{-4}

13) In water: $K_{sp} = [Fe^{3+}][OH^-]^3 = 27S^4 = 4 \times 10^{-38}$, $S = 2.0 \times 10^{-10}$ M

In 3.0 M NaOH: $(S)(3.0)^3 = 4 \times 10^{-38}$, $S = 1.\underline{5} \times 10^{-39}$ M

Solubility lowered by: $2.0\times10^{-10} \div 1.\underline{5} \times 10^{-39} =$ $1.\underline{3} \times 10^{29}$

14) After dilution: $[Ag^+] = 0.015$ M, $[SO_4^{2-}] = 0.014$ M

$Q > K_{sp}?$, $[Ag^+]^2[SO_4^{2-}] > K_{sp}?$, $(0.015)^2(0.014) > 1.6\times10^{-5}?$,

$3.2\times10^{-6} > 1.6\times10^{-5}?$ No, Ag_2SO_4 <u>will not precipitate</u>.

CHAPTER 20

25)

$$25 \text{ g NaHCO}_3 \times \frac{1 \text{ mol NaHCO}_3}{84.01 \text{ g NaHCO}_3} = 0.30 \text{ moles NaHCO}_3$$

$$0.75 \text{ L} \times \frac{0.20 \text{ mol HCl}}{\text{L}} = 0.15 \text{ moles HCl (or NaOH)}$$

a) $HCO_3^- + H^+ \rightleftarrows H_2CO_3$ leaves 0.15 moles HCO_3^- and forms 0.15 moles H_2CO_3: a buffer so:

$$[H^+] = K_a \times \frac{[H_2CO_3]}{[HCO_3^-]} = 4.3\times10^{-7} \times \frac{0.15}{0.15} = 4.3\times10^{-7} \text{ M}, \quad \boxed{\text{pH} = 6.37}$$

b) $HCO_3^- + OH^- \rightleftarrows CO_3^{2-} + H_2O$ leaves 0.15 moles HCO_3^- and forms 0.15 moles CO_3^{2-}: a buffer so:

$$[H^+] = K_a \times \frac{[HCO_3^-]}{[CO_3^{2-}]} = 4.8\times10^{-11} \times \frac{0.15}{0.15} = 4.8\times10^{-11}, \quad \boxed{\text{pH} = 10.32}$$

26) $$\frac{[H_2PO_4^-]}{[HPO_4^{2-}]} = \frac{[H^+]}{K_a} = \frac{4.0 \times 10^{-8}}{6.2 \times 10^{-8}} = \boxed{0.65}$$

$$\frac{[H_2CO_3]}{[HCO_3^-]} = \frac{[H^+]}{K_a} = \frac{4.0 \times 10^{-8}}{4.3 \times 10^{-7}} = \boxed{0.093}$$

27) $$\frac{x^2}{0.015 - x} = 6.8\times10^{-4} \quad \text{so} \quad x^2 + 6.8\times10^{-4}x - 1.0\times10^{-5} = 0$$

$$x = \frac{-(6.8\times10^{-4}) \pm \sqrt{(6.8\times10^{-4})^2 - 4(1)(-1.0\times10^{-5})}}{2(1)} = \boxed{0.0028 \text{ M} = [F^-]}$$

$Q > K_{sp}?$, $[Ca^{2+}][F^-]^2 > K_{sp}?$, $(0.050)(0.0028)^2 > K_{sp}?$,

$4.0 \times 10^{-7} > 4.0 \times 10^{-11}?$ Yes, CaF_2 <u>will precipitate</u>.

28)

$$M = \frac{10.00 \text{ g} \times \dfrac{1 \text{ mol}}{60.06 \text{ g}}}{0.1000 \text{ kg} \times \dfrac{1 \text{ L}}{1.025 \text{ kg}}} = 1.707 \text{ M urea}$$

also pOH = 6.75

so $[OH^-] = 1.7\underline{8} \times 10^{-7}$ M

$$\underset{1.707 \text{ M}}{NH_2CONH_2(aq)} + H_2O(\ell) \rightleftarrows \underset{1.7\underline{8}\times10^{-7} \text{ M}}{NH_2CONH_3^+(aq)} + \underset{1.7\underline{8}\times10^{-7} \text{ M}}{OH^-(aq)}$$

$$K_b = \frac{(1.7\underline{8} \times 10^{-7})^2}{1.707} = \boxed{1.9 \times 10^{-14}}$$

ANSWERS TO CHAPTER TESTS

CHAPTER 1 TEST

1) a) -44.742 b) 44.069224 c) -4.6121458 d) 5.9344721
   e) 12.023743 g) 7/20 h) 7/15

2) a) z^{-2} b) y^8 c) 64

3) a) 3 b) -20 c) -2 d) 4cd/3abe e) 2 f) 8 g) 3.2 h) 3.5, -1

4) $K = Fr^2/q_1q_2$ 5) t = 57.030251 minutes

CHAPTER 2 TEST

1) a) 7.835×10^6 b) 2.18×10^{-5} c) 0.000319 d) 503,000

2) 3.853590×10^{-3} 3) 6.1916×10^5 4) 6.366×10^{-3}

5) 66.3 seconds 6) 27.1% sodium 7) 27,200 registered 8) 6770 this year

CHAPTER 3 TEST

1) 3; 2; 4; 2 (be conservative); 4

2) 1.3; 1.30; 1.296; 1.2961; 1.29605; 1.296055

3) 2.0×10^{-3} (0.0082 has 2 sf)

4)
$$\begin{array}{r} 106.054 \\ 35.60 \\ 0.01428 \\ 2.3311 \\ \hline 143.99938 \end{array} = 144.00$$

5) 8×10^{-4} g, 8×10^{-7} kg 6) 2.78 lb 7) 7.98×10^3 cm, 7.98×10^4 mm

8) 6.05×10^4 m 9) 16.1 ft 10) 356 L 11) 13.9 m^3 12) 0.786 L

13) 5°C 14) 180°F, 355 K 15) 1.13 g/mL 16) 2.65×10^3 g

CHAPTER 5 TEST

1) a) 6.407 b) -25.316 2) a) 3.1×10^9 b) 1.2×10^{-6}

3) -37.15 (assumes 3 and 15 are exact numbers) 4) 2.8×10^7 5) 0.44

CHAPTER 6 TEST

1) a) Si b) Sr c) Au d) potassium e) phosphorus f) chlorine

2)

symbol	Z	A	p	n
^{46}Ti	22	46	22	24
^{55}Fe	26	55	26	29
^{127}Xe	54	127	54	73
^{60}Ni	28	60	28	32

3) a)

	family
C:	Si, Ge, Sn, Pb
Mg:	Be, Ca, Sr, Ba, Ra
Sb:	N, P, As, Bi
Hg:	Zn, Cd

b)

period
Li, Be, B, N, O, F, Ne
Na, Al, Si, P, S, Cl, Ar
Rb to Xe
Cs to Rn, also Ce to Lu

4) 63.55 amu 5) 26.0 amu

CHAPTER 7 TEST

chlorine, Cl_2
helium, He
phosphorus tribromide, PBr_3
disilicon hexafluoride, Si_2F_6
calcium ion, Ca^{2+}
sulfide ion, S^{2-}
iodic acid, HIO_3
sulfurous acid, H_2SO_3
phosphate ion, $PO_4{}^{3-}$
perchlorate ion, $ClO_4{}^-$
hydrogen carbonate ion, $HCO_3{}^-$
ammonium phosphate, $(NH_4)_3PO_4$
iron(II) nitrate, $Fe(NO_3)_2$
potassium hypobromite, KBrO
sodium hydrogen sulfite, $NaHSO_3$
potassium fluoride, KF

Fe, iron
N_2, nitrogen
S_2F_2, disulfur difluoride
OF_2, oxygen difluoride
Br^-, bromide ion
Fe^{3+}, iron(III) ion
HBr, hydrobromic acid
HNO_3, nitric acid
$SO_4{}^{2-}$, sulfate ion
OH^-, hydroxide ion
$BrO_3{}^-$, bromate ion
Na_2S, sodium sulfide
Na_2CO_3, sodium carbonate
$Mg(H_2PO_4)_2$, magnesium dihydrogen phosphate
$PbCl_2$, lead(II) chloride
$Cu(IO_4)_2$, copper(II) periodate

CHAPTER 8 TEST

1) a) 54.00 g/mol b) 46.07 g/mol c) 149.10 g/mol

2) 5.53×10^{23} O_3 molecules 3) 947 g $Cu(NO_3)_2$

4) a) 58.80% C, 9.870% H, 31.33% O b) 43.38% Na, 11.33% C, 45.29% O

4) 0.894 mol HOOCCOOH 6) 57.89 g $Al(NO_3)_3$ 7) a) NH_4CO_2 b) $N_2H_8C_2O_4$

CHAPTER 9 TEST

1)

most complicated:	$NaCl(aq)$ +	$H_2O(\ell)$ →	$H_2(g)$ +	$Cl_2(g)$ +	1 $NaOH(aq)$
balance Na + O:	1 $NaCl(aq)$ +	1 $H_2O(\ell)$ →	$H_2(g)$ +	$Cl_2(g)$ +	1 $NaOH(aq)$
balance H + Cl:	1 $NaCl(aq)$ +	1 $H_2O(\ell)$ →	½ $H_2(g)$ +	½ $Cl_2(g)$ +	1 $NaOH(aq)$
clear fractions:	2 $NaCl(aq)$ +	2 $H_2O(\ell)$ →	1 $H_2(g)$ +	1 $Cl_2(g)$ +	2 $NaOH(aq)$

2)

most complicated:	$KMnO_4(aq)$ +	$K_2C_2O_4(aq)$ →	1 $K_2MnO_4(aq)$ +	$CO_2(g)$
balance Mn:	1 $KMnO_4(aq)$ +	$K_2C_2O_4(aq)$ →	1 $K_2MnO_4(aq)$ +	$CO_2(g)$
balance K:	1 $KMnO_4(aq)$ +	½ $K_2C_2O_4(aq)$ →	1 $K_2MnO_4(aq)$ +	$CO_2(g)$
balance C + O:	1 $KMnO_4(aq)$ +	½ $K_2C_2O_4(aq)$ →	1 $K_2MnO_4(aq)$ +	1 $CO_2(g)$
clear fractions:	2 $KMnO_4(aq)$ +	1 $K_2C_2O_4(aq)$ →	2 $K_2MnO_4(aq)$ +	2 $CO_2(g)$

3) 2 $Na(s)$ + $Cl_2(g)$ → 2 $NaCl(aq)$ - Combination
4) $Fe_2S_3(s)$ → 2 $Fe(s)$ + 3 $S(s)$ - Decomposition
(Alternatively, 8 $Fe_2S_3(s)$ → 16 $Fe(s)$ + 3 $S_8(s)$ because sulfur is octatomic.)
5) 2 $Al(OH)_3(s)$ → $Al_2O_3(s)$ + 3 $H_2O(\ell)$ - Decomposition
6) 2 $C_2H_6(g)$ + 7 $O_2(g)$ → 4 $CO_2(g)$ + 6 $H_2O(\ell)$ - Combustion
7) 2 $C_2H_5COOH(\ell)$ + 7 $O_2(g)$ → 6 $CO_2(g)$ + 6 $H_2O(\ell)$ - Combustion
8) 2 $Al(s)$ + 3 $Ni(NO_3)_2(aq)$ → 3 $Ni(s)$ + 2 $Al(NO_3)_3(aq)$ - Single Replacement
9) $Pb(NO_3)_2(aq)$ + 2 $NaCl(aq)$ → $PbCl_2(s)$ + 2 $NaNO_3(aq)$ - Double Replacement (Pptn)
10) $KCN(aq)$ + $HCl(aq)$ → $HCN(aq)$ + $KCl(aq)$ - Double Replacement (Weak Acid Formation)
11) $Na_2CO_3(aq)$ + $BaCl_2(aq)$ → 2 $NaCl(aq)$ + $BaCO_3(s)$ - Double Replacement (Precipitation)
12) $HNO_3(aq)$ + $KOH(aq)$ → $KNO_3(aq)$ + $H_2O(\ell)$ - Double Replacement (Neutralization)

CHAPTER 10 TEST

1) 13.0 g C_3H_7OH 2) 19.0 g O_2 3) 91.2% yield 4) 79.5 g CS_2

5) 2.95 g $Mg(OH)_2$ produced, 1.3 g NaOH remain 6) 5.02% I_2

CHAPTER 11

1) 2.042×10^{-18} J 2) $n_i = 2$, $n_o = 4$

3) longest: infrared > green > indigo > ultraviolet > gamma :shortest

4) $2s^22s^22p^63s^23p^64s^23d^{10}4p^65s^24d^{10}5p^3$, $[Kr]5s^24d^{10}5p^3$, valence: $5s^25p^3$

5) excited, m effect

CHAPTER 12 TEST

1) 25.6 torr = 0.0337 atm = 3.41 kPa 2) 15.0 atm

3) 0.152 L 4) 1.81 L 5) 105 L 6) 78.2 g/mol

7) 33.9 g/mol 8) 1.07×10^3 L O_2 9) 0.286 g $KClO_3$ 10) 21.9 L SO_2

11) 153 torr 12) SO_2 1.118 times faster than SO_3 13) 34 g/mol

CHAPTER 13 TEST

1) 1.1 kJ 2) 1.4 kg 3) 0.0446 J/g•°C 4) 81 g Ni 5) 2.45 kJ/g

6) 25 kJ 7) 5.56×10^3 kJ 8) -1669 kJ/mol 9) 2058.3 kJ/mol

CHAPTER 14 TEST

1) 6.24% $BaBr_2$ 2) 8.0×10^1 mL soln 3) $X_{CH_3OH} = 0.510$; $X_{H_2O} = 0.490$

4) 339 g KNO_3 5) 85.3 g KI 6) 2.57 mmol $KMnO_4$

7) 17.5 m $CO(NH_2)_2$; 51.2% $CO(NH_2)_2$ 8) 72.0% H_2SO_4; 26.3 m H_2SO_4

9) 1.43 M NaOH 10) 4.20 g PbI_2 11) 39.4 mL Na_2CO_3

12) 0.501 M NaOH 13) 0.837 M $NaHCO_3$

CHAPTER 15 TEST

1) strong, major: 2 $Al^{3+}(aq)$ + 3 $SO_4^{2-}(aq)$, minor: none
2) strong, major: $H^+(aq)$ + $HSO_4^-(aq)$, minor: $SO_4^{2-}(aq)$ (from dissociation of HSO_4^-)
3) weak, major: $HNO_2(aq)$, minor: $H^+(aq)$ + $NO_2^-(aq)$
4) weak, major: $NH_3(aq)$, minor: $NH_4^+(aq)$ + $OH^-(aq)$

5) $Al(NO_3)_3(aq) + 3\ NaOH(aq) \rightarrow Al(OH)_3(s) + 3\ NaNO_3(aq)$ - Precipitation
$Al^{3+}(aq) + 3\ NO_3^-(aq) + 3\ Na^+(aq) + 3\ OH^-(aq) \rightarrow Al(OH)_3(s) + 3\ Na^+(aq) + 3\ NO_3^-aq)$
$Al^{3+}(aq) + 3\ OH^-(aq) \rightarrow Al(OH)_3(s)$

6) $2\ NH_4Cl(aq) + Na_2SO_4(aq) \rightarrow (NH_4)_2SO_4(aq) + 2\ NaCl(aq)$ - No Reaction - All Soluble

7) $Ca(OH)_2(s) + 2\ HNO_3(aq) \rightarrow 2\ H_2O(\ell) + Ca(NO_3)_2(aq)$ - Neutralization
$Ca(OH)_2(s) + 2\ H^+(aq) + 2\ NO_3^-(aq) \rightarrow 2\ H_2O(\ell) + Ca^{2+}(aq) + 2\ NO_3^-(aq)$
$Ca(OH)_2(s) + 2\ H^+(aq) \rightarrow 2\ H_2O(\ell) + Ca^{2+}(aq)$

8) $2\ HBr(aq) + Zn(s) \rightarrow H_2(g) + ZnBr_2(aq)$ - Single Replacement
$2\ H^+(aq) + 2\ Br^-(aq) + Zn(s) \rightarrow H_2(g) + Zn^{2+}(aq) + 2\ Br^-(aq)$
$2\ H^+(aq) + Zn(s) \rightarrow H_2(g) + Zn^{2+}(aq)$

9) $Na_2SO_3(aq) + 2\ HCl(aq) \rightarrow H_2O(\ell) + SO_2(aq) + 2\ NaCl(aq)$ - Weak Acid H_2SO_3 Forms/Decomp
$2\ Na^+(aq) + SO_3{}^{2-}(aq) + 2\ H^+(aq) + 2\ Cl^-(aq) \rightarrow H_2O(\ell) + SO_2(aq) + 2\ Na^+(aq) + 2\ Cl^-(aq)$
$SO_3{}^{2-}(aq) + 2\ H^+(aq) \rightarrow H_2O(\ell) + SO_2(aq)$

10) $Ba(s) + 2\ H_2O(\ell) \rightarrow H_2(g) + Ba(OH)_2(aq)$ - Single Replacement
$Ba(s) + 2\ H_2O(\ell) \rightarrow H_2(g) + Ba^{2+}(aq) + 2\ OH^-(aq)$

11) $NH_4Br(aq) + KOH(aq) \rightarrow NH_3(aq) + H_2O(\ell) + KBr(aq)$ - Weak Base Formation
$NH_4{}^+(aq) + Br^-(aq) + K^+(aq) + OH^-(aq) \rightarrow NH_3(aq) + H_2O(\ell) + K^+(aq) + Br^-(aq)$
$NH_4{}^+(aq) + OH^-(aq) \rightarrow NH_3(aq) + H_2O(\ell)$

12) $2\ Al(s) + 3\ Cu(NO_3)_2(aq) \rightarrow 2\ Al(NO_3)_3(aq) + 3\ Cu(s)$ - Single Replacement
$2\ Al(s) + 3\ Cu^{2+}(aq) + 6\ NO_3{}^-(aq) \rightarrow 2\ Al^{3+}(aq) + 6\ NO_3{}^-(aq) + 3\ Cu(s)$
$2\ Al(s) + 3\ Cu^{2+}(aq) \rightarrow 2\ Al^{3+}(aq) + 3\ Cu(s)$

CHAPTER 16 TEST

1) a) +4 b) +3 c) +4 d) +3

2) a) C is oxidized from +3 to +4. $H_2C_2O_4$ is the reducing agent.
As is reduced from +5 to +3. H_3AsO_4 is the oxidizing agent.

3) a) $4\ e^- + 4\ H^+ + SiO_2 \rightarrow Si + 2\ H_2O$ b) $H_2O + 2\ e^- + BrO^- \rightarrow Br^- + 2\ OH^-$

4) a)
$$6\ e^- + 6\ MnO_4{}^- \rightarrow 6\ MnO_4{}^{2-}$$
$$8\ OH^- + CH_3OH \rightarrow 6\ H_2O + CO_3{}^{2-} + 6\ e^-$$
$$8\ OH^- + 6\ MnO_4{}^- + CH_3OH \rightarrow 6\ MnO_4{}^{2-} + 6\ H_2O + CO_3{}^{2-}$$

b)
$$3\ H_2C_2O_4 \rightarrow 6\ CO_2 + 6\ H^+ + 6\ e^-$$
$$16\ H^+ + 6\ e^- + 2\ MoO_4{}^{2-} \rightarrow 2\ Mo^{3+} + 8\ H_2O$$
$$10\ H^+ + 3\ H_2C_2O_4 + 2\ MoO_4{}^{2-} \rightarrow 6\ CO_2 + 2\ Mo^{3+} + 8\ H_2O$$

c)
$$4\ Zn \rightarrow 4\ Zn^{2+} + 8\ e^-$$
$$NO_3{}^- + 8\ e^- + 10\ H^+ \rightarrow NH_4{}^+ + 3\ H_2O$$
$$4\ Zn + NO_3{}^- + 10\ H^+ \rightarrow 4\ Zn^{2+} + NH_4{}^+ + 3\ H_2O$$

CHAPTER 17 TEST

1) 98.0 torr 2) 2.4°C 3) 2.5×10^2 g/mol 4) i = 1.01 (1% ionized) 5) 1.80×10^4 g/mol

CHAPTER 18 TEST

1) $K = \frac{[CO]^4}{[C_2H_6]^2[O_2]^7}$ 2) K = 0.011 3) RIGHT 4) positive

5) More favorable 6) More CO_2 7) $[Cl_2] = [PCl_3] = 0.032$ M, $[PCl_5] = 0.018$ M

8) $[CO_2] = [H_2] = 0.089$ M, $[CO] = [H_2O] = 0.051$ M

CHAPTER 19 TEST

1) 1.6×10^{-5} 2) 8.3×10^{-9}

3) a) 1.3×10^{-26} mol/L = 3.0×10^{-24} g/L b) 1.3×10^{-4} mol/L = 0.036 g/L

4) 3.6×10^{-6} mol/L = 4.9×10^{-4} g/L 5) $Q = 7.0 \times 10^{-4} > 8.3 \times 10^{-5}$. Ppt.

CHAPTER 20 TEST

1) $[H^+] = 4.1 \times 10^{-4}$ M, $[OH^-] = 2.4 \times 10^{-11}$ M, pH = 3.39, pOH = 10.61

2) $[OH^-] = 1.7 \times 10^{-5}$ M, $[H^+] = 6.0 \times 10^{-10}$ M, pH = 9.22, pOH = 4.78

3) 6.1×10^{-5} 4) pH = 2.89, 4.6% ionized 5) 12.38

6) 10.41 7) 4.80 8) 4.28 9) 0.67/1

10) 2.7×10^2 mL 7.4 M NH_3 and 1.7×10^2 g NH_4Cl, or quantities in same ratio.

INDEX

Note: (t) after a page number indicates a table.

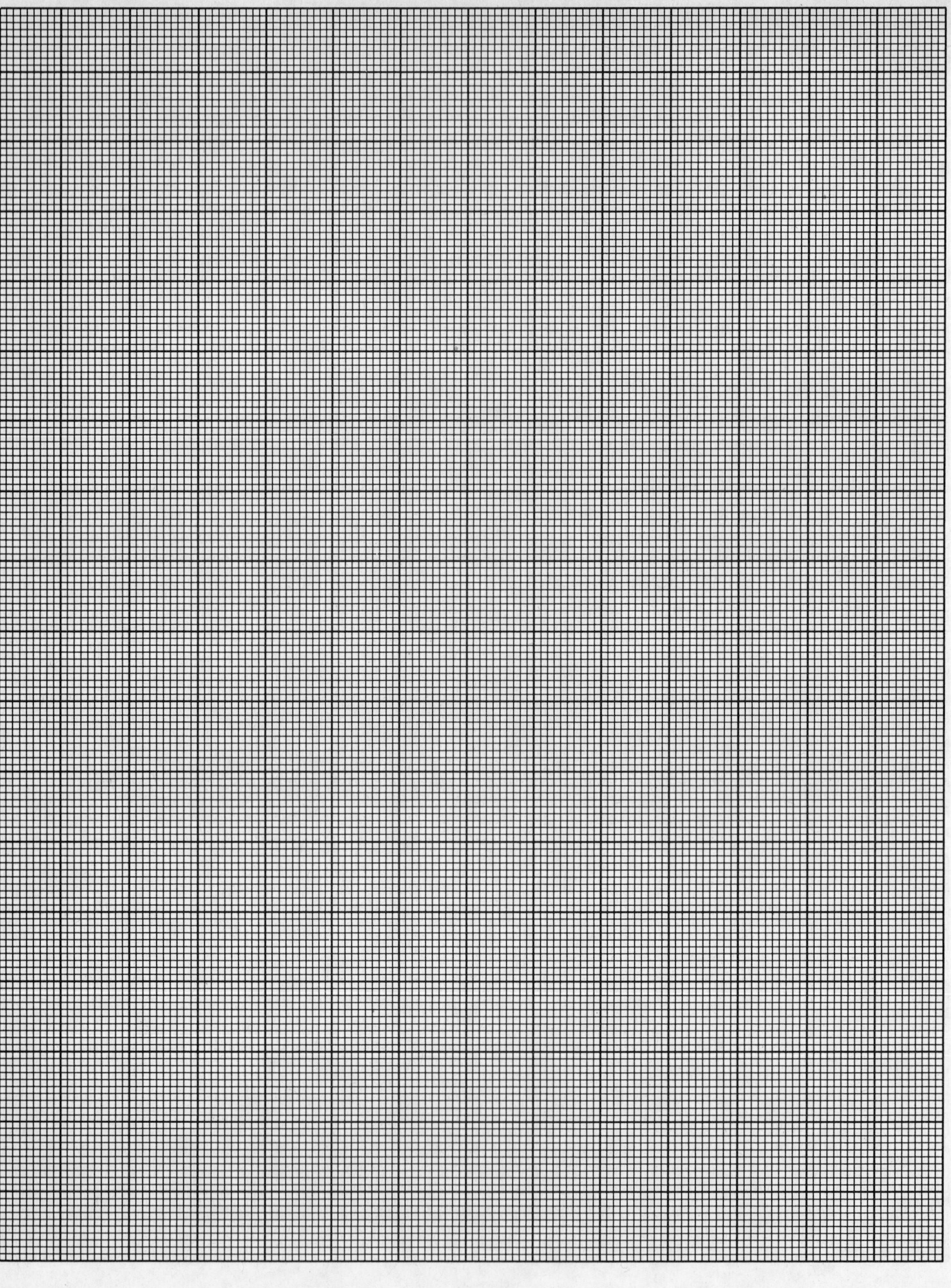